Circuit Analysis:
A Systems Approach

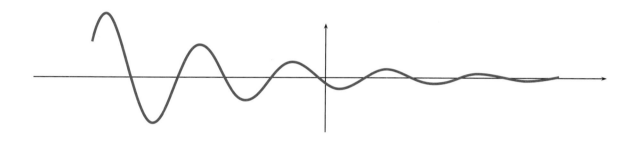

Russell M. Mersereau
Georgia Institute of Technology

Joel R. Jackson
Georgia Institute of Technology

PEARSON

Prentice
Hall

Upper Saddle River, NJ 07458

Library of Congress Cataloging-in-Publication Data

Mersereau, Russell M.
  Circuit analysis : a systems approach / Russell M. Mersereau, Joel R. Jackson
    p. cm.
  ISBN 0–13–093224–8
  1. Electronic circuit design.   I. Jackson, Joel R.

TK7867.M39 2006
621.319'2—dc22   2005048933

Vice President and Editorial Director, ECS: *Marcia Horton*
Executive Managing Editor: *Vince O'Brien*
Managing Editor: *David A. George*
Production Editor: *Daniel Sandin*
Director of Creative Services: *Paul Belfanti*
Art Director: *Jayne Conte*
Art Editor: *Gregory Dulles*
Cover Designer: *Bruce Kenselaar*
Manufacturing Manager: *Alexis Heydt-Long*
Manufacturing Buyer: *Lisa McDowell*
Senior Marketing Manager: *Holly Stark*

MATLAB is a registered trademark of The MathWorks, Inc., 3 Apple Hill Drive, Natick, MA 07160-2098.

© 2006 Pearson Education, Inc.
Pearson Prentice Hall
Upper Saddle River, NJ 07458

Printed in the United States of America

10   9   8   7   6   5   4   3   2   1

ISBN   0-13-093224-8

Pearson Education Ltd., *London*
Pearson Education Australia Pty. Ltd., *Sydney*
Pearson Education Singapore, Pte. Ltd.
Pearson Education North Asia Ltd., *Hong Kong*
Pearson Education Canada, Inc., *Toronto*
Pearson Educación de Mexico, S.A. de C.V.
Pearson Education—Japan, *Tokyo*
Pearson Education Malaysia, Pte. Ltd.
Pearson Education, Inc., *Upper Saddle River, New Jersey*

To Martha—RMM

To Danielle, Alexander, and Christine—JRJ

Contents

4   Operational Amplifiers   150

7   System Functions   253

Foreword

From the beginning of electrical engineering education, *Electric Circuits* was the first course that students encountered in their prospective field. There were good reasons for this:

1. The study of linear circuits introduced students to the use of mathematics as a language for describing physical laws and for thinking about engineering problems;
2. Linear-circuit analysis was a good way to teach problem solving techniques of general applicability;
3. The subject laid useful groundwork for subsequent courses such as electronics, signals and systems, communications, and control.

These three basic principles remain valid today when choosing a first course for the much broader field of electrical and computer engineering (ECE). However, in the School of ECE at Georgia Tech, we feel that there are good reasons for making the linear-circuits course the *second course* following an introductory course on signal processing taken in the first semester of the sophomore year. Mersereau and Jackson's elegant text was developed with this structure in mind.

The reasons for choosing to teach signal processing first are the above (1–3), which are equally true with signal processing replacing linear circuits, and the following:

4. The study of signal processing makes a strong connection to modern day digital computation as a means for implementing systems;

5. Widely available, powerful computers and easy-to-use software such as MATLAB make it possible to motivate beginning engineers with interesting hands-on applications of the mathematics that they study in the classroom.

These reasons acknowledge the increasing importance of digital implementations of engineering systems, and recognize that significant applications of circuit knowledge, while still exceedingly important, usually cannot occur until later in an ECE program through labs involving expensive test equipment.

Losing first position at Georgia Tech and in many other ECE programs is far from bad news for the teaching of linear circuits. Indeed, there are many benefits that result from the fact that students have gained some understanding of such topics as complex numbers and their use in representing sinusoidal signals, the notion of time-domain and frequency-domain representations, and the concepts of frequency response and system function. Mersereau and Jackson have exploited these benefits in structuring a linear circuits book that emphasizes a systems approach to circuits and includes material on analog filters, their frequency responses, and their implementation with op-amp circuits. Their text begins in a traditional way with four chapters that introduce the important concepts of circuit theory such as circuit element definitions, KVL and KCL, methods for writing equilibrium equations, basic circuit theorems, and the concepts of energy and power in circuits. These concepts, which are specific to circuit theory, are developed using mostly resistive-circuit examples. Then, in a departure from most classical linear-circuits texts, they introduce the Laplace transform, which is used throughout the rest of the text for solving the differential equations that represent circuit behavior and as the basis for the impedance concept in networks. This approach would work even when the circuits course comes first, but it has the distinct advantage that in a second course many of the time-domain/frequency-domain concepts that students find difficult are being seen in a new light and are being *reinforced* rather than introduced as new material. Furthermore, the Laplace transform fills in a part of the signals and systems picture that is not covered in our text *Signal Processing First*.

We congratulate our Georgia Tech colleagues on producing such a well-written, student-friendly textbook. It will make a perfect companion to our *Signal Processing First* text. Together these two texts provide the basis for a complete introduction to linear circuits and both discrete-time and continuous-time signals and systems. We hope that our colleagues in other schools will view the publication of this text as an opportunity to try our approach and add their own innovations to what we have done.

JAMES H. MCCLELLAN
RONALD W. SCHAFER
MARK A. YODER

Preface

Historically, courses in linear circuits, lasting anywhere from one to three semesters, have formed the foundation of the electrical engineering curriculum. In the twenty-first century, this material is still important, but few programs can devote more than a single semester to it. Furthermore, the role of a basic circuits course is changing: While it still needs to explain circuit models and help students develop intuition about how circuits behave in preparation for more advanced electronics courses, it is more important that it develop a student's ability to reason clearly and think critically than that it present him or her with a collection of currently useful techniques. Computers can relieve the tedium of hand calculations, but only the enlightened student can tell whether the results make sense.

This text is designed to be covered in a single semester. It takes the point of view that circuits accept input signals, such as the output of a sensor, and convert them into output signals. It stresses understanding how a circuit responds to different types of input signals. Thus, the text places a strong emphasis on the transient behavior of a circuit when a discontinuity occurs in the input and on its frequency-response behavior. Because of the limitation of the text to material that can be covered in a single semester, it omits many topics that are often covered in more encyclopedic texts. For example, there is little treatment of the time-domain solution of differential equations; instead, these are solved by using Laplace transforms. Similarly, there are no discussions of two-port networks, three-phase circuits, or electrical history. A more controversial omission is any discussion of network-simulation tools, such as PSPICE. In addition to the fact that a single semester does not allow time to do justice to this topic, this omission is also pedagogical. We strongly believe that, if used too early, these simulation tools can inhibit developing intuition about how circuits behave. PSPICE has its place in electronics courses, but we do not believe that it belongs in a first course in circuit analysis.

At Georgia Tech, where this text was developed, a linear-circuits course is intended to be taught in the second semester of the second year to all electrical engineering and computer engineering majors. It serves as a prerequisite to courses in microelectronic circuits in electromechanical and electromagnetic energy conversions, and in systems and controls. In these respects, it is similar to circuit-analysis courses anywhere. Unlike most other programs, however, circuit analysis is not the first electrical and computer engineering course that our students are required to take. That role is played by the course *Introduction to Signal Processing,* which students normally take in the beginning of their second year. In that course, students are exposed to discrete-time and continuous-time signal processing, sampling, filtering, Fourier transforms, z-transforms, and the designing and implementing of discrete-time signal-processing systems by using MATLAB. As a result, students entering our circuit-analysis course are already (somewhat) comfortable with the manipulation of complex numbers and functions, with phasor representations of sinusoidal signals, with switching between time-domain and transform-domain representations of signals and systems, and with the fact that many systems can be described usefully by their impulse responses, frequency responses, and system functions. Since all of these topics are included in this text, it can still be used in the more traditional setting, but our desire to couple *Linear Circuits* to *Introduction to Signal Processing* has guided our selection of topics and our approach.

In addition to teaching students about how circuits behave, a major role of any circuits course is to teach students how to solve problems, particularly problems that require mapping familiar concepts to new situations. As a result, each of the chapters includes a number of worked examples, and each chapter is followed by a number of problems. These problems are generally grouped into four categories. The first group is called *Drill Problems*. These problems are very similar to the worked examples. Indeed, they are cross-referenced to specific examples. Worked solutions to the Drill Problems are included on the web site. The second and largest group is called *Basic Problems*. These are reasonably straightforward, although not always trivial, extensions of the examples and explanations in the text. Answers, but not solutions, to some of these problems appear at the back of the text. The next group is called *Advanced Problems*. As the name suggests, these problems require more thought, they usually do not resemble the exercises, and they might require pulling together multiple concepts. The final group is called *Design Problems*. As their name suggests, these involve some aspect of the design process—coming up with a circuit that meets a certain specification. Admittedly, the designs in the early chapters are somewhat contrived, but they become increasingly relevant in the later chapters. The web site includes additional aids to help with problem solving, including animated solutions and video clips illustrating the thought process that lies behind the solutions of certain problems.

Certain skills associated with circuits require more than a simple understanding; they need to become second nature. These include identifying circuit elements connected in series or parallel, finding the equivalent resistance of series and parallel connections of elements, applying voltage and current dividers, and relating the locations of the poles and zeros of a circuit to the frequency response and the transient behavior of a circuit. To aid in the development of these skills, a number of randomized, timed drilling programs are also included on the web site.

Chapters 1–3 deal almost exclusively with circuits constructed from only resistors and sources, in order to explain how Kirchhoff's current and voltage laws and the v–i relations of the elements define the solution of a circuit without introducing the complication of differential equations. Chapter 4 is a short interlude that discusses linear operational-amplifier circuits and methods for their analysis. It develops a collection of useful operational-amplifier circuits.

Chapters 5–7 are concerned with the Laplace-domain analysis of circuits containing reactive elements. After the transform is developed in Chapter 5, a procedure is developed for mapping a circuit containing reactive elements to the Laplace domain. In that domain, we have another environment in which Kirchhoff's current law, Kirchhoff's voltage law, and Ohm's law hold. Thus, in the Laplace domain, a circuit with reactive elements can be analyzed, just like a resistive circuit. More importantly, all of the methods from the earlier chapters are immediately applicable. As a by-product of the Laplace-domain analysis, we introduce the system function and show how it can be used to characterize a circuit and predict its transient behavior to discontinuous input signals.

Chapters 8–10 discuss the frequency-domain behavior of circuits. It begins with a discussion of the steady-state behavior of a circuit when the input signal is a sinusoid. We show that the complex amplitudes of the sinusoidal signals within a circuit once again obey KCL, KVL, and Ohm's law and thus represent another domain in which the familiar analysis methods can be exploited. Sinusoidal-steady-state analysis is generalized to introduce the frequency response, and the interrelationship between the frequency response, the system function, and the transient behavior of a circuit is discussed in some detail. Finally, Chapter 10 looks at the specific problem of designing and implementing lowpass, highpass, and bandpass filter circuits. It also discusses frequency and impedance scaling as straightforward methods for modifying circuit behavior.

There is undoubtedly more material in this text than will fit into most one-semester courses. Furthermore, each individual instructor will have certain topics that he or she chooses to emphasize or omit. At Georgia Tech we often omit some or all of those sections that are marked by an asterisk (*) in the Contents. These include Sections 2-1-1, 2-3-3, 3-5, 6-4-2, 6-5-1, and 7-4. Chapter 10 is also occasionally omitted or abridged. In addition, the material in Sections 5-1-1 and 5-1-2 on sinusoidal signals, Section 9-1 on Fourier series, Section 9-2-1 on the Fourier transform, and Appendix A on complex numbers are all presented in the prerequisite course that is based on *Signal Processing First*. Therefore, these topics are omitted as well.

RUSSELL M. MERSEREAU
JOEL R. JACKSON

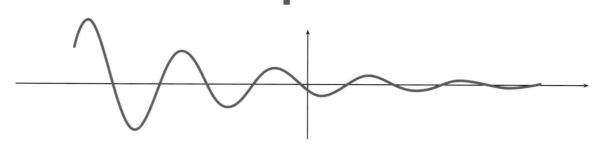

CHAPTER 1

Circuit Elements and Models

Objectives

By the end of this chapter, you should be able to do the following:

1. *Write equations that relate the potential difference across the terminals of an element to the current flowing through it.*

2. *Write a sufficient set of KCL (Kirchhoff's current law) and KVL (Kirchhoff's voltage law) equations that will define all of the constraints on the element voltages and the currents in a circuit.*

3. *Find the complete solution (all of the voltages and all of the currents) in a circuit containing only resistors and sources.*

4. *Recognize the role of circuit models as approximations of physical behavior, with limitations that might be ignorable, depending upon the circumstances.*

1-1   Introduction

This text develops techniques that can be used to understand the behavior of certain types of interconnected electrical devices. For the most part, this development is done by using idealized models that are expressed mathematically. These models impose constraints on the voltages and currents that appear in the circuit: constraints imposed by the *types of elements* in the circuit, and constraints imposed by the *way in which they are connected*. All of these constraints can be expressed mathematically in a way that captures the underlying physics. They interrelate the various voltages and currents associated with the elements and sources. By writing a sufficient number of independent equations, we can calculate all the voltage and current waveforms. These underlying constraints are deceptively easy to state, but they are very powerful, and the problems to which they can be applied can become very complex.

Why study circuit theory? There are at least three reasons. First, models that closely approximate many electronic devices provide a means for analyzing and designing circuits constructed from these devices. Second, many of the principles that we will learn by studying these models allow us to design and analyze realizations of key components of various electrical systems, such as filters, amplifiers, and mixers. Finally, the most expansive reason for studying circuit theory is that it can be used to understand the behavior a wider variety of systems whose inputs and outputs are related by sets of differential equations. Circuit-theory methods are regularly applied to the design and analysis of nonelectrical systems in mechanical, aerospace, and chemical engineering, and even more far-reaching applications have been found in economics and physiology. The web site contains additional information on some of these applications.

EXTRA INFORMATION: *Neurology Application*

This chapter is largely concerned with the definitions and underpinnings that form the basis for the analysis and design of all circuits. We begin by defining the basic circuit elements and sources from which our models are constructed. Then, we introduce the two basic laws that underlie all of circuit analysis. All of the material that follows this chapter is merely an exploration of the implications of these two basic laws. The chapter concludes with a more thorough discussion of the role of mathematical models, their range of applicability, and their limitations.

1-2   Network Elements

A large class of electrical devices can be modelled by using one or more two-terminal *elements*. Figure 1-1 shows a generic element whose two terminals are shown as the two lines representing wires that are attached

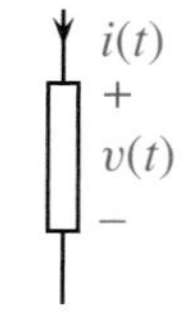

Fig. 1-1. A graphical representation for a generic two-terminal network element, showing the reference directions for the voltage, $v(t)$, and current, $i(t)$.

to the device. These terminals allow the element to be connected to other elements in a circuit. Associated with every element are two measurable physical quantities: a *voltage* or *potential difference* measured between the two terminals, and a *current* flowing through the element. We assume that these quantities are familiar from your courses in physics.

The current, $i(t)$, measures the rate of flow of electrical charge through the device; that is,

$$i(t) = \frac{dq(t)}{dt}, \tag{1.1}$$

where $q(t)$ is charge, which varies as a function of time. The basic measurement unit of current is the *ampere (A)*, where one ampere (1A) is equal to one coulomb/second. Although we know that current flow in conductors results from the motion of negatively charged electrons, the universally adopted convention is to treat current flow as a movement of *positive* electrical charge. This means that the actual movement of the electrons is in the direction opposite from the current, but this will not affect our analysis. We will always treat current flow as a net movement of positive charge, regardless of the physical phenomena that take place.

Positive current will be treated as a flow of positive electric charge.

Whenever positive and negative charges are separated, energy is expended (work is performed) and an electric field is generated. This energy can be recovered when the charges are brought back together. Thus, separated electrical charges possess potential energy. The *voltage,* $v(t)$, is the energy per unit charge that is created by the separation, or

$$v(t) = \frac{dU(t)}{dq(t)}, \qquad (1.2)$$

where $U(t)$ is the energy. More specifically, the voltage associated with an element is the energy required to move a unit charge from one of its terminals, marked by a $^-$, to the other, marked by a $^+$. The measurement unit for voltages is the *volt (V)*, where $1\ V = 1$ joule/coulomb.

As our formulas for the voltage and current suggest, the voltages and currents in a circuit are usually not constant, but instead vary with time. In our mathematical model for an element, we have assigned variables $v(t)$ to the voltage and $i(t)$ to the current; t represents time. We shall explicitly retain the time dependence of these variables unless we know in advance that they are constant.

🌐 EXTRA INFORMATION: *Voltage and Current*

A fairly extensive understanding of electrical circuits can be derived from mathematical models that approximate the idealized macroscopic behavior of real electrical elements. The first of these is a statement about the current in an element:

> *The current that enters a two-terminal element through one terminal is equal to the current that leaves it through the other.*

This is equivalent to saying that there is no net accumulation of charge in an element.

Every element imposes some kind of constraint between its voltage $v(t)$ and its current $i(t)$; this constraint is known as the *v–i characteristic* of the element. It is also called the *element relation*. By whichever name, it is a mathematical relationship (equation) that expresses $v(t)$ in terms of $i(t)$ (or vice versa) and that varies from element to element.

The scales of the voltages and currents that engineers encounter in different contexts vary over many orders of magnitude. Cross-country power transmission lines often have potentials with respect to the ground of hundreds of kilovolts; potential differences measured at the input antenna in your cell phone might be measured in nanovolts. To deal with these wide variations, we shall regularly make use of the *standard SI system of prefixes*, summarized in Table 1-1. Whenever a quantity is quoted without its units, the understanding should be that it is measured in its base units, which are volts for

Table 1-1. SI prefixes.

Multiple	Prefix	Symbol
10^{24}	yotta	Y
10^{21}	zetta	Z
10^{18}	exa	E
10^{15}	peta	P
10^{12}	tera	T
10^{9}	giga	G
10^{6}	mega	M
10^{3}	kilo	k
10^{-3}	milli	m
10^{-6}	micro	μ
10^{-9}	nano	n
10^{-12}	pico	p
10^{-15}	femto	f
10^{-18}	atto	a
10^{-21}	zepto	z
10^{-24}	yocto	y

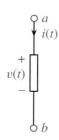

Fig. 1-2. Labelling conventions for the voltages and currents associated with an element.

voltages and amperes for currents. Whenever we wish to express a quantity in modified units, those will be indicated explicitly by using the prefixes in Table 1-1. For example, we might say $v(t) = 3\text{mV}$ and $i(t) = 1.4\mu\text{A}$, but $v(t) = 3$ should be understood to be the same as $v(t) = 3\text{V}$.

1-2-1   Sign Conventions for Defining Element Variables

Consider the generic two-terminal element shown in Figure 1-2. Its two terminals are labelled a and b. It is important to understand that current can flow both ways through an element and that $i(t)$ is a *signed quantity*. Stating that there is a current of 3A flowing from terminal a to terminal b is equivalent to stating that there is a current of -3A flowing from terminal b to terminal a.

> *Changing the sign of a current changes its direction.*

To give meaning to the variable $i(t)$, we assign a reference direction for the current by means of the arrow next to the variable $i(t)$. The meaning of the arrow is as follows: When $i(t)$ is positive, the current flows in the direction of the arrow; when it is negative, the current flows in the

opposite direction. Without the reference direction, the meaning of $i(t)$ would be ambiguous. For the element in Figure 1-2, when $i(t)$ is positive, a positive current flows from terminal a to terminal b; when $i(t)$ is negative, the positive flow is from b to a. The purpose of the reference direction is merely to give meaning to the variable; it in no way limits the current. Remember that the current $i(t)$ can be either positive or negative; indeed, usually it will be positive for some values of t and negative for others.

We defined the voltage $v(t)$ as the work required to move a unit charge from the terminal marked by a $-$, in this case terminal b, to the terminal marked by a $+$, in this case terminal a. That work, and the voltage $v(t)$, however, are also signed quantities. If the work is positive, the voltage will be positive, and a unit charge at terminal a will have a higher potential energy than it does at terminal b; if the work is negative, the voltage will also be negative, and the unit charge will have a higher potential energy at terminal b than it does at terminal a. The change in potential energy that results from moving a unit charge from terminal b to terminal a is the negative of the change in potential energy that results from moving the charge in the opposite direction.

> *Reversing the polarity of the terminals (i.e., interchanging the + and − signs) is equivalent to changing the sign of the voltage.*

The $+$ and $-$ signs define the reference direction for $v(t)$. If $v(t)$ is positive, a charge has higher potential energy at the $+$ terminal than it does at the $-$ one; if it is negative, the opposite is true. If the terminals were not so labelled, the definition of the voltage would also be ambiguous.

It is customary to coordinate the reference directions for the voltage and current for an element so that the reference direction for the current flows into the positive $+$ terminal of the device, as in Figure 1-1. This is sometimes called the passive sign convention, because when $i(t)$ and $v(t)$ are both positive the device will be absorbing energy. For us, it is the only sign convention that we shall use, so we shall regularly refer to this labelling as our *default sign convention*. If this sign convention is used consistently, there is no need to label the reference directions for both the current and voltage, because either can be inferred from the other. This will be our implied convention when the sense of only the voltage or the current is indicated. It is important to stress that the purpose of assigning reference directions to the voltages and currents is merely to give meaning to the voltage and current variables associated with an element; it in no way restricts the values of those voltages to be positive or negative, as we shall see. We will come back to this point later in the chapter.

If a current $i(t)$ flows into the + terminal of an element, as in Figure 1-2, the *instantaneous power* absorbed by the element, $P_{\text{inst}}(t)$, is defined as

$$P_{\text{inst}}(t) = \frac{dU(t)}{dt} = \left(\frac{dU(t)}{dq(t)}\right)\left(\frac{dq(t)}{dt}\right) = v(t)i(t).$$

The fundamental unit of power is the *watt (W)*. The instantaneous power can be either positive or negative. When it is positive, the element *absorbs* power; when it is negative, the element *supplies* power.

Notice that the definition of the instantaneous power is critically dependent on the consistent definitions of the voltage and current variables for an element. This is also the case for the element relations that relate $v(t)$ and $i(t)$ for an element.

Table 1-2 lists many of the common ideal elements and sources that we shall encounter, along with their graphical symbols and v–i (element) relations. The next three sections look at the first three of these in some detail: ideal resistors, ideal capacitors, and ideal inductors. We shall examine sources later in the chapter.

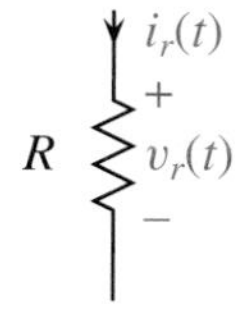

$$v_r(t) = Ri_r(t)$$
$$i_r(t) = \frac{1}{R}\,v_r(t) = Gv_r(t)$$

Fig. 1-3. An ideal resistor and its v–i relations.

1-2-2 Resistors

The mathematically simplest of the linear-circuit elements is the *resistor*, shown in Figure 1-3. When defined according to the default sign convention, the element voltage $v_r(t)$ and current $i_r(t)$ in an ideal resistor are proportional to each other:

$$v_r(t) = Ri_r(t). \tag{1.3}$$

This relationship is commonly known as *Ohm's law*.

> *Ohm's law: For an ideal resistor, the instantaneous value of the voltage is proportional to the instantaneous value of the current.*

The reference directions of the voltage and current become very important here. For (1.3) to be valid, the reference direction for the current must flow into the + terminal of the resistor. For a resistor, current always flows from the terminal with the higher potential to the one with the lower. When the + terminal has the higher potential, both $v_r(t)$ and $i_r(t)$ are positive; when the − terminal has the higher potential, both $v_r(t)$ and $i_r(t)$ are negative; but, in both cases, (1.3) is valid.

The constant of proportionality, R, in (1.3) is called the *resistance* of the resistor. It represents the degree to which the element resists the flow of current. If R is large, the resistor resists the flow of current considerably.

Table 1-2. Circuit Elements and Sources

	Symbol	v–i Relation	See Section
Resistor	$i(t)$ R $+$ $v(t)$ $-$	$v(t) = Ri(t)$	1-2-2
Inductor	$i(t)$ L $+$ $v(t)$ $-$	$v(t) = L\dfrac{di(t)}{dt}$	1-2-3
Capacitor	$i(t)$ C $+$ $v(t)$ $-$	$i(t) = C\dfrac{dv(t)}{dt}$	1-2-4
Voltage Source	$v_s(t)$ $i(t)$ $+|$ $+$ $v(t)$ $-$	$v(t) = v_s(t)$ for all $i(t)$	1-3
Current Source	$i_s(t)$ $i(t)$ $\rightarrow$ $+$ $v(t)$ $-$	$i(t) = i_s(t)$ for all $v(t)$	1-3
Voltage-Controlled Voltage Source	$av_c(t)$ $i(t)$ $+|$ $+$ $v(t)$ $-$	$v(t) = av_c(t)$ for all $i(t)$	1-5
Voltage-Controlled Current Source	$gv_c(t)$ $i(t)$ $\rightarrow$ $+$ $v(t)$ $-$	$i(t) = gv_c(t)$ for all $v(t)$	1-5
Current-Controlled Voltage Source	$ri_c(t)$ $i(t)$ $+|$ $+$ $v(t)$ $-$	$v(t) = ri_c(t)$ for all $i(t)$	1-5
Current-Controlled Current Source	$bi_c(t)$ $i(t)$ $\rightarrow$ $+$ $v(t)$ $-$	$i(t) = bi_c(t)$ for all $v(t)$	1-5

When R is small, it offers little resistance. Resistances are measured in *ohms*, abbreviated as Ω, where

$$1\Omega = 1 \text{ volt/ampere.}$$

A resistance of 1Ω is quite small for most circuit applications. Larger resistances are more often encountered in electrical circuits. These are often measured in kilohms (kΩ) and megohms (MΩ), which correspond to $10^3\Omega$ and $10^6\Omega$, respectively.

 TOOL: *Ohm's Law Calculator*

Equation (1.3) expresses the voltage in terms of the current. This equation can be inverted to express the current in terms of the voltage when that is more convenient:

$$i_r(t) = \frac{1}{R}v_r(t) = Gv_r(t). \qquad (1.4)$$

The reciprocal of the resistance, G, is called the *conductance*. The name was chosen because, for a high value of the conductance, the resistor conducts current very well. Values of conductance are measured in *siemens (S)*:

$$1\text{S} = 1 \text{ ampere/volt.}$$

A common way of presenting a resistor's element equation, $v(t) = Ri(t)$, is graphically, as shown in Figure 1-4. Notice that the graph of the v–i relation for an ideal resistor is a straight line that passes through the origin. It is for this reason that this type of resistor is also referred to as a *linear* resistor. As the graph implies, the linear behavior of an ideal resistor extends over the whole range of voltages and currents. Actual resistors that you would put into a circuit are less nearly ideal. They might behave like linear resistors only over some limited range of voltages and currents. Furthermore, their resistance might vary with external factors, such as temperature.

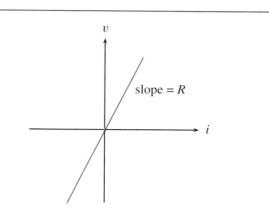

Fig. 1-4. Graphical depiction of the instantaneous v–i relation for an ideal resistor with a resistance R. The voltage is always proportional to the current.

 DEMO: *Interactive Resistance Graph*

To derive an expression for the instantaneous power absorbed by a resistor, we begin with the definition of the instantaneous power,

$$P_{\text{inst}}(t) = v_r(t)i_r(t),$$

and substitute Ohm's law (either (1.3) or (1.4)). This gives

$$P_{\text{inst}}(t) = Ri_r^2(t) = Gv_r^2(t).$$

From either expression, we notice that, for a resistor, $P_{\text{inst}}(t)$ is always nonnegative. From this, we conclude that

Resistors always absorb power.

 FLUENCY EXERCISE: *Ohm's law*

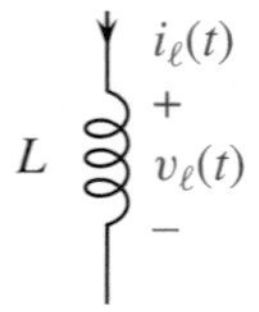

$$i_\ell(t)$$
$$+$$
$$L \quad v_\ell(t)$$
$$-$$

$$v_\ell(t) = L \frac{di_\ell(t)}{dt}$$

$$i_\ell(t) = \frac{1}{L} \int_{t_0}^{t} v_\ell(\beta)\, d\beta + i_\ell(t_0)$$

Fig. 1-5. An ideal inductor and the mathematical description of its behavior.

1-2-3   Inductors

The second of the three basic linear-circuit elements of interest is the *inductor*, the graphical symbol for which is given in Figure 1-5. For an ideal inductor, the voltage is proportional to the first derivative of the current:

$$v_\ell(t) = L \frac{di_\ell(t)}{dt}. \tag{1.5}$$

The parameter L is called the *inductance*, which is measured in *henrys (H)*, where

$$1\mathrm{H} = 1\ (\text{volt} \cdot \text{second})/\text{ampere}. \tag{1.6}$$

A 1H inductance is extremely large. Inductances measured in millihenries (mH) or microhenries (μH) (10^{-3}H and 10^{-6}H, respectively) are more common in circuits.

 DEMO: *Interactive Inductor Graph*

Equation (1.5) is useful when the current is the independent variable and the voltage is the dependent variable. For those occasions on which the roles of the two variables are reversed, it might be convenient to express the v–i relation of the inductor so that the inductor current is a function of its voltage. The inductor voltage

is proportional to the derivative of the current, so we can compute the inductor current by integrating the voltage:

$$i_\ell(t) = \frac{1}{L} \int_{t_0}^{t} v_\ell(\beta)\, d\beta + i_\ell(t_0). \tag{1.7}$$

The time t_0 corresponds to a reference time at which observations are begun (often chosen to be $t = 0$), and $i_\ell(t_0)$ is called the *initial value of the inductor current*.

The instantaneous power absorbed by an inductor is

$$P_{\text{inst}}(t) = v_\ell(t)i_\ell(t). \tag{1.8}$$

If we substitute (1.5), this equation can be rewritten as

$$P_{\text{inst}}(t) = L \frac{di_\ell(t)}{dt} i_\ell(t)$$
$$= \frac{d}{dt}\left[\frac{1}{2}Li_\ell^2(t)\right].$$

Integrating this result from t_0 to t gives an expression for the total *energy* that flows into the inductor during this interval:

$$U_\ell(t) = \int_{t_0}^{t} \frac{d}{d\tau}\left[\frac{1}{2}Li_\ell^2(\tau)\right] d\tau$$
$$= \frac{1}{2}Li_\ell^2(t) - \frac{1}{2}Li_\ell^2(t_0). \tag{1.9}$$

Notice that $P_{\text{inst}}(t)$ can be either positive or negative—that is, energy can flow either into or out of an inductor. We also notice that the total energy flow into the inductor from t_0 to t depends only on the values of the current at the ends of the interval and not on how the current varies within that interval. If we select t_0 to be a time such that

$i_\ell(t_0) = 0$, then we can say that the total energy stored in the inductor is

$$U_\ell(t) = \frac{1}{2} L i_\ell^2(t). \tag{1.10}$$

Any inductor that has a nonzero current flowing through it stores energy. This energy is stored in a magnetic field associated with the inductor.

The definition of inductance comes from the relation

$$\psi = Li,$$

where ψ is the **flux** created by the magnetic field associated with the inductor structure. This straight-line relationship means that the inductance is constant and independent of the magnetic flux. Air-core inductors, which are constructed as hollow coils of conducting wire, come closest to approximating this ideal linear behavior. Essentially, we can model any device that stores energy in a magnetic field, including a straight length of perfectly conducting wire, as an inductor. Inductors can take a variety of forms, varying from coils of wire to the printed metal spirals on the semiconductor gallium arsenide that are used as inductors for radio frequency integrated circuits (RFICs). The voltage is related to the flux through the relation

$$v_\ell(t) = \frac{d\psi(t)}{dt}, \tag{1.11}$$

which reduces to (1.5) when the inductance is constant.

1-2-4 Capacitors

The third basic linear-circuit element is the ideal *capacitor*. A capacitor is a physical device that stores energy in an *electric* field. For an ideal capacitor the voltage and current are again related by derivatives, but the roles of the voltage and current variables are reversed from the case that we saw for an inductor. With the voltage $v_c(t)$ and current $i_c(t)$ being defined under our normal convention, as shown in Figure 1-6, for the capacitor,

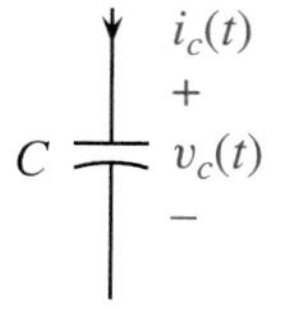

$$v_c(t) = \frac{1}{C} \int_{t_0}^{t} i_c(\beta)\, d\beta + v_c(t_0)$$

$$i_c(t) = C \frac{dv_c(t)}{dt}$$

Fig. 1-6. An ideal capacitor with the conventions for its voltage and current. The v–i relations are also included.

---

the current is proportional to the first derivative of the voltage:

$$i_c(t) = C \frac{dv_c(t)}{dt}. \tag{1.12}$$

The constant of proportionality, C, is called the *capacitance*, which is measured in *farads (F)*:

$$1\text{F} = 1 \text{ coulomb/volt}. \tag{1.13}$$

Again, the basic unit is quite large and is rarely seen. The capacitances more commonly encountered are measured in microfarads ($1\mu\text{F} = 10^{-6}\text{F}$) or picofarads ($1\text{pF} = 10^{-12}\text{F}$).

The capacitor voltage, $v_c(t)$, can be computed from the capacitor current, $i_c(t)$, by integration:

$$v_c(t) = \frac{1}{C} \int_{t_0}^{t} i_c(\beta)\, d\beta + v_c(t_0). \tag{1.14}$$

The quantity $v_c(t_0)$ is called the *initial capacitor voltage*.

 DEMO: *Interactive Capacitor Graph*

Like an inductor, a capacitor can have either a positive or a negative instantaneous power—that is, it can (temporarily) absorb energy or release it. The *instantaneous power* absorbed by a capacitor can be computed by using the definition:

$$P_{\text{inst}}(t) = v_c(t)i_c(t)$$

$$= C\frac{dv_c(t)}{dt}v_c(t) \tag{1.15}$$

$$= \frac{d}{dt}\left[\frac{1}{2}Cv_c^2(t)\right]. \tag{1.16}$$

Treating the capacitor the same way that we treated the inductor earlier, we see that the *total energy flow* into the capacitor between time t_0 and time t is given by

$$U_c(t) = \int_{t_0}^{t}\frac{d}{d\tau}\left[\frac{1}{2}Cv_c^2(\tau)\right]d\tau$$

$$= \frac{1}{2}Cv_c^2(t) - \frac{1}{2}Cv_c^2(t_0) \tag{1.17}$$

and that the total energy stored in the capacitor is

$$U_c(t) = \frac{1}{2}Cv_c^2(t). \tag{1.18}$$

This energy is stored in an electric field.

A capacitor can be constructed out of two conducting surfaces separated by a gap filled with some non-conducting material. Charge accumulates on the surfaces, and energy is stored in the electric field created between them. These are not the only devices that act like capacitors, however. Anywhere that there is an electric field, there will be some capacitance. For example, semiconductor junctions formed by abutting p-type and n-type materials exhibit capacitance, as do the plasma membranes of electrically excitable cells, such as neurons.

The physical principle that underlies a capacitor is the fact that the electric charge stored on a pair of closely separated conducting plates is proportional to the voltage across them. The capacitance is defined as the proportionality constant. Thus, for an ideal capacitor,

$$q(t) = Cv(t),$$

where $q(t)$ is the charge on one of the plates of the capacitor, C is the capacitance, and $v(t)$ is the applied voltage across the capacitor. One way to think of this is that, if we apply a constant voltage V across a capacitor by using a battery, a charge $+Q$ will appear on the plate connected to the positive terminal of the battery and the charge $-Q$ will appear on the plate connected to the negative terminal of the battery. The total charge on a capacitor is zero, which means that there will be (it is hoped) a negligible electric field outside the capacitor; the electric field is confined to the volume between the plates.

1-2-5   Short Circuits

You probably already have an intuitive notion of what a *short circuit* is. In circuit terms, it is a circuit element whose two terminals are connected by a perfectly conducting wire. The voltage across the terminals of a short circuit is always zero, regardless of the value of the current (which, in general, is nonzero). It corresponds to the limiting case of a resistor whose resistance is zero and whose conductance is infinite. Figure 1-7 shows an example of a short circuit.

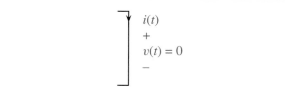

Fig. 1-7. A short circuit. The voltage $v(t)$ is zero, regardless of the value of the current $i(t)$.

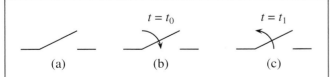

Fig. 1-8. An open circuit. The current $i(t)$ is zero, regardless of the value of the voltage $v(t)$.

1-2-6 Open Circuits

An *open circuit* can also be considered as a degenerate circuit element, one whose two terminals are not connected. It is the opposite of a short circuit; the current $i(t)$ through an open circuit is always zero, regardless of the voltage across the terminals. It thus corresponds to the limiting case of a resistor with an infinite resistance and a zero conductance. An example of an open circuit is shown in Figure 1-8.

1-2-7 Switches

Switches are familiar devices from your own experience. They can be used to turn devices on or off. The most familiar examples are mechanical devices in which an electrical connection is either made or broken by the position of the switch, but electrical switches made from semiconductors are also common. The graphical representation of a switch, which is shown in Figure 1-9, suggests the mechanical interpretation. An ideal switch,

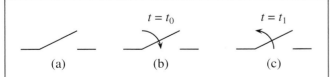

Fig. 1-9. (a) A generic switch. (b) A switch that is closed at time $t = t_0$. (c) A switch that is opened at time $t = t_1$.

which we shall use in our mathematical models, corresponds to a short circuit when it is closed, but to an open circuit when it is open.

1-3 Independent Sources

Independent sources provide a mechanism for getting one or more signals (and power) into a circuit. They come in two types: *voltage sources* and *current sources*. In circuit diagrams, these are represented as in Figure 1-10. Notice that, for both, there is no need to specify the reference directions for the voltage or current explicitly, since they are implied by the device symbols themselves.

A *voltage source* with a voltage of $v_s(t)$ maintains that voltage across its terminals no matter how much current flows through the device. This means that the instantaneous v–i characteristic for a voltage source is a horizontal line, as drawn in Figure 1-11.

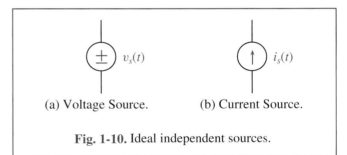

(a) Voltage Source.   (b) Current Source.

Fig. 1-10. Ideal independent sources.

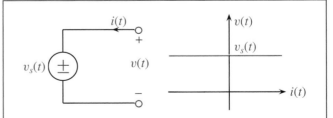

Fig. 1-11. The graph on the right shows the instantaneous v–i characteristic of the voltage source shown on the left.

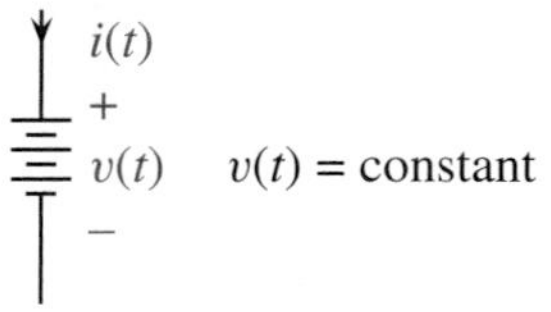

$i(t)$

$+$

$v(t)$ $v(t) = \text{constant}$

$-$

Fig. 1-12. An ideal battery. A battery is a voltage source whose source voltage is constant, regardless of the current.

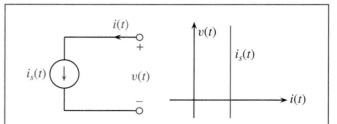

Fig. 1-13. The graph on the right shows the instantaneous v–i characteristic of the current source shown on the left.

An *ideal battery* is a special case of a voltage source for which $v_s(t) = \text{constant}$, regardless of the current drawn from it. Batteries are often drawn as in Figure 1-12. This, of course, is not a full and accurate description of the performance of an actual battery. Common sense tells us that a battery cannot supply an infinite amount of current. If the current drawn is large enough, we will discharge the battery and the voltage across it will go to zero. As is the case with all physical components, however, if we restrict the element variables—the current, in this case—to the required range of operation, then our approximation of a real battery by an ideal one will be sufficiently accurate to allow us to use it in circuit designs. Always be mindful, however, of the fact that it is important to know the range of operation of a device and, to the extent possible, to keep the device within that preferred region of operation.

A *current source* with a source current of $i_s(t)$ maintains its current as $i_s(t)$ no matter what voltage appears across the terminals. The graph of the instantaneous v–i characteristic of a current source is a vertical line in the $(v$–$i)$-plane, as drawn in Figure 1-13.

Many of the analysis techniques that we shall develop involve temporarily turning off one or more of the independent sources. For a voltage $v_s(t)$ on a voltage source to be "turned off" means setting $v_s(t) = 0$. The result is a device that has a potential difference of zero volts between its terminals, regardless of the current that flows through it: A voltage source that is turned off is

equivalent to a *short circuit* or a length of perfectly conducting wire. When we turn off a current source, we set $i_s(t) = 0$. The result is a device through which no current flows, regardless of the voltage across its terminals: A current source that is turned off is equivalent to an *open circuit*.

1-4   Kirchhoff's Laws

Elements and sources by themselves are not very interesting, and there is not a great deal that we can do with them until we connect them together; then we can do some very interesting and useful things. We connect elements and sources by joining their terminals; the connection points themselves are called *nodes*. When two or more terminals are connected at a node, current can flow freely from any element to any other. A node does not have to be an isolated point. If several elements are connected by short circuits, the entire set of connections, including the short circuits, constitutes the node. Figure 1-14 shows a circuit with its nodes encircled by dashed lines.

Any arrangement of elements and sources connected at their terminals is called a network or circuit.

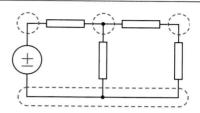

Fig. 1-14. A circuit with its nodes indicated by dashed lines.

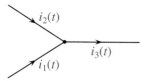

Fig. 1-15. Elements joined at a node to illustrate Kirchhoff's current law.

The voltages and currents associated with the elements and sources in a network are governed by three types of constraints. First, there are the v–i (element) relations, such as we have already seen for resistors, inductors, and capacitors. Those relate the potential differences between the terminals of the elements to the currents passing through them. The element relations are properties of the elements alone and do not depend upon how the elements are connected together. The other two types of constraints are embodied in Kirchhoff's laws. These depend only on the connections and have nothing to do with the elements themselves. The first of these is a constraint on the currents in the network known as ***Kirchhoff's current law (KCL)***. The second is a constraint on the voltages known as ***Kirchhoff's voltage law (KVL)***.

1-4-1 Kirchhoff's Current Law (KCL)

Kirchhoff's current law states the following:

> *The sum of all the currents entering a connection point, or node, is equal to zero.*

Physically, this is a statement that, in the ideal networks of our models, electric charge cannot accumulate at a node. It is a statement about the conservation of charge: every electron that leaves an element or source must then flow into one of the other elements or sources to which it is connected. If we invoke a fluid analogy, where we think of electrical current as water flowing in a closed system of pipes (i.e., one with no leaks), then the amount of water flowing into some junction in the system is equal to the amount of water flowing out of that junction. For the special case of the node illustrated in Figure 1-15, KCL means that

$$i_1(t) + i_2(t) - i_3(t) = 0.$$

The minus sign appears in front of $i_3(t)$ because this is a current whose reference direction *leaves* the node; thus, $-i_3(t)$ is the current entering the node through this branch. If we multiply both sides of this equation by -1, we get

$$-i_1(t) - i_2(t) + i_3(t) = 0.$$

This equation states that the sum of all the currents *leaving* the node must also be zero; KCL equations can be written either way.

Example 1-1   Writing KCL Equations

To illustrate the process of writing KCL equations, consider the circuit shown in Figure 1-16, which contains three elements and a voltage source. The three nodes where the elements and sources are connected are enclosed by dashed circles and labeled *a*, *b*, and *c*. We have also defined four currents associated with these

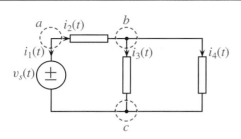

Fig. 1-16. A circuit with three nodes and four currents to illustrate the writing of KCL equations.

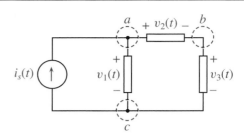

Fig. 1-17. Another circuit with three nodes, to illustrate writing KCL equations when voltage labels are used and the default sign convention is implied.

devices, for which the reference directions have been assigned arbitrarily. Since there are three nodes, we can write three KCL equations, although, as we shall see shortly, only two of these are independent. To be consistent, we write each equation by summing the currents that *enter* the node:

> *node a:* $-i_1(t) - i_2(t) = 0$
>
> *node b:* $i_2(t) - i_3(t) - i_4(t) = 0$
>
> *node c:* $i_1(t) + i_3(t) + i_4(t) = 0.$

As mentioned earlier, it is very important to pay attention to the reference directions, which determine the signs in these equations. Each current whose reference direction is *into* a particular node has a plus sign, and each current whose reference direction is *out of* a particular node receives a minus sign.   ■

 To check your understanding, try Drill Problems P1-1 and P1-2, then compare your solutions with the ones on the web site.

 WORKED SOLUTION: *Writing KCL equations*

Example 1-2   Writing KCL Equations by Using the Default Sign Convention

Labelling all of the voltages and currents in a circuit can often lead to a very cluttered drawing. If the default sign convention is rigidly followed, it is sufficient to supply reference directions for *either* the voltages *or* the currents, since the reference directions for the other variables are then implied.

Recall the default sign convention, which assigns the positive reference direction for a current associated with an element to be the direction that enters the terminal marked by a $^+$. When we need to write KCL equations for a circuit diagram on which only the voltages have been labelled, the procedure now involves two steps: we use the indicated voltages to define the reference directions for the currents, and then we use the reference directions for the currents to write the KCL equations.

In the circuit in Figure 1-17, the voltages across the three elements have been labelled, but not the currents. Instead, the current reference directions are implied by the reference directions of the voltages. Define current $i_1(t)$ to be the current flowing into the $^+$ terminal of the element with voltage $v_1(t)$, and define $i_2(t)$ and $i_3(t)$, similarly, as the currents flowing into the $^+$ terminals of the other two elements. If we again write the KCL equations by summing the currents that *enter* the nodes of the circuit, we get the following three KCL equations:

> *node a:* $i_s(t) - i_1(t) - i_2(t) = 0$
>
> *node b:* $i_2(t) - i_3(t) = 0$
>
> *node c:* $-i_s(t) + i_1(t) + i_3(t) = 0.$ ∎

 To check your understanding, try Drill Problem P1-3.

Example 1-3   Writing KCL Equations by Using Voltages

When the circuit elements are resistors, we can take what we learned in the previous example one step further and write the KCL equations directly in terms of the voltage variables, by incorporating Ohm's law. This allows us to avoid defining the current variables altogether. Since the effort required to solve a series of equations is directly related to their number, reducing the number of variables usually reduces the number of equations that must be solved.

To illustrate this approach, consider the circuit in Figure 1-18. We approach this problem exactly as we did the previous one; the current $i_1(t)$ flows into the $+$ terminal of the resistor with the voltage $v_1(t)$, etc., but,

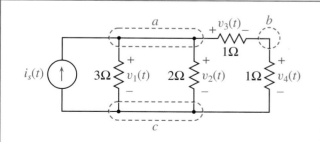

Fig. 1-18. A circuit containing three nodes at which we wish to write KCL equations in terms of the voltage variables and the source current.

instead of writing $i_1(t)$, we incorporate the statement of Ohm's law for this 3Ω resistor:

$$i_1(t) = \frac{v_1(t)}{3}.$$

The three KCL equations at the three nodes are then given by

> *node a:* $i_s(t) - \dfrac{v_1(t)}{3} - \dfrac{v_2(t)}{2} - \dfrac{v_3(t)}{1} = 0$
>
> *node b:* $\dfrac{v_3(t)}{1} - \dfrac{v_4(t)}{1} = 0$
>
> *node c:* $-i_s(t) + \dfrac{v_1(t)}{3} + \dfrac{v_2(t)}{2} + \dfrac{v_4(t)}{1} = 0.$

Even though these equations are written in terms of the voltages, it is important to remember that they are still KCL equations, which are statements about the conservation of currents.   ∎

 To check your understanding, try Drill Problem P1-4.

 MULTIMEDIA TUTORIAL: *KCL Equations*

Although we presented KCL as a constraint on the currents at a connection point, or node, it can also be applied in more general situations. A more general statement of KCL states,

> *The sum of all the currents that enter any closed portion of a circuit is equal to zero.*

 DEMO: *Kirchhoff's Current Law*

To explain what we mean by this statement, and also to see why it is implied by the statements of KCL applied at the nodes of a circuit, consider the portion

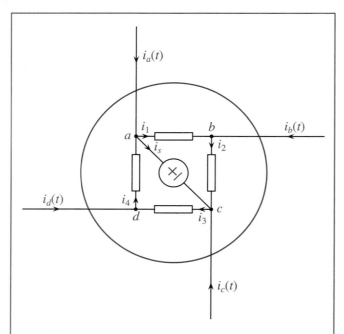

Fig. 1-19. A portion of a large circuit that is isolated from the remainder of the network by the closed surface shown as a large colored circle. The only connections to the remainder of the circuit are through the four wires with currents $i_a(t)$, $i_b(t)$, $i_c(t)$, and $i_d(t)$. The sum of all the currents intersecting the surface satisfy the generalized KCL.

of a more complete circuit illustrated in Figure 1-19. In that figure, the colored circle isolates a portion of a network containing, in this case, the four nodes a, b, c, and d, along with several elements and one source. This subnetwork is connected to the remainder of the circuit by wires carrying the currents $i_a(t)$, $i_b(t)$, $i_c(t)$, and $i_d(t)$, respectively. These are the only currents that enter the subnetwork from other parts of the circuit. Applying the more general statement of KCL to the currents that cross this surface implies the relation

$$i_a(t) + i_b(t) + i_c(t) + i_d(t) = 0. \qquad (1.19)$$

To show that this equation is implied by our earlier statement of Kirchhoff's current law, we write KCL equations at all of the nodes within the subnetwork— that is, at all of the nodes located within the circle. For this circuit, this involves writing equations at nodes a, b, c, and d. To be consistent, we sum the currents *entering* the nodes for all four equations:

node a: $i_4(t) - i_1(t) - i_s(t) + i_a(t) = 0$

node b: $i_1(t) - i_2(t) + i_b(t) = 0$

node c: $i_2(t) - i_3(t) + i_s(t) + i_c(t) = 0$

node d: $i_3(t) - i_4(t) + i_d(t) = 0.$

Adding the four equations together gives

$$i_a(t) + i_b(t) + i_c(t) + i_d(t) = 0,$$

which corresponds to our more general statement of KCL.

It should be clear that this result is true in general, and not just for this specific network, since every element that is wholly contained within the subnetwork connects exactly two nodes in that subnetwork. The reference direction of the current associated with such an element will necessarily enter one of the nodes and leave the other. Thus, if the equations are written consistently, in the KCL equation for one node that current will appear with a plus sign and in the KCL equation for the other node it will appear with a minus sign. When the KCL equations are added, these currents cancel. Those currents that intersect the separating surface are all attached to the subnetwork at only a single node inside the surface, which means that they appear in only one KCL equation. When the KCL equations are added, these terms do not cancel.

Two further observations can be made about using this more general statement of KCL. First,

> *It is sufficient to write KCL equations only at the connection points or nodes in a network.*

These incorporate all of the constraints on the currents in the circuit. Writing KCL equations across more general surfaces does not add additional information, because, as we have just seen, these equations can be derived from the KCL equations at the nodes inside the surface. Thus, normally, we shall choose to write KCL equations only at nodes.

Actually, we can go one step further:

> *If a network has a total of n nodes, it is sufficient to write KCL equations at only n − 1 of them.*

The KCL equation at the remaining node is not independent of the others and provides no additional information. This fact follows from the above derivation. Imagine drawing a surface that encircles the entire circuit and writing a KCL equation for each enclosed node, i.e., for every node in the circuit. If these equations are written consistently—that is, if in each case, we sum the currents entering the node (or the currents leaving the node)—then the sum of all the nodal KCL equations will be identically zero. This means that any particular equation is equal to −1 times the sum of all the others. The reader is encouraged to verify this fact, using the examples in this section.

1-4-2 Kirchhoff's Voltage Law (KVL)

Kirchhoff's current law constrains the currents in a circuit; Kirchhoff's voltage law constrains the voltages. It states,

> *The sum of the element and source voltages around any closed path in a circuit is equal to zero.*

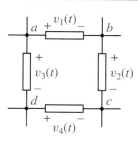

Fig. 1-20. A portion of a network containing four nodes used to illustrate Kirchhoff's voltage law (KVL).

If we consider the clockwise path **abcd** in the portion of a circuit shown in Figure 1-20, then KVL implies the following constraint on the voltages:

$$v_1(t) + v_2(t) - v_4(t) - v_3(t) = 0. \qquad (1.20)$$

It is important to pay attention to the reference directions for the voltage variables. In traversing the path, if the $^+$ terminal of an element or source is encountered before the $^-$ terminal, the voltage gets a positive sign; if the $^-$ terminal is encountered first, the voltage gets a minus sign.

KVL allows us to attribute a quantity called the *potential* to each node. The potential is the sum of the voltages along any path that ends at the node and begins at a designated reference node. Again, we refer to the portion of a circuit in Figure 1-20 as an example. We let node d be the reference node, and we assign to it a potential of zero. We denote the potentials at the other three nodes as $e_a(t), e_b(t),$ and $e_c(t)$. Then, if we compute the potential at node b as the accumulated voltages over the path connecting nodes d and b that passes through node a, we have

$$e_b(t) = v_3(t) - v_1(t).$$

From the statement of KVL in (1.20), however,

$$v_3(t) - v_1(t) = v_2(t) - v_4(t).$$

Thus, we can also say that

$$e_b(t) = -v_4(t) + v_2(t),$$

which is the sum of the accumulated voltages over the path that connects the same two nodes, but which passes through node c.

> *Because of KVL, the sum of the accumulated voltages over any path that connects two particular nodes is independent of the path!*

This makes the definition of the potential unambiguous.

A two-terminal element is necessarily connected to two nodes in a circuit. The voltage associated with the element is thus equal to the difference between the potentials of the two nodes to which it is connected. For our example,

$$v_1(t) = e_a(t) - e_b(t)$$

$$v_2(t) = e_b(t) - e_c(t)$$

$$v_3(t) = e_a(t) - e_d(t)$$

$$v_4(t) = e_d(t) - e_c(t).$$

For this reason the term *potential difference* is used synonymously with voltage, and the voltage of an element is equal to the potential of its $^+$ terminal minus the potential of its $^-$ terminal.

To become more familiar with the process of writing KVL equations, let us revisit the examples that we saw earlier.

 DEMO: *Kirchhoff's Voltage Law*

Example 1-4   Writing KVL Equations

As a first example, consider the circuit in Figure 1-21, which is the same one that we encountered in

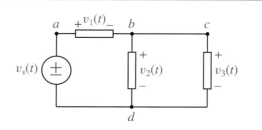

Fig. 1-21. A circuit with three elements and an independent source, used to illustrate the writing of KVL equations.

Example 1-1, except that this time we specifically indicate the voltages.

This network contains three simple closed paths (closed paths that do not cross themselves). By using the node labels, we can call these paths *abda*, *bcdb*, and *abcda*. These yield the following three KVL equations:

> *path abda:* $v_1(t) + v_2(t) - v_s(t) = 0$
>
> *path bcdb:* $v_3(t) - v_2(t) = 0$
>
> *path abcda:* $v_1(t) + v_3(t) - v_s(t) = 0.$

Remember that the voltage associated with a short circuit is zero. Thus, there is no potential difference between the points in the circuit labelled b and c.   ∎

 To check your understanding, try Drill Problems P1-5 and P1-6.

Example 1-5   **Writing KVL Equations by Using the Default Sign Convention**

When only the currents are labelled on a circuit drawing and the default sign convention is implied, finding the reference directions for the voltages in the KVL equations involves using the current labels to identify the $^+$ and $^-$

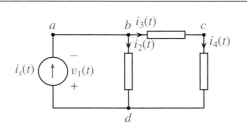

Fig. 1-22. A circuit to illustrate the writing of KCL equations when only the currents are labelled.

terminals of the elements. This is straightforward, but it does require an extra level of care.

In the circuit in Figure 1-22, the currents are labelled, but not the voltages. Notice that we have introduced the variable $v_1(t)$ to represent the voltage across the current source. This is a necessary and important step. Beginning students often assume that this voltage is zero, but this is a serious mistake.

> *The potential difference across the terminals of a current source, like the current through a voltage source, is almost never zero, because the sources usually supply power.*

As in the previous example, there are again three simple closed paths: *abda*, *bcda*, and *abcda*. If $v_2(t)$, $v_3(t)$, and $v_4(t)$ are the three implied, but unlabelled (except by the default sign convention), element voltages, the three KVL equations are as follows:

path abda: $\quad v_2(t) + v_1(t) = 0$

path bcdb: $\quad v_3(t) + v_4(t) - v_2(t) = 0$

path abcda: $\quad v_3(t) + v_4(t) - v_1(t) = 0.$

Notice that, when we traverse the path and we go in the direction of one of the current reference arrows, that

voltage gets a plus sign; whenever we go in a direction that is opposite to one of the reference arrows, the corresponding voltage gets a minus sign.   ∎

To check your understanding, try Drill Problem P1-7.

WORKED SOLUTION: *Writing KVL equations*

Example 1-6   **Writing KVL Equations by Using Currents**

If a circuit contains only resistors and independent sources, we can use Ohm's law to express the voltages directly in terms of the current variables. This allows both the KCL and KVL equations to be written in terms of the same reduced set of variables. This is similar to what we did in Example 1-3, except that there we wrote KCL equations in terms of voltage variables and here we are writing KVL equations in terms of current variables. To illustrate the procedure, we use the same circuit as for that earlier example. It is redrawn in Figure 1-23 with the current variables indicated. Notice, as in the previous example, that we must again introduce a voltage variable to account for the initially unknown (and nonzero) potential difference across the terminals

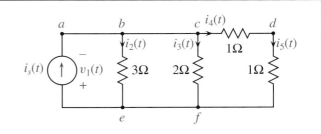

Fig. 1-23. A resistive circuit to demonstrate writing KCL equations in terms of the current variables and the voltage across the current source.

of the current source. For this circuit, there are six simple paths on which we could write KVL equations. As we shall see shortly, these are not all independent, but since the goal here is to become familiar with the procedure of writing KVL equations, we shall go ahead and write all six equations:

$$\textbf{\textit{path abea:}} \quad 3i_2(t) + v_1(t) = 0$$

$$\textbf{\textit{path bcfeb:}} \quad -3i_2(t) + 2i_3(t) = 0$$

$$\textbf{\textit{path cdfc:}} \quad -2i_3(t) + 1i_4(t) + 1i_5(t) = 0$$

$$\textbf{\textit{path abcfea:}} \quad 2i_3(t) - v_1(t) = 0$$

$$\textbf{\textit{path bcdfeb:}} \quad 1i_4(t) + 1i_5(t) - 3i_2(t) = 0$$

$$\textbf{\textit{path abcdfea:}} \quad 1i_4(t) + 1i_5(t) - v_1(t) = 0.$$

Writing these equations is not difficult, but there are many opportunities for careless errors, both with the signs and with the repeated application of Ohm's law. (*Caveat lector!*) ■

 To check your understanding, try Drill Problem P1-8.

For most circuits, there are many possible closed paths that can be used to write KVL equations, but not all of these provide *independent* constraints on the voltages, so some of these equations are redundant. As an example, consider the network in Figure 1-24. If we write a KVL equation for the voltages encountered on the closed clockwise path *abeda* (path #1), we get

$$v_{ab}(t) + v_{be}(t) - v_{de}(t) - v_{ad}(t) = 0. \quad (1.21)$$

Similarly, path *bcfeb* (path #2) gives rise to the KVL equation

$$v_{bc}(t) + v_{cf}(t) - v_{ef}(t) - v_{be}(t) = 0. \quad (1.22)$$

These two equations are independent because (1.22) cannot be derived from (1.21). The longer path *abcfeda*,

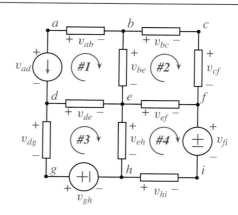

Fig. 1-24. A network containing a number of elements and sources and four meshes.

which contains portions of both paths #1 and #2, however, yields a KVL equation,

$$v_{ab}(t) + v_{bc}(t) + v_{cf}(t) - v_{ef}(t) - v_{de}(t) - v_{ad}(t) = 0, \quad (1.23)$$

that contains no new information; (1.23) is equal to the sum of (1.21) and (1.22).

This leads to two rather obvious questions: how many independent KVL equations can we write and how can we find the appropriate paths with the least effort? For *planar networks* these questions have straightforward answers.[1] A planar network or circuit is a circuit that can be drawn (or redrawn) on a sheet of paper with no elements or wires crossing each other. In a sense, these circuits are thus two dimensional (i.e., planar), while nonplanar circuits are three dimensional. All of the examples that we have seen so far are planar, as, indeed, are most of the examples that we shall encounter in the remainder of this text.

 FLUENCY EXERCISE: *Planar Networks*

---

[1] Some of the methodologies that we shall develop in Chapter 2 will enable us to analyze nonplanar networks.

Define a *mesh* as a simple closed path that does not encircle any elements or sources. In Figure 1-24, Path #1 and Path #2 are meshes, but path *abcfeda* is not, because it encircles the element that connects node *b* to node *e*. Figure 1-24 contains four meshes, which are the paths that encircle the four "window panes" indicated by the four colored circular arrows.

> *For a planar network, we can obtain a minimal set of independent KVL equations by writing one KVL equation for each path that is associated with a mesh in the network.*

This selection procedure is based on two facts: (1) the KVL equations written on the meshes are independent; and (2) the KVL equations on any other paths can be derived from the mesh equations and are thus redundant. It is important to note that this choice of KVL equations is sufficient, but it is not necessary; there are other paths that we could choose to derive minimal sets of independent KVL equations. These alternative sets of equations, however, are all equivalent to using the mesh equations; they impose equivalent constraints on the voltages in the circuit and they result in the same solution.

How do we know that the mesh KVL equations are independent? For the network shown in Figure 1-24, each mesh equation involves at least one element that is not present in the other equations, so clearly each equation cannot be derived from the others. Unfortunately, this situation does not always occur. The proof that the mesh equations are nonetheless always independent for all planar circuits is left as an exercise for the reader. (See Problem P1-50.)

To demonstrate the sufficiency of the mesh equations, again consider the network in Figure 1-24. The four KVL equations derived from closed clockwise paths around the meshes are as follows:

#1: $v_{ab}(t) + v_{be}(t) - v_{de}(t) - v_{cd}(t) = 0$ (1.24)

#2: $v_{bc}(t) + v_{cf}(t) - v_{ef}(t) - v_{be}(t) = 0$ (1.25)

#3: $v_{de}(t) + v_{eh}(t) - v_{gh}(t) - v_{dg}(t) = 0$ (1.26)

#4: $v_{ef}(t) + v_{fi}(t) - v_{hi}(t) - v_{eh}(t) = 0.$ (1.27)

The path *abcfeda*, which we saw earlier, encircles the union of meshes 1 and 2, and its KVL equation (1.23) is the sum of the equations for those meshes. Because we took care to write our KVL equations by proceeding around the paths in a consistent direction (here clockwise), an element such as $v_{be}(t)$ that is encircled by the nonmesh path appears in one mesh equation with a positive sign and in the other equation with a negative sign. When the equations for adjacent meshes are added, the element voltages from any encircled elements cancel. Similarly, we could also write a KVL equation around the outermost path *abdfihgda*, but it can readily be seen that this corresponds to the sum of all four mesh equations; the voltages from all four of the encircled elements cancel similarly.

1-4-3   Solving for the Element Variables in a Circuit

The *complete solution* of a circuit is a set of time-varying voltages and currents such that the *v–i* relations of all of the elements and sources are satisfied, the currents satisfy KCL at all of the nodes and across all possible closed surfaces in the circuit, and the voltages satisfy KVL on all possible closed paths. When the circuit contains only resistors and sources, those equations are linear *algebraic* equations. If the equations are independent and the number of equations equals the number of unknowns, the solution is unique.

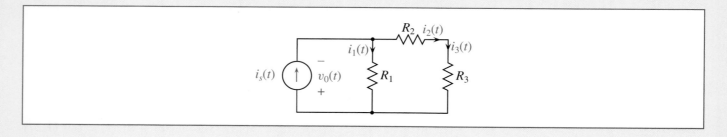

A Confusing Issue: Selecting Reference Directions

Many students who are new to the study of circuits find the process of defining reference directions for the voltages and currents to be intimidating. Their confusion is often accompanied by a question such as "How do we know which way the currents are going if we haven't solved the circuit yet?" This might be caused by thinking about voltages as unsigned quantities. In fact, most voltages and currents in most circuits vary as functions of time and the direction of current flow in a wire or an element will change from one time instant to another. (Think about the 60 Hz sinusoidal voltages in the power sockets in your house.) The choice of reference directions for the voltages and currents is completely arbitrary. Once we have chosen those directions, the signs of the resulting variables, when properly interpreted, will tell us the directions of the current flows at any instant and the polarities of the voltages. Now that we are in a position to solve simple resistive circuits, we can verify this fact.

Consider the simple resistive circuit in the figure at the top of this page.

The circuit has three nodes and two meshes, so we can write two KCL equations and two KVL equations using the four variables $i_1(t)$, $i_2(t)$, $i_3(t)$, and $v_0(t)$. We can incorporate Ohm's law directly to avoid having to define new variables for the voltages. The two KCL equations at the two upper nodes are

$$i_1(t) + i_2(t) = i_s(t)$$
$$i_2(t) - i_3(t) = 0.$$

and the two KVL equations are

$$v_0(t) + R_1 i_1(t) = 0$$
$$-R_1 i_1(t) + R_2 i_2(t) + R_3 i_3(t) = 0.$$

From the second of these four equations, we have $i_2(t) = i_3(t)$ and from the third $v_0(t) = -R_1 i_1(t)$. The remaining two equations are then

$$i_1(t) + i_2(t) = i_s(t)$$
$$-R_1 i_1(t) + (R_2 + R_3) i_2(t) = 0.$$

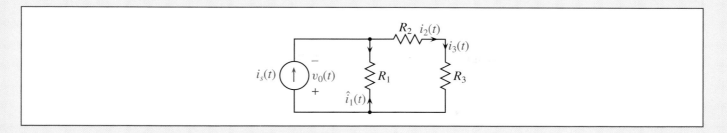

If we multiply the first of these equations by R_1 and add the result to the second, we have

$$(R_1 + R_2 + R_3)i_2(t) = R_1 i_s(t),$$

from which it follows that

$$i_2(t) = \frac{R_1}{R_1 + R_2 + R_3} i_s(t) \qquad (1.28)$$

$$= i_3(t). \qquad (1.29)$$

Using this result, we see that

$$i_1(t) = \frac{R_2 + R_3}{R_1 + R_2 + R_3} i_s(t)$$

$$v_0(t) = -\frac{R_1(R_2 + R_3)}{R_1 + R_2 + R_3} i_s(t).$$

Suppose now that we had originally defined $i_1(t)$ as going in the opposite direction as in the figure at the top of this page.

With these variables the KCL and KVL equations would be

$$-\hat{i}_1(t) + i_2(t) = i_s(t)$$

$$i_2(t) - i_3(t) = 0$$

$$v_0(t) - R_1 \hat{i}_1(t) = 0$$

$$R_1 \hat{i}_1(t) + R_2 i_2(t) + R_3 i_3(t) = 0.$$

These can be solved in a manner similar to the earlier set to give the solution

$$\hat{i}_1(t) = -\frac{(R_2 + R_3)}{R_1 + R_2 + R_3} i_s(t)$$

$$i_2(t) = i_3(t) = \frac{R_1}{R_1 + R_2 + R_3} i_s(t)$$

$$v_0(t) = -\frac{R_1(R_2 + R_3)}{R_1 + R_2 + R_3} i_s(t).$$

Notice that $v_0(t)$, $i_2(t)$, and $i_3(t)$ are the same as for the earlier solution and that $\hat{i}_1(t) = -i_1(t)$. A positive value for $i_1(t)$ represents a flow of current from top to bottom through the resistor, as does a negative value for $\hat{i}_1$; both solutions say the same thing. Thus, either choice for the reference direction for $i_1(t)$ (or any of the other element variables) gives the same solution for the circuit, when properly interpreted.

In Chapter 2, we shall show the following:

For a planar network it is sufficient to write

   1. *A v–i relation for each element,*

   2. *A KCL equation at all but one of the nodes of the circuit, and*

   3. *A KVL equation over the closed path defined by each mesh in the circuit,*

to find the complete solution of the circuit.

If this strategy is followed, the number of equations will always equal the number of unknowns. For the moment, however, we simply use this method to find the solution of a simple circuit containing only resistors and sources, without having proved its sufficiency.

Example 1-7   Finding the Complete Solution of a Circuit

Let us begin by finding the complete solution for the simple circuit shown in Figure 1-25, which contains a single voltage source and three resistors. There are four voltages and four currents, but the potential difference across the terminals of the independent voltage source, $v_s(t)$, is assumed to be known. The remaining voltage and current variables, $v_1(t), v_2(t), v_3(t), i_1(t), i_2(t), i_3(t)$, and $i_s(t)$ are initially unknown. The circuit contains three

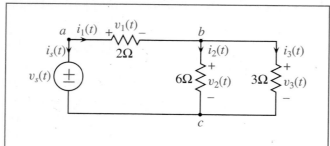

Fig. 1-25. A simple network to be analyzed.

elements, three nodes, and two meshes. Thus, following our proposed methodology, we write three v–i relations, two independent KCL equations, and two independent KVL equations. This corresponds to a total of seven equations in the seven variables.

The element relations for the three resistors allow us to express three of the unknown currents in terms of three of the voltages:

$$i_1(t) = v_1(t)/2 \qquad (1.30)$$
$$i_2(t) = v_2(t)/6 \qquad (1.31)$$
$$i_3(t) = v_3(t)/3. \qquad (1.32)$$

We must write KCL equations at only two of the three nodes if they are to be independent. Since it does not matter which two we select, let us arbitrarily choose to write them at nodes *a* and *b*. Summing the currents entering the nodes, we get

node a: $-i_s(t) - i_1(t) = 0$ (1.33)
node b: $i_1(t) - i_2(t) - i_3(t) = 0.$ (1.34)

Finally, writing KVL equations on the clockwise paths around the two meshes gives

left mesh: $-v_s(t) + v_1(t) + v_2(t) = 0$ (1.35)
right mesh: $-v_2(t) + v_3(t) = 0.$ (1.36)

We can now solve these seven equations in seven unknowns to find the complete solution by expressing each unknown voltage and current in terms of $v_s(t)$. Substituting (1.30), (1.31), (1.32), (1.33) and (1.36) into (1.34) and (1.35) reduces the number of equations to two in the two unknowns $v_1(t)$ and $v_2(t)$. After simplifying, these become

node b: $v_1(t) - v_2(t) = 0$
left mesh: $v_1(t) + v_2(t) = v_s(t).$

By solving these two equations algebraically, we get

$$v_1(t) = v_2(t) = \frac{1}{2}v_s(t).$$

Now that we know $v_1(t)$ and $v_2(t)$, it is straightforward to substitute for the other variables:

[from(1.36)] $v_3(t) = v_2(t) = v_s(t)/2$

[from(1.30)] $i_1(t) = v_1(t)/2 = v_s(t)/4$

[from(1.31)] $i_2(t) = v_2(t)/6 = v_s(t)/12$

[from(1.32)] $i_3(t) = v_2(t)/3 = v_s(t)/6$

[from(1.33)] $i_s(t) = -i_1(t) = -v_s(t)/4.$

We see that the current $i_1(t)$ flowing through the 2Ω resistor is divided into two streams—one that flows through the 6Ω resistor and one that flows through the 3Ω resistor. The magnitudes of those two currents $i_2(t)$ and $i_3(t)$ are inversely proportional to their respective resistances (i.e., the larger component of the current flows through the resistor with the smaller resistance). We also notice in this circuit, which contains only resistors and a single independent voltage source, that none of the voltages in the circuit is larger in magnitude than the source voltage itself. More importantly, however, we are able to demonstrate, at least for one circuit, that our procedure for finding the complete solution works!

This method of solution, which we shall later more formally name the *exhaustive method*, results in a fairly large number of equations in an equally large number of unknowns. It often tells us more about the circuit than we need to know. In later chapters we shall develop techniques that are faster and simpler, but they all originate from the insight gained from the exhaustive method. ■

 To check your understanding, try Drill Problems P1-9, P1-10, and P1-11.

Even though computers can be used to solve the linear equations that define the complete solution, there are nonetheless decided advantages to reducing the number of variables and equations involved. We shall look at some formal methods for deriving efficient network formulations in Chapter 2, but there are a couple of obvious simplifications that can be made that are evident in this example. These are the subject of the next section.

1-4-4   Elements Connected in Series and Parallel

If a node connects *exactly* two elements, those elements are said to be **connected in series**. In the circuit of Figure 1-25, which was used for the previous example, the voltage source and the 2Ω resistor are connected in series. Figure 1-26 shows two generic elements that are connected in series. Notice that the node connects exactly two elements. If we write a KCL equation at that node, it implies that

$$i_1(t) - i_2(t) = 0 \implies i_1(t) = i_2(t). \tag{1.37}$$

Thus,

> *The currents flowing through two elements connected in series are the same.*

(*Note:* if the reference directions for the two currents are different, the currents will be the negatives of each other.) We can use this fact to remove one unknown and one equation from the set of equations that must be solved to find the complete solution of any circuit that contains two

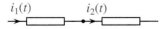

Fig. 1-26. Two elements connected in series.

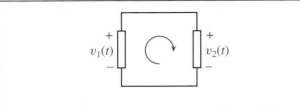

Fig. 1-27. Two elements connected in parallel.

elements connected in series. We do this by equating the currents of the two elements that are connected in series and omitting the KCL equation at the node where they are joined.

> *Whenever we detect two elements connected in series in a circuit, we can reduce the number of current variables and the number of KCL equations by one.*

Similarly, if a closed path around a mesh passes through *exactly* two elements, those elements are said to be *connected in parallel*. The 6Ω and 3Ω resistors in the previous example are connected in parallel. Figure 1-27 shows a generic parallel connection. Notice that, when two elements are connected in parallel, they connect the same pair of nodes. Writing a KVL equation on the closed path defined by the mesh between the two parallel elements yields

$$-v_1(t) + v_2(t) = 0 \implies v_1(t) = v_2(t).$$

(This is not surprising; since they connect the same pair of nodes, they have the same potential difference!) Thus, we can conclude the following:

> *Whenever two elements are connected in parallel, the same voltage appears across both of them.*

> *For each pair of parallel elements in a network, we can reduce the number of variables and the number of equations by one by equating the two element voltages and removing the KVL equation for the mesh created by the two parallel elements.*

Identifying series and parallel connections can sometimes be difficult for beginning students. We will have much more to say about series and parallel connections, and how we can exploit them to simplify circuits, in Chapter 3. For now, we will be content with a couple of additional examples.

Example 1-8   Exploiting Series and Parallel Connections

The circuit in Figure 1-28 contains four resistors and a current source, but two of the resistors are connected in series and the other two resistors and the current source are connected in parallel. Were we not to exploit these connections, we would need to write nine independent equations to solve for the nine unknown variables (voltages and currents for the four resistors and the potential difference across the current source). This would involve writing four v–i relations for the resistors, two KCL equations at two of the three nodes (which are

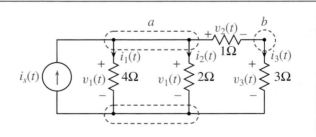

Fig. 1-28. A network containing both series and parallel connections.

circled), and three KVL equations on the paths defined by the three meshes. We observe, however, that the potential difference across the terminals of the current source and the voltages of the 4Ω and 2Ω resistors are the same and that the same current flows through both the 3Ω and 1Ω resistors. This reduces the number of unknowns to six. We can find these by writing four v–i relations, one KCL equation, and one KVL equation:

$$v\text{–}i \ \#1: \quad i_1(t) = v_1(t)/4 \tag{1.38}$$

$$v\text{–}i \ \#2: \quad i_2(t) = v_1(t)/2 \tag{1.39}$$

$$v\text{–}i \ \#3: \quad v_2(t) = 1 i_3(t) \tag{1.40}$$

$$v\text{–}i \ \#4: \quad v_3(t) = 3 i_3(t) \tag{1.41}$$

$$node \ a: \quad i_1(t) + i_2(t) + i_3(t) = i_s(t) \tag{1.42}$$

$$right \ mesh: \quad -v_1(t) + v_2(t) + v_3(t) = 0. \tag{1.43}$$

There are many approaches that will lead to a solution. One is to use (1.39), (1.40), and (1.41) to express $v_1(t)$, $v_2(t)$, and $v_3(t)$ in terms of $i_2(t)$ and $i_3(t)$. From (1.38) and (1.39) we know that $i_2(t) = 2i_1(t)$. Substituting all of this information into the last two equations reduces them to

$$3i_1(t) + i_3(t) = i_s(t)$$

$$-4i_1(t) + 4i_3(t) = 0.$$

The solution is

$$i_1(t) = i_s(t)/4 \qquad v_1(t) = i_s(t)$$

$$i_2(t) = i_s(t)/2 \qquad v_2(t) = i_s(t)/4$$

$$i_3(t) = i_s(t)/4 \qquad v_3(t) = 3i_s(t)/4.$$

∎

The next example demonstrates a simplification that is possible when we do not need all of the element variables. It also shows that we can solve for quantities that are not element variables.

Example 1-9   Ignoring Some Variables and Creating Others

The circuit in Figure 1-29 contains four resistors and a current source. The goal here is to learn the dependence of the voltage $v_{\text{out}}(t)$ on the source current, $i_s(t)$. This differs from our earlier examples in two important respects: first, we do not care about all the variables in the network; second, the variable of interest is not an element voltage, but instead is simply the difference in electrical potential between two nodes in the circuit that are not connected by a single element. The first of these differences simplifies both the problem setup and its solution; the second complicates it slightly. To deal with the second problem, we will need to solve for the voltages across the four resistors. Once these are known, we can then write an additional KVL equation over a "closed" path involving two of the resistors and the open circuit across which the output voltage is defined.

Examining the circuit we see that it contains four nodes and two meshes, but that some of these equations can be eliminated quickly. For example, in the central branch, resistor R_1 is connected in series with R_2; and, in the right branch, resistor R_2 is connected in series with R_1. We can use this fact to eliminate two current variables

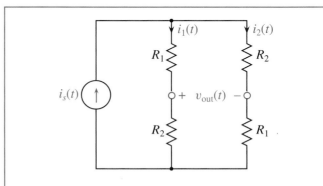

Fig. 1-29. A circuit for which the voltage of interest is not an element voltage.

and two KCL equations. Furthermore, a KVL equation around the left mesh serves only to compute the value of the voltage across the current source, but this value is not needed, so we can avoid this equation also. The v–i relations for the resistors (Ohm's law) can be also applied by inspection. This leaves a single KCL equation for the upper (or lower) node and a single KVL equation for the right mesh. As variables, we use the two currents $i_1(t)$ and $i_2(t)$:

$$i_1(t) + i_2(t) = i_s(t) \tag{1.44}$$

$$-(R_2 + R_1)i_1(t) + (R_2 + R_1)i_2(t) = 0. \tag{1.45}$$

Solving for $i_1(t)$ and $i_2(t)$ gives

$$i_1(t) = i_2(t) = i_s(t)/2. \tag{1.46}$$

Now that we know the currents, we write a final KVL equation around a closed path that involves the open circuit between the terminals where $v_{\text{out}}(t)$ is defined and the two resistors at the bottom of the circuit:

$$v_{\text{out}}(t) + R_1 i_2(t) - R_2 i_1(t) = 0.$$

After we substitute the result from (1.46), we have our answer:

$$v_{\text{out}}(t) = \frac{R_2 - R_1}{2} i_s(t).$$

Observe again that the output is proportional to the input.

∎

1-4-5   Circuits with Inductors and/or Capacitors

When a circuit contains one or more inductors and capacitors (collectively called *reactive elements*), the equations that define the complete solution are not all linear algebraic equations, since some of the element relations contain derivatives. The resulting set of coupled differential equations still guarantees a unique solution, provided that the initial values of the capacitor voltages and initial values of the inductor currents are known. Finding the complete solution, however, requires that we solve those differential equations.

Manipulating the differential equations and solving them greatly complicates the analysis procedure. For this reason, we will put off looking at circuits containing inductors and capacitors until Chapter 6. In the meantime, we shall look at resistive circuits in order to develop some insight into how circuits behave and to develop some techniques for simplifying the analysis procedure. We shall also develop a key tool, the Laplace transform, that will greatly help us to manipulate and solve circuits that contain reactive elements. Before putting these circuits aside, however, we first look at one simple example containing a capacitor.

Example 1-10   A Circuit with a Capacitor

The circuit in Figure 1-30 contains a voltage source, two resistors, and a capacitor. As variables, there are three element currents and three element voltages. There is no need to define a separate variable for the current in the voltage source, since that current is clearly the negative of $i_1(t)$. Let $v_1(t)$ be the voltage across the terminals of the element with current $i_1(t)$, etc., using the default sign convention.

For this circuit, we can write three element relations,

$$v_1(t) = (1\,\Omega)\,i_1(t) = i_1(t) \tag{1.47}$$

$$v_2(t) = (2\,\Omega)\,i_2(t) = 2i_2(t) \tag{1.48}$$

$$i_3(t) = \frac{1}{2}\frac{dv_3(t)}{dt}, \tag{1.49}$$

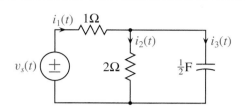

Fig. 1-30. A circuit containing a two resistors and a capacitor.

two KVL equations,

$$v_2(t) = v_3(t) \qquad (1.50)$$

$$v_1(t) + v_2(t) = v_s(t), \qquad (1.51)$$

and one KCL equation,

$$i_1(t) - i_2(t) - i_3(t) = 0. \qquad (1.52)$$

The first four of these equations allow us to express $v_3(t)$ and the three currents in terms of $v_1(t)$ and $v_2(t)$. Substituting for these variables into (1.51) and (1.52) gives

$$v_1(t) + v_2(t) = v_s(t)$$

$$v_1(t) - \frac{1}{2}v_2(t) - \frac{1}{2}\frac{dv_2(t)}{dt} = 0.$$

Subtracting the second equation from the first and multiplying by 2 eliminates $v_1(t)$ and results in the single differential equation

$$\frac{dv_2(t)}{dt} + 3v_2(t) = 2v_s(t), \qquad (1.53)$$

which must be solved for $v_2(t)$. To solve this equation requires that we know the value of $v_2(t)$ at one time instant. Once $v_2(t)$ is determined, all of the remaining element variables can be found, if needed. ∎

To check your understanding, try Drill Problem P1-12.

WORKED SOLUTION:

1-5 Dependent Sources

A *dependent source* is a voltage or current source whose source (controlled) variable depends on some other voltage or current in the network (the controlling variable). Because there are two types of sources and two types of controlling variables, there are the four different types of dependent sources indicated in Figure 1-31.

The current (voltage) in a dependent voltage (current) source is not constrained by the source. In this respect, a dependent source behaves like an independent source, but dependent and independent sources differ in the way that they are used. Independent sources are used primarily as a means for injecting signals into a network (i.e., they are used to provide inputs or power to a circuit); dependent sources are used primarily to model certain types of devices. For example, models for bipolar transistors include, as their dominant element, a current-controlled current source. Similarly, models

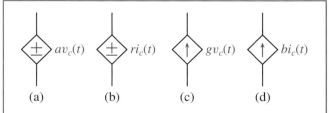

Fig. 1-31. Four types of dependent sources. (a) Voltage-controlled voltage source (VCVS). (b) Current-controlled voltage source (CCVS). (c) Voltage-controlled current source (VCCS). (d) Current-controlled current source (CCCS).

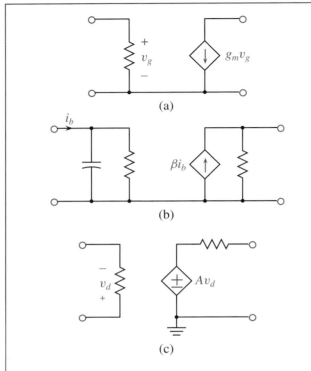

Fig. 1-32. Low-frequency models for some three-terminal active devices that incorporate dependent sources as their key component. (a) Field-effect transistor. (b) Bipolar *npn* transistor. (c) Operational amplifier.

for field-effect transistors include voltage-controlled current sources and models for operational amplifiers include voltage-controlled voltage sources. Figure 1-32 shows some low-frequency models for these devices. Analyzing circuits containing dependent sources is fairly straightforward. One approach is to treat them like independent sources in terms of setting up KVL and KCL equations, and then to incorporate the dependence (controlling) relation as a sort of v–i characteristic for the device. We use this approach in the following example.

Example 1-11   Working with a Dependent Source

To illustrate the procedure, we shall work out the value of the voltage $v_2(t)$ in the network shown in Figure 1-33. Notice that the current through the dependent source is proportional to the current that flows through the 5Ω resistor. We begin by looking at this circuit just as we do any other. It contains three nodes and two meshes, but one of the KCL equations simply tells us that the current through the voltage source is the negative of the current flowing through the 5Ω resistor and one of the mesh KVL equations tells us that the voltage across the dependent source is $v_2(t)$. Thus, if we ignore these trivial equations, we need write only one KCL equation and one KVL equation in addition to the v–i relation for the resistors. Because our goal is to figure out the value of $v_2(t)$, we would like $v_2(t)$ to be one of our independent variables; we choose $i_1(t)$ to be the other, since it is the controlling variable for the dependent source.

The current flowing downward through the 2Ω resistor is equal to $v_2(t)/2$ by Ohm's law. Therefore, a KCL equation written by summing the currents entering node a gives the equation

$$i_1(t) - \frac{v_2(t)}{2} + \frac{i_1(t)}{3} = 0, \qquad (1.54)$$

which incorporates the constraint relation for the dependent source. The KVL equation written for the

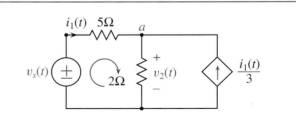

Fig. 1-33. A simple network containing a dependent source.

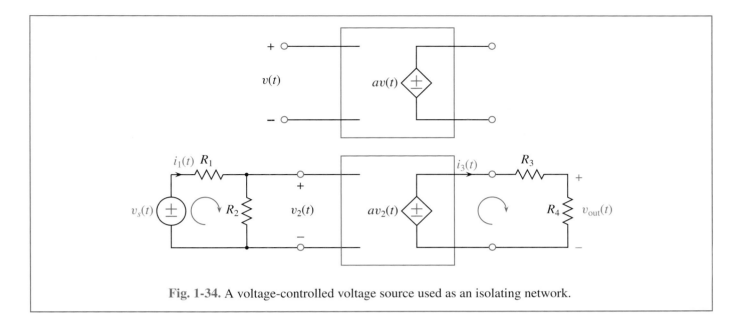

Fig. 1-34. A voltage-controlled voltage source used as an isolating network.

clockwise path around the left mesh gives the additional equation

$$-v_s(t) + 5i_1(t) + v_2(t) = 0. \qquad (1.55)$$

We now have two independent equations in the two unknowns $i_1(t)$ and $v_2(t)$. Because $v_2(t)$ is the variable of interest, we use (1.54) to solve for $i_1(t)$:

$$i_1(t) = \frac{3}{8}v_2(t).$$

We then substitute this result into (1.55), yielding

$$\frac{15}{8}v_2(t) + v_2(t) = v_s(t)$$

or

$$v_2(t) = \frac{8}{23}v_s(t).$$

∎

 To check your understanding, try Drill Problems P1-13 and P1-14.

 MULTIMEDIA TUTORIAL: *Dependent Sources*

Example 1-12   Isolating Networks

Figure 1-34 shows a voltage-controlled voltage source used to isolate two parts of a circuit. Isolating networks greatly simplify the design of circuits having specific behavior because they allow us to design subnetworks in those designs independently of one another. The terminals at the input of this isolating network behave like an open circuit drawing no current.

The open circuit at the input to the isolating network in the circuit at the bottom of Figure 1-34 means that all of the current that flows through R_1 will also flow through R_2. If we denote that current as $i_1(t)$, then a KVL equation written around the leftmost mesh reveals

$$-v_s(t) + R_1 i_1(t) + R_2 i_1(t) = 0,$$

or

$$i_1(t) = \frac{1}{R_1 + R_2} v_s(t).$$

Furthermore, since

$$v_2(t) = R_2 i_1(t),$$

it follows that

$$v_2(t) = \frac{R_2}{R_1 + R_2} v_s(t).$$

Notice that $v_2(t)$ depends on only R_1 and R_2, which are part of the left subnetwork; it does not depend upon the values of R_3, R_4, or the gain a, which are part of the right subnetwork. The dependent source isolates the left subnetwork from the right one.

In the right subnetwork, we observe that the same current flows through R_3 and R_4. Call this current $i_3(t)$. Then a KVL equation around the right mesh gives

$$-av_2(t) + R_3 i_3(t) + R_4 i_3(t) = 0,$$

or

$$i_3(t) = \frac{a}{R_3 + R_4} v_2(t) .$$

Because $v_{\text{out}}(t)$ is measured across R_4, we have

$$v_{\text{out}}(t) = R_4 i_3(t) = \frac{a R_4}{R_3 + R_4} v_2(t)$$

$$= \frac{a R_4}{(R_3 + R_4)} \cdot \frac{R_2}{(R_1 + R_2)} \cdot v_s(t).$$

■

Figure 1-35 shows the four types of ideal isolating networks. In Chapter 4, we show how some of these can be realized via operational amplifiers.

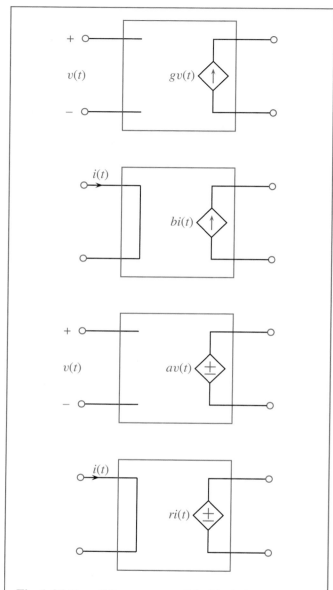

Fig. 1-35. Four different types of ideal isolating networks.

1-6 The Role of Models

Kirchhoff's laws and ideal resistors, inductors, capacitors, and sources provide mathematical models that can be used to understand the behavior of interconnected electrical devices. They provide a simplified description of the underlying physics. Maxwell's equations, which, in principle, describe all electrical phenomena, represent a more accurate model that can be applied to the electric and magnetic fields within and between components. They are important for understanding and creating more precise models for real-world devices and components, such as antennas, generators, and transformers. Analyzing a complex circuit by using Maxwell's equations, however, is a formidable task.

The simpler models that we have derived in this chapter ignore the facts that electron speed is necessarily finite and that the underlying electrical and magnetic fields are not always confined to the elements themselves. These idealized models allow us to obtain acceptable solutions for most problems, including those for which our knowledge of the underlying physics is incomplete or for which the effort required to obtain an exact solution is more than we choose to invest or is unnecessary. If our task is to select an amplifier component for a more nearly complete audio system, it is not necessary to know all of the electrical details at every location within the amplifier. Instead, it is sufficient to know only a few important facts about how the amplifier will perform when it is connected to the other components of the system. These details might include the maximum amount of power that the amplifier can produce and the range of input amplitudes and frequencies that can be amplified without producing a distorted output.

A useful engineering alternative to a complete and comprehensive analysis is a mathematical model that incorporates a number of simplifying assumptions, the *right* number of simplifying assumptions. Though this can be subject to some debate, it is the aim of physics to incorporate all possible physical phenomena in as succinct a manner as possible. It is the essence of good engineering to be able to sift through the myriad possible effects and focus on only those that are essential to the design at hand. When a model is *eigenstadt sympatico harmonius* (has precisely the right degree of detail—no more, no less), then using it in analysis will provide insight into the original problem that we might otherwise find difficult to obtain. It takes a fair amount of engineering maturity to know which physical effects are essential and which ones may be neglected. The ability to do this with aplomb is the essence of the art of engineering.

To illustrate, consider a resistor—not an ideal one, but one that you might actually insert into a circuit. Under some circumstances, the ideal model, which obeys Ohm's law, adequately approximates the real device. In these cases, we should use it, since it is quite simple. How well the resistor approximates the ideal depends upon several factors. For example, the modelled relationship between the voltage and current for the resistor that is valid when the voltages and currents are small might not be valid when they are large. Furthermore, because the behavior of inductors and capacitors is frequency dependent, effects that are negligible in one frequency range might become significant in another. The behavior of a real component might also vary with the manner of its construction. For example, a resistor is typically constructed out of a length of a highly resistive material, such as a thin film or a piece of carbon. In such cases, charge tends to collect on the end surfaces of the material, which causes the device to acquire some of the properties of a capacitor. Furthermore, the current flowing through the leads creates a magnetic field, so the leads themselves have a small amount of inductance. Finally, there is also some stray capacitance between the leads of the device if these are close together. If we were to capture all of these effects in a model, we might end up with the model for a real resistor shown in Figure 1-36. Clearly, this model

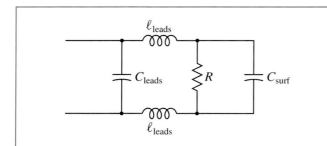

Fig. 1-36. A model for a real resistor that incorporates the inductance of the leads, the end surface capacitance, and the capacitance between the leads of the device.

is more complicated and harder to use than the simple one, but we could use it if we had to: Since the model is still based on our ideal elements, using it will not create a need to change our methodology.

How do we know which model we should use? Those voltages and currents that we can calculate from our models should approximate the values that we would measure in an actual circuit, but we must recognize that all models have limitations. It is very important that we compare the predictions based on our models against experimental measurements to assess their validity and efficacy. When our measurements disagree with our predictions, it could be appropriate to reexamine the simplifying assumptions behind the models or to replace them with more accurate—and necessarily more complex—models.

The elements relations that we have defined are not the only idealizations that we have made. Kirchhoff's laws also ignore certain effects. Different layouts of the components on a printed circuit board, for example, often perform quite differently because of the interactions of the electric and magnetic fields associated with the elements, which Kirchhoff's laws ignore. Indeed, layout requires knowledge of many areas, including not only circuit theory, but also electromagnetics,

thermodynamics, and, to some extent, economics.

This text focuses on the study of mathematical models. It does not concern itself with the very important problem of going from physical devices to those models. That process involves a detailed knowledge of the physics of materials and it incorporates concepts from electromagnetic fields. Although we shall not focus on the generation of models of actual physical devices in this text, it is an essential and important endeavor. Perhaps the best way to obtain an intuitive feel for the limitation of a model is by also understanding the basic principles used to develop the model. We must always keep in mind that the powerful techniques that we will develop yield useful information only when the models we are using provide accurate representations of the actual devices.

 EXTRA INFORMATION: *Engineering Models*

1-7   Chapter Summary

1-7-1   Important Points Introduced

- A circuit is a connection of elements and sources.

- Every two-terminal element has a voltage variable and a current variable.

- The current that enters a two-terminal element through one terminal is equal to the current that leaves it through the other.

- SI system of units:
    Basic unit for voltage = volt (V),
    Basic unit for current = ampere (A),
    Basic unit for power = watt (W),
    Basic unit for resistance = ohm (Ω),
    Basic unit for inductance = henry (H),
    Basic unit for capacitance = farad (F).

- The power absorbed by an element is $P_{\text{inst}}(t) = v(t)i(t)$.

- Element relations for ideal resistors—Eq. (1.3), inductors Eq. (1.5), and capacitors Eq. (1.12).

- Resistors always absorb power.

- Inductors and capacitors store energy in their magnetic and electric fields, respectively.

- Independent sources are used to introduce waveforms and/or power into a circuit.

- There is no voltage across the terminals of a short circuit, although current can flow through it.

- No current flows through an open circuit, although there can be a voltage across it.

- A switch is a device that has two states—one in which it looks like an open circuit and one in which it looks like a short circuit.

- The sum of all currents entering any node is zero (KCL).

- The sum of the currents that enter any closed potion of a circuit is zero.

- If a network has a total of n nodes, it is sufficient to write KCL equations at only $n - 1$ of them.

- The sum of the potential differences along any closed path in a circuit is zero (KVL).

- We can obtain a minimal set of independent KVL equations by writing one KVL equation for each path that defines a mesh in the network.

- Two elements (or an element and a source) are connected in series if the same current (or its negative) flows through both elements.

- Two elements (or an element and a source) are connected in parallel if they connect the same pair of nodes.

- Dependent sources behave like independent sources, except that the source waveform is a function of some other voltage or current variable in the circuit.

1-7-2 New Abilities Acquired

You should now be able to do the following:

(1) Assign voltages and currents to all of the elements and sources in a circuit.

(2) Relate the voltage to the current flowing into an ideal resistor, inductor, or capacitor.

(3) Compute the power absorbed by any element or source.

(4) Write a KCL equation for any node or enclosing boundary in a circuit.

(5) Write a KVL equation for any closed path in a circuit.

(6) Find the complete solution of a circuit containing resistors and sources.

(7) Identify elements in a circuit that are connected in series and/or parallel.

1-7-3 Links

The web site contains a number of tools and interactive demonstrations relating to element relations, Kirchhoff's laws, and networks.

- **TOOLS:** Ohm's law calculator

- **DEMONSTRATIONS:** As listed

The web site also contains supplemental information on selected topics, fluency exercises, multimedia tutorials, and solved problems.

1-8 Problems

1-8-1 Drill Problems

All drill problems have solutions on the web site.

P1-1 For the circuit in Figure P1-1, write a KCL equation at every node in the circuit in terms of the current variables whose reference directions are given. This problem is similar to Example 1-1.

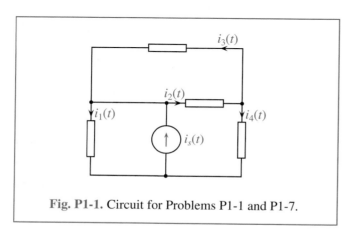

Fig. P1-1. Circuit for Problems P1-1 and P1-7.

P1-2 Write a sufficient set of KCL equations for the circuit in Figure P1-2. This problem is similar to Example 1-1.

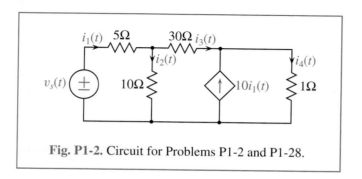

Fig. P1-2. Circuit for Problems P1-2 and P1-28.

P1-3 The circuit in Figure P1-3 contains four elements and a voltage source. Write a sufficient set of KCL equations to fully constrain all the currents in the circuit. You may exploit any obvious series and parallel connections to reduce the number of variables. Notice that only the voltages are labelled, although you should write your equations using current variables. You should assume that the reference directions for the currents are consistent with the default sign convention. Furthermore, let the current flowing into the element with voltage $v_1(t)$ be $i_1(t)$, etc. This problem is similar to Example 1-2.

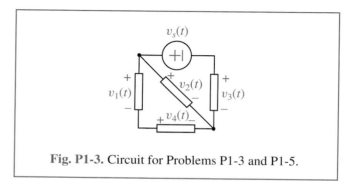

Fig. P1-3. Circuit for Problems P1-3 and P1-5.

P1-4 For the circuit in Figure P1-4, write a sufficient set of KCL equations in terms of the voltage variables that are labelled. You should incorporate Ohm's law for the resistors. Exploit any obvious series and parallel connections to reduce the number of equations and variables. This problem is similar to Example 1-3.

P1-5 Write a KVL equation for every simple closed path in the circuit in Figure P1-3 in terms of the voltages whose reference directions are given in that figure. This problem is similar to Example 1-4.

P1-6 Write a sufficient set of KVL equations to incorporate the constraints on the voltages over all closed paths for the circuit in Figure P1-6. This problem is similar to Example 1-4.

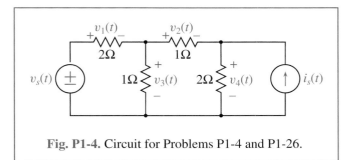

Fig. P1-4. Circuit for Problems P1-4 and P1-26.

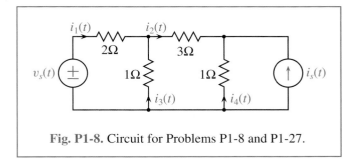

Fig. P1-8. Circuit for Problems P1-8 and P1-27.

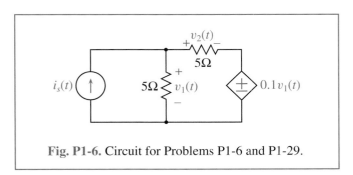

Fig. P1-6. Circuit for Problems P1-6 and P1-29.

P1-9 This problem is similar to Example 1-7.

(a) How many nodes are present in the circuit in Figure P1-9?

(b) How many meshes are present in that circuit?

(c) Write a KCL equation at *every* node in the circuit in terms of the indicated current and voltage variables.

(d) Write a KVL equation for every mesh in the circuit. Write those equations, using the indicated current variables, by incorporating Ohm's law for each resistor.

P1-7   Write a sufficient set of KVL equations to incorporate all of the voltage constraints for the circuit in Figure P1-1. Since only the currents are labelled, you should assume that the reference directions for the element voltages are consistent with the default sign convention. (Let the voltage of the element with current $i_1(t)$ be $v_1(t)$, etc.) Define a potential difference across the terminals of the current source. This problem is similar to Example 1-5.

P1-8 For the circuit in Figure P1-8, write a sufficient set of KVL equations to constrain all of the element variables in terms of the current variables that are indicated. You should incorporate Ohm's law for the resistors. Use any obvious series and parallel connections to reduce the number of equations and variables. This problem is similar to Example 1-7.

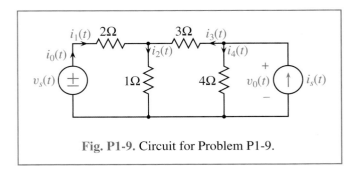

Fig. P1-9. Circuit for Problem P1-9.

P1-10 This problem is similar to Example 1-7.

(a) How many meshes are present in the circuit in Figure P1-10?

(b) How many nodes are present in that circuit?

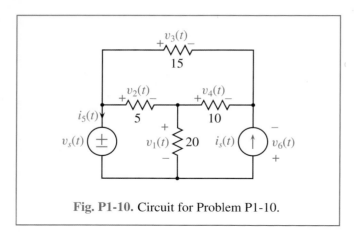

Fig. P1-10. Circuit for Problem P1-10.

(c) Write a KVL equation at every mesh in the circuit in terms of the indicated voltage and current variables.

(d) Write a KCL equation for *every* node in the circuit. Write those equations, using the indicated voltage variables, by incorporating Ohm's law for each resistor. All resistances are measured in ohms.

P1-11 This problem is similar to Example 1-7.

(a) For the network in Figure P1-11, write a set of equations that expresses all of the constraints on the element and source voltages and currents that are implied by the network model.

(b) Solve those equations for $i_2(t)$.

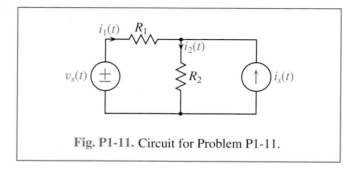

Fig. P1-11. Circuit for Problem P1-11.

P1-12 Find the differential equation that relates the current $i_R(t)$ to the source current $i_s(t)$ in the circuit of Figure P1-12. This problem is similar to Example 1-10.

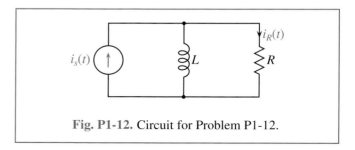

Fig. P1-12. Circuit for Problem P1-12.

P1-13 The circuit in Figure P1-13 contains a voltage-controlled current source. Write a sufficient set of KCL equations to specify the current constraints in the circuit, using the variables $i_1(t)$, $i_2(t)$, and $i_3(t)$.

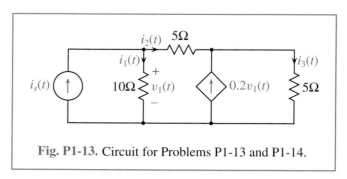

Fig. P1-13. Circuit for Problems P1-13 and P1-14.

P1-14 Find the values of the currents for the circuit in Figure P1-13. This problem is similar to Example 1-11.

1-8-2 Basic Problems

P1-15 Assume that the current flowing through a device is $i(t)$ and that the voltage across its terminals is $v(t)$, where these waveforms are sketched in Figure P1-15. Mathematically, these are given by

$$v(t) = \begin{cases} 1, & 0 < t < 1 \\ 0, & \text{otherwise} \end{cases}$$

$$i(t) = \begin{cases} 1 - e^{-3t}, & 0 \leq t < 1 \\ e^{-3(t-1)}, & 1 \leq t \\ 0, & \text{otherwise} \end{cases}.$$

(a) Calculate and sketch the power $P_{\text{inst}}(t)$ absorbed by the device as a function of time.

(b) Compute the total amount of energy that the device absorbs.

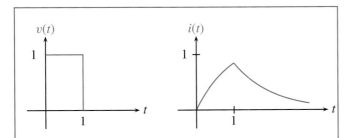

Fig. P1-15. Voltage and current waveforms for Problem P1-15.

P1-16 The current source in the network shown in Figure P1-16a has the time dependence shown in Figure P1-16b.

(a) Sketch $v_r(t)$.

(b) Sketch $v_\ell(t)$.

(c) Assuming $v_c(0) = 0$, sketch $v_c(t)$.

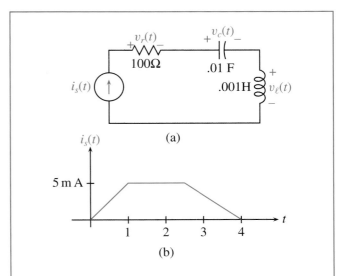

Fig. P1-16. Circuit and source waveform for Problem P1-16.

P1-17 The voltage source in the circuit shown in Figure P1-17a has the source waveform shown in Figure P1-17b.

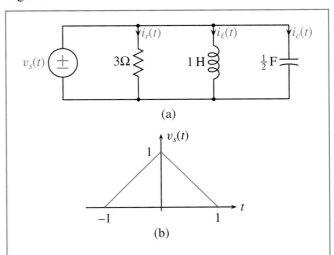

Fig. P1-17. Circuit and source waveform for Problem P1-17.

(a) Sketch $i_r(t)$.

(b) Sketch $i_c(t)$.

(c) Sketch $i_\ell(t)$. Assume that $i_\ell(-\infty) = 0$.

P1-18 The voltage waveform for the voltage source in the network of Figure P1-18 is

$$v_s(t) = \begin{cases} \sin 2\pi (100)t, & t \geq 0 \\ 0, & t < 0 \end{cases}.$$

(a) Determine $i_r(t)$.

(b) Assuming $i_\ell(0) = 0$, calculate $i_\ell(t)$.

(c) Find $i_c(t)$.

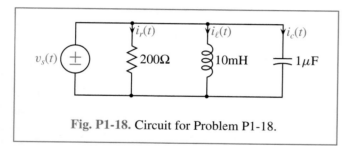

Fig. P1-18. Circuit for Problem P1-18.

P1-19 In the circuit shown in Figure P1-19, the current source waveform is

$$i_s(t) = \begin{cases} 5\cos(50t), & t > 0 \\ 0, & t < 0 \end{cases}.$$

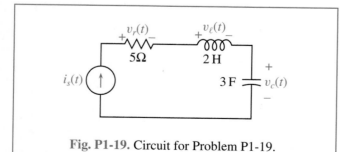

Fig. P1-19. Circuit for Problem P1-19.

(a) Figure out $v_r(t)$.

(b) Figure out $v_\ell(t)$.

(c) Find $v_c(t)$ if $v_c(0) = 0$.

P1-20 For the circuit in Figure P1-20, the current source waveform is

$$i_s(t) = \begin{cases} e^{-t} - e^{-2t}, & t > 0 \\ 0, & t < 0 \end{cases}.$$

It is known that $v_c(-\infty) = 0$.

(a) What is $v_r(t)$?

(b) What is $v_\ell(t)$?

(c) What is $v_c(t)$?

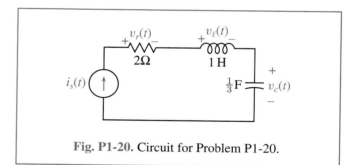

Fig. P1-20. Circuit for Problem P1-20.

P1-21

(a) The voltage across the terminals of a 1F capacitor is

$$v(t) = \begin{cases} \frac{1}{2}(1 - \cos \pi t), & 0 < t < 2 \\ 0, & \text{otherwise} \end{cases}.$$

This waveform is shown in Figure P1-21.

  (i) Sketch the current flowing through the device, $i(t)$.

  (ii) Sketch the energy stored in the device as a function of t.

  (iii) For what values of t is the device supplying power?

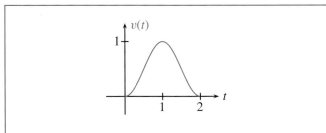

Fig. P1-21. Waveform for Problem P1-21.

(iv) For what values of t is the device absorbing power?

(b) Repeat the questions asked in part (a) if $v(t)$ is the voltage across the terminals of a 1H inductor.

P1-22

(a) The voltage, $v_\ell(t)$ measured between the terminals of an ideal 2H inductor is

$$v_\ell(t) = 6e^{-2t} + 3e^{-3t}\,\text{mV}.$$

The inductor current, $i_\ell(t)$, at $t = 0$ is zero. Find $i_\ell(t)$.

(b) Repeat for the voltage waveform shown in Figure P1-22.

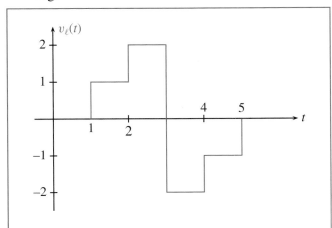

Fig. P1-22. Voltage waveform for Problem P1-22.

P1-23 The source waveform $v_s(t)$ applied to the circuit in Figure P1-23a is shown in Figure P1-23b. Sketch the current $i(t)$ flowing through the capacitor as a function of time.

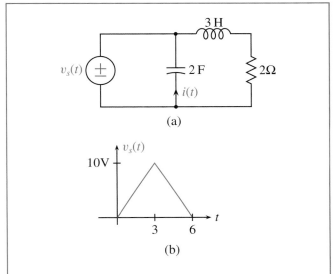

Fig. P1-23. (a) Circuit for Problem P1-23; (b) voltage-source waveform $v_s(t)$.

P1-24

(a) Write the KCL equations that constrain the currents at all of the nodes of the network in Figure P1-24.

(b) Write the KVL equations that constrain the voltages for all of the meshes in that same network.

P1-25

(a) Write the KCL equations that constrain the currents at all of the nodes of the network in Figure P1-25. The time dependence of the currents has been suppressed to limit clutter.

(b) Write the KVL equations that constrain the voltages for all of the meshes in that same network.

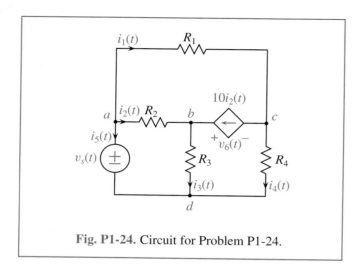

Fig. P1-24. Circuit for Problem P1-24.

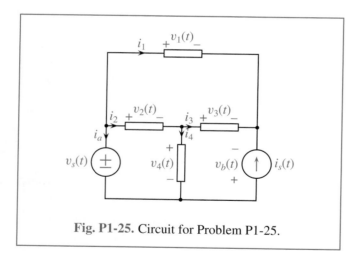

Fig. P1-25. Circuit for Problem P1-25.

P1-26 Find the values of the voltages from the complete solution of the circuit in Figure P1-4.

P1-27 Find the values of the currents from the complete solution of the circuit in Figure P1-8.

P1-28 Find the values of the currents for the circuit in Figure P1-2.

P1-29 Find the values of the voltages in the circuit in Figure P1-6.

P1-30 In the circuit of Figure P1-30 express $v(t)$ as a function of $v_{s_1}(t)$, $v_{s_2}(t)$, and $i_s(t)$.

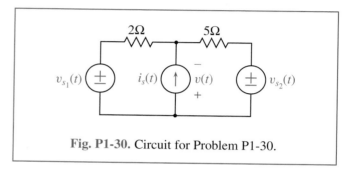

Fig. P1-30. Circuit for Problem P1-30.

P1-31 In the circuit in Figure P1-31, both source waveforms (and all of the element variables) are constant. Compute the values of i_1, v_1, i_2, and v_2.

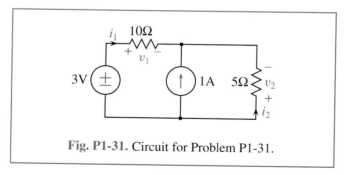

Fig. P1-31. Circuit for Problem P1-31.

P1-32 Consider the circuit in Figure P1-32.

(a) What is $i_1(t)$ (in terms of $v_s(t)$)?
(b) What is $v_2(t)$?

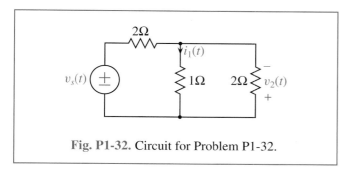

Fig. P1-32. Circuit for Problem P1-32.

P1-33 Solve the circuit in Figure P1-33 for $v_1(t)$, $v_2(t)$, and $v_3(t)$.

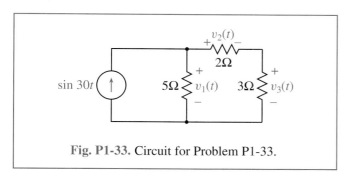

Fig. P1-33. Circuit for Problem P1-33.

P1-34 For the circuit of Figure P1-34, express $v(t)$ in terms of $i_s(t)$.

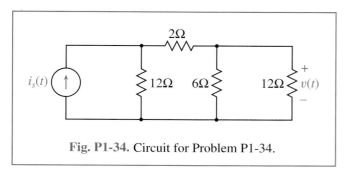

Fig. P1-34. Circuit for Problem P1-34.

P1-35

(a) Express the current $i(t)$ in terms of $v_s(t)$ in the circuit of Figure P1-35 when the switch is in the *open* position.

(b) Find that same current when the switch is *closed*.

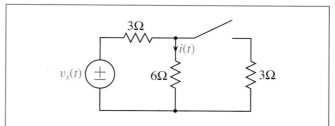

Fig. P1-35. Circuit for Problem P1-35, which contains a switch.

P1-36

(a) Express the voltage $v(t)$ in terms of $v_s(t)$ in the circuit of Figure P1-36 when the switch is *closed*.

(b) Find that same voltage when the switch is *open*.

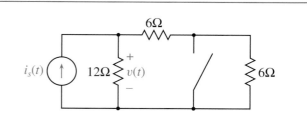

Fig. P1-36. Circuit for Problem P1-36, which contains a switch.

P1-37

(a) If a 1-kΩ resistor and a 2-kΩ resistor are connected in series, which will absorb more power?

(b) If a 1-kΩ resistor and a 2-kΩ resistor are connected in parallel, which will absorb more power?

P1-38 Two 1-μF capacitors are connected to a battery. Will they store more energy if they are connected in series or in parallel?

P1-39 The circuit in Figure P1-39 contains a current-controlled voltage source. Write a sufficient set of KVL equations to specify the element-voltage constraints over all closed paths in the circuit, using the variables $v_1(t)$, $v_2(t)$, $v_3(t)$ and $v_s(t)$. ($i_1(t)$ is the current flowing into the + terminal of the 5Ω resistor.)

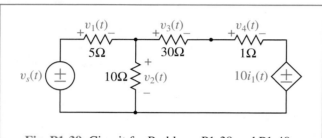

Fig. P1-39. Circuit for Problems P1-39 and P1-40.

P1-40 Find the values of the voltages in the circuit in Figure P1-39.

P1-41 Find the current $i_1(t)$ for the circuit shown in Figure P1-41. (You will need to define some additional element variables.)

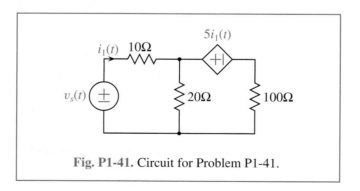

Fig. P1-41. Circuit for Problem P1-41.

P1-42 Figure out the voltage $v(t)$ in the circuit in Figure P1-42.

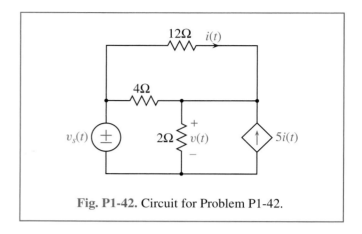

Fig. P1-42. Circuit for Problem P1-42.

P1-43 For the circuit in Figure P1-43,

(a) compute the power *absorbed* by the independent voltage source;

(b) compute the power *absorbed* by the dependent current source.

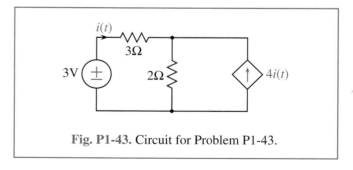

Fig. P1-43. Circuit for Problem P1-43.

P1-44 In the circuit in Figure P1-44, which contains a voltage-dependent current source, compute $v(t)$.

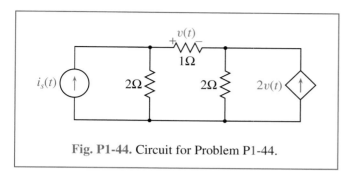

Fig. P1-44. Circuit for Problem P1-44.

P1-45 Find $v(t)$, the potential difference between the two indicated nodes, for the circuit in Figure P1-45. All resistances are measured in ohms.

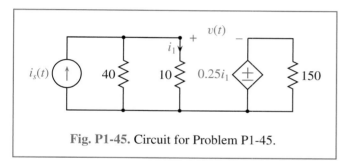

Fig. P1-45. Circuit for Problem P1-45.

1-8-3 Advanced Problems

P1-46 In the circuit of Figure P1-46, both inductors have no current flowing through them at $t = 0$.

(a) If

$$i_s(t) = \begin{cases} \sin 3t, & t \geq 0 \\ 0, & t < 0 \end{cases},$$

what is $v(t)$?

(b) If

$$v_s(t) = \begin{cases} e^{-3t}, & t \geq 0 \\ 0, & t < 0 \end{cases},$$

what is $i(t)$?

(c) With both sources turned on, as indicated in (a) and (b), what is $\hat{i}(t)$?

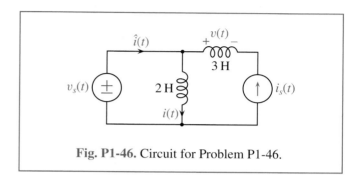

Fig. P1-46. Circuit for Problem P1-46.

P1-47 Consider the circuit in Figure P1-47(a). The waveforms corresponding to the source signals are given graphically in Figure P1-47(b).

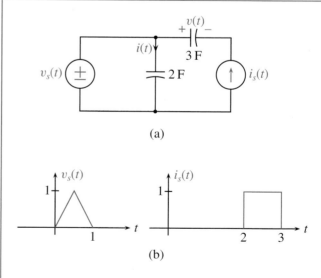

(a)

(b)

Fig. P1-47. Figure for Problem P1-47: (a) circuit (b) voltage and current source waveforms.

(a) Compute the current $i(t)$. Express your result as a graph of the current versus time.

(b) Compute the voltage $v(t)$. Express your result as a graph of the voltage versus time.

(c) Draw a graph of the instantaneous power absorbed by the voltage source versus time.

(d) Draw a graph of the instantaneous power absorbed by the current source versus time.

P1-48 Find the value of the resistance R in the circuit in Figure P1-48 at which the instantaneous power absorbed by that resistor, P_{inst} is maximized. *Hint:* To find the value of R for which $P_{inst}(R)$ is maximized, set the derivative of $P_{inst}(R)$ to zero.

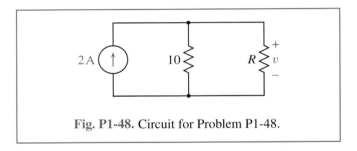

Fig. P1-48. Circuit for Problem P1-48.

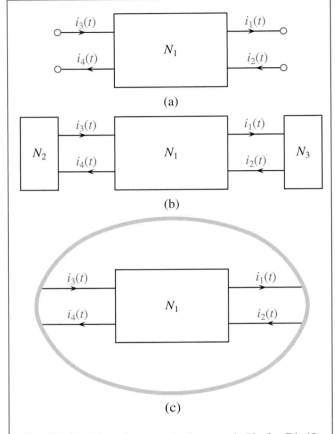

Fig. P1-49. (a) A four-terminal network N_1 for P1-49. (b) N_1 connected to two two-terminal networks. (c) N_1 embedded into a more general network.

P1-49 Consider the four-terminal network N_1 shown in Figure P1-49(a).

(a) When network N_1 is connected to the two subnetworks N_2 and N_3 as shown in Figure P1-49(b), what is the relation between currents $i_1(t)$ and $i_2(t)$?

(b) Does the result that you derived in (a) apply to $i_1(t)$ and $i_2(t)$ when N_1 is embedded in a larger (but unknown) network as shown in Figure P1-49(c)? Explain.

P1-50 In the text, it was claimed that the KVL equations formed on the closed paths that encircle the meshes in a planar network are independent. In this problem, we prove this claim. A planar network is one that can be drawn on a piece of paper with none of the wires crossing each other. Our proof (and our methodology) is limited to planar networks because these are the only ones for which the concept of a mesh is defined. (Nonplanar circuits can be solved via the node method, which will be described in Chapter 2.)

(a) Our proof proceeds by *induction*. Recall the basis behind these proofs.

(i) Verify that the statement is true for $\ell = 1$.

(ii) Assume that it is valid for $\ell = k$.

(iii) Prove that, if it is valid for $\ell = k$, then it must also be true for $\ell = k + 1$.

For this problem, ℓ corresponds to the number of meshes in a simplified circuit formed from the original by removing some of the elements. Begin by removing all of the interior elements from the network, leaving a single path going around the outside of the original circuit, and write a KVL equation around that path. Since there is only a single equation, which is nontrivial, it is independent. This corresponds to Step 1 above. Next, assume that several elements have been reinserted into the circuit so that the partially completed circuit now contains k meshes for whose paths the KVL equations are independent. This corresponds to Step 2. Finally, we add enough additional elements back into the circuit to create an additional mesh. This can be caused only by dividing one of the meshes from the kth stage, called the parent, into two meshes at the next stage, called the children. Let I equal the sum of the voltages across the inserted elements, let A equal the sum of the voltages from the parent that go to child #1, and let B equal the sum of the voltage drops that go to child #2.

(i) Write a KVL equation for the path corresponding to the parent in terms of A, B, and I.

(ii) Write two KVL equations for the paths corresponding to the children in terms of A, B, and I.

(iii) Show that these two equations are independent, in that neither can be derived from the other.

(iv) Show that neither of these equations can be derived from the other meshes in the circuit at this level.

This completes the inductive proof.

(b) Use the constructive procedure of your proof to verify that the KVL equations for the meshes in the network in Figure P1-50 are independent.

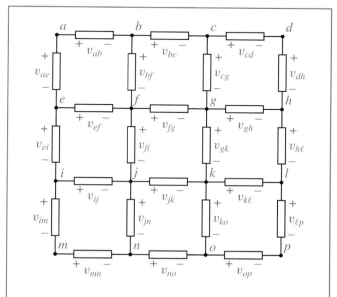

Fig. P1-50. A network containing a number of elements for proving the independence of mesh equations in Problem P1-50.

P1-51   In Problem P1-50, a method of proof was outlined for showing that the KVL equations derived from the paths surrounding the meshes in a planar circuit are independent. In this problem, we tackle the KCL equations. Prove that the KCL equations written at all but one of the nodes of a circuit are also independent— that is, that no one of the equations can be derived from the others. As with the KVL equations, this can also be proved by induction. Begin with a single enclosing surface that surrounds all of the nodes of the circuit but one, and write a single KCL equation for that subnetwork. Then divide that subnetwork (or one of the eligible subnetworks when later there is more than one) into two parts, each of which contains at least one node, and show that the two KCL equations derived from the surfaces surrounding those subnetworks are independent of each other and from the remaining equations. Continue this procedure until the number of subnetworks generated is equal to one less than the number of nodes and each subnetwork contains a single node.

P1-52 Consider the circuit in Figure P1-52.

(a) In that circuit, only one voltage and one current have been labelled, since these are the only ones that we shall eventually want. In order to analyze the circuit, however, we need to assign (explicitly or implicitly) voltages and currents for all of the elements and a voltage across the terminals of the current source. On a drawing of the circuit, indicate such a complete set of variables. Adhere to the default sign convention.

(b) Using your variables in addition to those that are indicated, write the following set of equations: one KVL equation at each mesh, one KCL equation at each node, and one element relation for each resistor.

(c) Solve your equations for the quantities $v(t)$ and $i(t)$. Since the other element variables are not requested, you should eliminate them first in solving your equations. Notice that you might not need all of the equations that you wrote in part (c). Your answer will be in the form $v(t) = K_1 i_s(t)$, $i(t) = K_2 i_s(t)$.

P1-53 Compute $v(t)$ and $i(t)$ in the network shown in Figure P1-53.

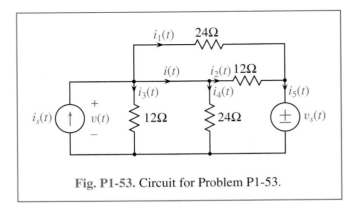

Fig. P1-53. Circuit for Problem P1-53.

P1-54 The circuit in Figure P1-54 contains a two-terminal nonlinear element N that satisfies the v–i relation

$$v(t) = \begin{cases} 9 - i^2(t), & 0 \le i(t) \le 9 \\ 0, & \text{otherwise} \end{cases} .$$

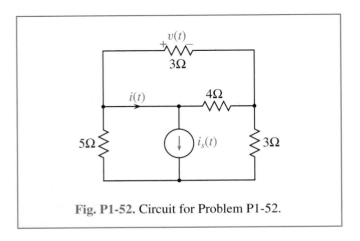

Fig. P1-52. Circuit for Problem P1-52.

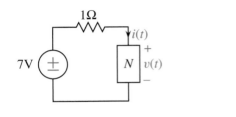

Fig. P1-54. Circuit containing a nonlinear element for Problem P1-54.

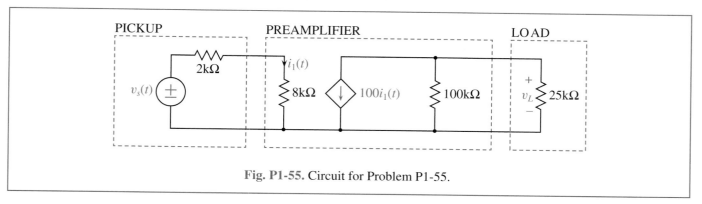

Fig. P1-55. Circuit for Problem P1-55.

Discover two possible values for the current, $i(t)$. *Note:* Although you have not seen any examples involving nonlinear elements, Kirchhoff's laws still apply to the circuit and Ohm's law still applies to the resistor.

P1-55 In the center of Figure P1-55 is a model of a one-transistor preamplifier that is used to amplify the output of a low-amplitude magnetic pickup and drive a 25 kΩ load. Express the voltage $v_L(t)$ measured across the load in terms of $v_s(t)$.

P1-56 Two resistors connected in series act like a single resistor. Similarly, two resistors connected in parallel behave like a single resistor. In this problem, we derive these basic results. We shall use them extensively in Chapter 3.

(a) Consider two resistors connected in series and connected across a voltage source, as in Figure P1-56(a). Show that the current flowing through them is proportional to the source voltage.

(b) This implies that, from the point of view of the voltage source, the series connection of resistors is equivalent to a single resistor, as is shown in Figure P1-56(b). Express R_{eq} in terms of R_1 and R_2.

(c) We can similarly consider two resistors connected in parallel across a voltage source, as in Figure P1-56(c). Show again that the current flowing through them is proportional to the source voltage.

(d) Find the equivalent resistance of the parallel combination as you did in part (b).

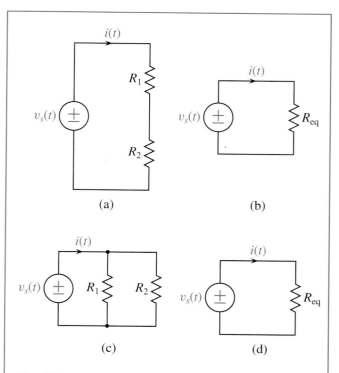

Fig. P1-56. (a)–(b) Two resistors connected in series and their equivalent resistance. (c)–(d) Two resistors connected in parallel and their equivalent resistance.

P1-57 Consider the circuit shown in Figure P1-57.

(a) This circuit contains three elements; thus, there are six element variables, all of which are constant, since the voltage source is constant. The currents are labelled, and the voltages across the terminals of the three resistors are implied by the default sign convention. Write a set of six linear equations in the variables v_1, v_2, v_3, i_1, i_2, and i_3 that specify the complete solution. These should take the form of three element relations, one KCL equation, and two KVL equations.

(b) Solve the set of equations from (a) to compute the values of the element variables.

(c) Evaluate the power absorbed by all of the elements and sources. Show that the total power absorbed in the resistors is equal to the total power supplied by the source.

   The net power in any circuit must always be zero—that is, the total power absorbed must always equal the total power supplied. This problem demonstrates that fact for one particular circuit. We shall prove the general case in Chapter 2.

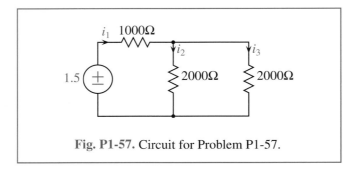

Fig. P1-57. Circuit for Problem P1-57.

P1-58 Find the differential equation that relates the current $i(t)$ to the source voltage $v_s(t)$ in the circuit of Figure P1-58.

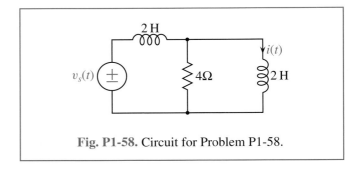

Fig. P1-58. Circuit for Problem P1-58.

1-8-4   Design Problems

P1-59 Find the minimum and maximum values of the resistance R in the circuit of Figure P1-59 such that the following two conditions will be met:

(a) $i \geq 25\,\text{mA}$

(b) $P_{\text{inst}} \leq 500\,\text{mW}$.

Here, P_{inst} is the instantaneous power absorbed by the resistor.

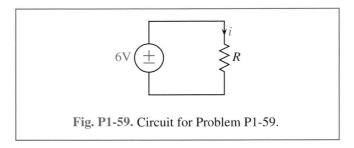

Fig. P1-59. Circuit for Problem P1-59.

P1-60

(a) Express the voltage v in the circuit in Figure P1-60 as a function of the resistance R.

(b) Compute the maximum and minimum values of R such that both of the following conditions will be true:

   (i) $v \geq 10\,\text{V}$
   (ii) $P_{\text{inst}} \geq 5\,\text{W}$.

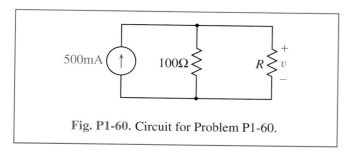

Fig. P1-60. Circuit for Problem P1-60.

P1-61 Resistors are specified by their maximum power rating as well as by their resistance. For example, a resistor with a power rating of 1W can dissipate 1W of power indefinitely. The larger the power rating, the bulkier and more expensive the resistor, so good design practice dictates that the power ratings should be large enough to handle the load, but no larger than necessary. If the resistors in the circuit in Figure P1-61 are available in power ratings of 1 W, $\frac{1}{2}$ W, $\frac{1}{4}$ W, and $\frac{1}{10}$ W, specify the power ratings needed.

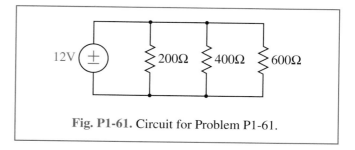

Fig. P1-61. Circuit for Problem P1-61.

C H A P T E R 2

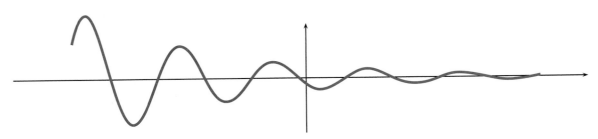

Writing Circuit Equations

Objectives

By the end of this chapter, you should be able to do the following:

1. *Find the complete solution of a circuit using the exhaustive, node, and mesh methods.*

2. *Put a set of linear equations, such as the ones that define the constraints of a resistive circuit, into matrix-vector form.*

3. *Find the element variables in a circuit, using the method of superposition of sources.*

In the last chapter, we saw that, for some planar circuits containing only resistors and sources, we could find the complete solution by writing v–i relations for all of the elements, KCL equations at all but one of the nodes, and KVL equations on closed paths around all of the meshes.

This brute-force approach is a useful starting point, but it leaves many questions unanswered, not the least of which is, will it always work? It can also produce a very large number of equations to be solved to find the complete solution. A more serious limitation of the brute-force approach, however, is that it provides very little insight into the behavior of a circuit.

This chapter seeks answers to three questions. Will our procedures for finding the solution of a circuit always work? How can we reduce the number of equations that we need to write? And how should the variables in those equations be chosen? We demonstrate that the first of these questions has an affirmative answer. Later we show that we can reduce the number of equations that we have to write if we do not need to find all of the variables in the circuit. By choosing the independent variables of the problem carefully, however, it is often possible to reduce the number of equations still further, so that, in effect, we only have to write *either* the KCL equations *or* the KVL equations. This is because the choice of the variables guarantees that the other equations are satisfied by construction. This choice of variables leads to two popular formal techniques for analyzing circuits—node analysis and mesh analysis. The chapter concludes with a development and demonstration of these two techniques. Along the way, we also discover a few more simplifying techniques and useful properties of circuits.

All three of the formal methods that we develop, however, are still brute-force methods. They still provide very little insight that can be used in design, and they are still tedious. The circuit simplifications that we develop later in Chapter 3 go a long way toward remedying these difficulties. The formal methods are still very important, however, for the following reasons: (1) We can prove that they will work. Thus, they serve as a starting point for other important simplification and analysis techniques. (2) They serve as our back-up for those situations when ad hoc methods fail or when we lack the insight to select the appropriate simplifications.

2-1   The Exhaustive Method for Writing a Sufficient Set of Circuit Equations

The circuit analysis method that we used in Chapter 1 is one that we call the *exhaustive method*, although it also has other names. It could, with some justification, also be called the exhausting method, because, of all the methods we will consider, it results in the largest number of equations in the largest number of variables. We shall usually choose to use other methods, ones that involve fewer variables and provide more insight, but the exhaustive method nevertheless does have several things going for it, not the least of which is that its equations are the easiest to write. In addition, we can prove that the method is guaranteed to produce a solution; as a result, we shall establish the adequacy of later methods by tying

them to the guarantees of the exhaustive method. For simple circuits involving only a few elements, it is often the solution method of choice. As a starting point, let us formally restate the method.

EXHAUSTIVE METHOD:

1. *Define a voltage variable and a current variable for each element in the circuit. Define a voltage variable for each (independent and dependent) current source and a current variable for each voltage source.*
2. *Write a v–i relation for each element.*
3. *Write a KVL equation for each mesh.*
4. *Write a KCL equation for all but one of the nodes. (It does not matter which one is excluded.)*
5. *Solve the resulting set of equations.*

If the circuit has e elements and s sources, Step 1 of the procedure defines $2e + s$ variables. Steps 2, 3, and 4 allow us to write $2e + s$ independent linear equations in those variables, a fact that we shall demonstrate shortly. For a resistive circuit, the e element relations in Step 2 are quite simple. We can use these e statements of Ohm's law to eliminate immediately either the resistor voltages or the resistor currents from the set of variables. If we do this, we are left with a set of $e + s$ equations in $e + s$ unknowns that need to be solved. Is this set of equations sufficient?

2-1-1   Proof of Sufficiency of the Method*

If we write a KCL equation at all but one of the nodes of a planar circuit, write a KVL equation at all of its meshes, and write a v–i relation for each element, can we be sure that we will have enough equations to specify the complete solution of the circuit? This is the question to which this section finds an affirmative answer. It exploits a fundamental property of mathematical graphs (of which circuits are a special case) that constrains the numbers of nodes, meshes, and branches (connections between nodes) that a planar circuit can have. For those willing to accept the method without justification, this section can be omitted without disturbing the flow of the material.

We begin with a few definitions. As we have already seen, a *planar network* is a circuit that we can draw on a flat sheet of paper without any elements crossing each other. All the networks that we have drawn so far are planar, and most of those that we shall see in what is to follow are planar also.[1] The exhaustive method, as we have stated it, is applicable only to planar graphs. Let e be the number of *elements* in the circuit, s be the number of *sources* (independent or dependent), ℓ be the number of *meshes*, and n be the number of *nodes*. A *branch* is a connection between two nodes. In a circuit, a branch can be either an element or a source; we let $b = e + s$ denote the total number of branches.

The numbers of nodes, meshes, and branches in any planar graph are not independent; knowing any two of these specifies the other. To show this, we define a *tree* as a connected network (or graph) that contains no closed paths. A tree can be constructed from any circuit by selectively removing elements. Figure 2-1 shows an example of a circuit and one possible tree that can be obtained from it. Because there are no closed paths, no currents can flow in a tree. Notice that in a tree there is exactly one path connecting any two nodes. We can construct a tree by building the network in the following fashion. We begin with the isolated nodes from the network, and we construct the tree by inserting some of

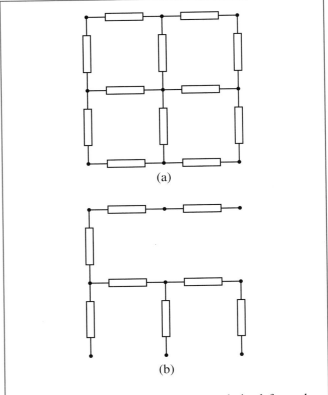

Fig. 2-1. (a) A network. (b) A tree derived from the network in (a).

the branches of the complete network one at a time. For the first branch, we select any branch from the circuit, which must connect two nodes. Then we select one of the isolated nodes and add a branch from the original circuit that will attach that node to the growing tree. We stop when there are no more isolated nodes. At this point we have a tree, plus several branches remaining to be inserted. We have not yet created any closed paths, because every time that we added a branch, one terminal of that branch was attached to a node that was previously isolated and that now is attached only to the single element. Notice that we have created only one out of

---

[1] Planar networks are the only ones for which our definition of a mesh is applicable. For dealing with the nonplanar case, see Problems P2-52 and P2-53.

many possible trees; but all of the tree possibilities will contain exactly $n - 1$ branches, where n is the number of nodes. This is because the first branch attached to the growing tree decreased the number of isolated nodes by two and each additional insertion reduced the number of isolated nodes remaining by one until we ran out of isolated nodes.

 DEMO: *Trees*

Once we have the tree, we add the remaining branches to the growing network one at a time. When the first of these additional branches is added, a network with one mesh is created, because there are now two paths that connect the two nodes to which it is attached—one through the new element and one through the original tree. The addition of another element either creates a second mesh in the same fashion or divides a previously constructed mesh into two smaller meshes. As this procedure is continued, the addition of each new branch adds one to the number of meshes, either by creating a new mesh or by dividing an existing mesh into two. The first $n - 1$ branches, which make up the tree, did not create any meshes, but each branch added after the tree was created adds exactly one additional mesh. Thus, the number of meshes created is $b - (n - 1)$. This must be the total number of meshes in the complete network. Therefore,

$$\ell = b - n + 1$$

or

$$b = n + \ell - 1. \qquad (2.1)$$

As an example, notice that, for the network in Figure 2-1(a), there are 12 branches, nine nodes, and four meshes, which agrees with the formula. You are also encouraged to look at any of the other planar networks in this text and verify that this formula is true for all of them also.

Now let us use this result to verify that the exhaustive method will produce enough independent equations to be able to specify the complete solution. After we complete Step 1 of the procedure, each of the e elements in the circuit has two unknown quantities associated with it, a voltage variable and a current variable, while each of the s sources has one unknown, the voltage in the case of a current source or the current in the case of a voltage source. By adding these up, we can see that the total number of unknowns is

$$\# \text{ unknowns } = 2e + s.$$

In order to find these unknowns, we write a number of linear equations, each of which imposes a constraint. There are $n - 1$ KCL equations, ℓ KVL equations, and e element relations. Thus, the total number of equations is

$$\# \text{ equations } = (n - 1) + \ell + e.$$

Equation (2.1), however, tells us how the number of nodes and meshes is related to the number of elements and sources, since $b = e + s = (n - 1) + \ell$. Substituting this relation, we see that the total number of equations is

$$\# \text{ equations } = 2e + s.$$

Since the number of independent equations is equal to the total number of unknowns, we are guaranteed that the exhaustive method will find the complete solution.

 MULTIMEDIA TUTORIAL: *Exhaustive Method*

2-1-2   Examples of the Method

Example 2-1   **The Exhaustive Method**

As a simple first example, consider the circuit in Figure 2-2, which contains three resistors and a current source. The first step is to define seven variables: a voltage and a current for each of the three resistors and a voltage (potential difference) across the terminals of the current

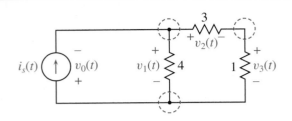

Fig. 2-2. A simple circuit to be analyzed via the exhaustive method. The circuit contains three nodes and two meshes (Example 2-1). All resistances are measured in ohms.

source. The four voltages are indicated on the figure. Let $i_1(t)$ be the current flowing into the $+$ terminal of the 4 Ω resistor, $i_2(t)$ be the current flowing into the $+$ terminal of the 3 Ω resistor, and $i_3(t)$ be the current flowing into the $+$ terminal of the 1 Ω resistor. The total number of variables is seven.

Step 2 of the procedure says that we should write element relations for all of the elements. Since in this case there are three elements (the resistors), we write three statements of Ohm's law:

$$v_1(t) - 4i_1(t) = 0$$
$$v_2(t) - 3i_2(t) = 0$$
$$v_3(t) - i_3(t) = 0.$$

The circuit contains three nodes and two meshes. Therefore, we should write two KCL equations and two KVL equations. For the KCL equations written at the two upper nodes in the circuit, we have

$$i_1(t) + i_2(t) = i_s(t)$$
$$-i_2(t) + i_3(t) = 0;$$

for the KVL equations written on the paths defined by the two meshes, we have

$$v_0(t) + v_1(t) = 0$$
$$-v_1(t) + v_2(t) + v_3(t) = 0.$$

To solve these equations, we notice first that the top KVL equation is the only equation involving $v_0(t)$. If we ignore it for the moment, we are left with six equations in six unknowns. Using the three v–i relations to eliminate the voltage variables gives for the remaining three equations

$$i_1(t) + i_2(t) = i_s(t)$$
$$-i_2(t) + i_3(t) = 0$$
$$-4i_1(t) + 3i_2(t) + i_3(t) = 0.$$

From the second of these equations, $i_2(t) = i_3(t)$. Applying this fact to the other two equations gives

$$i_1(t) + i_2(t) = i_s(t)$$
$$-i_1(t) + i_2(t) = 0.$$

Solving these yields the result

$$i_1(t) = i_2(t) = i_3(t) = 0.5 \, i_s(t).$$

Knowing the currents then gives

$$v_1(t) = 4i_1(t) = 2 \, i_s(t),$$
$$v_2(t) = 3i_2(t) = 1.5 \, i_s(t),$$
$$v_3(t) = 1i_3(t) = 0.5 \, i_s(t),$$
$$v_0(t) = -v_1(t) = -2.0 \, i_s(t).$$

∎

 To check your understanding, try Drill Problem P2-1.

 WORKED SOLUTION: *Exhaustive Method*

Example 2-2   **A Second Example of the Exhaustive Method**

As a second and somewhat more complicated example, consider the circuit in Figure 2-3, which contains four resistors, a voltage source, and a current source. To solve this circuit by using the exhaustive method will require setting up and solving 10 linear equations in 10 unknowns. The unknowns are the voltages and currents associated with the four resistors, the current through the voltage source, $i_0(t)$, and the potential difference across the terminals of the current source, $v_0(t)$. The resistor currents are shown. Assume that the four voltages $v_1(t)$, $v_2(t)$, $v_3(t)$, and $v_4(t)$ are defined under the default sign convention on the appropriate resistors. The circuit contains four nodes and three meshes. Therefore, the ten equations will consist of four element relations, three KCL equations, and three KVL equations.

Writing the equations is straightforward. The four element relations are

$$v_1(t) - 4i_1(t) = 0$$
$$v_2(t) - 3i_2(t) = 0$$

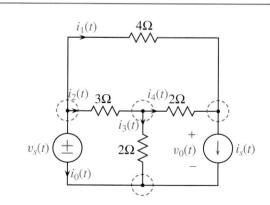

Fig. 2-3. Another circuit to illustrate the Exhaustive Method (Example 2-2).

$$v_3(t) - 2i_3(t) = 0$$
$$v_4(t) - 2i_4(t) = 0.$$

If we write KCL equations at the three nodes that line up horizontally in the middle of the circuit, we have

$$i_0(t) + i_1(t) + i_2(t) = 0$$
$$-i_2(t) + i_3(t) + i_4(t) = 0$$
$$i_1(t) + i_4(t) = i_s(t).$$

Finally, writing KVL equations over the three meshes gives

$$v_1(t) - v_2(t) - v_4(t) = 0$$
$$v_2(t) + v_3(t) = v_s(t)$$
$$v_0(t) - v_3(t) + v_4(t) = 0.$$

Notice that $i_0(t)$ and $v_0(t)$ each occurs in only one equation. These equations can be ignored for the time being, since each merely provides values for these two variables in terms of the others. Using the element variables to eliminate the voltages as variables in the remaining equations reduces the number of equations and unknowns to four:

$$-i_2(t) + i_3(t) + i_4(t) = 0$$
$$i_1(t) + i_4(t) = i_s(t)$$
$$4i_1(t) - 3i_2(t) - 2i_4(t) = 0$$
$$3i_2(t) + 2i_3(t) = v_s(t). \qquad (2.2)$$

Unfortunately, finding the solution of the final four equations in four unknowns is algebraically tedious. It can be approached with a pencil and paper, using familiar methods from algebra, or it can be done on a scientific calculator or computer. (Section 2-3 will discuss how

MATLAB can be used to solve equations like these.) By whatever means, the currents are seen to be

$$i_1(t) = \frac{1}{12} v_s(t) + \frac{4}{9} i_s(t)$$

$$i_2(t) = \frac{1}{6} v_s(t) + \frac{2}{9} i_s(t)$$

$$i_3(t) = -\frac{1}{12} v_s(t) + \frac{5}{9} i_s(t)$$

$$i_4(t) = \frac{1}{4} v_s(t) - \frac{1}{3} i_s(t),$$

which means that the element voltages are

$$v_1(t) = 4i_1(t) = \frac{1}{3} v_s(t) + \frac{16}{9} i_s(t)$$

$$v_2(t) = 3i_2(t) = \frac{1}{2} v_s(t) + \frac{2}{3} i_s(t)$$

$$v_3(t) = 2i_3(t) = -\frac{1}{6} v_s(t) + \frac{10}{9} i_s(t)$$

$$v_4(t) = 2i_4(t) = \frac{1}{2} v_s(t) + \frac{2}{3} i_s(t).$$

Finally, from the two equations that we omitted earlier,

$$v_0(t) = v_3(t) - v_4(t) = -\frac{2}{3} v_s(t) + \frac{4}{9} i_s(t)$$

and

$$i_0(t) = -i_1(t) - i_2(t) = -\frac{1}{4} v_s(t) - \frac{2}{3} i_s(t).$$

Exhausting!   ■

 To check your understanding, try Drill Problem P2-2.

2-2   Supernodes and Supermeshes

In each of the two examples of the last section, we were able to simplify the solution slightly by observing that some of the equations (one in the first example and two in the second) could be decoupled from the others. This reduced the size of the set of equations that had to be solved simultaneously. In both cases, the equations that could be set aside were those that involved source variables (voltages across the terminals of current sources and currents flowing through voltage sources). There was nothing special about these two examples; we can always decouple the equations involving the source variables from the others, but to do this we could have to write some KCL equations that are associated with isolated subnetworks rather than with simple nodes, and we might have to write some KVL equations on closed paths that are not meshes. To simplify the development in Sections 2-2-1 and 2-2-2, we assume initially that our circuits do not contain dependent sources. This is only temporary, and we will look at circuits that contain them in Section 2-2-3.

2-2-1   Supernodes

Let us first consider those equations that contain as variables the currents that flow through voltage sources. We observe that all of these equations must be KCL equations, since these variables cannot be part of the element relations for the resistors and the KVL equations contain only voltage variables. Each voltage source must necessarily connect two nodes. Figure 2-4 shows an example of a voltage source in a circuit that connects two nodes labelled a and b. The current flowing through the source is i. If KCL equations are written at both of these nodes, those equations are

$$node\ a:\quad i_1 + i_2 - i = 0 \qquad (2.3)$$

$$node\ b:\quad i_3 + i_4 + i = 0. \qquad (2.4)$$

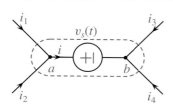

Fig. 2-4. A section of a network containing a voltage source that connects two nodes.

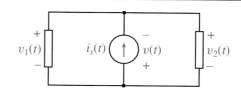

Fig. 2-5. A current source separating two meshes.

Adding the two equations together gives

$$i_1 + i_2 + i_3 + i_4 = 0. \qquad (2.5)$$

This equation does not involve the source variable i. It is, however, the KCL equation for the surface that encloses the voltage source and the two nodes to which it is attached! By replacing the pair of equations (2.3) and (2.4) by the single equation (2.5), we remove one variable (i) and reduce the number of equations by one. A surface that encircles a voltage source and its two attached nodes, such as this one, is called a *supernode*.

In the preceding argument, it was assumed that the exhaustive method wrote KCL equations at both a and b. Suppose instead that node b is the single node at which a KCL equation is not written. In this case the KCL equation at node a,

$$i_1 + i_2 - i = 0,$$

is the only equation that involves the variable i and it only serves to tell us that

$$i = i_1 + i_2.$$

Thus, unless we need to know the value of i, we can ignore this equation, thereby again reducing both the number of variables and the number of equations by one. If we do need the value of i, we can use the above equation *after* we have solved for i_1 and i_2.

If we turn all of the voltage sources into supernodes and write KCL equations at all but one of the non-encircled nodes or supernodes, we will have eliminated all of the voltage source currents as variables and reduced the number of equations by the number of voltage sources.

2-2-2   Supermeshes

Next, we need to decouple the equations that involve the voltages associated with current sources. The only equations that contain these voltages are KVL equations.

There are two cases that we need to consider. First, consider the case where the current source separates two meshes, as in Figure 2-5.

The potential difference between the terminals of the current source is $v(t)$. The KVL equations for the two meshes are

$$\textit{left mesh:} \quad -v_1(t) - v(t) = 0 \qquad (2.6)$$

$$\textit{right mesh:} \quad v(t) + v_2(t) = 0. \qquad (2.7)$$

Adding the two equations together gives

$$-v_1(t) + v_2(t) = 0. \qquad (2.8)$$

Eq. (2.8) is the KVL equation of the closed path formed from the union of these two meshes that are separated by the current source. It is, of course, a valid KVL equation and it is consistent with (2.6) and (2.7). A path such as this one that is formed from two adjacent meshes separated by a current source is called a *supermesh*. Replacing (2.6)

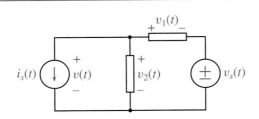

Fig. 2-6. A circuit containing a current source that does not separate two meshes.

unknowns are element voltages and e of them are element currents. e of the equations are element relations and the remaining e are KCL and KVL equations.

The modifications to the exhaustive method that incorporate supernodes and supermeshes to eliminate source variables define the *simplified exhaustive method*. Even though we have yet to incorporate dependent sources, we shall formally state the method now. This statement is valid regardless of whether there are dependent sources.

and (2.7) by (2.8) reduces the number of equations and the number of variables by one ($v(t)$).

The second case that we need to consider occurs when a circuit contains a current source that does not separate two meshes, as in Figure 2-6. This will occur when the current source lies on the outer "boundary" of the circuit. To give these sources a name, we shall call them *exterior current sources*. In this case, the KVL equation written over the left mesh,

$$-v(t) + v_2(t) = 0,$$

is the only one that contains the variable $v(t)$. Ignoring that equation (unless we need the value of $v(t)$) again reduces the number of equations and the number of variables by one.

If we define a supermesh for every pair of adjacent meshes (or supermeshes) that are separated by a current source and then write a KVL equation for each mesh or supermesh whose path does not contain an exterior current source, we will have decoupled all of the current source voltages as variables and reduced the number of equations by the number of current sources.

By using both supernodes and supermeshes, we eliminate *all* of the source variables as unknowns and reduce the number of equations by the number of sources. If the circuit contains e elements, the resulting system of equations contains $2e$ equations in $2e$ unknowns. e of the

---

SIMPLIFIED EXHAUSTIVE METHOD:

1. *Create a supernode around each voltage source and the nodes to which it is attached.*
2. *Create a supermesh from each pair of adjacent meshes that are separated by a current source.*
3. *Define a voltage variable and a current variable for each element (not the sources) in the circuit.*
4. *Write a v–i relation for each element.*
5. *Write a KVL equation for each mesh or supermesh whose path does not contain an exterior current source.*
6. *Write a KCL equation for all but one of the nodes or supernodes.*
7. *Solve the resulting set of equations for the element variables.*
8. *If any of the source variables are required, solve for each of them after the element variables have been determined.*

Either of these methods—the exhaustive method or the simplified exhaustive method—is guaranteed to find the solution of any planar circuit. The complete solution can be found by solving even fewer linear equations by using the node method and the mesh method. These procedures are developed later in the chapter.

Example 2-3   Using the Simplified Exhaustive Method to Write KCL and KVL Equations

Let us illustrate the procedure we have just developed using the simple circuit sketched in Figure 2-7, which we solved via the exhaustive method in Example 2-1.

Since this circuit contains three resistors, there are six element variables: the currents through the three resistors, and the three associated voltages. The circuit does not contain any voltage sources, so we do not need to create any supernodes, but it does contain a current source. Since that source is located on the exterior edge of the circuit, we will not write a KVL equation on the path corresponding to the left mesh. This means that we need to write two KCL equations and one KVL equation. To limit the number of additional equations, we use only the voltage variables and then use Ohm's law and the default sign convention to express the element currents in terms of the voltages ($i = v/R$).

The KVL equation written around the right mesh is

$$-v_1(t) + v_2(t) + v_3(t) = 0.$$

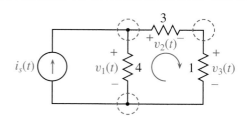

Fig. 2-7. A simple circuit to illustrate how the simplified exhaustive method can be used to construct a minimal set of KVL and KCL equations. All resistances are measured in ohms.

From KCL written at the upper left node, we have

$$\frac{1}{4}v_1(t) + \frac{1}{3}v_2(t) = i_s(t),$$

and from KCL at the upper right node, we have

$$\frac{1}{3}v_2(t) - v_3(t) = 0.$$

Solving these three linear equations in three unknowns gives the solution

$$v_1(t) = 2\,i_s(t); \quad v_2(t) = \frac{3}{2}\,i_s(t); \quad v_3(t) = \frac{1}{2}\,i_s(t).$$

The three element currents can be computed from the three voltages:

$$i_1(t) = \frac{1}{2}\,i_s(t); \quad i_2(t) = \frac{1}{2}\,i_s(t); \quad i_3(t) = \frac{1}{2}\,i_s(t).$$

We observe that every voltage in the circuit and every current is proportional to the single source signal, $i_s(t)$. This is the same solution that we got in Example 2-1, except that, in that example, we set up and solved seven equations in seven unknowns. (We very quickly reduced those to three equations in three unknowns that were comparable to the equations above.) ∎

 To check your understanding, try Drill Problem P2-3.

Example 2-4   Example 2-2 Revisited

The circuit in Figure 2-8 reexamines the slightly more complex example that we saw in Example 2-2. Since the circuit contains a voltage source, we create a supernode that encircles it. The circuit also contains a current source, but that source is located on the exterior boundary of the circuit, so its effect will be to render unnecessary the KVL equation on its mesh. There are three nonencircled nodes or supernodes, as shown on the drawing, so we will need to write two KCL equations in addition to

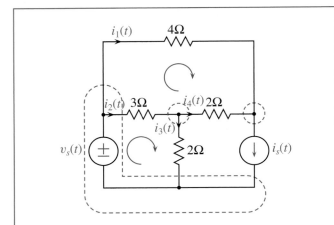

Fig. 2-8. Another circuit to illustrate writing a minimal set of KCL and KVL equations.

 To check your understanding, try Drill Problem P2-4.

2-2-3 Dependent Sources

To simplify our discussion to this point, we have assumed that our networks were free of any dependent sources. Now it is time to remove that restriction. Fortunately, this is not difficult.

Dependent sources behave like independent ones; they have one variable that is controlled (the voltage in a dependent voltage source or the current in a dependent current source) and one variable that is uncontrolled (the current in a dependent voltage source or the voltage in a dependent current source). The value of the *uncontrolled* variable is determined by the solution of the complete circuit. It can be found once the element variables are known by writing a single KCL or KVL equation. The value of the *controlled* variable is linked to another voltage or current in the circuit. As a result, the addition of a dependent source does not introduce any new variable into the network whose value needs to be computed initially. Thus, it does not create the need for any additional equations. We cannot, of course, simply ignore the dependent sources; we need to account for them in our equations, but it is important to note that *we do not need to add any more equations or variables to do this with the simplified exhaustive method.*

The simplest way to deal with a dependent source is to treat it initially as if it were an independent source. We create a supernode around each of the dependent voltage sources and we create a supermesh encircling any of the interior dependent current sources. We then write a KVL equation around the meshes or supermeshes in the circuit that do not contain (independent or dependent) current sources and a KCL equation at all but one of the nodes or supernodes. For the dependent sources, we next substitute the controlling relation, which must be expressed in terms

the two KVL equations on the two remaining meshes. Together with the four v–i relations for the resistors, these will be enough to specify completely the eight element variables associated with the four elements. (With the exhaustive method, we had to write three KCL equations and three KVL equations.) As in the previous example, we can incorporate Ohm's law directly into the writing of some of the equations to help reduce their number. We use the resistor currents as our variables, as we did in Example 2-2, being careful to adhere to the default sign convention. We arbitrarily select the supernode as the node to be ignored when writing KCL equations. The four equations are as follows:

$$left\ node:\quad -i_2(t) + i_3(t) + i_4(t) = 0$$
$$right\ node:\quad i_1(t) + i_4(t) = i_s(t)$$
$$top\ mesh:\quad 4i_1(t) - 3i_2(t) - 2i_4(t) = 0$$
$$bottom\ mesh:\ 3i_2(t) + 2i_4(t) = v_s(t).$$

Since these are the same four equations that we had in equation (2.2) in Example 2-2, the solution is obviously the same as well.   ■

of element voltages and/or element currents. We illustrate this procedure in the following example.

Example 2-5   Simplified Exhaustive Method with a Dependent Source

Consider the circuit drawn in Figure 2-9. If we create a supernode that encircles the voltage source and the two nodes to which it is attached, there will be two remaining isolated nodes, each of which is connected to one of the terminals of the 2Ω resistor. Thus, we need to write two KCL equations. Since the center and right mesh are separated by a (dependent) current source, we create a supermesh from them and write two KVL equations—one around the left mesh and one around the supermesh. We also write four element relations for the four resistors. This results in a total of eight equations in the four resistor voltages and the four resistor currents. In this circuit, since all of the elements are resistors, the element relations are simple:

$$v_1(t) = 6i_1(t)$$
$$v_2(t) = 3i_2(t)$$
$$v_3(t) = 2i_3(t)$$
$$v_4(t) = 4i_4(t).$$

We can substitute immediately for the element voltages and write both the KVL and KCL equations in terms of the current variables alone. Doing this reduces the number of equations that we have to solve formally.

The first KVL equation is written around the left mesh and uses the current variables. Remember to be careful with the sign conventions.

$$6i_1(t) + 3i_2(t) = v_s(t). \tag{2.9}$$

The second KVL equation is written over the supermesh that encircles the dependent current source:

$$-3i_2(t) + 2i_3(t) + 4i_4(t) = 0. \tag{2.10}$$

Since the network has three node/supernodes, we need to write KCL equations at two of them. We choose to write them at the two isolated nodes. For the left node, we get

$$i_1(t) - i_2(t) - i_3(t) = 0, \tag{2.11}$$

and for the right one,

$$i_3(t) - i_4(t) = -\tfrac{1}{2}i_2(t)$$
$$\implies \tfrac{1}{2}i_2(t) + i_3(t) - i_4(t) = 0. \tag{2.12}$$

These are our four equations in four unknowns. By putting all of the variables on one side of the equations and the source terms on the other, we have the equations in a form that software packages, such as MATLAB, can accept to provide a ready solution. It is also a form that makes a pencil-and-paper solution easier. By whatever method, for this problem the solution is

$$i_1(t) = 7/60 \ v_s(t) = 0.1167 \ v_s(t)$$
$$i_2(t) = 1/10 \ v_s(t) = 0.1000 \ v_s(t)$$
$$i_3(t) = 1/60 \ v_s(t) = 0.0167 \ v_s(t)$$
$$i_4(t) = 1/15 \ v_s(t) = 0.0667 \ v_s(t).$$

Section 2-3 reviews methods for writing linear equations, such as these, in matrix–vector form and shows how MATLAB or similar packages can be used to solve these equations. ∎

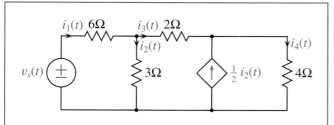

Fig. 2-9. An example of a network with one independent and one dependent source.

To check your understanding, try Drill Problem P2-5.

specifies its column. The special case of an $m \times 1$ matrix is called a *column vector*. Column vectors take the form

$$b = \begin{bmatrix} b_1 \\ b_2 \\ \vdots \\ b_m \end{bmatrix}.$$

2-3   Solving Circuit Equations

If a planar, resistive circuit contains e elements (resistors), then finding the complete solution of the network via the simplified exhaustive method involves setting up and solving a system of e linear equations in e unknowns when Ohm's law is used to eliminate either the voltage variables or the current variables before the equations are solved. We shall shortly explore some other techniques that reduce the number of equations further, but setting up and solving a set of linear equations is a key component of all of these methods. This section looks at methods for writing and manipulating the linear equations in matrix–vector form. This frames the problem in a form where a programmable calculator or a software package, such as MATLAB, can deal with it. In doing so, we will also make a key observation about circuits that will prove to be particularly useful. We begin by reviewing some properties of matrices.

A $1 \times n$ matrix is called a *row vector*. The vector

$$c = [\, c_1 \ c_2 \ \cdots \ c_n \,]$$

is an example of a row vector. Notice that, for row and column vectors, a single-subscript notation is sufficient to represent the elements. Our convention will be to use lowercase boldface letters to denote vectors and uppercase boldface letters to denote more general matrices.

      EXTRA INFORMATION: *Matrix Equations*

2-3-2   Matrix Operations

Matrix Addition: Two matrices can be added only if they have the same dimensions (the same numbers of rows and columns.) If two matrices A and B are of the same size, we define their sum, $C = A + B$, as the matrix formed from the sums of their corresponding elements. That is,

$$c_{ik} = a_{ik} + b_{ik}.$$

Since row and column vectors are special cases of matrices, the same procedure can be used for summing vectors, provided that they are the same size.

2-3-1   Matrices

An $m \times n$ *matrix* is a rectangular array of numbers or functions constructed to have m rows and n columns. The numbers that make up the array are called the *elements* of the matrix, and the numbers m and n are called its *dimensions*. As an example, the matrix

$$A = \begin{bmatrix} \alpha_{11} \ \alpha_{12} \ \alpha_{13} \ \alpha_{14} \\ \alpha_{21} \ \alpha_{22} \ \alpha_{23} \ \alpha_{24} \\ \alpha_{31} \ \alpha_{32} \ \alpha_{33} \ \alpha_{34} \end{bmatrix}$$

is a 3×4 matrix consisting of the elements α_{ik}, where i specifies the row in which the element is located and k

Scalar Multiplication: Multiplication of a matrix A by a scalar α, denoted by $D = \alpha A$, is done by multiplying each element in the matrix by the scalar. Thus, if a_{ik} is the $(i, k)^{th}$ element of A and d_{ik} is the $(i, k)^{th}$ element of D, then

$$d_{ik} = \alpha a_{ik}.$$

Matrix Multiplication: The product of two matrices $E = AB$ is defined only if the two matrices have compatible dimensions; the number of columns in A must equal the number of rows in B. If A is an $m \times n$ matrix and B is an $n \times p$ matrix, then the product E will be $m \times p$. The element e_{ik} of E is computed from the formula

$$e_{ik} = \sum_{\ell=1}^{n} a_{i\ell} b_{\ell k},$$

which is the inner product of the ith row of A with the kth column of B. Usually, matrix multiplication is *not* commutative (i.e. $AB \neq BA$). In fact, if $m \neq p$, the product BA is not even defined. Even when $m = p$, however, these two products are generally different.

Matrix multiplication, like scalar multiplication, obeys the distributive law with respect to addition. Thus,

$$A(B + C) = AB + AC$$

and

$$(D + E)F = DF + EF.$$

Matrix Inverse: The rows of a matrix are *linearly independent* if none of them can be expressed as a linear combination of the others. This means that if a_i is the ith row of a matrix A, which contains m rows, there are no constants c_k such that

$$a_i = \sum_{\substack{j=1 \\ j \neq i}}^{m} c_j a_j.$$

If the rows of the matrix A are linearly independent and if the number of rows, m, of A is equal to the number of its columns, then A is *invertible*. The inverse of A, denoted as A^{-1}, is a matrix that satisfies the matrix equation

$$A^{-1}A = AA^{-1} = I,$$

where I is the $m \times m$ *identity matrix*. An identity matrix is a square matrix whose elements $i_{k\ell}$ satisfy

$$i_{k\ell} = \begin{cases} 1, & k = \ell \\ 0, & k \neq \ell. \end{cases}$$

Example 2-6   **Matrix Inverse**

The matrix

$$A = \begin{bmatrix} 1 & 0 & -1 \\ 1 & 1 & 0 \\ 1 & 0 & 1 \end{bmatrix}$$

has the inverse

$$A^{-1} = \begin{bmatrix} 0.5 & 0 & 0.5 \\ -0.5 & 1 & -0.5 \\ -0.5 & 0 & 0.5 \end{bmatrix}.$$

It is straightforward to verify that

$$AA^{-1} = \begin{bmatrix} 1 & 0 & -1 \\ 1 & 1 & 0 \\ 1 & 0 & 1 \end{bmatrix} \begin{bmatrix} 0.5 & 0 & 0.5 \\ -0.5 & 1 & -0.5 \\ -0.5 & 0 & 0.5 \end{bmatrix} = \begin{bmatrix} 1 & 0 & 0 \\ 0 & 1 & 0 \\ 0 & 0 & 1 \end{bmatrix}.$$

We also notice that

$$A^{-1}A = \begin{bmatrix} 0.5 & 0 & 0.5 \\ -0.5 & 1 & -0.5 \\ -0.5 & 0 & 0.5 \end{bmatrix} \begin{bmatrix} 1 & 0 & -1 \\ 1 & 1 & 0 \\ 1 & 0 & 1 \end{bmatrix} = \begin{bmatrix} 1 & 0 & 0 \\ 0 & 1 & 0 \\ 0 & 0 & 1 \end{bmatrix}.$$

■

Any text in linear algebra will discuss a number of efficient techniques for determining the inverse of an invertible matrix. We shall not discuss any of these here because we do not need them. Computational packages such as MATLAB contain functions for inverting matrices and we shall simply use them whenever we need to invert a matrix.

2-3-3 Representing Linear Equations in MATLAB*

The system of m linear equations in n unknowns

$$a_{11}x_1 + b_{12}x_2 + \cdots + a_{1n}x_n = b_1$$

$$a_{21}x_1 + b_{22}x_2 + \cdots + a_{2n}x_n = b_2$$

$$\vdots$$

$$a_{m1}x_1 + b_{m2}x_2 + \cdots + a_{mn}x_n = b_m$$

can be written in matrix–vector form as

$$Ax = b$$

if we define

$$A = \begin{bmatrix} a_{11} & a_{12} & \dots & a_{1n} \\ a_{21} & a_{22} & \dots & a_{2n} \\ \vdots & \vdots & \ddots & \vdots \\ a_{m1} & a_{m2} & \dots & a_{mn} \end{bmatrix}$$

as the $m \times n$ matrix of coefficients,

$$x = \begin{bmatrix} x_1 \\ x_2 \\ \vdots \\ x_n \end{bmatrix}$$

as an $n \times 1$ column vector of variables, and

$$b = \begin{bmatrix} b_1 \\ b_2 \\ \vdots \\ b_m \end{bmatrix}$$

as an $m \times 1$ column vector of constants. The set of equations has a unique solution if the matrix of coefficients is invertible.

The default operations in MATLAB are matrix operations. To enter a series of linear equations, such as the ones above, one simply enters the matrix of coefficients A and the column vector of constants from the right-hand side, b. The solution can then be found by using the MATLAB construction

```
x=A\b
```

Example 2-7   Solving Equations by Using MATLAB

As a simple example, we can use MATLAB to solve the simple set of three linear equations in three unknowns

$$1.4x_1 + 2.3x_2 + 3.7x_3 = 6.5$$

$$3.3x_1 + 1.6x_2 + 4.3x_3 = 10.3$$

$$2.5x_1 + 1.9x_2 + 4.1x_3 = 8.8.$$

We solve these by entering the 3×3 matrix of coefficients, entering the 3×1 column vector of coefficients from the right-hand sides, and then solving the equations, using the "backslash" command. This is shown in the following transcription.

```
>> A = [1.4 2.3 3.7; 3.3 1.6 4.3;...
2.5 1.9 4.1]

A =
1.4000     2.3000     3.7000
3.3000     1.6000     4.3000
2.5000     1.9000     4.1000

>> b = [6.5; 10.3; 8.8]

b =
6.5000
10.3000
8.8000

>> x=A\b

x =
```

```
1.0000
-1.0000
2.0000
```

```
>>
```

The commands are the expressions entered following the prompt >>. The values of the expressions are echoed by MATLAB. Those echoes are suppressed if a command is followed by a semicolon (;). To enter a matrix (or vector), a space is used to delineate entries on a row and a semicolon is used to mark the end of a row.   ■

 DEMO: *Solving Equations with MATLAB*

2-3-4   Matrix Descriptions of Resistive Circuits

All of the circuit examples to this point have had generic waveforms, such as $v_s(t)$ or $i_s(t)$, as their independent sources. This allowed us to keep our results completely general, and it also allowed us to make a key observation about resistive circuits: all of the element-variable waveforms are linear combinations of the source waveforms. When the inputs are generic, we can still solve the resulting equations by using MATLAB, but we need to manipulate them a bit more first.

To illustrate the approach, we use the network in Figure 2-10 as a working example. This network contains two independent sources and five resistors. Thus, there are 10 element variables, whose values can be found from two KVL equations (one on the right mesh and one on the supermesh that goes around the current source), three KCL equations (at, for example, the supernode that encircles the voltage source and two of the three remaining isolated nodes), and the five v–i relations. The element relations allows us to eliminate the voltage variables immediately and to write both the KCL and KVL equations in terms of the five current variables

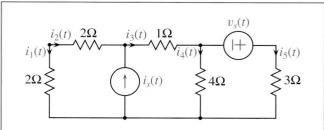

Fig. 2-10. A resistive network with two independent sources.

that are indicated. Anticipating putting these equations into matrix–vector form, we write them with the current variables on one side (the left) and the sources on the other (the right).

We begin with the KVL equations written on the supermesh and the right mesh:

$$-2i_1(t) + 2i_2(t) + i_3(t) + 4i_4(t) = 0$$

$$-4i_4(t) + 3i_5(t) = v_s(t).$$

Then we add the KCL equations at the two nodes attached to the horizontal 2Ω resistor and the supernode:

$$-i_1(t) - i_2(t) = 0$$

$$i_2(t) - i_3(t) = -i_s(t)$$

$$i_3(t) - i_4(t) - i_5(t) = 0.$$

Next, we put these equations into matrix–vector form:

$$\boldsymbol{C}\boldsymbol{i}(t) = \boldsymbol{s}(t). \tag{2.13}$$

$\boldsymbol{C}$ is the 5×5 matrix of coefficients, $\boldsymbol{i}(t)$ is a 5×1 column vector containing the current variables, and $\boldsymbol{s}(t)$ is a 5×1 column vector of independent source terms from the right-hand sides. Notice that the coefficients of the matrix $\boldsymbol{C}$ are all constants; they do not vary with time.

For our example, (2.13) takes the form

$$
\begin{bmatrix}
-2 & 2 & 1 & 4 & 0 \\
0 & 0 & 0 & -4 & 3 \\
-1 & -1 & 0 & 0 & 0 \\
0 & 1 & -1 & 0 & 0 \\
0 & 0 & 1 & -1 & -1
\end{bmatrix}
\underbrace{\begin{bmatrix}
i_1(t) \\
i_2(t) \\
i_3(t) \\
i_4(t) \\
i_5(t)
\end{bmatrix}}_{\boldsymbol{i}(t)}
=
\underbrace{\begin{bmatrix}
0 \\
v_s(t) \\
0 \\
-i_s(t) \\
0
\end{bmatrix}}_{\boldsymbol{s}(t)}. \quad (2.14)
$$

The rows of the $\boldsymbol{C}$ and $\boldsymbol{s}(t)$ matrices correspond to the five equations, and the columns of the $\boldsymbol{C}$ matrix correspond to the coefficients of the individual variables. That is, the first column of $\boldsymbol{C}$ contains the coefficients of $i_1(t)$ in the five equations, the second column contains the coefficients of $i_2(t)$, etc. Notice that we must explicitly include a coefficient value of zero for each current that does not appear in a particular equation.

The right-hand side of (2.14) is still not in a form that we can use. We can manipulate it into such a form, however, by rewriting it as the sum of column vectors, each of which is proportional to one of the source waveforms:

$$
\begin{bmatrix}
-2 & 2 & 1 & 4 & 0 \\
0 & 0 & 0 & -4 & 3 \\
-1 & -1 & 0 & 0 & 0 \\
0 & 1 & -1 & 0 & 0 \\
0 & 0 & 1 & -1 & -1
\end{bmatrix}
\begin{bmatrix}
i_1(t) \\
i_2(t) \\
i_3(t) \\
i_4(t) \\
i_5(t)
\end{bmatrix}
=
\begin{bmatrix}
0 \\
1 \\
0 \\
0 \\
0
\end{bmatrix} v_s +
\begin{bmatrix}
0 \\
0 \\
0 \\
-1 \\
0
\end{bmatrix} i_s.
$$
$$(2.15)$$

If we define $\boldsymbol{s}_v$ and $\boldsymbol{s}_i$ as the two (constant) column vectors on the right-hand side of (2.15), namely,

$$
\boldsymbol{s}_v = \begin{bmatrix} 0 \\ 1 \\ 0 \\ 0 \\ 0 \end{bmatrix} \quad \text{and} \quad \boldsymbol{s}_i = \begin{bmatrix} 0 \\ 0 \\ 0 \\ -1 \\ 0 \end{bmatrix},
$$

then the system of equations can be written as

$$
\boldsymbol{C}\boldsymbol{i}(t) = \boldsymbol{s}_v \, v_s(t) + \boldsymbol{s}_i \, i_s(t). \quad (2.16)
$$

Since this system of equations possesses a solution, $\boldsymbol{C}$ is invertible, and we can write the solution in the form

$$
\boldsymbol{i}(t) = \boldsymbol{C}^{-1}\boldsymbol{s}_v \, v_s(t) + \boldsymbol{C}^{-1}\boldsymbol{s}_i \, i_s(t)
$$
$$
= \boldsymbol{i}_v \, v_s(t) + \boldsymbol{i}_i \, i_s(t),
$$

where we have defined

$$
\boldsymbol{i}_v = \boldsymbol{C}^{-1}\boldsymbol{s}_v \quad \text{and} \quad \boldsymbol{i}_i = \boldsymbol{C}^{-1}\boldsymbol{s}_i.
$$

Notice that $\boldsymbol{i}_v$ and $\boldsymbol{i}_i$ are constant column vectors and that they can be found by solving the following two sets of linear equations:

$$
\boldsymbol{C}\boldsymbol{i}_v = \boldsymbol{s}_v; \quad \boldsymbol{C}\boldsymbol{i}_i = \boldsymbol{s}_i.
$$

Because both sets of equations have the same coefficient matrix $\boldsymbol{C}$, we can solve both of them at the same time by creating a matrix $\boldsymbol{s}$ with two columns, one containing $\boldsymbol{s}_v$ and the other containing $\boldsymbol{s}_i$. The resulting solution vector $\boldsymbol{i}$ will also contain two columns, one containing $\boldsymbol{i}_v$ and the other containing $\boldsymbol{i}_i$. (See the MATLAB documentation.) This is preferable to computing $\boldsymbol{C}^{-1}$ explicitly, since it requires less computation if properly programmed. The complete solution can be found by using the following MATLAB dialog:

```
>> C=[-2 2 1 4 0; 0 0 0 -4 3; ...
-1 -1 0 0 0; 0 1 -1 0 0; ...
0 0 1 -1 -1];
>> s=[0 0; 1 0; 0 0; 0 -1; 0 0];
>> i=C\s

i =

-0.0851      0.4043
0.0851      -0.4043
0.0851       0.5951
-0.1064      0.2553
0.1915       0.3404
```

Therefore, the five currents are

$$i_1(t) = -0.0851\ v_s(t) + 0.4043\ i_s(t)$$

$$i_2(t) = 0.0851\ v_s(t) - 0.4043\ i_s(t)$$

$$i_3(t) = 0.0851\ v_s(t) + 0.5951\ i_s(t)$$

$$i_4(t) = -0.1064\ v_s(t) + 0.2553\ i_s(t)$$

$$i_5(t) = 0.1915\ v_s(t) + 0.3404\ i_s(t).$$

Example 2-8   Using MATLAB to Find Circuit Variables

To illustrate the procedure with a second example, consider the circuit in Figure 2-11. Since the two current sources lie on the outer boundary of the circuit, only one KVL equation is necessary (on the center mesh). We also need to write KCL equations at two of the three nodes. Using the three resistor currents as variables and selecting the two upper nodes makes the three equations

$$\textit{KVL:} \quad -2i_1(t) + i_2(t) + 3i_3(t) = 0$$

$$\textit{KCL1:} \quad i_1(t) + i_2(t) = i_s(t)$$

$$\textit{KCL2:} \quad i_2(t) + 2i_2(t) - i_3(t) = 0.$$

Again, all of the variables have been placed on the left-hand sides, and the independent source terms have been placed on the right. In matrix–vector form, these become

$$\begin{bmatrix} -2 & 1 & 3 \\ 1 & 1 & 0 \\ 0 & 3 & -1 \end{bmatrix} \begin{bmatrix} i_1(t) \\ i_2(t) \\ i_3(t) \end{bmatrix} = \begin{bmatrix} 0 \\ 1 \\ 0 \end{bmatrix} i_s(t).$$

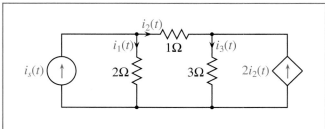

Fig. 2-11. Circuit for Example 2-6.

The solution obtained from MATLAB is

$$i_1(t) = 0.8333\ i_s(t)$$

$$i_2(t) = 0.1667\ i_s(t)$$

$$i_3(t) = 0.5000\ i_s(t).$$

■

 WORKED SOLUTION: *Circuit Variables in MATLAB*

 To check your understanding, try Drill Problem P2-6.

2-3-5   Superposition of Independent Sources

In addition to providing a means to enable us to use MATLAB to find the element variables in a circuit, the matrix representation of the circuit equations allows us to make several useful observations about circuits. One of these is an analysis method known as *superposition of sources*.

Suppose that a network is excited by more than one independent source, such as the network in Figure 2-10, which has two independent sources. Then the source vector $s(t)$—the vector from the right-hand sides of the equations that contains the source terms—can be separated into components from each of the sources. In the example of Figure 2-10, when the equations that defined the solution were put into matrix–vector form, we saw that the right-hand side could be written in the form

$$s(t) = \begin{bmatrix} 0 \\ v_s(t) \\ 0 \\ -i_s(t) \\ 0 \end{bmatrix} = \begin{bmatrix} 0 \\ 1 \\ 0 \\ 0 \\ 0 \end{bmatrix} v_s(t) + \begin{bmatrix} 0 \\ 0 \\ 0 \\ -1 \\ 0 \end{bmatrix} i_s(t)$$

$$\stackrel{\triangle}{=} s_v v_s(t) + s_i i_s(t).$$

Thus,

$$\boldsymbol{C}\boldsymbol{i}(t) = \boldsymbol{s}_v v_s(t) + \boldsymbol{s}_i i_s(t).$$

The vector of solutions also divides into two components:

$$\begin{aligned}
\boldsymbol{i}(t) &= \boldsymbol{C}^{-1}\boldsymbol{s}(t) \\
&= \boldsymbol{C}^{-1}[\boldsymbol{s}_v\, v_s(t) + \boldsymbol{s}_i\, i_s(t)] \\
&= \boldsymbol{C}^{-1}\boldsymbol{s}_v\, v_s(t) + \boldsymbol{C}^{-1}\boldsymbol{s}_i\, i_s(t) \\
&= \boldsymbol{i}_v\, v_s(t) + \boldsymbol{i}_i\, i_s(t). \quad\quad (2.17)
\end{aligned}$$

Here,

$$\boldsymbol{i}_v \triangleq \boldsymbol{C}^{-1}\boldsymbol{s}_v$$

$$\boldsymbol{i}_i \triangleq \boldsymbol{C}^{-1}\boldsymbol{s}_i.$$

We used this relationship in the previous section to describe how to obtain a solution by using MATLAB. Now, let us look at it more closely. It says that the value of each current (and each voltage) in the network contains two components, one of which is proportional to the voltage source and one of which is proportional to the current source. Notice that the value of $\boldsymbol{s}_i$ is independent of the value of $\boldsymbol{s}_v$, and vice versa. This means that we can turn off $\boldsymbol{s}_i$ (replace the current source by an open circuit) and solve for $\boldsymbol{i}_v$. We can then turn off the voltage source (replace it by a short circuit) and solve for $\boldsymbol{i}_i$. The resulting vectors $\boldsymbol{i}_v$ and $\boldsymbol{i}_i$ define the solution to the original circuit. This property is known as *the source superposition principle.* Source superposition is clearly possible for any network that has multiple independent sources. For a network with N independent sources, the output can be expressed as the sum of N components from the sources acting independently. Source superposition implies that we can solve for the network variables by applying the sources one at a time and then summing the results. Chapter 3 presents some techniques for network simplification that will make this approach particularly useful. *Source superposition cannot be used*

for dependent sources, however, since turning off a dependent source would affect the coefficient matrix $\boldsymbol{C}$.

Example 2-9   Source Superposition

As an example of how source superposition can be applied, consider the network in Figure 2-12(a), which contains two independent sources. Because this network is fairly simple, it is not difficult to solve it for the current $i(t)$. We need to write one KVL equation over the path

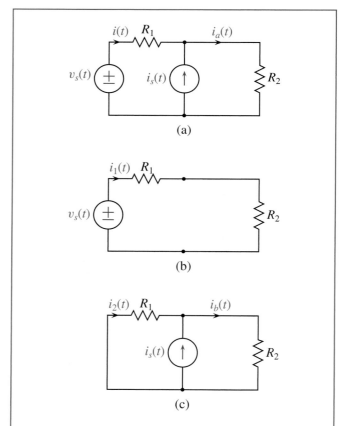

Fig. 2-12. An illustration of source superposition. (a) A network containing two independent sources. (b) The network in (a) with the current source turned off. (c) The network in (a) with the voltage source turned off.

defined by the supermesh and one KCL equation on either the supernode or the isolated node. Either KCL equation gives

$$i(t) - i_a(t) = -i_s(t).$$

Applying KVL around the supermesh and using the currents as the variables results in the equation

$$R_1 i(t) + R_2 i_a(t) = v_s(t).$$

From these, we can solve for $i(t)$ by eliminating $i_a(t)$:

$$i(t) = \frac{1}{R_1 + R_2} [v_s(t) - R_2 i_s(t)], \qquad (2.18)$$

which contains one component that is proportional to each source.

As an alternative, we can consider the two networks shown in Figure 2-12(b) and (c). In Figure 2-12(b), the current source from the original network is turned off; in Figure 2-12(c), the voltage source is turned off. If we solve for $i_1(t)$ in the network in Figure 2-12(b) and for $i_2(t)$ in the network in Figure 2-12(c), then, by source superposition, we must have

$$i(t) = i_1(t) + i_2(t).$$

For the circuit in Figure 2-12(b), the current $i_1(t)$ flows through both resistors R_1 and R_2, which are connected in series. Applying KVL around the loop, we have

$$R_1 i_1(t) + R_2 i_1(t) = v_s(t)$$

or

$$i_1(t) = \frac{1}{R_1 + R_2} v_s(t).$$

In part (c) of the figure, the voltage source is replaced by a short circuit. If we let the current through the resistor R_2 be denoted by $i_b(t)$, then, from KCL at the upper node,

$$i_b(t) = i_2(t) + i_s(t),$$

and from KVL applied around the outer path,

$$R_1 i_2(t) + R_2 i_b(t) = 0.$$

We can solve these two equations for $i_2(t)$ by eliminating $i_b(t)$:

$$i_2(t) = -\frac{R_2}{R_1 + R_2} i_s(t).$$

Then

$$
\begin{aligned}
i_1(t) + i_2(t) &= \frac{1}{R_1 + R_2} [v_s(t) - R_2 i_s(t)] \\
&= i(t),
\end{aligned}
$$

which agrees with (2.18). ∎

 WORKED SOLUTION: *Source Superposition*

 To check your understanding, try Drill Problem P2-7.

Two additional observations can be made from (2.17): (1) For a resistive circuit, all of the element variables will be equal to a superposition of the source waveforms; (2) When all of the independent sources are turned off, all of the element variables will be zero.

2-4   The Node Method

The exhaustive method and the simplified exhaustive method are systematic procedures for writing sets of linear equations that can be solved to find the complete solution of a planar resistive network. The simplified exhaustive method results in a complete set of $2e$ equations in the $2e$ element variables, when e is the number of elements in the network. As we saw, however, the matrix of coefficients in those equations is fairly sparse. This suggests that we might be able to reduce the number of equations if we use some of the simpler equations to reduce the number of variables. One possibility is to exploit the simplicity of the element relations when the elements are resistors, since, for a resistor,

$$v_k(t) = R_k i_k(t).$$

Clearly, if we know all of the resistor currents, we can trivially determine all of the resistor voltages or vice versa. This suggests that we might use Ohm's law to express all of the element voltages in terms of the element currents, solve a reduced set of e equations to calculate the e currents, and then use the calculated current values to find the voltages. We did this in many of the examples.

Actually, we can go farther than this. If we choose our variables carefully, we can incorporate both the element relations and the KVL equations into the KCL equations, to reduce the number of variables and equations to $n - 1$, where n is the number of isolated nodes or supernodes in the circuit after supernodes have been created at the voltage sources. This approach is called the *node method*. It is the subject of this section. Alternatively, we can incorporate the element relations and the KCL equations into the KVL equations. This reduces the number of variables and equations to ℓ, where ℓ is the number of meshes or supermeshes that do not contain exterior current sources in the network. This latter approach is called the *mesh method*; it is the subject of the next

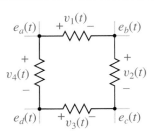

Fig. 2-13. A mesh extracted from a larger network with the node potentials indicated.

section. Which approach is to be preferred depends upon a number of factors, one of which is the relative values of n and ℓ. All other factors being equal, we would normally prefer the approach that leads to the smaller number of equations to be solved, particularly if we are solving those equations by hand.

The key to the node method is identifying an alternative set of fewer than e variables from which all of the element voltages and currents can be calculated. One way to do this is to find a set of variables that implicitly incorporates the KVL constraints. One such set is the set of *node potentials*, the electrical potentials associated with each node measured with respect to a reference node. To see how these have the desired attributes, consider the mesh extracted from a larger circuit in Figure 2-13, where the node potentials are denoted by $e_a(t)$, $e_b(t)$, $e_c(t)$, and $e_d(t)$. The voltage across each element is the difference in potential between its two terminals. Thus, for example,

$$v_1(t) = e_a(t) - e_b(t).$$

Recall that, in Chapter 1, we argued that the voltage associated with an element is the potential at the $^+$ terminal minus the potential at the $^-$ terminal. If we sum the voltages around the mesh, we obtain

$$v_1(t) + v_2(t) - v_3(t) - v_4(t)$$
$$= [e_a(t) - e_b(t)] + [e_b(t) - e_c(t)]$$
$$- [e_d(t) - e_c(t)] - [e_a(t) - e_d(t)] = 0.$$

As we traverse any path in the network, the potential at the current node is equal to the potential at the starting node plus the accumulated potential differences of the elements that have been encountered. When we return to the starting node, the potential must be the same as the initial potential, which means that the sum of the accumulated voltages must be zero.

> *When we use the node potentials as variables, KVL will always be satisfied automatically for all closed paths!*

If we know all of the node potentials, we can compute all of the element voltages as their differences; once we know all of the element voltages, we can calculate the element currents by using Ohm's law. Thus, knowing the node potentials is enough to specify the complete solution. The next example illustrates the process of computing element voltages and currents from the node potentials.

Example 2-10   **Finding Element Variables from Node Potentials**

The circuit in Figure 2-14 contains three nodes, with node potentials $e_a(t)$, $e_b(t)$, and $e_c(t)$, measured with respect to some reference, as indicated. The voltage across the terminals of the 1Ω resistor, $v_1(t)$, is simply the difference between two node potentials. Thus,

$$v_1(t) = e_a(t) - e_c(t).$$

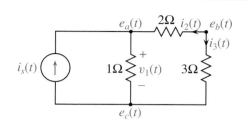

Fig. 2-14. A simple circuit with three nodes (Example 2-7).

The current $i_2(t)$, which flows through the 2Ω resistor, has its reference direction defined as pointing from node b to node a. Therefore,

$$v_2(t) = e_b(t) - e_a(t)$$

and

$$i_2(t) = \frac{1}{2}[e_b(t) - e_a(t)].$$

Similarly,

$$i_3(t) = \frac{1}{3}[e_b(t) - e_c(t)].$$

Every voltage in the circuit and every current can be expressed in terms of the three node potentials $e_a(t)$, $e_b(t)$, and $e_c(t)$ and in terms of $i_s(t)$.   ∎

When we defined the node potentials, we were very careful to state that they were a measure of the electrical potential *measured with respect to a reference node.* This is because we cannot compute the absolute values of these potentials, only their relative values. We can see this by observing that, if we were to add a constant value to each node potential, this would not affect the voltages across the terminals of any of the elements or their currents. Therefore, it is customary to arbitrarily set one of the node potentials to zero and measure all of the remaining potentials relative to it. This reference node is designated as the *ground*; on a circuit diagram, it is indicated by the symbol shown in Figure 2-15. The ground node is frequently chosen to be the one that is omitted when writing KCL equations, but this is not necessary; any of the *n* nodes (or supernodes) of the circuit can be designated as the ground and any node can be the one omitted when writing KCL equations.

When the network contains voltage sources, there will be supernodes created. We shall examine what to do in these cases in one of the examples that follows. First, however, we state the node method as a formal procedure.

THE NODE METHOD:

1. *Create a supernode encircling each independent or dependent voltage source and the two nodes to which it is attached.*

2. *Select one of the nodes of the circuit as the ground node.*

3. *Define n − 1 node potential variables at the remaining nodes or supernodes of the circuit.*

4. *Set up KCL equations at n − 1 of the nodes/supernodes in the network. The currents in these equations must be expressed in terms of the node potentials.*

5. *Solve those n − 1 equations for the n − 1 node potentials.*

6. *Calculate the element voltages and currents of interest from the node potentials.*

The whole procedure will be clearer after we look at a few examples.

Fig. 2-15. A ground node.

 FLUENCY EXERCISE: *Node Potentials*

Example 2-11   The Node Method with Current Sources

This first example, the circuit drawn in Figure 2-16, looks at the node method for a circuit that does not contain any voltage sources. This is the simplest case, because, for such networks, there are no supernodes. The four nodes are indicated by colored dashed lines. Three of the nodes are labelled *a*, *b*, and *c*; the fourth is identified as the ground. Let the node potentials on the nonground nodes be denoted by $e_a(t)$, $e_b(t)$, and $e_c(t)$, respectively. (The potential at the ground node is zero.) Notice that, when we use the node potentials, it is not necessary to indicate the reference directions for the element voltages

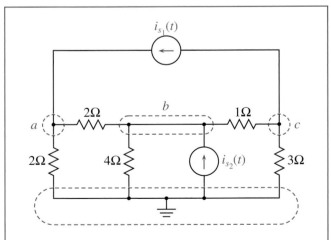

Fig. 2-16. An example of a circuit with no voltage sources to illustrate the use of the node method.

and currents, unless we desire the value of a particular one. The voltage across the 2Ω resistor measured from node a to node b, for example, is simply $e_a(t) - e_b(t)$ and the current that flows through that resistor from node a to node b is simply $[e_a(t) - e_b(t)]/2$. Similarly, the current flowing in that resistor in the direction from node b to node a is $[e_b(t) - e_a(t)]/2 = -[e_a(t) - e_b(t)]/2$, which is the negative of the current flowing in the opposite direction.

The next step, Step 4 of the formal procedure, involves writing three KCL equations by summing the currents that leave (or enter) nodes a, b, and c. (Although it is not required, it is recommended that all of the equations be written the same way, i.e., by summing the currents entering each node or by summing the currents leaving each node.) These need to be written by using the node potentials as variables. Summing the currents that leave node a, we have

$$\tfrac{1}{2}[e_a(t) - 0] + \tfrac{1}{2}[e_a(t) - e_b(t)] = i_{s_1}(t).$$

After we regroup the terms, this simplifies to

$$e_a(t) - \tfrac{1}{2}e_b(t) = i_{s_1}(t). \qquad (2.19)$$

Moving on to node b, we write

$$\tfrac{1}{2}[e_b(t) - e_a(t)] + \tfrac{1}{4}[e_b(t) - 0] + [e_b(t) - e_c(t)] = i_{s_2}(t),$$

which simplifies to

$$-\tfrac{1}{2}e_a(t) + \tfrac{7}{4}e_b(t) - e_c(t) = i_{s_2}(t). \qquad (2.20)$$

Finally, at node c, we get the final KCL equation,

$$[e_c(t) - e_b(t)] + \tfrac{1}{3}[e_c(t) - 0] = -i_{s_1}(t),$$

which reduces to

$$-e_b(t) + \tfrac{4}{3}e_c(t) = -i_{s_1}(t). \qquad (2.21)$$

Notice that there is a regular structure to these equations if they are written at the nonground nodes and if they are written consistently. In the KCL equation for node k, each of the current terms from the elements includes a contribution from node potential $e_k(t)$. Furthermore, all of these terms have the same algebraic sign, whereas all of the terms involving other node potentials in that equation occur with the opposite sign.

We now have three equations in three unknowns. The next step is to solve them. This is facilitated by first putting them into matrix–vector form:

$$\begin{bmatrix} 1 & -\tfrac{1}{2} & 0 \\ -\tfrac{1}{2} & \tfrac{7}{4} & -1 \\ 0 & -1 & \tfrac{4}{3} \end{bmatrix} \begin{bmatrix} e_a(t) \\ e_b(t) \\ e_c(t) \end{bmatrix} = \begin{bmatrix} 1 \\ 0 \\ -1 \end{bmatrix} i_{s_1}(t) + \begin{bmatrix} 0 \\ 1 \\ 0 \end{bmatrix} i_{s_2}(t).$$

Then we use MATLAB (or a calculator or hand calculation) to get the solution for the node potentials:

$$e_a(t) = \tfrac{5}{6}i_{s_1}(t) + \tfrac{2}{3}i_{s_2}(t)$$

$$e_b(t) = -\tfrac{1}{3}i_{s_1}(t) + \tfrac{4}{3}i_{s_2}(t) \qquad (2.22)$$

$$e_c(t) = -i_{s_1}(t) + i_{s_2}(t). \qquad (2.23)$$

Notice that, once again, each independent source produces a component in each of the variables.

We can use the node potentials to find any of the other variables in the circuit. For example, the current flowing from node a to node b through the 2Ω resistor is

$$i_{ab}(t) = \frac{1}{2}[e_a(t) - e_b(t)] = \frac{7}{12}i_{s_1}(t) - \frac{1}{3}i_{s_2}(t).$$

Using the node method required that we solve three linear algebraic equations in three unknowns. The simplified exhaustive method would have required solving five equations in five unknowns; the exhaustive method would have required solving seven equations in seven unknowns.   ∎

 To check your understanding, try Drill Problems P2-8 and P2-9.

The node method is valid for all circuits, but circuits that contain voltage sources can be trickier to deal with,

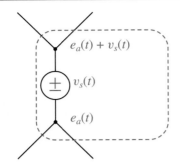

Fig. 2-17. A supernode containing a voltage source has two node potentials, but these are not independent. Their difference is equal to the source voltage.

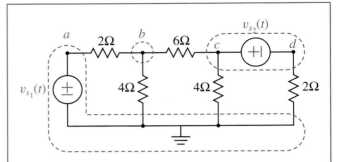

Fig. 2-18. An example to illustrate the use of the node method when voltage sources are present (Example 2-12).

because they contain supernodes. Consider the supernode in Figure 2-17. It contains two nodes within it, but their two node potentials are not independent, since the difference of these potentials must be equal to the source voltage, $v_s(t)$. Thus, we treat one of the node potentials as an unknown, $e_a(t)$ in Figure 2-17, and we write the other as $e_a(t) + v_s(t)$, where $v_s(t)$ is the source voltage. There is only one variable associated with the supernode, and we write only one KCL equation for it. In the writing of that equation, however, some of the currents will be calculated by using the node potential $e_a(t)$ and some will use $e_a(t) + v_s(t)$, depending on which of the two nodes receives the current. (If one of the nodes in the supernode is the ground node, then $e_a(t) = 0$.) The whole procedure is illustrated in the next example.

Example 2-12   The Node Method with Voltage Sources

To illustrate how to use the node method when there are voltage sources, consider the network in Figure 2-18, which contains two of them. After Step 1 of the procedure, we have two supernodes and one isolated node, which are indicated by the colored dashed lines. To aid in the discussion, the isolated node is node b, node a is one of the two nodes within the left supernode (the

other is the ground), and nodes c and d are the two nodes within the other supernode. Although the complete circuit contains four different nonzero node potentials, denoted as $e_a(t)$, $e_b(t)$, $e_c(t)$, and $e_d(t)$, only two of these are independent. We can see this by looking at the network diagram. Notice that

$$e_a(t) = v_{s_1}(t), \qquad (2.24)$$

because $v_{s_1}(t)$ represents the difference in potential between node a and the ground. Similarly,

$$e_c(t) = e_d(t) + v_{s_2}(t). \qquad (2.25)$$

For this network, we can treat $e_b(t)$ and either $e_d(t)$ or $e_c(t)$ as the independent variables and write our KCL equations in terms of these. We shall write equations at node b and at the supernode enclosing nodes c, d, and the associated voltage source. As before, we sum the currents *leaving* the nodes (and surfaces).

At node b, we write

$$\tfrac{1}{2}\left[e_b(t) - v_{s_1}(t)\right] + \tfrac{1}{4}e_b(t)$$
$$+ \tfrac{1}{6}\left\{e_b(t) - \left[e_d(t) + v_{s_2}(t)\right]\right\} = 0.$$

Here we used (2.24) and (2.25) to eliminate $e_a(t)$ and $e_c(t)$ as variables. This equation can be simplified to

$$\tfrac{11}{12}e_b(t) - \tfrac{1}{6}e_d(t) = \tfrac{1}{2}v_{s_1}(t) + \tfrac{1}{6}v_{s_2}(t). \qquad (2.26)$$

At the surface corresponding to the supernode containing nodes c and d of the original circuit, three currents cross the surface. Thus,

$$\tfrac{1}{6}\left\{[e_d(t) + v_{s_2}(t)] - e_b(t)\right\}$$
$$+ \tfrac{1}{4}\left[e_d(t) + v_{s_2}(t)\right] + \tfrac{1}{2}e_d(t) = 0.$$

Notice that, for the currents that originate from node c, the node potential is $e_c(t) = e_d(t) + v_{s_1}(t)$, and for those that originate at node d, $e_d(t)$ is the node potential. This equation can now be simplified to

$$-\tfrac{1}{6}e_b(t) + \tfrac{11}{12}e_d(t) = -\tfrac{5}{12}v_{s_2}(t). \qquad (2.27)$$

Solving (2.26) and (2.27) gives

$$e_b(t) = 0.5641\, v_{s_1}(t) + 0.1026\, v_{s_2}(t) \qquad (2.28)$$
$$e_d(t) = 0.1026\, v_{s_1}(t) - 0.4359\, v_{s_2}(t). \qquad (2.29)$$

From $e_b(t)$ and $e_d(t)$, we can compute any of the other voltages and currents in the network.

The important facts to remember when we use the node method with voltage sources present are (1) that some of the surfaces at which we write KCL equations are not nodes, but supernodes, and (2) that the node potential for the node at one terminal of a voltage source is tied to the potential of the node at the other terminal by the source voltage.   ∎

 To check your understanding, try Drill Problem P2-10.

When the node method is used to solve a circuit containing a dependent source, there is an additional step that is required. Consider a current-controlled voltage source whose source voltage is governed by the relation

$$v_s(t) = r i(t),$$

where $i(t)$, the controlling variable, is some other current in the circuit. Since the node method requires that we write KCL equations by using only the node potentials as variables, we need to first express $i(t)$ in terms of the node potentials in the circuit before writing the KCL equations. This step is performed in the next example.

Example 2-13   Node Method with a Dependent Source

As a final example of the node method, consider the network in Figure 2-19. This contains both a voltage source and a dependent current source. When a dependent source is present, it is necessary to restate the dependence relation in terms of the variables of the problem, which in this case are the node potentials.

The circuit contains one isolated node and one supernode, so we will need only one unknown node potential and one KCL equation. In this network, the quantity of interest is the voltage across the load resistor, R_L, which is denoted as $v_{out}(t)$. Since the value of this voltage is the same as the node potential, we denote the node potential as $v_{out}(t)$. The potential at the node that connects the voltage source with resistor R_s is $v_s(t)$.

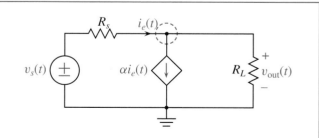

Fig. 2-19. An example of the node method with a dependent source.

Now that the node potentials have been labelled, we can express the current $i_e(t)$ in terms of them as

$$i_e(t) = \frac{1}{R_s}[\,v_s(t) - v_{out}(t)].$$

The current through the dependent source is, therefore, equal to $\alpha i_e(t) = \alpha[\,v_s(t) - v_{out}(t)]/R_s$. Writing a KCL equation at the circled node by summing the currents that leave that node gives

$$\frac{1}{R_s}[v_{out}(t) - v_s(t)] + \frac{1}{R_L}v_{out}(t)$$

$$= -\alpha\frac{1}{R_s}[\,v_s(t) - v_{out}(t)].\quad (2.30)$$

Solving for $v_{out}(t)$ gives

$$v_{out}(t) = \frac{\frac{1-\alpha}{R_s}}{\frac{1-\alpha}{R_s} + \frac{1}{R_L}}v_s(t)$$

$$= \frac{1}{1 + \frac{R_s}{R_L(1-\alpha)}}v_s(t).$$

∎

 To check your understanding, try Drill Problem P2-11.

2-5   The Mesh Method

The mesh method is very similar to the node method, except that the roles of voltages and currents are reversed. The latter uses the definitions of the node potentials to guarantee that KVL is satisfied for all closed paths. Then, finding the complete solution becomes equivalent to finding a solution of the KCL equations, once those are expressed in terms of the node potentials. The mesh method makes use of a different set of variables that guarantees that KCL is satisfied for all nodes. It is then sufficient merely to set up and solve the resulting

set of KVL equations on the appropriate meshes and supermeshes. The mesh method produces the same values for all the element variables that the node method does, but it results in fewer equations to be solved when the number of meshes/supermeshes is fewer than the number of nodes/supernodes ($\ell \leq n - 1$); when this condition is not true, the node method requires fewer equations.

How can we find the appropriate variables? We are looking to repeat the success of the node method, but with the roles of KVL and KCL, the roles of nodes and meshes, and the roles of voltages and currents reversed. With the node method, a node potential was a *voltage* that was associated with each *node*, such that each element *voltage* in the network could be expressed as the difference of two node potentials. For the mesh method, we would like to find a variable that is a *current* that we can associate with each *mesh*, such that all of the element *currents* can be expressed as differences of these new variables. We call these variables **mesh currents**, or **circulating currents**. In the network shown in Figure 2-20, they are shown as the currents $i_\alpha(t)$, $i_\beta(t)$, and $i_\gamma(t)$ that appear in the centers of the meshes.

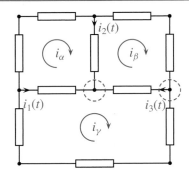

Fig. 2-20. A network with three meshes and their associated mesh currents. (The time dependence of the mesh currents has been suppressed.)

We defined the mesh currents in Figure 2-20 to have a clockwise direction. This is not required, and for some problems, it could be more convenient for one or more of them to be defined to be counterclockwise. For those elements not on the exterior boundary of the network, the element current is the sum or difference of the mesh currents of the meshes separated by the element. If the reference direction of the element current flows in the direction of the mesh current, it is added (receives a + sign); if it is in the opposing direction, it is subtracted (receives a − sign). For example, for the currents $i_1(t)$, $i_2(t)$, and $i_3(t)$ in Figure 2-20,

$$i_1(t) = i_\gamma(t) - i_\alpha(t)$$

$$i_2(t) = i_\alpha(t) - i_\beta(t)$$

$$i_3(t) = i_\beta(t) - i_\gamma(t).$$

For the elements on the perimeter, the element currents are equal either to the appropriate mesh currents or to their negatives.[2]

If the currents in the network are specified by using the mesh currents as variables, then KCL is satisfied automatically. For example, consider the network in Figure 2-20. If we sum the currents entering the dashed node in the center of the network, we see that

$$i_1(t) + i_2(t) + i_3(t) = [i_\gamma(t) - i_\alpha(t)]$$
$$+ [i_\alpha(t) - i_\beta(t)] + [i_\beta(t) - i_\gamma(t)] = 0. \quad (2.31)$$

It is similar for the node on the right. The current entering the node through the upper element is $i_\beta(t)$, the current entering through the lower element is $-i_\gamma(t)$, and that entering through the horizontal element is $i_\gamma(t) - i_\beta(t)$. These three currents also sum to zero.

─────────────

[2]For symmetry, we could define an additional mesh corresponding to the exterior of the network. We could then assign a (counterclockwise) mesh current with a value of zero to this mesh, which would serve as a "ground mesh." Then all element currents could be expressed as a sum or difference of two mesh currents and the mesh method would be a closer parallel to the node method.

> *When we use the mesh currents as variables, KCL is satisfied automatically at all nodes!*

Once we know all of the mesh currents, we can compute all of the element currents, and once we know all of the element currents, we can calculate all of the element voltages, using Ohm's law.

Example 2-14 Finding Element Variables from Mesh Currents

The simple circuit in Figure 2-21 contains two meshes and two mesh currents, labelled $i_\alpha(t)$ and $i_\beta(t)$. It is straightforward to express the indicated element variables in terms of $i_\alpha(t)$ and $i_\beta(t)$. The voltage $v_1(t)$ associated with the 3Ω resistor is

$$v_1(t) = 3i_\alpha(t),$$

since the reference direction for the current (by the default sign convention) is in the same direction as the mesh current. On the other hand,

$$v_3(t) = -5i_\beta(t),$$

because the reference direction for this current is opposite to the direction of the mesh current. Finally,

$$i_2(t) = i_\alpha(t) - i_\beta(t).$$

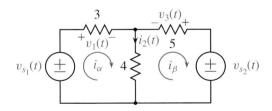

Fig. 2-21. A circuit with element variables to be expressed in terms of its two mesh currents. All resistances are measured in ohms.

Since this current is associated with an element that separates two meshes, it is the difference of two mesh currents. ∎

The paths over which we write KVL equations will be either meshes or supermeshes; as in the simplified exhaustive method, we also omit meshes or supermeshes that contain exterior current sources. Just as the node method requires special care when a circuit contains voltage sources, so the mesh method requires special care when the circuit contains current sources. As before, we address the details of this method with examples after first formally stating the mesh method as a procedure.

THE MESH METHOD:

1. *Create a supermesh from any pair of adjacent meshes that are separated by an independent or dependent current source.*

2. *Define a mesh current for each mesh or supermesh that does not contain a current source.*

3. *Write ℓ KVL equations, each of which is written over one of the paths over which the mesh currents mesh have been defined. The voltages in these equations must be expressed in terms of the mesh currents.*

4. *Solve those ℓ equations for the ℓ mesh currents.*

5. *Compute the element currents and voltages of interest from the mesh currents.*

Example 2-15   **The Mesh Method with Voltage Sources**

As a first example of the method, let's consider the network in Figure 2-18, which we solved earlier by the node method. This circuit is redrawn in Figure 2-22 to show the mesh currents explicitly. We have defined three mesh currents, because the circuit contains three meshes and no current sources.

The next step of the procedure is to write three KVL equations. As we prefer to do, we write these with the element variables on the left-hand side and the source terms on the right. For the left mesh, the KVL equation can be written as

$$2i_\alpha(t) + 4[i_\alpha(t) - i_\beta(t)] = v_{s_1}(t).$$

After regrouping terms, we find that this simplifies to

$$6i_\alpha(t) - 4i_\beta(t) = v_{s_1}(t).$$

Similarly, for the center and right meshes, the KVL equations are

$$4[i_\beta(t) - i_\alpha(t)] + 6i_\beta(t) + 4[i_\beta(t) - i_\gamma(t)] = 0$$
$$4[i_\gamma(t) - i_\beta(t)] + 2i_\gamma(t) = -v_{s_2}(t).$$

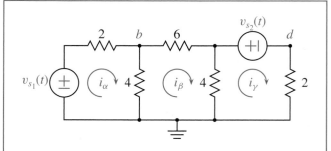

Fig. 2-22. A redrawing of the network of Figure 2-18 with the three mesh currents identified. All resistances are measured in ohms.

We can combine these three KVL equations into a single matrix–vector equation

$$
\begin{bmatrix} 6 & -4 & 0 \\ -4 & 14 & -4 \\ 0 & -4 & 6 \end{bmatrix}
\begin{bmatrix} i_\alpha(t) \\ i_\beta(t) \\ i_\gamma(t) \end{bmatrix}
=
\begin{bmatrix} 1 \\ 0 \\ 0 \end{bmatrix} v_{s_1}(t)
+
\begin{bmatrix} 0 \\ 0 \\ -1 \end{bmatrix} v_{s_2}(t),
$$

whose solution is

$$i_\alpha(t) = 0.2179\, v_{s_1}(t) - 0.0513\, v_{s_2}(t)$$
$$i_\beta(t) = 0.0769\, v_{s_1}(t) - 0.0769\, v_{s_2}(t)$$
$$i_\gamma(t) = 0.0513\, v_{s_1}(t) - 0.2179\, v_{s_2}(t).$$

Since we previously solved this network with the node method, it is instructive to compare these answers with the ones in (2.28) and (2.29). The potential at node b relative to the ground is the same as the voltage across the leftmost 4Ω resistor. This can be computed, using the mesh currents, as

$$e_b(t) = 4[i_\alpha(t) - i_\beta(t)]$$
$$= 0.5641 v_{s_1}(t) + 0.0126 v_{s_2}(t)$$

and the potential at node d is the voltage across the rightmost 2Ω resistor, which is

$$e_d(t) = 2i_\gamma(t) = 0.1026 v_{s_1}(t) - 0.4359 v_{s_2}(t).$$

Both of these agree with the earlier values, as indeed they must.

The node method requires solving two equations in two unknowns, whereas the mesh method requires solving three equations in three unknowns, although the equations were easier to write with the mesh method than with the node method. Clearly, either method can be used, and both give the same result. Which method should be selected is largely a matter of personal taste. If the equations will be solved by hand, the method that yields fewer equations is probably to be preferred; if a computer or calculator will be used to solve the equations, the one that simplifies writing the equations might be the better candidate. ∎

To check your understanding, try Drill Problem P2-12.

When a circuit contains current sources, supermeshes might be generated and/or meshes containing exterior current sources will be ignored. In either case, there will be fewer KVL equations and fewer independent mesh currents than there are meshes in the circuit. The next example shows how to deal with this situation.

Example 2-16   The Mesh Method with Current Sources

The mesh method is complicated by the presence of current sources, just as the node method was complicated by the presence of voltage sources. To explore this case, let us reexamine the network in Figure 2-16, using the mesh method. The circuit is redrawn in Figure 2-23, with mesh currents shown in each of the four meshes of the complete network.

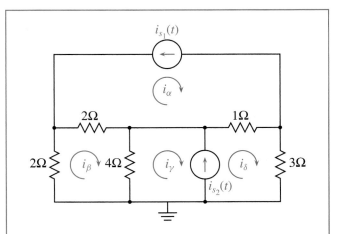

Fig. 2-23. A repetition of the network of Figure 2-16 to illustrate the mesh method. The time dependence of the mesh currents has been suppressed for clarity.

Even though we have indicated a mesh current for each mesh of the complete network, there are only two independent mesh currents, and we will write only two KVL equations. The independent mesh currents are $i_\beta(t)$ and either $i_\gamma(t)$ or $i_\delta(t)$. The other two mesh currents shown in the complete circuit are constrained by these and the current sources. If we look closely at the circuit drawing, we notice that

$$i_\alpha(t) = -i_{s_1}(t)$$
$$i_\delta(t) = i_\gamma(t) + i_{s_2}(t). \tag{2.32}$$

In writing the KVL equations, it is often helpful to define a mesh current for each mesh in the circuit, as we have done, and then to replace those that are not independent by their equivalent values in a second step. In this case, this means adding the mesh currents $i_\alpha(t)$ and $i_\delta(t)$. Then, once these equations have been written, we substitute (2.32) to reduce the number of unknowns. Writing KVL around the left mesh gives

$$2i_\beta(t) + 2[i_\beta(t) - i_\alpha(t)] + 4[i_\beta(t) - i_\gamma(t)] = 0.$$

Setting $i_\alpha(t) = -i_{s_1}(t)$, this becomes

$$2i_\beta(t) + 2[i_\beta(t) + i_{s_1}(t)] + 4[i_\beta(t) - i_\gamma(t)] = 0. \tag{2.33}$$

For the supermesh, we have

$$4[i_\gamma(t) - i_\beta(t)] + [i_\delta(t) - i_\alpha] + 3i_\delta(t) = 0.$$

After we substitute $i_\alpha(t) = -i_{s_1}(t)$ and $i_\delta(t) = i_\gamma(t) + i_{s_2}(t)$, this becomes

$$4[i_\gamma(t) - i_\beta(t)] + [i_\gamma(t) + i_{s_2}(t) + i_{s_1}(t)]$$
$$+ 3[i_\gamma(t) + i_{s_2}(t)] = 0. \tag{2.34}$$

We can collect terms to put these equations into our more standard form:

$$8i_\beta(t) - 4i_\gamma(t) = -2i_{s_1}(t)$$
$$-4i_\beta(t) + 8i_\gamma(t) = -i_{s_1}(t) - 4i_{s_2}(t).$$

Or, in matrix–vector form,

$$\begin{bmatrix} 8 & -4 \\ -4 & 8 \end{bmatrix} \begin{bmatrix} i_\beta(t) \\ i_\gamma(t) \end{bmatrix} = \begin{bmatrix} -2 \\ -1 \end{bmatrix} i_{s_1}(t) + \begin{bmatrix} 0 \\ -4 \end{bmatrix} i_{s_2}(t).$$

The solution is

$$i_\beta(t) = -0.4167\,i_{s_1}(t) - 0.3333\,i_{s_2}(t)$$
$$i_\gamma(t) = -0.3333\,i_{s_1}(t) + 0.6667\,i_{s_2}(t).$$

We leave it as an exercise for the reader to verify that this solution is identical to the one we obtained for this circuit via the node method in (2.23).   ■

To check your understanding, try Drill Problem P2-13.

Circuits that contain dependent sources require an extra step when you are using the mesh method for analysis, just as the node method required an extra step. Before we can write down the KVL equations, we need to rewrite the constraint expression for the source in terms of the mesh currents. The next example illustrates the procedure.

Example 2-17   Mesh Method with a Dependent Source

As a final example of the mesh method, let us reconsider the circuit with a dependent source that we analyzed via the node method. This is redrawn in Figure 2-24. Although we have drawn the network with two mesh currents, the circuit contains a single supermesh. Thus, there will be only one KVL equation and one of the mesh currents depends upon the other. Looking at the current through the dependent source, we see that

$$i_\alpha(t) - i_\beta(t) = \alpha i_e(t).$$

Furthermore, the controlling variable, $i_e(t)$ can easily be written in terms of $i_\alpha(t)$, since

$$i_e(t) = i_\alpha(t).$$

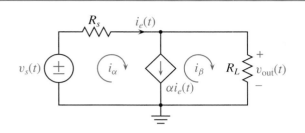

Fig. 2-24. An example of the mesh method with a dependent source.

Putting these two relations together gives

$$i_\alpha(t) - i_\beta(t) = \alpha i_\alpha(t)$$

$$\Longrightarrow i_\alpha(t) = \frac{1}{1-\alpha} i_\beta(t). \qquad (2.35)$$

Now we can write the KVL equation around the outer loop, which corresponds to the supermesh:

$$R_s i_\alpha(t) + R_L i_\beta(t) = v_s(t).$$

Substituting for $i_\alpha(t)$ yields

$$R_s \left[\frac{1}{1-\alpha} i_\beta(t) \right] + R_L i_\beta(t) = v_s(t),$$

or

$$i_\beta(t) = \frac{1}{R_L + \frac{R_s}{1-\alpha}} v_s(t).$$

Now that we know the mesh current, we can use it to calculate the output voltage:

$$v_{\text{out}}(t) = R_L i_\beta(t) = \frac{R_L}{R_L + \frac{R_s}{1-\alpha}} v_s(t)$$

$$= \frac{1}{1 + \frac{R_s}{R_L(1-\alpha)}} v_s(t).$$

For this example, both the node method and the mesh method required solving only a single equation in a single unknown.   ∎

          WORKED SOLUTION: *Mesh Method*

          To check your understanding, try Drill Problem P2-14.

Both the node method and the mesh method are systematic procedures for finding the complete solution of a resistive network. We shall see later that they also serve as useful tools for more general networks. For simple networks, such as the final example above, however, ad hoc methods can be just as easy. In the next chapter, we shall look at a number of techniques for simplifying networks to make them easier to understand and easier to analyze.

2-6   Conservation of Power

The net power absorbed in any circuit is zero.

This law follows as a direct consequence of Kirchhoff's laws, by using the definitions of the node potentials or the mesh currents. (See Problem P2-57.) This result is included in this chapter because it illustrates an application of node potentials that is different from simply analyzing a circuit.

For every watt of power absorbed in the resistors in a circuit, a watt must be supplied, either by the sources or by the other elements. If $v_k(t)$ is the voltage across the terminals of the kth element or source and $i_k(t)$ is the current flowing into its $^+$ terminal, then the formal statement of the conservation of power for a circuit with N elements (or sources) is

$$\sum_{k=1}^{N} v_k(t) i_k(t) = 0. \qquad (2.36)$$

In this section, we demonstrate this result for the specific circuit shown in Figure 2-25, although the result is completely general.

A Confusing Issue: Proper Treatment of Sources

Most errors made by beginning students when they use the node and mesh methods occur because they treat the source variables improperly. Consider the following circuit, in which all of the resistances are measured in ohms.

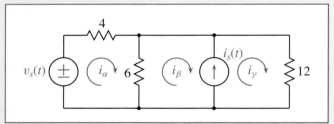

If the mesh method is applied as described in Section 2-5, we would write two KVL equations—one around the mesh on the left and one over the supermesh formed from the other two meshes with a path that encircles the current source. Because of the current source, mesh currents $i_\beta(t)$ and $i_\gamma(t)$ are not independent: $i_\gamma(t) - i_\beta(t) = i_s(t)$; hence, $i_\gamma(t) = i_\beta(t) + i_s(t)$. The two KVL equations are

$$- v_s(t) + 4i_\alpha(t) + 6[i_\alpha(t) - i_\beta(t)] = 0$$

$$6[i_\beta(t) - i_\alpha(t)] + 12[i_\beta(t) + i_s(t)] = 0.$$

Algebraic simplification turns these into

$$10i_\alpha(t) - 6i_\beta(t) = v_s(t)$$

$$-6i_\alpha(t) + 18i_\beta(t) = -12i_s(t),$$

and their solution is readily seen to be

$$i_\alpha(t) = \tfrac{1}{8} v_s(t) - \tfrac{1}{2} i_s(t)$$

$$i_\beta(t) = \tfrac{1}{24} v_s(t) - \tfrac{5}{6} i_s(t),$$

from which all of the other variables in the circuit can be calculated.

Errors frequently occur when students try to write KVL equations on all three meshes. Doing so requires assigning a potential difference across the terminals of the current source. **Remember, this voltage is almost never zero, nor is it equal to** $i_s(t)$. To get the correct solution in this situation, it is necessary to assign a variable to this voltage and then solve for it. For this example, let $v(t)$ be the voltage across the terminals of the current source (+ sign at the top). The three KVL equations are then

$$- v_s(t) + 4i_\alpha(t) + 6[i_\alpha(t) - i_\beta(t)] = 0$$

$$6[i_\beta(t) - i_\alpha(t)] + v(t) = 0$$

$$-v(t) + 12i_\gamma(t) = 0.$$

To these we must add the constraint between i_β and i_γ:

$$i_\gamma(t) - i_\beta(t) = i_s(t).$$

These four equations reduce to the ones that we derived earlier and, of course, they have the same (correct) solution. While this approach can be made to work, it results in more equations to solve and probably makes errors more likely. It is for this reason that we prefer writing a minimal set of equations by creating the appropriate supermeshes and supernodes, which eliminates the need to define any source variables.

As a final note, observe that the potential difference across the current source is $v(t) = 12i_\gamma(t)$ from the third KVL equation. Thus, $v(t) = 12[i_\beta(t) + i_s(t)] = 12[\tfrac{1}{24} v_s(t) + \tfrac{1}{6} i_s(t)] = \tfrac{1}{2} v_s(t) + 2 i_s(t)$, **which is not zero!**

Similar errors occur with the node method if improper assumptions are made about the current flowing through the voltage sources. The best way to deal with this situation is to follow the procedure as we have presented it in Section 2-4. Alternatively, the correct solution can be obtained if auxiliary variables are defined for these currents and then solved for along with the node potentials.

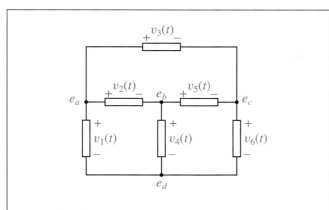

Fig. 2-25. Circuit to demonstrate conservation of power.

Let P be the total net power:

$$P = \sum_{k=1}^{6} v_k i_k \qquad (2.37)$$

$$= v_1 i_1 + v_2 i_2 + v_3 i_3 + v_4 i_4 + v_5 i_5 + v_6 i_6.$$

We need to show that $P = 0$. In writing (2.37), we have suppressed the dependence on time to simplify the resulting expressions. Using Figure 2-25, we next express each of the voltages in terms of the four indicated node potentials:

$$P = (e_a - e_d)i_1 + (e_a - e_b)i_2 + (e_a - e_c)i_3$$
$$+ (e_b - e_d)i_4 + (e_b - e_c)i_5 + (e_c - e_d)i_6.$$

Then we regroup the terms by node potential, instead of by current. This gives the alternative expression

$$P = e_a(i_1 + i_2 + i_3)$$
$$+ e_b(-i_2 + i_4 + i_5)$$
$$+ e_c(-i_3 - i_5 + i_6)$$
$$+ e_d(-i_1 - i_4 - i_6). \quad (2.38)$$

Now we write KCL equations for the four nodes, summing the currents that leave the node in each case and paying close attention to the default sign convention:

$$\begin{aligned}
\textit{node a:} \quad & i_1 + i_2 + i_3 = 0 \\
\textit{node b:} \quad & -i_2 + i_4 + i_5 = 0 \\
\textit{node c:} \quad & -i_3 - i_5 + i_6 = 0 \\
\textit{node d:} \quad & -i_1 - i_4 - i_6 = 0.
\end{aligned}$$

Notice that the expression that multiplies e_a in (2.38) is the sum of the currents that leave node a, which must be zero because of KCL. Similarly, each of the other node potentials is multiplied by the sum of the currents leaving its particular node. Therefore,

$$P = e_a \cdot 0 + e_b \cdot 0 + e_c \cdot 0 + e_d \cdot 0 = 0,$$

which establishes the result. *The net power absorbed in any circuit is zero.*

Notice that this result depends only upon KVL and KCL. It places no restrictions on the elements, which may be resistors, inductors, capacitors, sources, or some other elements not yet defined.

 DEMO: *Conservation of Power*

Example 2-18   Conservation of Power

To demonstrate the conservation of power, consider the circuit in Figure 2-26. This is a simple circuit, but we allow it to have arbitrary source waveforms. The first step is to solve the circuit. The circuit has one supernode, one isolated node, and one supermesh, so either the node method or the mesh method will require only a single equation. Let us use the node method. Let the bottom node be the ground and the one at the other end of the 2Ω resistor have node potential $v_1(t)$. Then, writing a KCL equation, we get

$$\frac{1}{2}v_1(t) + \frac{1}{4}[v_1(t) - v_s(t)] = i_s(t).$$

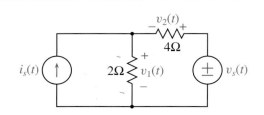

Fig. 2-26. Circuit to demonstrate conservation of power.

Thus,

$$\frac{3}{4}v_1(t) = \frac{1}{4}\,v_s(t) + i_s(t)$$

$$\Longrightarrow v_1(t) = \frac{1}{3}\,v_s(t) + \frac{4}{3}\,i_s(t).$$

We can use $v_1(t)$ to find the other three element variables:

$$v_2(t) = \ v_s(t) - v_1(t) = \frac{2}{3}\,v_s(t) - \frac{4}{3}\,i_s(t)$$

$$i_1(t) = \frac{1}{2}v_1(t) = \frac{1}{6}\,v_s(t) + \frac{2}{3}\,i_s(t)$$

$$i_2(t) = \frac{1}{4}v_2(t) = \frac{1}{6}\,v_s(t) - \frac{1}{3}\,i_s(t).$$

Next, we need to compute the power absorbed in the four elements. The power absorbed in the vertical resistor is

$$P_1 = v_1(t)i_1(t)$$

$$= \frac{1}{18}v_s^2(t) + \frac{8}{9}i_s^2(t) + \frac{4}{9}\,v_s(t)\,i_s(t),$$

and the power absorbed in the horizontal resistor is

$$P_2 = v_2(t)i_2(t)$$

$$= \frac{1}{9}v_s^2(t) + \frac{4}{9}i_s^2(t) - \frac{4}{9}\,v_s(t)\,i_s(t).$$

Thus, the total power absorbed in the two resistors is

$$P_1 + P_2 = \frac{1}{6}v_s^2(t) + \frac{4}{3}i_s^2(t).$$

The voltage across the terminals of the current source is $-v_1(t)$ if we adhere to the default sign convention, and its current is $4i_s(t)$. Therefore, the power absorbed by the current source is

$$P_3 = -v_1(t)\,i_s(t)$$

$$= -\frac{1}{3}\,v_s(t)\,i_s(t) - \frac{4}{3}i_s^2(t).$$

The current *entering* the $^+$ terminal of the voltage source is $-i_2(t)$ and its voltage is $v_s(t)$. Thus, the power absorbed by the voltage source is

$$P_4 = -i_2(t)\,v_s(t)$$

$$= -\frac{1}{6}v_s^2(t) + \frac{1}{3}\,v_s(t)\,i_s(t).$$

Therefore, the total power absorbed by the two sources is

$$P_3 + P_4 = -\frac{1}{6}v_s^2(t) - \frac{4}{3}i_s^2(t) = -P_1 - P_2,$$

so that

$$P_1 + P_2 + P_3 + P_4 = 0,$$

and the net power absorbed is zero. Notice that the two sources have a negative absorbed power, which means that they are supplying the power that is absorbed by the resistors. ∎

 To check your understanding, try Drill Problem P2-15.

2-7   Chapter Summary

2-7-1   Important Points Introduced

- A planar network is a circuit that can be drawn on a sheet of paper without any elements or wires crossing each other.
- The exhaustive method for finding the complete solution finds every voltage and current, including the unconstrained source variables, by solving $2e+s$ linear equations in $2e+s$ unknowns, where e is the number of elements and s is the number of sources.
- For a planar network, there is a constraint between the number of branches, b, the number of nodes, n, and the number of meshes, ℓ: $b = n + \ell - 1$.
- To calculate the values of the element variables in a circuit without solving for the unconstrained source variables, it is sufficient to create a supernode around each voltage source and a supermesh surrounding each interior current source, then to write a v–i relation for each element, a KCL equation at every node or supernode but one, and a KVL equation for every mesh or supermesh in the circuit that does not contain an exterior current source.
- The simplified exhaustive method finds the values of all of the element voltages and currents by solving $2e$ linear equations in $2e$ unknowns.
- When the node potentials are used as the independent variables in a circuit, there is no need to write KVL equations; KVL is automatically satisfied on any closed path.
- Using the node method, we need solve only $n - 1$ equations in $n - 1$ node potentials, where n is the number of nodes or supernodes remaining after supernodes have been created around each of the voltage sources.
- When the mesh currents are used as the independent variables in a circuit, there is no need to write KCL equations; KCL is automatically satisfied at all nodes.
- Using the mesh method, we need solve only ℓ equations in ℓ mesh currents, where ℓ is the number of meshes or supermeshes that do not contain exterior current sources.
- The net power absorbed in a circuit is always zero.

2-7-2   New Abilities Acquired

You should now be able to do the following:

(1) Identify the nodes or supernodes in the circuit where a minimal set of KCL equations can be written.

(2) Use supermeshes to identify the closed paths in the circuit where a minimal set of KVL equations can be written.

(3) Set up and solve the set of linear equations that defines the complete solution of a circuit using either the exhaustive method or the simplified exhaustive method.

(4) Put a set of linear equations, such as the ones that define the constraint equations of a resistive circuit, into matrix–vector form.

(5) Solve a set of linear equations, using MATLAB.

(6) Use superposition of sources to find the complete solution of a circuit with multiple independent sources.

(7) Express all element voltages and currents in terms of the node potentials.

(8) Find the complete solution of a circuit, using the node method.

(9) Express all element voltages and currents in terms of a set of mesh currents.

(10) Find the complete solution of a circuit using the mesh method.

(11) Compute the total amount of power absorbed and supplied in a circuit.

2-8   Problems

2-8-1   Drill Problems

All drill problems have solutions on the web site.

P2-1 Solve for the element voltages in the circuit in Figure P2-1, using the exhaustive method. This problem is similar to Example 2-1.

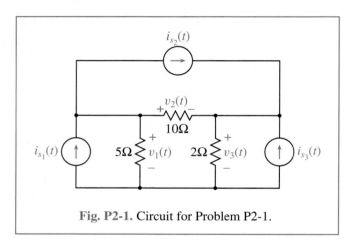

Fig. P2-1. Circuit for Problem P2-1.

P2-2 This problem, which is similar to Example 2-2, is concerned with the circuit in Figure P2-2. It should be solved by using the exhaustive method.

(a) Define a complete set of variables for this circuit.

(b) How many nodes are present in this circuit?

(c) How many meshes are present in this circuit?

(d) Write a sufficient set of KCL equations for this circuit. Your equations should be written in terms of the indicated variables.

(e) Write a sufficient set of KVL equations. These should also be written in terms of the indicated variables.

(f) Solve these equations for $i_1(t)$.

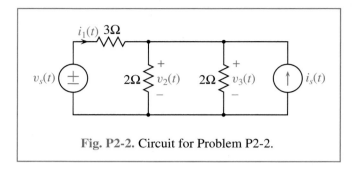

Fig. P2-2. Circuit for Problem P2-2.

P2-3 Find the voltages $v_1(t)$, $v_2(t)$, and $v_3(t)$ for the circuit in Problem P2-1, using the simplified exhaustive method (Figure P2-1). This is similar to Example 2-3.

P2-4 Define an appropriate set of variables and solve for $i_1(t)$ and $v_2(t)$ in the circuit of Figure P2-2 using the simplified exhaustive method.

P2-5 This problem is similar to Example 2-5. For the circuit in Figure P2-5,

(a) Write a minimal set of KCL equations to specify the element variables in the circuit. These should be expressed in terms of the variables indicated on the figure. Define supernodes as appropriate.

(b) Write a minimal set of KVL equations to specify the element variables in the circuit. These should be expressed in terms of the indicated variables.

(c) Put these equations into matrix–vector form and solve for $v_1(t)$ and $i_2(t)$.

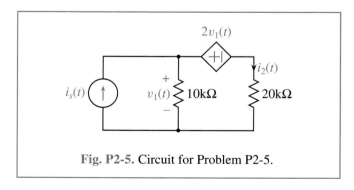

Fig. P2-5. Circuit for Problem P2-5.

P2-6 This problem is concerned with the three networks, shown in Figure P2-6. For each of the networks, the goal is to solve for the element voltages and currents. The uncontrolled source variables are not needed.

(a) Identify the closed paths in the circuit that can be used to write a minimal set of KVL equations.

(b) Identify the nodes and supernodes in the circuit that can be used to write a minimal set of KCL equations.

(c) Write the set of linear equations associated with the paths and closed surfaces that you found in (a) and (b).

(d) Put these equations into matrix–vector form and solve them for the indicated variables, using MATLAB.

P2-7 The purpose of this problem is to demonstrate the superposition principle. We shall use the circuit in Figure P2-7.

(a) Defining an appropriate set of variables where necessary, use one of the familiar methods to express $v(t)$ in terms of $v_s(t)$ and $i_s(t)$.

(b) Now turn off the voltage source by replacing it by a short circuit. Solve the resulting circuit for the voltage $v(t)$.

(c) Next, turn the voltage source back on, and turn off the current source by replacing it by an open circuit. Solve the resulting circuit for the voltage $v(t)$.

(d) Show that your result in part (a) is equal to the sum of your results in parts (b) and (c).

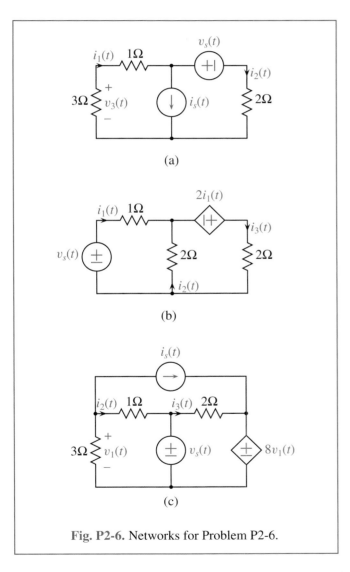

(a)

(b)

(c)

Fig. P2-6. Networks for Problem P2-6.

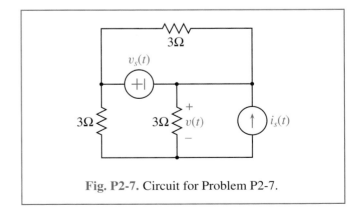

Fig. P2-7. Circuit for Problem P2-7.

P2-8

(a) For the circuit in Figure P2-8, what is the minimum number of KCL equations that need to be written to specify the complete solution?

(b) Select one of the nodes of the circuit as the ground node and label it. Define a supernode around each of the voltage sources. Define an appropriate set of independent node potentials, and indicate these on the circuit drawing. Label all of the non-independent node potentials as well. Next, write a KCL equation at each of the nonground nodes or supernodes in terms of the node potentials (and the source waveforms).

(c) Put your equations into matrix–vector form with the node potentials as unknowns. You do not need to solve the equations.

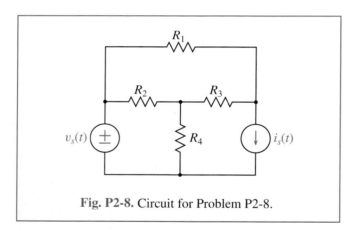

Fig. P2-8. Circuit for Problem P2-8.

P2-9

(a) Write KCL equations at nodes a and b in the circuit in Figure P2-9, using their node potentials $e_a(t)$ and $e_b(t)$ as variables.

(b) Solve your equations for $e_a(t)$ and $e_b(t)$.

(c) What is $i(t)$?

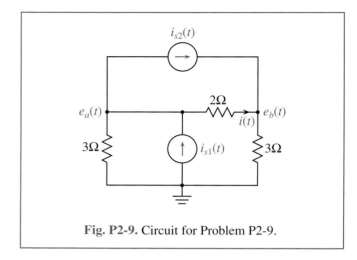

Fig. P2-9. Circuit for Problem P2-9.

P2-10

(a) Denote the potentials at nodes a, b, and c in the circuit in Figure P2-10 as $e_a(t)$, $e_b(t)$, and $e_c(t)$. Express $e_a(t)$ and $e_c(t)$ in terms of $e_b(t)$ and the source voltages.

(b) Define a supernode surrounding each voltage source. How many KCL equations will have to be written?

(c) Write a KCL equation at the supernode that encircles nodes a, b, and the voltage source on the left.

(d) Solve for $e_b(t)$.

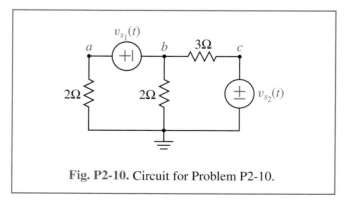

Fig. P2-10. Circuit for Problem P2-10.

P2-11

(a) Using the node potentials $e_a(t)$ and $v(t)$ in Figure P2-11 as variables, write KCL equations at nodes a and b.

(b) Solve your equations for $v(t)$.

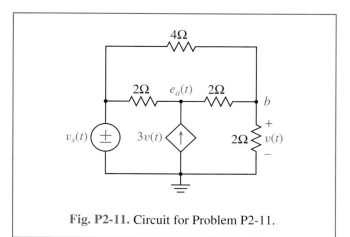

Fig. P2-11. Circuit for Problem P2-11.

P2-12

(a) Write KVL equations at the two meshes in the circuit in Figure P2-12, using the mesh currents $i_\alpha(t)$ and $i_\beta(t)$ as variables.

(b) Solve your equations for $i_\alpha(t)$ and $i_\beta(t)$.

(c) What is $v(t)$?

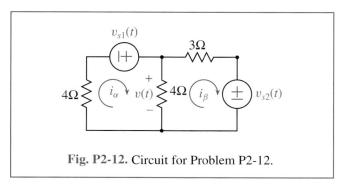

Fig. P2-12. Circuit for Problem P2-12.

P2-13

(a) Denote the mesh current in the lower left mesh in the circuit in Figure P2-13 as $i_\alpha(t)$. Express each of the other mesh currents in that circuit in terms of $i_\alpha(t)$ and the source currents.

(b) Write a KVL equation on the closed path that encircles the current source $i_{s1}(t)$ in terms of $i_\alpha(t)$ and the source currents.

(c) Solve for $i_\alpha(t)$.

(d) Use your result to compute $i(t)$.

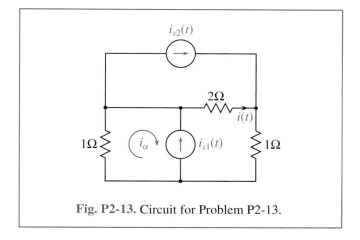

Fig. P2-13. Circuit for Problem P2-13.

P2-14

(a) For the circuit in Figure P2-14, what is the minimum number of KVL equations that need to be written to specify the solution of the circuit?

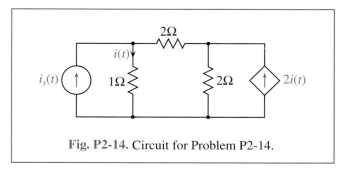

Fig. P2-14. Circuit for Problem P2-14.

(b) Define a number of mesh currents equal to your minimum number of equations from part (a). Write a sufficient set of KVL equations in terms of the mesh currents and the source waveforms.

(c) Solve your equations, and use the results to find $i(t)$.

P2-15 In the circuit in Figure P2-15, the voltage source is a DC source with $v_s = 60V$, and the current source supplies a constant current of 3mA.

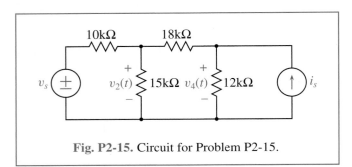

Fig. P2-15. Circuit for Problem P2-15.

(a) Compute the voltages v_2 and v_4.

(b) Compute the amount of power supplied by each source.

(c) Compute the amount of power absorbed by each of the four resistors.

(d) Verify that the total power supplied by the two sources is equal to the total power absorbed by the four resistors.

2-8-2   Basic Problems

P2-16 In the circuit in Figure P2-16, the only variables of interest are $i(t)$ and $v(t)$.

(a) Write a minimal set of KCL equations in terms of $i(t)$ and $v(t)$ to specify the solution of the circuit.

(b) Write a similar set of KVL equations.

(c) Solve these for $i(t)$ and $v(t)$.

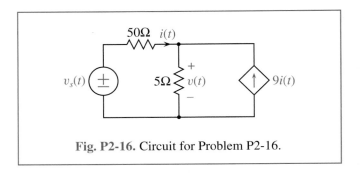

Fig. P2-16. Circuit for Problem P2-16.

P2-17 Set up a set of minimal KCL and KVL equations to find the element currents in the circuit of Figure P2-17, using the simplified exhaustive method.

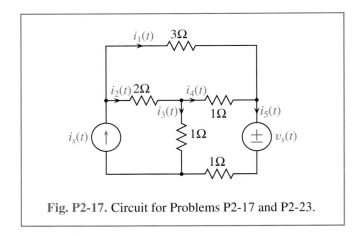

Fig. P2-17. Circuit for Problems P2-17 and P2-23.

P2-18 For the circuit in Figure P2-18,

(a) Write a minimal set of KVL equations to specify the complete solution. These should be expressed in terms of the variables indicated on the figure.

(b) Write a minimal set of KCL equations to specify the complete solution. These should also be expressed in terms of the indicated *voltage* variables.

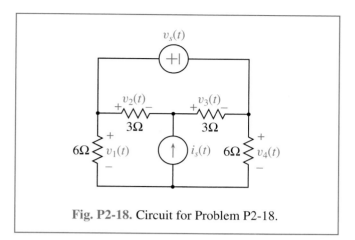

Fig. P2-18. Circuit for Problem P2-18.

(c) Put these equations into matrix–vector form and solve for the four voltages $v_1(t)$, $v_2(t)$, $v_3(t)$, and $v_4(t)$.

P2-19 For the circuit in Figure P2-19,

(a) Write a minimal set of KCL equations to specify the complete solution. These should be expressed in terms of the variables indicated on the figure.

(b) Write a minimal set of KVL equations to specify the complete solution. These should be expressed in terms of the indicated *current* variables.

(c) Put these equations into matrix–vector form and solve for the four currents $i_1(t)$, $i_2(t)$, $i_3(t)$, and $i_4(t)$.

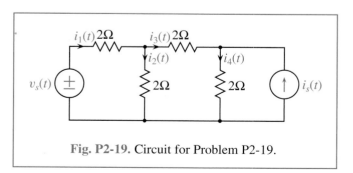

Fig. P2-19. Circuit for Problem P2-19.

P2-20

(a) In the circuit in Figure P2-20, what is the minimum number of KCL equations that you need to write to complete the solution?

(b) What is the minimum number of KVL equations that you will need to write to complete the solution?

(c) Write the KCL equations, using the current variables defined on the figure.

(d) Write the KVL equations in terms of the currents and $v_s(t)$ by incorporating the element relations.

(e) Write the complete set of KCL and KVL equations that is sufficient to calculate $i_1(t)$, $i_2(t)$, $i_3(t)$, $i_4(t)$, and $i_5(t)$, and put them into matrix–vector form.

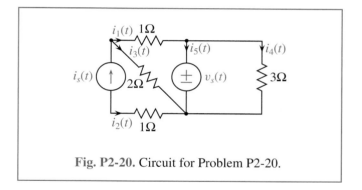

Fig. P2-20. Circuit for Problem P2-20.

P2-21   This problem will solve the circuit in Figure P2-21 via the exhaustive method by using both voltage and current variables. In addition to the three voltage variables that are indicated, define current variables $i_1(t)$, $i_2(t)$, and $i_3(t)$ flowing through the three resistors, with reference directions defined according to the default sign conventions. Let $i_1(t)$ be the current through the resistor with voltage $v_1(t)$, etc.

(a) Write a sufficient set of KCL equations to constrain the currents. Write these equations in terms of the current variables $i_1(t)$, $i_2(t)$, and $i_3(t)$.

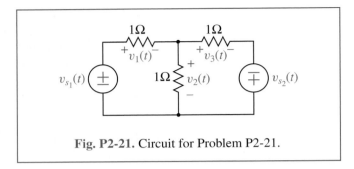

Fig. P2-21. Circuit for Problem P2-21.

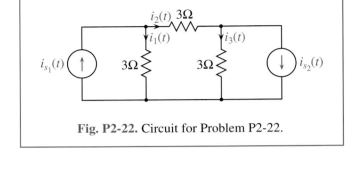

Fig. P2-22. Circuit for Problem P2-22.

P2-24 In the circuit in Figure P2-24, the voltage $v(t)$ is the variable of interest.

(b) Write a sufficient set of KVL equations to constrain the voltages. Write these equations in terms of the voltage variables $v_1(t)$, $v_2(t)$, and $v_3(t)$.

(c) Write the element relations for the three resistors. Write these with both the voltage and current variables on the left-sides of the equations, since these are both element variables.

(d) Put these equations into matrix–vector form. Define the vector of variables as

$$x = \begin{bmatrix} v_1(t) \\ v_2(t) \\ v_3(t) \\ i_1(t) \\ i_2(t) \\ i_3(t) \end{bmatrix}.$$

Solve these equations for the complete set of element variables.

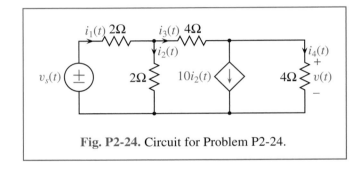

Fig. P2-24. Circuit for Problem P2-24.

(a) Show that, once the variables $i_1(t)$, $i_2(t)$, $i_3(t)$, and $i_4(t)$ are known, then all of the other variables in the circuit are known. You can do this by expressing all of the other voltages and currents in terms of these.

(b) Indicate on a drawing of the circuit a sufficient set of paths and nodes (or supernodes) on which KVL and KCL equations should be written to solve for the four indicated variables.

(c) Write the KVL and KCL equations that you identified in part (b).

(d) Solve your equations for $v(t)$. Your answer should be in terms of $v_s(t)$.

P2-22   Repeat Problem P2-21 for the circuit in Figure P2-22.

P2-23 Find the currents $i_1(t), i_2(t), i_3(t), i_4(t)$, and $i_5(t)$ for the circuit in Problem P2-17 (Figure P2-17).

P2-25 The circuit in Figure P2-25 is to be analyzed by using the simplified exhaustive method.

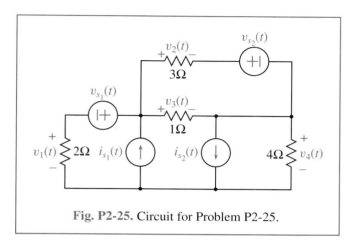

Fig. P2-25. Circuit for Problem P2-25.

(a) Indicate on a drawing of the original circuit the closed paths on which the KVL equations should be written.

(b) Also indicate on that drawing a sufficient set of surfaces (minimal number) on which the KCL equations should be written. These may correspond to nodes or to supernodes.

(c) Write the KVL and KCL equations corresponding to your selections in parts (a) and (b). These should be written in terms of the indicated voltages only, by incorporating Ohm's law when appropriate. You do not need to solve the equations, but you should verify that the number of equations is equal to the number of unknowns.

P2-26 Use the simplified exhaustive method to analyze the circuit in Figure P2-26.

(a) Indicate on a drawing of the original circuit the closed paths on which the KVL equations should be written.

(b) Also indicate on that drawing a sufficient set of surfaces (minimal number) on which the KCL equations should be written. They may correspond to nodes or to supernodes.

(c) Write the KVL and KCL equations corresponding to your selections in parts (a) and (b). These should be written in terms of the indicated currents only, by incorporating Ohm's law when appropriate.

(d) Solve your equations, using MATLAB, to express the current $i_1(t)$ in terms of the two source waveforms $v_s(t)$ and $i_s(t)$.

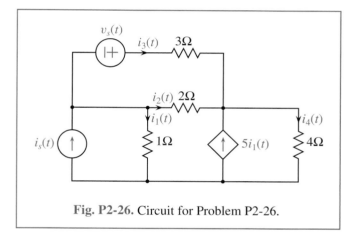

Fig. P2-26. Circuit for Problem P2-26.

P2-27 Use the simplified exhaustive method to solve for the voltage $v(t)$ in the circuit in Figure P2-27.

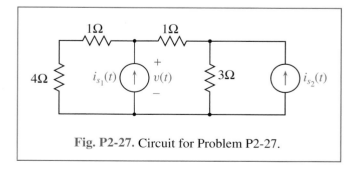

Fig. P2-27. Circuit for Problem P2-27.

P2-28

(a) For the circuit in Figure P2-28, write a minimal
sufficient set of KVL equations that will provide
independent constraints on the element voltages that
are labelled. Write your equations in terms of the
element voltages and the source signals only.

(b) For the same circuit, write a minimal sufficient set
of KCL equations that will provide independent
constraints on the same element voltages. Again,
write your equations in terms of the element voltages
and source signals only.

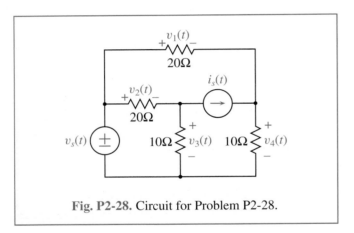

Fig. P2-28. Circuit for Problem P2-28.

P2-29 This problem analyzes the circuit in Figure P2-29.

(a) Write a minimal sufficient set of KCL equations that
will provide independent constraints on the element
currents that are labelled. Write your equations in
terms of the element currents and the source signals
only.

(b) For the same circuit, write a minimal sufficient set
of KVL equations that will provide independent
constraints on the same element currents. Again,
write your equations in terms of the element currents
and source signals only.

(c) Put your equations into matrix–vector form by
filling in the empty matrix and vector below.

$$
\begin{bmatrix} \end{bmatrix}
\begin{bmatrix} i_1(t) \\ i_2(t) \\ i_3(t) \\ i_4(t) \end{bmatrix}
=
\begin{bmatrix} \end{bmatrix}
v_s(t).
$$

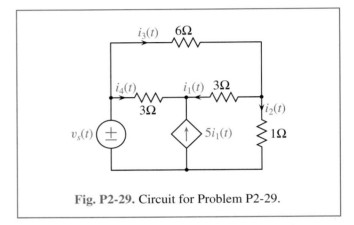

Fig. P2-29. Circuit for Problem P2-29.

P2-30

(a) For the circuit in Figure P2-30, write a sufficient set
of KCL equations to find the solution of the circuit.
These should be in terms of the variables $i_1(t)$, $i_2(t)$,
and $i_3(t)$.

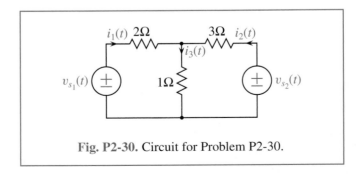

Fig. P2-30. Circuit for Problem P2-30.

(b) Write a sufficient set of KVL equations to find the solution. These should also be in terms of the variables $i_1(t)$, $i_2(t)$, and $i_3(t)$.

(c) Express your complete set of equations in matrix–vector form. You do not need to solve them.

P2-31 In the circuit in Figure P2-31, each of the current sources generates a constant current. Compute the values of the four resistor currents, i_1, i_2, i_3, and i_4.

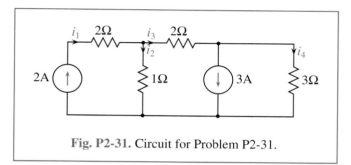

Fig. P2-31. Circuit for Problem P2-31.

P2-32

(a) Using the simplified exhaustive method and either the resistor voltages or the resistor currents as variables, write a set of three KCL and three KVL equations that will specify the solution of the circuit in Figure P2-32. $v(t)$ is the voltage across the terminals of the rightmost current source and the indicated open circuit.

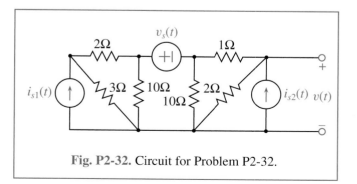

Fig. P2-32. Circuit for Problem P2-32.

(b) Express your equations in matrix–vector form.

(c) Use MATLAB to solve for your variables, and then use those to express $v(t)$. Your answer should be in the form $v(t) = K_1 i_{s1}(t) + K_2 i_{s2}(t) + K_3\, v_s(t)$.

P2-33 Use the node method to write a set of constraint equations for the network of Figure P2-33. Use as variables the voltages $e_a(t)$, $e_b(t)$, and $e_c(t)$. Do not solve the set of equations.

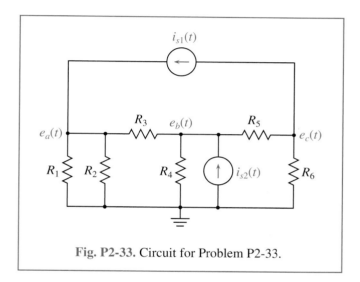

Fig. P2-33. Circuit for Problem P2-33.

P2-34 In this problem, we solve the circuit in Figure P2-34 by using the node method.

(a) Write the KCL equations at nodes a and b in terms of the node potentials at those nodes, $e_a(t)$ and $e_b(t)$.

(b) Put your equations into matrix–vector form by supplying the missing constants in the framework below.

$$\begin{bmatrix} & \\ & \end{bmatrix}\begin{bmatrix} e_a(t) \\ e_b(t) \end{bmatrix} = \begin{bmatrix} \\ \end{bmatrix} i_s(t) + \begin{bmatrix} \\ \end{bmatrix} v_s(t).$$

(c) Solve them for $e_a(t)$ and $e_b(t)$.

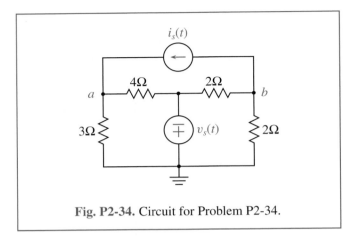

Fig. P2-34. Circuit for Problem P2-34.

P2-35 Find the voltage at each connection point in the circuit in Figure P2-35 with respect to the indicated ground.

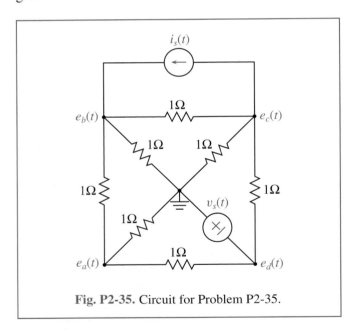

Fig. P2-35. Circuit for Problem P2-35.

P2-36 We wish to solve the circuit in Figure P2-36 by using the node method. Let $e_a(t)$ be the node potential at the indicated node.

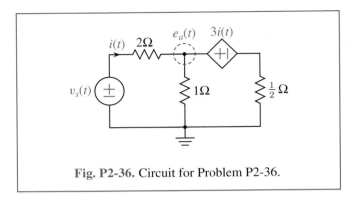

Fig. P2-36. Circuit for Problem P2-36.

(a) Express $i(t)$ in terms of $e_a(t)$ and $v_s(t)$.
(b) Write a KCL equation at the supernode that encircles the dependent voltage source. This equation should involve only the variables $e_a(t)$ and $v_s(t)$.
(c) Find $e_a(t)$.

P2-37

(a) Define and label an appropriate set of node potentials for the circuit in Figure P2-37.
(b) Write a set of KCL equations that can be solved for the node potentials that you defined in part (a), using the node potentials as variables.
(c) Solve your equations to express $v(t)$ in terms of $v_s(t)$.

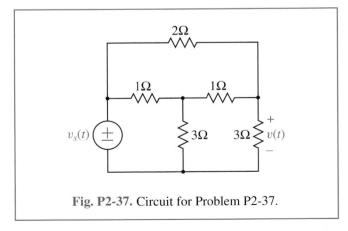

Fig. P2-37. Circuit for Problem P2-37.

P2-38 Consider the circuit with a CCVS shown in Figure P2-38.

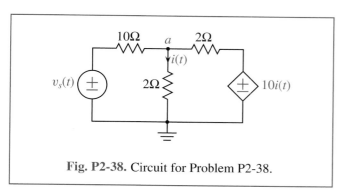

Fig. P2-38. Circuit for Problem P2-38.

(a) Redraw the circuit with the CCVS replaced by an equivalent VCVS that depends upon the node potential $e_a(t)$.

(b) Write a KCL equation at node a in terms of the variable $e_a(t)$.

(c) Solve for $e_a(t)$.

(d) What is the instantaneous power absorbed by the independent voltage source?

P2-39 Solve for $v(t)$ in terms of $i_s(t)$ in the circuit in Figure P2-39.

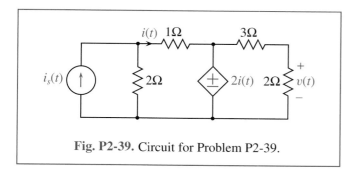

Fig. P2-39. Circuit for Problem P2-39.

P2-40 Find all of the element voltages and currents in the circuit of Figure P2-40, by using the mesh method. Be sure to identify the variables clearly.

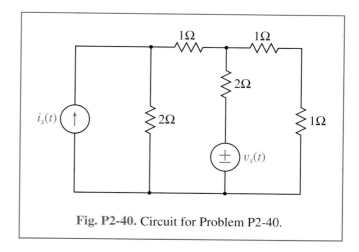

Fig. P2-40. Circuit for Problem P2-40.

P2-41

(a) What is the minimal number of KVL equations that need to be written if the mesh method is to be used to analyze the circuit in Figure P2-41?

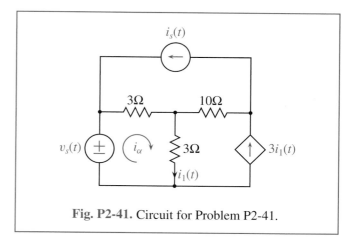

Fig. P2-41. Circuit for Problem P2-41.

(b) Write a KVL equation over the indicated mesh. Express your answer in terms of the mesh current $i_\alpha(t)$, $i_1(t)$, $v_s(t)$, and $i_s(t)$.

(c) Express $i_1(t)$ as a function of $i_\alpha(t)$.

(d) Write $i_1(t)$ as a function of $i_s(t)$ and $v_s(t)$.

P2-42 Solve for $v(t)$ in the circuit in Figure P2-42.

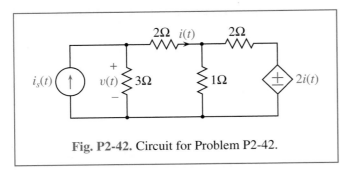

Fig. P2-42. Circuit for Problem P2-42.

P2-43 Consider the circuit with a CCVS shown in Figure P2-43.

(a) Redraw the circuit with the CCVS replaced by an equivalent CCVS that depends upon the mesh currents in the two meshes $i_\alpha(t)$ and $i_\beta(t)$.

(b) Write two KVL equations over paths defined by the meshes in terms of the variables $i_\alpha(t)$ and $i_\beta(t)$.

(c) Solve for $i(t)$.

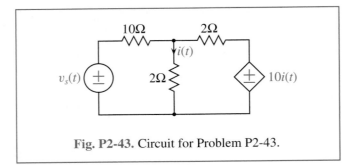

Fig. P2-43. Circuit for Problem P2-43.

P2-44

(a) Which method, the mesh method or the node method, will result in fewer equations to solve in order to find $v(t)$ in the circuit in Figure P2-44?

(b) Compute $v(t)$, using the method that you selected in (a).

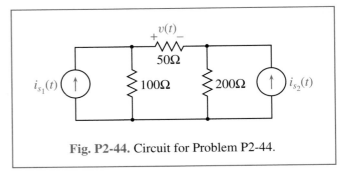

Fig. P2-44. Circuit for Problem P2-44.

P2-45

(a) For the circuit in Figure P2-45, express the mesh current in the lower right mesh in terms of $i_\alpha(t)$ and $i_\beta(t)$.

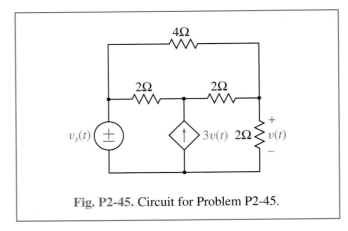

Fig. P2-45. Circuit for Problem P2-45.

(b) Redraw the circuit with the VCCS replaced by an equivalent CCCS that depends upon the mesh currents $i_\alpha(t)$ and $i_\beta(t)$. $i_\alpha(t)$ is the mesh current in the upper mesh of the circuit, and $i_\beta(t)$ is the mesh current associated with the lower left mesh.

(c) Write KVL equations over the upper mesh and the supermesh in terms of the variables $i_\alpha(t)$ and $i_\beta(t)$.

(d) Solve for $i_\alpha(t)$ and $i_\beta(t)$.

(e) What is $v(t)$?

P2-46 Find $v(t)$ in the circuit in Figure P2-46 by any method that you choose.

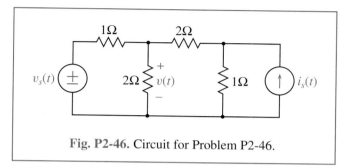

Fig. P2-46. Circuit for Problem P2-46.

P2-47 In the circuit in Figure P2-47, express $v(t)$ in terms of $v_s(t)$ and $i_s(t)$.

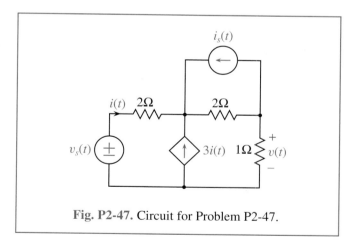

Fig. P2-47. Circuit for Problem P2-47.

P2-48 The node method and the mesh method are not the only systematic methods for finding the solution of a circuit, although they are the most popular. As an example of a different approach, consider the circuit in Figure P2-48.

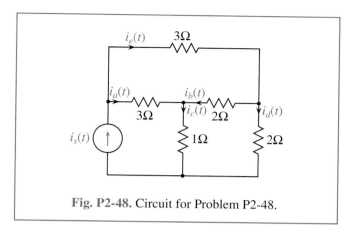

Fig. P2-48. Circuit for Problem P2-48.

(a) Show that all of the currents in the circuit (and, therefore, all of the voltages) can be expressed in terms of $i_a(t)$, $i_b(t)$, and $i_s(t)$; that is, express $i_c(t)$, $i_d(t)$, and $i_e(t)$ in terms of $i_a(t)$, $i_b(t)$, and $i_s(t)$.

(b) Write a set of two independent KVL equations over appropriate paths, using only $i_a(t)$, $i_b(t)$, and $i_s(t)$ as variables.

(c) Express your equations in matrix–vector form by filling in the missing entries in the equation below:

$$\begin{bmatrix} & \\ & \end{bmatrix} \begin{bmatrix} i_a(t) \\ i_b(t) \end{bmatrix} = \begin{bmatrix} \\ \end{bmatrix} i_s(t).$$

P2-49 Find the power delivered by each source in the circuit in Figure P2-49.

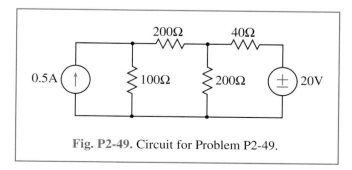

Fig. P2-49. Circuit for Problem P2-49.

2-8-3   Advanced Problems

P2-50 Solve for $i(t)$ in the circuit in Figure P2-50.

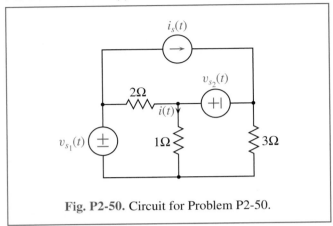

Fig. P2-50. Circuit for Problem P2-50.

P2-51 In the circuit in Figure P2-51,

(a) Compute the power supplied by the voltage source. (Note: The power supplied is the negative of the power absorbed.)

(b) Compute the power supplied by the current source.

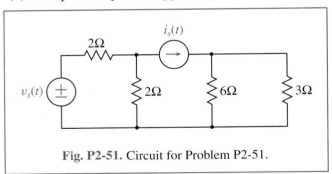

Fig. P2-51. Circuit for Problem P2-51.

P2-52 Our derivations of the simplified exhaustive method and the mesh method were limited to planar circuits. (The node method had no such restriction). In this problem, we extend the simplified exhaustive method to nonplanar circuits. An example of a nonplanar circuit is shown in Figure P2-52. All of the resistors in that circuit have resistance R.

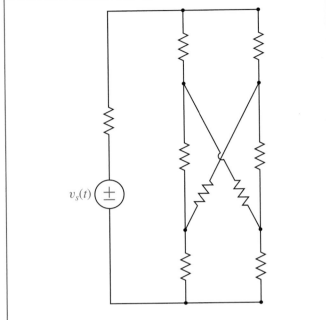

Fig. P2-52. Nonplanar circuit for Problems P2-52 and P2-53.

The procedure is summarized in the following steps

1. Create a supernode around each voltage source.

2. Write a v–i relation for each element and a KCL equation at all but one node/supernode, as before.

3. Remove sufficient branches to eliminate all branch crossings, to obtain a planar network, N.

4. Write a KVL equation at each mesh or supermesh of the resulting circuit that does not contain an exterior current source.

5. For each branch that was removed, form a closed path consisting of that branch and some subset of branches from N. Write a KVL equation on each of these paths.

6. Solve the resulting set of equations, using any appropriate method.

(a) On a drawing of the circuit, define as variables the nine resistor currents. Using those currents, write KCL equations at all but one of the nodes of the network. (There should be five of them.)

(b) Using the above procedure, write four KVL equations in terms of the current variables.

(c) Solve your equations for the resistor currents.

P2-53 Problem P2-52 outlined a procedure for extending the simplified exhaustive method to nonplanar circuits. A similar approach can be used to extend the mesh method. A mesh current is defined for each of the meshes, supermeshes and/or loops identified in steps 3 and 4. KVL equations are then written in terms of those mesh currents over this set of loops. Since the variables are mesh currents, there is no need to write KCL equations.

(a) Use the nonplanar extension of the mesh method to solve the circuit in Figure P2-52.

(b) Solve that same circuit by using the node method. Compute the resistor currents from the node potentials and show that the resulting values are the same as those derivable from the mesh currents that you computed in part (a).

P2-54 In our derivation of the mesh method, we stressed its duality with the node method: the similarity of the two methods if the roles of voltages and currents and of nodes and meshes are reversed. This problem lets you explore this issue further. Consider the network in Figure P2-54.

(a) Use the node method to construct the set of equations that must be solved to find the circuit solution. Let the ground node be the node that is omitted in writing your KCL equations. Express these equations in the form

$$C v(t) = s_1 v_{s_1}(t) + s_2 v_{s_2}(t),$$

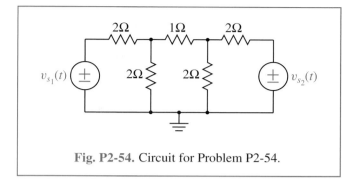

Fig. P2-54. Circuit for Problem P2-54.

where $v(t)$ is a vector of node potentials, s_1 and s_2 are column vectors of constants, and C is a constant matrix.

(b) Now design a *different* network containing two *current* sources with currents $i_{s_1}(t)$ and $i_{s_2}(t)$, such that the set of *mesh* equations that need to be solved to find the complete solution is

$$C i(t) = s_1 i_{s_1}(t) + s_2 i_{s_2}(t),$$

where $i(t)$ is the vector of mesh currents and where C, s_1, and s_2 are the same as in your solution in part (a).

(c) Solve your equations from part (b).

P2-55 In the circuit in Figure P2-55, express $v(t)$ in terms of $i_s(t)$.

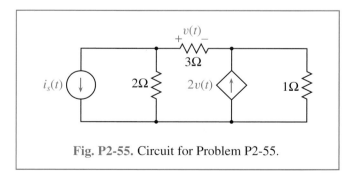

Fig. P2-55. Circuit for Problem P2-55.

P2-56 A thermistor is a sensor whose resistance R_T varies as a function of temperature. A plot of R_T vs. T for a particular (idealized) device is shown in Figure P2-56(a). A Wheatstone bridge, such as the one in Figure P2-56(b), is a circuit that can be used to convert a variable resistance into a variable voltage. Produce a plot of voltage v versus temperature T as T goes from 0° to 60°.

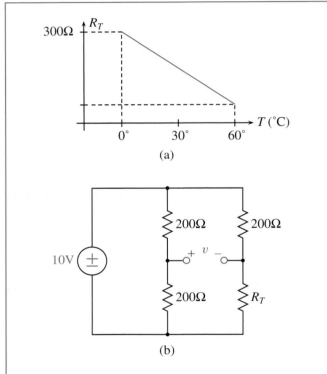

(a)

(b)

Fig. P2-56. (a) Plot of resistance vs. temperature for a particular thermistor. (b) Wheatstone bridge circuit for Problem P2-56.

P2-57 Section 2-6 demonstrated that the conservation of power was implied by KCL and KVL. That demonstration was done by using node potentials for the specific circuit in Figure 2.23. For the same circuit, produce a similar derivation using mesh currents.

2-8-4   Design Problems

P2-58 The circuit in Figure P2-58 contains a light-emitting diode, which is modelled as a 2Ω resistor. The diode will light if the current is greater than 10 mA, but will burn out if its current is greater than 15 mA. Choose the nominal value of the of the resistance R, if R can change by ±10% with changes in temperature.

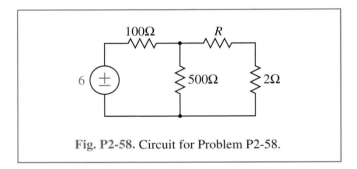

Fig. P2-58. Circuit for Problem P2-58.

P2-59 The circuit in Figure P2-59 has two sources, each of which is constant as a function of time. The voltage source has a value of 16V, and the current source has a value of imA. Find the value of i that will cause the voltage v, indicated in the circuit, to have a value of 4V.

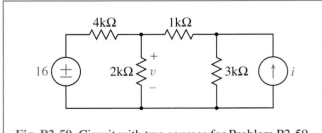

Fig. P2-59. Circuit with two sources for Problem P2-59.

P2-60 The circuit in Figure P2-60 represents a model of an amplifier that is driving a loudspeaker, which is modelled by the 16Ω resistor. The input to the amplifier is the signal $v_s(t) = 0.2\cos(1000t)$.

(a) Compute the value of the resistance R such that $v(t) = 20\cos(1000t)$.

(b) The power in the signal $x(t) = A\cos(\omega t)$ is $A^2/2$. Compute the ratio of the power in the signal $v(t)$ delivered to the loudspeaker to the power in the input signal $v_s(t)$.

(a) He vaguely remembers that he used the simplified exhaustive method to write the circuit equations. Assuming this to be the case, how many KCL and KVL equations should he have written?

(b) What are the KCL equations for the circuit?

(c) What are the KVL equations for the circuit?

(d) On the circuit drawing, label the four voltages and their reference directions.

(e) On the circuit drawing, label the four resistance values.

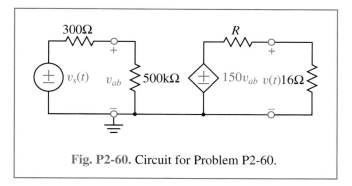

Fig. P2-60. Circuit for Problem P2-60.

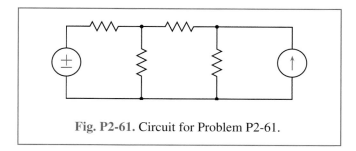

Fig. P2-61. Circuit for Problem P2-61.

P2-61 A distraught engineering student accidentally threw away a circuit design that he had been working on for some time. Rifling through his trash, he came upon the preliminary circuit in Figure P2-61 and the intermediate matrix–vector equation shown below.

$$\begin{bmatrix} \frac{1}{5} & 0 & 0 & \frac{1}{2} \\ 1 & 1 & 0 & -1 \\ -\frac{1}{5} & 1 & \frac{1}{3} & 0 \\ 0 & 1 & -1 & 0 \end{bmatrix} \begin{bmatrix} v_1(t) \\ v_2(t) \\ v_3(t) \\ v_4(t) \end{bmatrix} = \begin{bmatrix} 0 \\ 0 \\ 0 \\ -1 \end{bmatrix} v_s(t) + \begin{bmatrix} 1 \\ 0 \\ -1 \\ 0 \end{bmatrix} i_s(t).$$

He has no memory of the resistor values, which voltages are which, or what their reference directions should be. There are a few things that you should know about him: (1) he hates to do laundry, (2) he is not particularly systematic, and (3) he does not call home nearly as often as he should.

Subnetworks

Objectives

By the end of this chapter, you should be able to do the following:

1. *Identify and simplify circuits containing resistors connected in series and parallel.*

2. *Use voltage and current dividers to apportion the voltages and currents associated with resistors connected in series and parallel.*

3. *Find the Thévenin and Norton equivalents of two-terminal subnetworks and use those to simplify circuits.*

Frequently, we can isolate one part of a network that is connected to the remainder of a circuit at a single pair of terminals, as in Figure 3-1. Here the subnetwork N is connected to the remainder of the circuit E at the terminals labelled a and a'. The subnetwork could contain only a few elements, or it could be quite complex, with many elements, independent sources, and dependent sources.[1] If we do not care about all of the voltages and currents that lie within N, we can frequently save considerable analysis effort by modelling the subnetwork with a simpler circuit that contains fewer elements, but imposes the same constraints on the terminal variables $v(t)$ and $i(t)$. If done carefully, this can often either greatly simplify the circuit or simplify its analysis.

In essence, the two-terminal subnetwork N is a big element; it has a voltage $v(t)$ across its terminals and a current $i(t)$ passing through it[2], as shown in Figure 3-1. It also has a v–i characteristic that is created by its

---

[1] When it contains dependent sources, the controlling variables should also be located within N.

[2] KCL states that the current which enters the subnetwork through one terminal must be equal to the current which leaves it through the other.

106

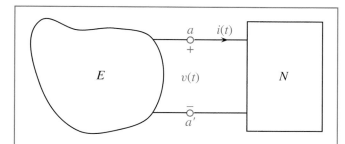

Fig. 3-1. A two-terminal network N connected to an external network E at a single pair of terminals.

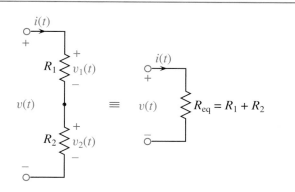

Fig. 3-2. Two resistors connected in series are network equivalent to a single resistor.

internal elements, sources, and connections. Two 2-terminal networks are called *network equivalent* if they have the same v–i characteristic.

This chapter explores two-terminal networks. It begins by looking at two-terminal subnetworks that contain only resistors and shows that these are network equivalent to a single resistor. Then we move on to subnetworks that contain both resistors and sources. Most of the chapter is concerned with network equivalence. We look at some generic forms for equivalent networks, and, through examples, we show how replacing parts of circuits with network equivalents can simplify both the analysis and the understanding of a circuit. Knowing how and when to introduce an equivalent circuit is more an art than a science, but it gets easier with experience. This chapter will touch on only some of the situations in which these substitutions are helpful. Later chapters will provide other opportunities to exploit network equivalent substitutions.

3-1 Resistor-Only Subnetworks

The simplest subnetworks are those that contain only resistors. Since each of these is equivalent to a single resistor, identifying situations in which to make this equivalent substitution can be particularly helpful.

3-1-1 Resistors in Series

The first of these simplifications is the observation that a two-terminal network consisting of two resistors connected in series is network equivalent to a single resistor. Consider the two-terminal network shown on the left side of Figure 3-2. Since the same current $i(t)$ flows through both resistors, by writing a KVL equation over the path defined by the two resistors and the external connection, we observe that

$$v(t) = v_1(t) + v_2(t)$$
$$= R_1 i(t) + R_2 i(t)$$
$$= (R_1 + R_2)i(t). \tag{3.1}$$

Thus, the voltage measured across the terminals is proportional to the current flowing into the $+$ terminal. Since this is the v–i relation that we associate with a single resistor, we can make the following observation.

> *Two resistors connected in series are network equivalent to a single resistor with a resistance equal to their sum.*
>
> $$R_{eq} = R_1 + R_2$$

It should be clear from this brief derivation that this result is true *only* when the resistors are connected in series; the derivation required that the currents flowing through both resistors must be the same. If we generalize this result to the case where more than two resistors are connected in series, they are also network equivalent to a single resistor with a resistance equal to their sum. This can be proved by extending the above derivation straightforwardly or by repeatedly applying the two-resistor result. Two inductors connected in series are also network equivalent to a single inductor and two capacitors connected in series are network equivalent to a single capacitor, although for capacitors the formula for the equivalent capacitance is somewhat different. These statements can be proven by derivations that are similar to the ones shown here for resistors. (See Problems P3-16 and P3-18.)

While we are looking at resistors connected in series, we can derive another useful result that tells how the total voltage $v(t)$ is apportioned among them. From (3.1), we can write for the two-resistor case that

$$i(t) = \frac{1}{R_1 + R_2} v(t).$$

Then

$$v_1(t) = R_1 i(t) = \frac{R_1}{R_1 + R_2} v(t)$$

$$v_2(t) = R_2 i(t) = \frac{R_2}{R_1 + R_2} v(t).$$

These results are known as the *voltage divider* relations. They state that the total voltage across a series connection of resistors is divided among those resistors in a manner that is proportional to their resistance. If N resistors with resistances R_i, $i = 1, 2, \ldots, N$ are connected in series, the voltage across the terminals of the ith resistor is

$$v_i(t) = \frac{R_i}{R_1 + R_2 + \cdots + R_N} v(t), \qquad (3.2)$$

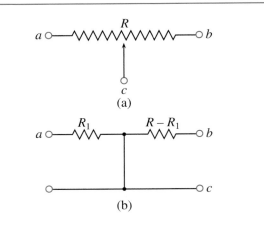

Fig. 3-3. (a) A potentiometer. (b) Equivalent circuit.

where $v(t)$ is the total voltage across the series combination.

 FLUENCY EXERCISE: *Voltage Dividers*

A *potentiometer*, shown in Figure 3-3, is a three-terminal device that is designed to exploit the voltage divider relation. In a potentiometer, the position of the contact at terminal c is adjustable by a knob (or screw). This affects the value of the resistance R_1 in the equivalent model shown in Figure 3-3(b). If the terminal at b is left as an open circuit, the two-terminal device with terminals a and c is a variable resistor whose resistance R_1 can vary between 0 and R ohms. When all three terminals are connected and the current flowing through the central terminal c is maintained at a level that is considerably lower than the current that flows through the other two terminals, a nearly constant resistance R is maintained between terminals a and b, and the potential $e_c(t)$ is continuously adjustable between $e_a(t)$ and $e_b(t)$. Potentiometers are commonly used in human interfaces such as volume controls, and they are also widely used as adjustable resistors. (See Problems P-3.14 and P-3.15.)

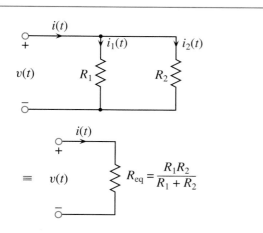

Fig. 3-4. Two resistors connected in parallel are equivalent to a single resistor.

EXTRA INFORMATION: *Resistance*

3-1-2   Resistors in Parallel

Two resistors connected in parallel, as in Figure 3-4, are also equivalent to a single resistor. For resistors connected in parallel, the voltages across the two resistors are each equal to $v(t)$. Therefore,

$$i_1(t) = \frac{1}{R_1} v(t)$$

$$i_2(t) = \frac{1}{R_2} v(t).$$

If we apply KCL at either of the nodes, we see that

$$i(t) = i_1(t) + i_2(t)$$
$$= \frac{1}{R_1} v(t) + \frac{1}{R_2} v(t)$$
$$= \left[\frac{1}{R_1} + \frac{1}{R_2} \right] v(t).$$

Notice that the current flowing into the two-terminal subnetwork is proportional to the voltage across its terminals. Thus, we have another configuration that is network equivalent to a single resistor. In this case, the equivalent resistance is seen to be

$$\frac{1}{R_{eq}} = \frac{1}{R_1} + \frac{1}{R_2}, \qquad (3.3)$$

which can be rewritten as

$$R_{eq} = \frac{R_1 R_2}{R_1 + R_2}.$$

The latter relation is easier to remember, but (3.3) is the form that generalizes when more than two resistors are connected in parallel. Recall that we defined the conductance, G, as the reciprocal of the resistance. Using conductances, (3.3) can be written as

$$G_{eq} = \frac{1}{R_{eq}} = G_1 + G_2. \qquad (3.4)$$

> *Two resistors connected in parallel are network equivalent to a single resistor with a conductance equal to the sum of their conductances*
>
> $$G_{eq} = G_1 + G_2.$$

As a special case, observe that, if two resistors with a resistance R are connected in parallel, they are equivalent to a single resistor with resistance $R/2$. When two resistors are connected in series, the equivalent resistance is greater than either one of them; when two resistors are connected in parallel, the equivalent resistance is smaller than either of them.

FLUENCY EXERCISE: *Resistors in Parallel*

TOOL: *Parallel Resistance Calculator*

A Confusing Issue: Identifying Series and Parallel Elements

The voltage and current dividers and the fact that elements connected in series or parallel are equivalent to single elements are important work savers. They can be used, however, *only* when elements are connected in series or parallel. Missing series and parallel connections can create extra work, but misidentifying series and parallel connections will produce wrong results.

Whether two elements are connected in series or parallel is a function of how they are connected, not of how they are drawn! Two elements are connected in series if and only if *all* of the current (every single electron) that passes through the first element also passes through the second and if *all* of the current in the second element also passes through the first. In the formal definition in Chapter 1, we stated that, if there was a node in the circuit that connected exactly two elements, then those two elements were connected in series. Two elements are also connected in series if there is a *supernode* (encircling any set of elements, not just a voltage source) in the circuit that connects exactly two elements. In the latter case, two elements can be connected in series even if they are not adjacent or not connected to each other. Similarly, we defined two elements as connected in parallel if they formed a two-element mesh; they are also connected in parallel if they form a two-element supermesh. Two elements connected in parallel must connect the same two nodes in the circuit. They must therefore have the same voltage. These elements do not have to be adjacent in a drawing of the circuit.

Consider the circuit at the top of the next column. Resistor R_1 is connected in series with the voltage source, since the current flowing through both devices is the same. It is also connected in

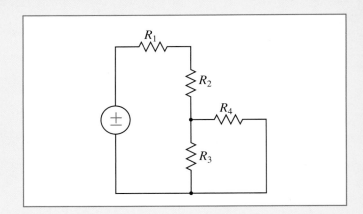

series with resistor R_2. Furthermore, the resistor R_2 is connected in series with the voltage source. Resistor R_3 is *not* connected in series with resistor R_2, because the presence of the resistor R_4 means that the currents through R_2 and R_3 are not the same. Resistors R_4 and R_3 are connected in parallel, since both elements connect the same pair of nodes. If this parallel connection is replaced by an equivalent resistor that resistor will be connected in series with resistor R_2 (and R_1 and the voltage source).

In the slightly more exotic circuit

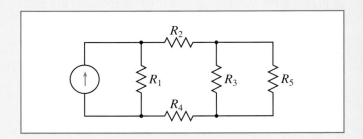

it is clear that the resistor with resistance R_3 is connected in parallel with resistor R_5. What is perhaps less clear, however, is the fact that the two resistors with resistances R_2 and R_4 are also connected in series!

We can generalize this result to the case where N resistors are connected in parallel. In this case, the parallel combination is still network equivalent to a single resistor and

$$G_{eq} = \frac{1}{R_{eq}} = G_1 + G_2 + \cdots + G_N.$$

Two or more inductors connected in parallel are network equivalent to a single inductor and two or more capacitors connected in parallel are network equivalent to a single capacitor, although here, too, the formula for the equivalent capacitance is different from the resistive and inductive cases. The derivations of these results are similar to the resistive case. (See Problems P3-17 and P3-19.)

For elements connected in parallel, the *current divider* relation describes how the individual currents are apportioned:

$$i_1(t) = \frac{1}{R_1}v(t) = \frac{1}{R_1}R_{eq}i(t)$$

$$= \frac{\frac{1}{R_1}}{\frac{1}{R_1} + \frac{1}{R_2}}i(t).$$

Thus,

$$i_1(t) = \frac{G_1}{G_1 + G_2}i(t) = \frac{R_2}{R_1 + R_2}i(t)$$

$$i_2(t) = \frac{G_2}{G_1 + G_2}i(t) = \frac{R_1}{R_1 + R_2}i(t).$$

Notice that, when we have two resistors in parallel, the larger resistance gets the smaller fraction of the current. If N resistors are connected in parallel, the current through the jth resistor, with conductance G_j, is

$$i_j(t) = \frac{G_j}{G_1 + G_2 + \cdots + G_N}i(t),$$

where $i(t)$ is the total current entering the terminals.

  FLUENCY EXERCISE: *Series and Parallel*

  FLUENCY EXERCISE: *Current Dividers*

Example 3-1   Using Network Equivalence to Simplify Circuits

The simplifications to a circuit that result from identifying series and parallel connections of elements can be very useful. Figure 3-5 shows an example of how the network equivalent of a circuit can be found by the successive replacements of series and parallel connections of resistors by their equivalents. Had we tried to solve this circuit via either the node method or the mesh method, we would have needed to set up and solve at least four equations in four unknowns to get the same result.

■

 To check your understanding, try Drill Problem P3-1.

Example 3-2   A Loaded Potentiometer

In Section 3-1-1 we introduced a potentiometer. Figure 3-6 shows a potentiometer being used as a variable gain control. The output voltage, $v_{out}(t)$ is the voltage applied to the $10k\Omega$ load resistor. The input signal is applied to the pair of terminals on the left. If the wiper of the potentiometer draws negligible current (corresponding to a nearly infinite load resistance), then the resistance seen by the input terminals will be nearly constant and $v_{out}(t)$ will vary linearly with the wiper position. Since the load resistance in this problem is only $10k\Omega$, however, the resistance seen at the input terminals will vary with the wiper position, and the output voltage, although still variable, will not vary linearly with position. To analyze these effects, we will use the

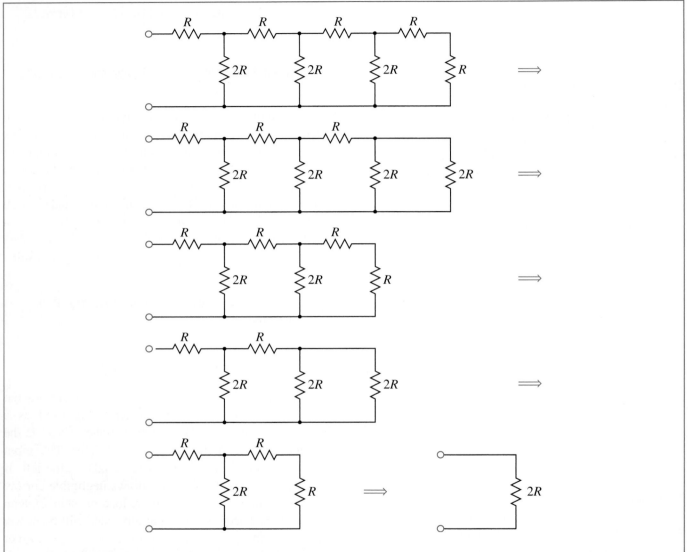

Fig. 3-5. An example of circuit simplification by the identification of series and parallel connections of resistors.

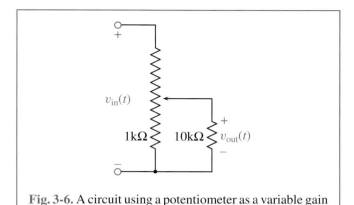

Fig. 3-6. A circuit using a potentiometer as a variable gain control.

potentiometer model given in Figure 3-3(b), with $R=1\,\text{k}\Omega$ and $R_1 = \theta R$, with $0 \le \theta \le 1$ denoting the wiper position. The equivalent problem is shown in Figure 3-7.

To calculate the equivalent resistance seen at the input terminals, observe that there is a resistance of $10\,\text{k}\Omega$ connected in parallel with a resistance of $10^3(1 - \theta)\,\Omega$ and that the equivalent resistance is connected in series with the $10^3\theta\,\Omega$ resistor. The equivalent resistance of the

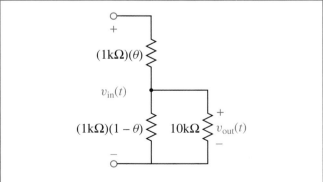

Fig. 3-7. The circuit of Figure 3-6, but with a model for the potentiometer substituted.

parallel connection is

$$R_{\text{par}} = \frac{(10^4)(10^3)(1 - \theta)}{10^4 + 10^3 - 10^3\theta} = 10^3 \left(\frac{1 - \theta}{1.1 - 0.1\theta} \right).$$

Therefore, the equivalent resistance seen at the terminals is

$$R_{\text{eq}} = 10^3 \left(\frac{1 - \theta}{1.1 - 0.1\theta} \right) + 10^3\theta$$

$$= 10^3 \left(\frac{1 + 0.1\theta - 0.1\theta^2}{1.1 - 0.1\theta} \right),$$

which varies from about $910\,\Omega$ to $1000\,\Omega$ over the range of the potentiometer.

To calculate the output voltage, we can use a voltage divider. We still have the lower portion of the potentiometer connected in parallel with the load resistor. This can be replaced by the equivalent resistance that we computed earlier. The output voltage is then

$$v_{\text{out}}(t) = v_{\text{in}}(t) \left(\frac{10^3 \left(\frac{1-\theta}{1.1-0.1\theta} \right)}{10^3\theta + 10^3 \left(\frac{1-\theta}{1.1-0.1\theta} \right)} \right)$$

$$= v_{\text{in}}(t) \left(\frac{1 - \theta}{1 + 0.1\theta - 0.1\theta^2} \right).$$

■

Source superposition, which we discussed in Chapter 2, becomes particularly effective when used to solve a circuit with the series and parallel connection simplifications and the voltage and current dividers. This is because turning off sources frequently creates many series and parallel combinations of the remaining elements. This is illustrated in the next example.

Example 3-3   Using Source Superposition

Consider the circuit shown in Figure 3-8(a), which can be solved by using a combination of source superposition, the series and parallel combination simplifications, and the voltage and current dividers.

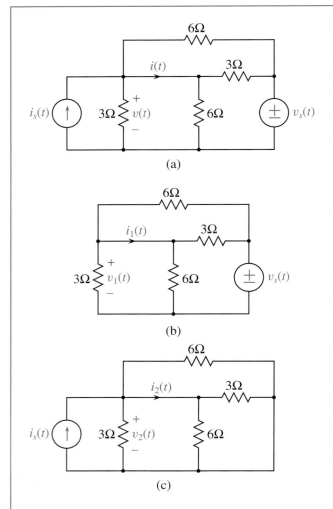

Fig. 3-8. (a) A circuit that can be solved by using circuit simplifications. (b) Circuit with the current source turned off. (c) Circuit with the voltage source turned off.

In this example, we are looking for two quantities: the voltage across the current source, $v(t)$, which is also the voltage across the vertical $3\,\Omega$ and $6\,\Omega$ resistors; and the current $i(t)$ that flows between those two parallel resistors. We observe first that the circuit has

two independent sources, which means that source superposition can be used. This implies that we can turn off the current source (resulting in the circuit of Figure 3-8(b)) and solve the simplified circuit for the quantities $v_1(t)$ and $i_1(t)$. Then we can turn off the voltage source (resulting in the circuit of Figure 3-8(c)) and solve that circuit for $v_2(t)$ and $i_2(t)$. Finally, we simply add these results together to get the solution to the original problem:

$$v(t) = v_1(t) + v_2(t)$$
$$i(t) = i_1(t) + i_2(t). \qquad (3.5)$$

The advantage of using source superposition is that the simpler circuits result in many more series and parallel combinations of elements that can be simplified further; the disadvantage of this approach is that we have to solve multiple circuits. With experience, the advantages usually outweigh the disadvantages. The challenge is to identify all of the series and parallel connections. This can often be confusing, particularly if we want to avoid redrawing the network constantly. For this problem, we want to emphasize the process, so to limit the degree of confusion we will redraw the circuit frequently.

Let us begin by solving for $v_1(t)$ in the circuit in Figure 3-8(b). Notice first that the two vertical resistors are connected in parallel, since each connects the same pair of nodes. For finding $v_1(t)$ (but not for later finding $i_1(t)$), we can replace them by a single resistor with a resistance of

$$\frac{(6\,\Omega)(3\,\Omega)}{6\,\Omega + 3\,\Omega} = 2\,\Omega.$$

The two horizontal resistors are also connected in parallel. They can, therefore, also be replaced by their equivalent resistance, which in this case is again $2\,\Omega$. Furthermore, the two $2\,\Omega$ equivalent resistors are now seen to be connected in series. This allows us to redraw

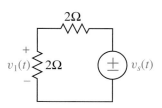

Fig. 3-9. The circuit of Figure 3.6b redrawn with the parallel resistors replaced by equivalents.

the circuit as shown in Figure 3-9. Using a voltage divider, we see that

$$v_1(t) = \frac{1}{2}v_s(t).$$

To compute the current $i_1(t)$, we proceed a little differently. This current is a little unusual in that it is a current that flows between connections within a single node. If we were to combine the parallel resistors, we would not be able to identify it. Instead, we apply KCL at the circled portion of the node shown in Figure 3-10. The potential at node b is $v_s(t)$ and at node a it is

$$e_a(t) = v_1(t) = \frac{1}{2}v_s(t).$$

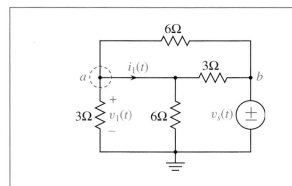

Fig. 3-10. Circuit used to find $i_1(t)$.

(Recall that we just solved for $v_1(t)$.) Since we know all the node potentials, we can compute all the currents. The current entering node a from the horizontal 6 Ω resistor is $[e_b(t) - e_a(t)]/6$, and the current leaving through the vertical 3 Ω resistor is $e_a(t)/3$. Therefore, by KCL,

$$i_1(t) = \frac{1}{6}[e_b(t) - e_a(t)] - \frac{1}{3}e_a(t) = \frac{1}{6}e_b(t) - \frac{1}{2}e_a(t)$$
$$= \frac{1}{6}v_s(t) - \frac{1}{4}v_s(t) = -\frac{1}{12}v_s(t).$$

Now we can turn off the voltage source and turn on the current source to figure out $v_2(t)$ and $i_2(t)$, as shown in Figure 3-8(c). Notice that all four resistors and the current source are connected in parallel and that $v_2(t)$ is the voltage across all of them. If we replace the vertical 6Ω and horizontal 3Ω resistors on the right by their equivalent 2Ω resistance, we observe that $i_2(t)$ is the current that would flow into that equivalent resistor. Figure 3-11 shows the circuit redrawn to reflect these observations.

Since a 3 Ω resistor in parallel with a 6 Ω resistor is equivalent to a 2 Ω resistor, a simple current divider reveals that the current $i_2(t)$ is one-half the current from the source:

$$i_2(t) = \frac{\frac{1}{2}}{\frac{1}{2} + \frac{1}{2}}i_s(t) = \frac{1}{2}i_s(t).$$

Finally, from Ohm's law,

$$v_2(t) = 2i_2(t) = i_s(t).$$

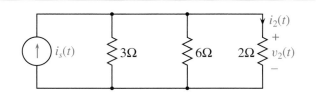

Fig. 3-11. A redrawing of the circuit in Figure 3-8(c) with an additional simplification.

These are all of the quantities that we need. Using (3.5), we get our final answer:

$$v(t) = v_1(t) + v_2(t) = \frac{1}{2}v_s(t) + i_s(t)$$

$$i(t) = i_1(t) + i_2(t) = -\frac{1}{12}v_s(t) + \frac{1}{2}i_s(t).$$

∎

 WORKED SOLUTION: *Source Superposition*

 To check your understanding, try Drill Problem P3-2.

3-2   The v–i Characteristics of Two-Terminal Networks

Consider a very general two-terminal subnetwork N, constructed from resistors and sources, that is connected to an external network E, as in Figure 3-1. The subnetwork N imposes a constraint, similar to an element relation, between the values of $v(t)$ and $i(t)$.[3]

Let us begin by reviewing the procedure that we would follow to find the complete solution of the entire network consisting of both E and N. We would generate a set of equations that constrain the values of the element variables, and then we would solve them. If we were using the exhaustive method, that set of equations would include element relations, KCL equations at an appropriate set of nodes or closed surfaces, and KVL equations on an appropriate set of closed paths. Now

---

[3]Notice that E is also a two-terminal subnetwork and that it imposes its own independent constraint on these two variables. The complete solution for the whole network corresponds to the values for $v(t)$ and $i(t)$ for which both of these constraints are satisfied simultaneously.

consider separating these equations into three groups: those equations involving voltages and currents that lie wholly within N, those that involve voltages and currents that lie wholly within E, and those KCL and KVL equations that involve voltages or currents from both N and E. (None of the element relations would fall into this latter group.) For the KCL equations involving currents from both N and E, we replace the sum of the currents entering terminal a from E by the quantity $i(t)$ and the sum of the currents entering terminal a' from E by the quantity $-i(t)$; for the KVL equations involving voltages from both N and E, we replace the sum of the voltages across the portions of the paths that lie wholly within E, but which connect terminals a and a', with the quantity $v(t)$. Next, we put aside the equations written in terms of variables that are entirely contained within E; now we are left with a modified subset of the original equations, written in terms of only the voltages and currents contained within N (along with the variables $v(t)$ and $i(t)$). Solving these equations allows us to obtain $v(t)$ as a function of $i(t)$, or $i(t)$ as a function of $v(t)$. If this argument sounds too formal, an example will perhaps make the procedure more specific.

Example 3-4   Finding the v–i characteristic

As a simple example, consider the two-terminal network shown in Figure 3-12.

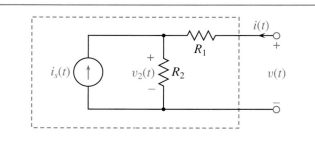

Fig. 3-12. An example of a two-terminal network.

Define $v_2(t)$ to be the voltage across the resistor with resistance R_2. For this simple network, we can write one KCL equation at the upper node, namely,

$$i_s(t) + i(t) = \frac{1}{R_2} v_2(t),$$

and one KVL equation over the path that runs from the $^+$ terminal, through the two resistors to the $^-$ terminal, and then through the external network back to the $^+$ terminal:

$$R_1 i(t) + v_2(t) = v(t).$$

This gives us two independent equations in the three variables $v(t)$, $i(t)$, and $v_2(t)$. Using the KCL equation to solve for $v_2(t)$ and then substituting this value into the KVL equation produces a single equation involving $v(t)$ and $i(t)$. This is the v–i relation that we want:

$$v(t) - R_1 i(t) = R_2 i(t) + R_2 i_s(t)$$

$$\implies v(t) = [R_1 + R_2] i(t) + R_2 i_s(t).$$

The graph of the instantaneous voltage as a function of the instantaneous current is a straight line as shown in Figure 3-13. All two-terminal networks that contain only resistors and sources have v–i relations of this form. ■

To check your understanding, try Drill Problem P3-3.

DEMO: v–i *Characteristic*

In analyzing a two-terminal network, it is tempting to look at the pair of terminals and treat it as an open circuit, but this temptation must be avoided. In fact, there is a nonzero current $i(t)$ that enters the network through the terminals. One way to avoid this trap is to attach a current source with a *generic* current $i(t)$ to the terminals, as in Figure 3-14(a). We then have a complete circuit that we can solve for the voltage $v(t)$ (as a function of $i(t)$). Alternatively, we could attach a voltage source with a voltage $v(t)$ and measure the current $i(t)$. This approach of attaching a source to a two-terminal network

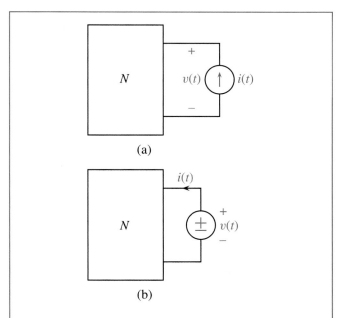

(a)

(b)

Fig. 3-14. Measuring the v–i characteristic of a two-terminal network by attaching the network to (a) a current source or (b) a voltage source.

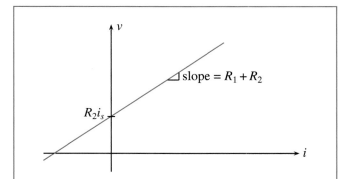

Fig. 3-13. The v–i characteristic for the two-terminal network in Figure 3-12.

is convenient because it turns the two-terminal network into a complete circuit, for which all of the standard analysis methods, such as the node method and the mesh method, can be used. It is an analytical approach, however, rather than an experimental one, because the source signals that are applied must be generic: we need to be able to learn the value of $v(t)$ for all $i(t)$ (or vice versa), and not simply for specific waveforms.

Recall that, when a network contains an independent source, one of the variables (the current in a current source or the voltage in a voltage source) is constrained to have a given functional form and the other assumes a value that is dictated by the other elements in the network. If we attach a current source with current $i(t)$ across the terminals and compute the resulting voltage $v(t)$, then $v(t)$ and $i(t)$ must be consistent with the v–i characteristic of the two-terminal network. This is why this approach works. It is demonstrated in the following example.

| **Example 3-5** | **Another Approach to Get the v–i Characteristic** |

As an example, consider the two-terminal network shown in Figure 3-15.

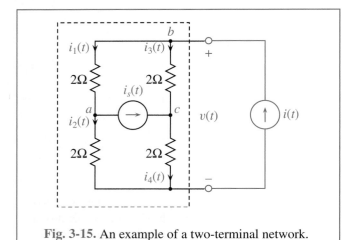

Fig. 3-15. An example of a two-terminal network.

Following the approach just outlined, we have added a current source to aid our analysis. This is shown in color because it is not part of the original network, which is contained within the dashed surface.

This network contains four nodes and two meshes, but the meshes are separated by a current source. To avoid defining and solving for the voltage across the terminals of that current source, we combine the two meshes into a supermesh and write a single KVL equation to go along with three KCL equations (written arbitrarily at nodes a, b, and c). If we let the currents through the resistors be our variables, we get

$$\begin{aligned}
\text{node } a: \quad & i_1(t) - i_2(t) = i_s(t) \\
\text{node } b: \quad & i_1(t) + i_3(t) = i(t) \\
\text{node } c: \quad & i_3(t) - i_4(t) = -i_s(t) \\
\text{KVL:} \quad & 2i_1(t) + 2i_2(t) - 2i_4(t) - 2i_3(t) = 0.
\end{aligned}$$

Notice that we have put $i(t)$ on the right-hand side of these equations with the other source terms. These equations are equivalent to the matrix equation

$$\begin{bmatrix} 1 & -1 & 0 & 0 \\ 1 & 0 & 1 & 0 \\ 0 & 0 & 1 & -1 \\ 2 & 2 & -2 & -2 \end{bmatrix} \begin{bmatrix} i_1(t) \\ i_2(t) \\ i_3(t) \\ i_4(t) \end{bmatrix} = \begin{bmatrix} 1 \\ 0 \\ -1 \\ 0 \end{bmatrix} i_s(t) + \begin{bmatrix} 0 \\ 1 \\ 0 \\ 0 \end{bmatrix} i(t),$$

from which we compute

$$\begin{aligned}
i_1(t) &= +0.5\, i_s(t) + 0.5\, i(t) \\
i_2(t) &= -0.5\, i_s(t) + 0.5\, i(t) \\
i_3(t) &= -0.5\, i_s(t) + 0.5\, i(t) \\
i_4(t) &= +0.5\, i_s(t) + 0.5\, i(t).
\end{aligned}$$

Now that we know the resistor currents, we can solve for the voltage across the terminals:

$$\begin{aligned}
v(t) &= 2i_3(t) + 2i_4(t) \\
&= 2i(t).
\end{aligned}$$

Since the voltage is proportional to the current, this particular network has the same effect on an external circuit as a single 2Ω resistor (i.e., it is network equivalent to a single 2Ω resistor).

It might seem surprising that the terminal behavior of this particular two-terminal network is not affected by its internal current source, but a closer look at the circuit reveals why this is so. From the superposition-of-sources principle, we can discover the dependence of $v(t)$ on $i_s(t)$ by leaving that source on and replacing $i(t)$ by an open circuit. This leaves the current source connected in parallel to two identical branches, each of which is a series connection of two 2Ω resistors. By a current divider, half of the current from the source will flow through one branch and half will flow through the other. This means that the two external terminals of the network will have the same node potential and that $v(t)$, which is the difference of those two node potentials, will be zero. Therefore, $v(t)$ does not contain a contribution from $i_s(t)$. This might not be the case, however, if the resistors did not have equal resistances.   ■

To check your understanding, try Drill Problem P3-4.

One of the major virtues of measuring the v–i characteristic of a two-terminal network by attaching a source across its terminals is that we can use our familiar methods for analyzing the circuit, including the node and mesh methods, without modification. Because we are solving the resulting circuit to find either the current through a voltage source or the voltage across a current source, however, we might have to perform a simple final computation after the normal variables in the circuit have been found. Indeed, such was the case in the previous example: Once we knew the values of all the element currents in the network, we had to perform an auxiliary computation to work out the potential difference across

the terminals. The next example uses the mesh method to find the v–i characteristic.

Example 3-6   Finding a v–i Characteristic by Using the Mesh Method

Before we conclude this section, let's look at the additional example in Figure 3-16, for which we would like to learn the v–i characteristic.

First, we should settle on our approach, which involves resolving questions such as: should we use the mesh method or would the node be better? Would it be easier to attach a voltage source and measure the current or to attach a current source and measure the voltage? A little thought before jumping in might save some effort in the long run. If we were to apply a voltage source and use the node method, we would need to write one KCL equation; if we were to apply a current source and use the mesh method, we would have to write one KVL equation. This reasoning suggests either of these approaches should require comparable effort. We shall arbitrarily, therefore, use the mesh method with an applied current source.

The center mesh is the one to be used to write the KVL equation. The mesh current in that mesh is the same as $i_0(t)$, so we simply make that identification. To aid in writing the KVL equation, we let the mesh current in the left mesh be $i(t)$ and we let the *counterclockwise* mesh

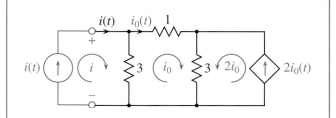

Fig. 3-16. Example of a two-terminal network illustrating the use of the mesh method to find the v–i characteristic. All resistances are measured in ohms.

current in the right mesh be $2i_0(t)$. KVL applied to the path gives

$$3[i_0(t) - i(t)] + i_0(t) + 3[i_0(t) + 2i_0(t)] = 0,$$

from which it follows that

$$i_0(t) = \frac{3}{13}i(t).$$

The voltage $v(t)$ is the potential difference across the terminals:

$$v(t) = 3[i(t) - i_0(t)] = \frac{30}{13}i(t).$$

Thus, this two-terminal network is network equivalent to a single resistor with a resistance of approximately 2.3Ω.

∎

 To check your understanding, try Drill Problem P3-5.

3-3   Thévenin Equivalent Networks

Whenever we replace a network containing many elements and sources with another one that contains fewer, our model becomes easier to work with and easier to understand. We have already seen this for examples of two-terminal networks that contain only resistors. In this section, we look at networks that contain both resistors and sources. The results of this section will later be extendable to more general networks containing inductors and capacitors in addition to resistors and sources.

 EXTRA INFORMATION: *Thévenin Equivalents*

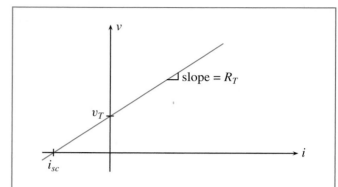

Fig. 3-17. The linear v–i characteristic that describes all two-terminal resistive networks.

3-3-1   Thévenin's Theorem for Resistive Networks

Almost all networks that contain only resistors and sources have v–i characteristics of the form

$$v(t) = R_T i(t) + v_T(t) \tag{3.6}$$

in which the instantaneous values of the terminal voltage and current satisfy a straight-line relationship, such as the one drawn in Figure 3-17.

This result is important and fairly straightforward to show.

 DEMO: *Resistive Equivalents*

Our approach will be the same as the one that we used in Chapter 2, in which we wrote out a complete set of element relations for all of the resistors in the network, a complete set of KCL equations at all but one of the nodes of the basic network, and a complete set of KVL equations at all of the meshes of the basic network. As before, we write our equations with the variables on the left-hand sides and all terms involving independent sources on the right. Dependent sources, when they exist in the circuit, modify the left sides of some of the equations. Because our network is a two-terminal network instead of a complete circuit, we place a current source across the terminals with a generic current $i(t)$.[4] The current in that current source is placed on the right sides of the appropriate KCL equations, since it is associated with an independent source. In addition, if the voltage $v(t)$ is not one of the element voltages, we add an additional KVL equation that relates that voltage to the element voltages in the network. In that equation, the term involving $v(t)$ is written on the left-hand side with the element variables and any source terms are written on the right.

Next we rewrite our equations in matrix–vector form, as we did in Section 2-3-4. They take the form

$$C v(t) = k_0 i(t) + \sum_{i=1}^{N} k_i s_i(t),$$

in which $v(t)$ is the column vector of element variables. If the two-terminal network contains b resistors, then there are $2b + 1$ variables, consisting of the b resistor voltages, b resistor currents, and $v(t)$. As a result, C is a $(2b + 1) \times (2b + 1)$ matrix and $v(t)$ is a $(2b + 1) \times 1$ column vector containing the variables. The signal $s_i(t)$ represents the ith independent source that is internal to the network (of which there are N), and the column vectors k_0 and k_i account for the constant coefficients of the various source terms in the various equations. Since the matrix C is invertible, we can write, for the complete solution,

$$v(t) = (C^{-1} k_0)i(t) + \sum_{i=1}^{N} (C^{-1} k_i)s_i(t).$$

---

[4]This will not work if the circuit is network equivalent to an isolated current source. In this case we cannot find a Thévenin equivalent, although we can find a Norton equivalent. (See Section 3-3-2.) This is an example of a network that does not have a v–i relation in the form of (3.6).

This represents a solution for all of the variables. Because we are interested in only $v(t)$, we extract this portion of the solution from the rest. Assuming that $v(t)$ was the $(2b + 1)^{th}$ variable, then if we define $u_v = [0, \ldots, 0, 1]$ we have $v(t) = u_v v(t)$. Thus,

$$v(t) = (u_v C^{-1} k_0)i(t) + \sum_{i=1}^{N} (u_v C^{-1} k_i)s_i(t).$$

This is in the form

$$v(t) = R_T i(t) + v_T(t), \tag{3.7}$$

where

$$R_T = u_v C^{-1} k_0 \tag{3.8}$$

and

$$v_T(t) = \sum_{i=1}^{N} (u_v C^{-1} k_i)s_i(t). \tag{3.9}$$

This result is known as *Thévenin's theorem* for resistive networks. It means that (almost) any resistive two-terminal network is network equivalent to a network of the form shown in Figure 3-18, which contains a single voltage source $v_T(t)$ connected in series with a resistor whose resistance is R_T.

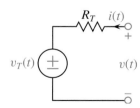

Fig. 3-18. A Thévenin equivalent network. Almost any two-terminal network containing only resistors and sources is network equivalent to a circuit of this form.

Although equations (3.8) and (3.9) serve as formulas for the Thévenin equivalent resistance and voltage, and were useful in our derivation, there are easier ways to compute numerical values for these quantities in actual circuits. Nonetheless, we observe that the Thévenin voltage, $v_T(t)$ is created by the independent sources contained within the circuit; *if there are no such sources, $v_T(t) = 0$ and the two-terminal network is network equivalent to a single resistor.* The Thévenin resistance, R_T, depends on the element values, the way that they are connected, and the dependent sources.

The Thévenin voltage and resistance can be computed by exploiting the fact that a two-terminal network, N, and its Thévenin equivalent are network equivalent, which means that any measurement that we can make

at the terminals of the two circuits will be the same in both circuits. Consider first the measurement depicted in Figure 3-19(a), where we simply measure the voltage across the isolated (open-circuited) terminals of the network. Since this is the value of $v(t)$ when $i(t) = 0$, it is called the *open-circuit voltage*. On the graph in Figure 3-17, it is the point where the v–i characteristic intersects the v (vertical) axis. If we measure the open-circuit voltage on the circuit diagram for the Thévenin equivalent (Figure 3-18), we see that $v_T(t) = v_{oc}(t)$. *Thus, the Thévenin source voltage is equal to the open-circuit voltage.*

Another measurement that we can make on both the network N and its Thévenin equivalent is a measurement of the *short-circuit current*. This is the current that flows into the device when the voltage across the terminals is zero. On the graph of Figure 3-18, it represents the intersection of the v–i characteristic with the current (horizontal) axis. Again, if we make this measurement on the Thévenin equivalent circuit in Figure 3-18, we see that

$$i_{sc}(t) = -\frac{1}{R_T}v_T(t) = -\frac{1}{R_T}v_{oc}(t).$$

From this, we obtain

$$R_T = -\frac{v_{oc}(t)}{i_{sc}(t)}. \tag{3.10}$$

This relation also follows from the graph, since knowledge of two points on a linear graph is sufficient to specify the graph.

 WORKED SOLUTION: *Thévenin Equivalent*

Example 3-7 **Using v_{oc} and i_{sc} to Find the Thévenin Equivalent**

As an example, let us find the Thévenin equivalent of the network shown in Figure 3-20(a).

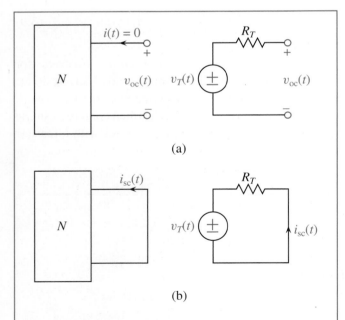

(a)

(b)

Fig. 3-19. The Thévenin resistance and voltage can be found from measurements of the (a) open-circuit voltage and (b) short-circuit current of a two-terminal network, N.

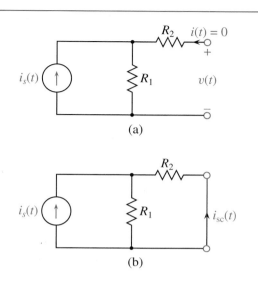

(a)

(b)

Fig. 3-20. (a) Network whose Thévenin equivalent is desired, configured to measure the open-circuit voltage. (b) Same network, configured to measure the short-circuit current.

To figure out the open-circuit voltage, $v_{oc}(t)$, we set $i(t) = 0$. Then all of the current from the current source flows through R_1. Since there is no voltage across R_2 when $i(t) = 0$,

$$v_{oc}(t) = R_1 i_s(t).$$

To discover the short-circuit current, $i_{sc}(t)$, we connect a short circuit across the terminals, as shown in Figure 3-20(b). This connects R_1 and R_2 in parallel, which means that we can find $i_{sc}(t)$ by using a current divider.

$$i_{sc}(t) = -\frac{\frac{1}{R_2}}{\frac{1}{R_1} + \frac{1}{R_2}} i_s(t)$$

$$= -\frac{R_1}{R_1 + R_2} i_s(t).$$

Notice that the sign on the short-circuit current is negative, because its reference direction is the same as

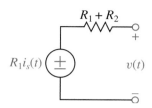

Fig. 3-21. The Thévenin equivalent for the network in Figure 3-20(a).

that of $i(t)$, which is defined to be entering the $+$ terminal of the two-terminal network. Thus,

$$R_T = -\frac{v_{oc}(t)}{i_{sc}(t)} = R_1 + R_2.$$

and the Thévenin equivalent network is shown in Figure 3-21. ∎

To check your understanding, try Drill Problem P3-6.

This approach to finding the Thévenin equivalent circuit finds the straight-line v–i characteristic by finding two points that lie on it. These points are defined by $v_{oc}(t)$ and $i_{sc}(t)$. An alternative is to specify the line by specifying one point—either $v_{oc}(t)$ or $i_{sc}(t)$—and the slope, which is R_T itself. One means for doing this was suggested by our derivation. We observe from (3.9) that, if all the independent sources in the network are turned off (set to zero), then the Thévenin equivalent voltage $v_T(t)$ is zero also; but, from (3.8), this operation has no effect on R_T. Thus, the network with all of its *independent* sources turned off is network equivalent to a single resistor whose resistance is R_T. This means that we can find the Thévenin equivalent resistance by replacing all of the independent current sources by open circuits, by replacing all of the independent voltage sources by short circuits, and by then finding the equivalent resistance at the terminals.

In the circuit of the previous example, we observe that, if we replace the current source in Figure 3-20(a) by an open circuit, the two-terminal network reduces to a resistor having resistance R_1 connected in series with a resistor having resistance R_2, so that $R_T = R_1 + R_2$.

At this point, we have three methods for computing the components of the Thévenin equivalent network:

Approach 1: Calculate $v_{oc}(t)$ and $i_{sc}(t)$. Then $v_T(t) = v_{oc}(t)$ and $R_T = -v_{oc}(t)/i_{sc}(t)$.

Approach 2: Calculate $v_{oc}(t)$ and R_T. Then $v_T(t) = v_{oc}(t)$.

Approach 3: Calculate $i_{sc}(t)$ and R_T. Then $v_T(t) = -i_{sc}(t)R_T$.

Which method to apply is largely a matter of personal taste, although, for circuits containing only independent sources and resistors, Approach 2 is probably the most popular. When a network contains both independent and dependent sources, Approach 1 is probably easier, but that method will fail when the network contains *only* dependent sources and resistors, since, in this case, $v_{oc}(t) = i_{sc}(t) = 0$ and we cannot find R_T by taking their ratio. Nonetheless, it is still true that $v_T(t) = 0$, but we generally have to resort to a more formal means to discover the v–i characteristic. We can attach a current source with a current $i(t)$ across the terminals and measure the voltage $v(t)$ that is induced by the v–i relation of the two-terminal network. It will be of the form

$$v(t) = R_T i(t);$$

hence, we can get the equivalent resistance from the constant of proportionality. This approach is illustrated in the next example.

 FLUENCY EXERCISE: *Thévenin Equivalents*

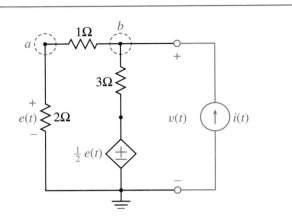

Fig. 3-22. A two-terminal network containing a dependent source for which we want to construct the Thévenin equivalent.

Example 3-8   **Finding the Thévenin Equivalent when there Are Dependent Sources**

To construct the Thévenin equivalent of the network in Figure 3-22, which contains a dependent source, we attach a current source with a current of $i(t)$ (shown as colored) and compute the voltage waveform $v(t)$.

Even though the mesh method, if applied to this circuit, would require only a single equation, we choose to use the node method because the key variables in the circuit are all voltages that we can identify with the node potentials. Specifically, the potential at node a is $e(t)$, which also controls the dependent source, and the potential at node b is the voltage $v(t)$ whose value we seek to learn. As usual with the node method, we begin by writing KCL equations at the nodes a and b:

node a: $\dfrac{1}{2}e(t) + [e(t) - v(t)] = 0$

$\Longrightarrow e(t) = \dfrac{2}{3}v(t)$

node b: $[v(t) - e(t)] + \dfrac{1}{3}\left[v(t) - \dfrac{1}{2}e(t)\right] = i(t)$

$\Longrightarrow e(t) = \dfrac{8}{7}v(t) - \dfrac{6}{7}i(t).$

Equating the two expressions for $e(t)$ gives us

$$\dfrac{2}{3}v(t) = \dfrac{8}{7}v(t) - \dfrac{6}{7}i(t)$$

$$\Longrightarrow v(t) = \dfrac{9}{5}i(t).$$

Since $v(t)$ is proportional to $i(t)$, this circuit is the equivalent of a single resistor with a resistance of $1.8\,\Omega$. The Thévenin voltage will always be zero when the two-terminal network does not contain any independent sources, regardless of whether it contains dependent sources. ∎

 To check your understanding, try Drill Problem P3-7.

3-3-2   Norton's Theorem for Resistive Networks

The previous section showed that (almost) any two-terminal network constructed from resistors, independent sources, and dependent sources has a v–i characteristic of the form

$$v(t) = R_T i(t) + v_T(t), \qquad (3.11)$$

which means that it is network equivalent to a series connection of an independent voltage source and a resistor. If we solve (3.11) for $i(t)$, we get the alternative representation

$$i(t) = \dfrac{1}{R_T}[v(t) - v_T(t)]$$

$$= \dfrac{1}{R_T}v(t) - \dfrac{1}{R_T}v_T(t)$$

$$= \dfrac{1}{R_T}v(t) + i_{\text{sc}}(t). \qquad (3.12)$$

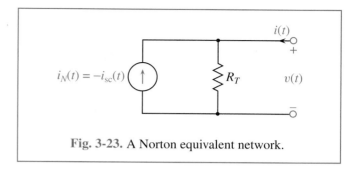

Fig. 3-23. A Norton equivalent network.

When it is written this way, we see that the v–i characteristic for the network N is equivalent to a *parallel* connection of a resistor and a *current* source, as shown in Figure 3-23. This network is known as the **Norton equivalent** of the network N. Notice that the values of the resistors in the Thévenin and Norton equivalent circuits are the same.

Most two-terminal circuits have both Thévenin and Norton equivalent circuits. There are two prominent exceptions. A two-terminal network consisting of an isolated current source has a Norton equivalent (actually, is its own Norton equivalent), but no Thévenin equivalent; and a circuit consisting of an isolated voltage source has a Thévenin equivalent, but no Norton equivalent. These two exceptions rarely cause problems, since these are not subnetworks that we would expect to be able to simplify, but the fact that these two exceptions exist often makes statements of simple circuit properties read like legal documents.

A network consisting of a voltage source connected in series with a resistor (Thévenin equivalent) is network equivalent to a current source connected in parallel with a resistor (its Norton equivalent). These two equivalent networks are shown in Figure 3-24.

The equivalence of these two structures can often be useful, through a process known as **source substitution**. Switching between these two structures can be extremely helpful in simplifying a circuit, particularly one

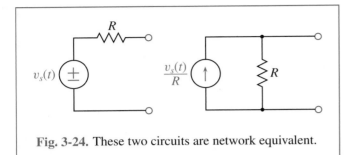

Fig. 3-24. These two circuits are network equivalent.

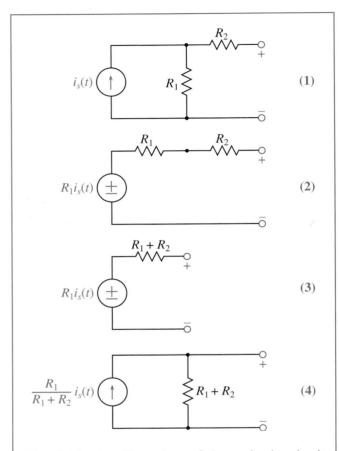

Fig. 3-25. An illustration of how simple circuit substitutions can be used to find the Thévenin and Norton equivalents for a circuit.

containing alternating series and parallel components known as *ladder networks*. (See the following two examples.) We could have used it, for example, to work Example 3-5 in Figure 3-20.

Example 3-9   Example 3-5 Revisited

In Example 3-5, we found the Thévenin equivalent of the two-terminal network in Figure 3-20 by finding $v_{oc}(t)$ and $i_{sc}(t)$. Figure 3-25 shows how we can accomplish the same goal by using source substitutions.

We first replace the parallel resistor and current source by a series resistor and voltage source. This produces the circuit shown as step 2 in that figure. The two resistors, R_1 and R_2, are now seen to be connected in series, and their series connection can be replaced by a single equivalent resistor. For this circuit, we are thus able to get the Thévenin (step 3) and Norton (step 4) equivalent circuits without the necessity of solving for the open-circuit voltage or the short-circuit current.

Be careful when computing the source waveform after making a source substitution. As a mnemonic device, remember from Ohm's law that amperes times ohms equals volts. Thus, the current source waveform in a Norton structure is *multiplied* by the resistance to produce

the voltage waveform in a Thévenin structure, but a *division* by the resistance is performed when we make the opposite substitution.   ∎

> **To check your understanding, try Drill Problem P3-8.**

> **FLUENCY EXERCISE:** *Equivalent Networks*

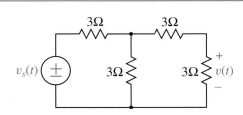

Fig. 3-26. A simple ladder network to be solved by using source substitution.

Example 3-10 Using Source Substitution to Analyze a Circuit

Consider the circuit in Figure 3-26. The goal here is to find the voltage $v(t)$.

Applying source substitution allows us to replace the voltage source connected in series with the 3 Ω resistor by a current source connected in parallel with a 3 Ω resistor. The current in that source is $v_s(t)/R_T = v_s(t)/3$ (amps). The resulting circuit is drawn in Figure 3-27.

The two 3 Ω resistors connected in parallel are network equivalent to a single $\frac{3}{2}$ Ω resistor, and that resistor in parallel with the current source is network equivalent to a voltage source connected in series with a resistor. The voltage in that source is equal to the current in the original current source ($\frac{v_s(t)}{3}$) multiplied by the resistance $\frac{3}{2}$ Ω.

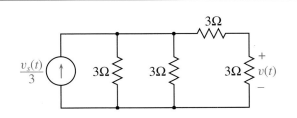

Fig. 3-27. The circuit of 3-26 after a source substitution replaced the voltage source and series resistor by a current source and parallel resistor.

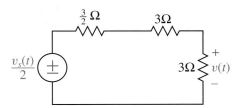

Fig. 3-28. The circuit of 3-27 after the two parallel resistors are combined and a second source substitution is performed.

This gives the circuit shown in Figure 3-28.

Finally, we can use a voltage divider to calculate $v(t)$.

$$v(t) = \frac{3}{\frac{3}{2} + 3 + 3} \cdot \frac{v_s(t)}{2} = \frac{1}{5} v_s(t).$$

∎

To check your understanding, try Drill Problem P3-9.

3-3-3 Other Equivalence Relations

In the previous few sections, we have seen several examples of circuits that have network equivalents that provide the same terminal behavior by means of a simpler structure. These and a number of other network equivalents are summarized in Table 3-1.

In addition to the series and parallel resistors that we have already seen and used and the Thévenin–Norton equivalence, we note that two voltage sources in series can be replaced by a single voltage source with a source voltage equal to their sum. The same is true for two current sources that are connected in parallel. These two examples can be seen either as simple applications of KVL and KCL, respectively, or as trivial consequences of source superposition. We also note that any element that is connected in parallel with a voltage source has the

Table 3-1. The two-terminal networks in the left-hand column are network equivalent to the corresponding ones in the right-hand column.

same terminal behavior as the voltage source by itself; the same is true also for any element that is connected in series with a current source. This table of network equivalents is not complete. We shall see other examples as we proceed.

3-4   Selecting an Analysis Method

At this point, we have built up a collection of several methods that can be used to analyze a circuit. There are systematic approaches—the exhaustive, node, and mesh methods—and methods based on simplifications, such as voltage and current dividers, superposition of sources, replacing series-connected and parallel elements by an equivalent element, and replacing subnetworks by their Thévenin and or Norton equivalents. In addition, there are yet more methods that we have not seen that can be helpful in certain situations. How are we to know which method to use for a particular situation? In the long run, as you gain experience, having many methods will make the task easier, but in the beginning they make it more confusing. Fortunately, there are no wrong methods; for a given circuit, almost all of the methods can be made to work, and all will produce the same answer. The only reward or penalty associated with any method is the amount of effort required to get the answer by that method. Furthermore, there is probably no best method for a particular circuit; the amount of effort that will be required by any method will depend upon the circuit and upon the circumstances. If you are quite comfortable with one particular method and less comfortable with others, you might choose to play it safe. This is a reasonable strategy, particularly if this results in fewer errors. If a computer or calculator is available that makes solving sets of linear equations straightforward, and if components are specified with numerical values, the systematic methods might be more appropriate than they would be if the resulting equations had to be solved by using a paper and pencil. If all that is needed is a

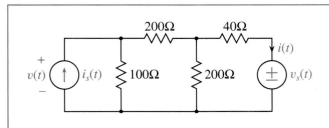

Fig. 3-29. Circuit to be analyzed by multiple methods.

single element variable, the methods that involve circuit simplifications are often the methods of choice, but if many voltages or currents in a circuit are required, these approaches begin to lose efficiency. A good way to reduce errors is to use one method to analyze the circuit and then use another to verify that the answer makes sense.

In order to compare the various methods for analyzing a circuit, consider the network in Figure 3-29. This particular circuit was chosen because most of the methods that we have seen can be applied to it. The goal is to compute $v(t)$, the voltage across the terminals of the current source, and $i(t)$, the current through the voltage source.

Example 3-11   The Exhaustive Method

Let the four resistor voltages be $v_1(t)$, $v_2(t)$, $v_3(t)$, and $v_4(t)$, defined from left to right. Furthermore, we shall let the $+$ terminal be the upper terminal on the two vertical resistors and the left terminal on the two horizontal ones. Since the current source lies on the exterior boundary of the circuit, we need to write two KVL equations, and since the circuit contains three nontrivial nodes, we need to write two KCL equations. We shall write all of these, using only the resistor voltages, by incorporating the four element relations into the KCL equations. The four equations are

$$KVL1: \quad -v_1(t) + v_2(t) + v_3(t) = 0$$

$$KVL2: \quad -v_3(t) + v_4(t) = -v_s(t)$$

$$KCL1: \quad \frac{1}{100}v_1(t) + \frac{1}{200}v_2(t) = i_s(t)$$

$$KCL2: \quad -\frac{1}{200}v_2(t) + \frac{1}{200}v_3(t) + \frac{1}{40}v_4(t) = 0.$$

In matrix–vector form, these become

$$\begin{bmatrix} -1 & 1 & 1 & 0 \\ 0 & 0 & -1 & 1 \\ 2 & 1 & 0 & 0 \\ 0 & -1 & 1 & 5 \end{bmatrix} \begin{bmatrix} v_1(t) \\ v_2(t) \\ v_3(t) \\ v_4(t) \end{bmatrix} = \begin{bmatrix} 0 \\ -1 \\ 0 \\ 0 \end{bmatrix} v_s(t) + \begin{bmatrix} 0 \\ 0 \\ 200 \\ 0 \end{bmatrix} i_s(t).$$

The solution for the four resistor voltages is

$$\begin{bmatrix} v_1(t) \\ v_2(t) \\ v_3(t) \\ v_4(t) \end{bmatrix} = \begin{bmatrix} 0.25 \\ -0.50 \\ 0.75 \\ -0.25 \end{bmatrix} v_s(t) + \begin{bmatrix} 70 \\ 60 \\ 10 \\ 10 \end{bmatrix} i_s(t).$$

For the two variables of interest, $v(t) = v_1(t)$ and $i(t) = \frac{1}{40}v_4(t)$, so we have

$$v(t) = \frac{1}{4}v_s(t) + 70i_s(t)$$

$$i(t) = -\frac{1}{160}v_s(t) + \frac{1}{4}i_s(t).$$

∎

Example 3-12   The Node Method

Figure 3-30 shows the circuit with all but one of its node potentials indicated; the remaining node at the $+$ terminal of the voltage source has a potential of $v_s(t)$. Writing KCL equations at the two circled nodes gives the following pair of equations in $e_a(t)$ and $e_b(t)$:

$$\frac{1}{100}e_a(t) + \frac{1}{200}[e_a(t) - e_b(t)] = i_s(t)$$

$$\frac{1}{200}[e_b(t) - e_a(t)] + \frac{1}{200}e_b(t) + \frac{1}{4}[e_b(t) - v_s(t)] = 0.$$

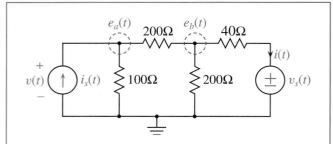

Fig. 3-30. Circuit of Figure 3-29 with node potentials indicated.

These can be simplified to

$$3e_a(t) - e_b(t) = 200i_s(t)$$

$$-e_a(t) + 7e_b(t) = 5v_s(t).$$

Their solution is readily seen to be

$$\begin{bmatrix} e_a(t) \\ e_b(t) \end{bmatrix} = \begin{bmatrix} 0.25 \\ 0.75 \end{bmatrix} v_s(t) + \begin{bmatrix} 70 \\ 10 \end{bmatrix} i_s(t).$$

In order to get the variables of interest, we observe that $v(t) = e_a(t)$ and that $i(t) = \frac{1}{40}[e_b(t) - v_s(t)] = \frac{1}{40}[-0.25v_s(t) + 10i_s(t)]$. Thus,

$$v(t) = \frac{1}{4}v_s(t) + 70i_s(t)$$

$$i(t) = -\frac{1}{160}v_s(t) + \frac{1}{4}i_s(t).$$

∎

Example 3-13   The Mesh Method

This same circuit can be analyzed by using the mesh method. Let the clockwise mesh currents in the center and right meshes be $i_\alpha(t)$ and $i_\beta(t)$, respectively. The two KVL equations are then

$$100[i_\alpha(t) - i_s(t)] + 200i_\alpha(t) + 200[i_\alpha(t) - i_\beta(t)] = 0$$

$$200[i_\beta(t) - i_\alpha(t)] + 40i_\beta(t) = -v_s(t).$$

These simplify to

$$5i_\alpha(t) - 2i_\beta(t) = i_s(t)$$

$$-5i_\alpha(t) + 6i_\beta(t) = -\frac{1}{40}v_s(t),$$

from which it follows that

$$\begin{bmatrix} i_\alpha(t) \\ i_\beta(t) \end{bmatrix} = \begin{bmatrix} -\frac{1}{400} \\ -\frac{1}{160} \end{bmatrix} v_s(t) + \begin{bmatrix} 0.30 \\ 0.25 \end{bmatrix} i_s(t).$$

Since $v(t) = 100[i_s(t) - i_\alpha(t)]$ and $i(t) = i_\beta(t)$, we have

$$v(t) = \frac{1}{4}v_s(t) + 70i_s(t)$$

$$i(t) = -\frac{1}{160}v_s(t) + \frac{1}{4}i_s(t).$$

∎

Example 3-14   Source Substitutions

Still another approach is to simplify the circuit systematically until the variables of interest can be written down by inspection. This approach is not as efficient for this circuit as it is for others, because the two variables of interest require different sets of substitutions. To find the current through the voltage source, we make the sequence of source substitutions shown in Figure 3-31, collapsing the circuit from the left. From the final simplification, we have

$$i(t) = \frac{1}{160}[40i_s(t) - v_s(t)]$$

$$= \frac{1}{4}i_s(t) - \frac{1}{160}v_s(t).$$

To solve for $v(t)$, we perform a similar series of simplifications, collapsing the circuit from the right. The successive steps are illustrated in Figure 3-32. From the final simplification, we have

$$v(t) = 70[i_s(t) + \frac{1}{280}v_s(t)]$$

$$= 70i_s(t) + \frac{1}{4}v_s(t).$$

∎

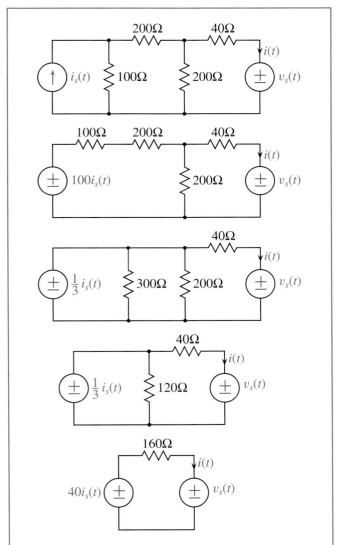

Fig. 3-31. Successive simplifications of the circuit of Figure 3-29 to solve for $i(t)$.

Example 3-15   Superposition of Sources

To conclude this comparison, we shall also solve the circuit by using superposition of sources. To do this, we

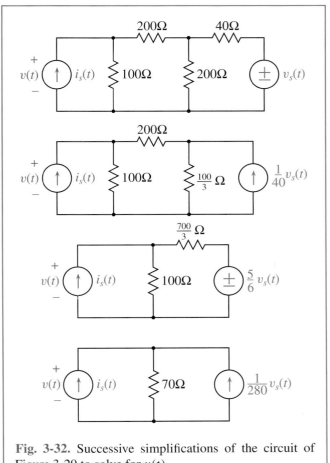

Fig. 3-32. Successive simplifications of the circuit of Figure 3-29 to solve for $v(t)$.

first set $i_s(t)$ to zero and solve for $v(t)$ and $i(t)$. Since these are merely the components of the two variables of interest caused by the voltage source, we shall call them $v_v(t)$ and $i_v(t)$, respectively. Then, we set $v_s(t)$ to zero, again solve for $v(t)$ and $i(t)$, and call the results $v_i(t)$ and $i_i(t)$. Finally, we add these two solutions together to get the true $v(t)$ and $i(t)$.

Setting $i_s(t)$ to zero is equivalent to replacing it by an open circuit. This results in the circuit shown in Figure 3-33.

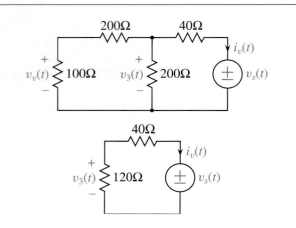

Fig. 3-33. (top) Circuit of Figure 3-29 with the current source turned off. (bottom) Same circuit after simplification created by replacing series and parallel connections of resistors by their equivalent values.

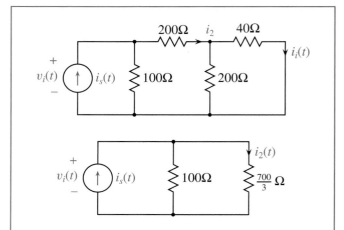

Fig. 3-34. (top) Circuit of Figure 3-29 with the voltage source turned off. (bottom) Same circuit after simplification created by replacing series and parallel connections of resistors by their equivalent values.

Applying a voltage divider to the upper circuit in that figure, we see that

$$v_v(t) = \frac{1}{3}v_3(t).$$

The lower circuit is a simplification that results from replacing the series connection of two resistors by its equivalent resistance, then recognizing that equivalent resistance is connected in parallel with the vertical 200Ω resistor. The equivalent resistance of this connection is 120Ω. Notice that the voltage across the terminals of this equivalent resistor is also $v_3(t)$. From another voltage divider applied to the simplified circuit, we observe that

$$v_3(t) = \frac{3}{4}v_s(t),$$

and that, therefore,

$$v_v(t) = \left(\frac{3}{4}\right)\left(\frac{1}{3}\right)v_s(t) = \frac{1}{4}v_s(t).$$

We also observe from the simplified circuit that

$$i_v(t) = -\frac{1}{160}v_s(t).$$

To complete the solution, we turn off the voltage source in the original circuit. Since turning off a voltage source replaces it by a short circuit, the circuit reduces to the one shown in Figure 3-34.

A current divider applied to the original circuit reveals that

$$i_i(t) = \frac{200}{240}i_2(t) = \frac{5}{6}i_2(t).$$

Having made this observation, we can now replace the parallel 40Ω and 200Ω resistors by their equivalent resistance ($\frac{100}{3}\Omega$), then observe that resistance is connected in series with the 200Ω resistor, and replace that combination by its equivalent, $\frac{700}{3}\Omega$. The simplified circuit is shown as the bottom circuit in Figure 3-34. Notice that current $i_2(t)$ passes through that equivalent resistance. A current divider applied to the simplified circuit yields

$$i_2(t) = \frac{100}{100 + \frac{700}{3}} i_s(t) = \frac{3}{10} i_s(t).$$

Thus,

$$i_i(t) = \left(\frac{5}{6}\right)\left(\frac{3}{10}\right) i_s(t) = \frac{1}{4} i_s(t).$$

The voltage across the terminals of the current source is found by replacing the parallel resistors by their equivalent resistance:

$$v_i(t) = \frac{(100)\left(\frac{700}{3}\right)}{100 + \frac{700}{3}} i_s(t) = 70 i_s(t).$$

Finally, we perform the superposition of the two partial solutions to get the now-familiar result

$$v(t) = v_v(t) + v_i(t) = \frac{1}{4} v_s(t) + 70 i_s(t)$$

$$i(t) = i_v(t) + i_i(t) = -\frac{1}{160} v_s(t) + \frac{1}{4} i_s(t).$$

∎

3-5   Graphical Analysis*

Frequently, the $v–i$ characteristic of a device is given in graphical rather than in analytical form.[5] This representation is particularly appropriate for many nonlinear elements. When one or more of the elements are specified graphically, it may be easier to solve the circuit graphically than by analytical means.

As an example, consider the circuit shown in Figure 3-35, which contains a nonlinear element.

_____

[5]A plot of $v(t)$ versus $i(t)$ can be given only when the value of $v(t)$ depends only on the instantaneous value of $i(t)$. Such is the case, for example, for resistors, but not for circuits containing inductors or capacitors.

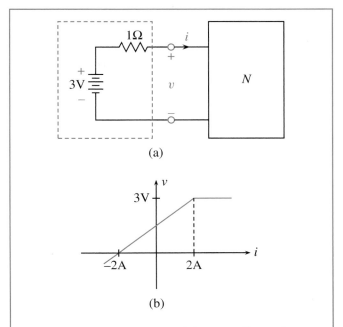

(a)

(b)

Fig. 3-35. (a) A network containing a nonlinear element, N. (b) The $v–i$ characteristic of the nonlinear element.

The $v–i$ relation in Figure 3-35(b) defines the constraint between v and i imposed by the network N. For the two-terminal network in the dashed box, which is network equivalent to a series connection of a resistor and a battery, the voltage and current satisfy the constraint

$$v = 3 - i. \qquad (3.13)$$

Notice that this current is defined as leaving the network through the positive terminal. This is necessary if we are going to use the same variables to describe both $v–i$ relations. Since the source is constant, neither v nor i varies as a function of time, t.

The solution to the circuit is that value of $[v, i]$ that satisfies both constraints simultaneously. When one of those constraints is specified graphically, a graphical

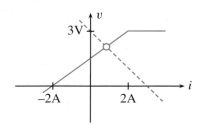

Fig. 3-36. The solution of a network consisting of two 2-terminal networks can be found from the intersections of the graphs of their v–i characteristics.

solution is often the easiest to obtain. We can find this by plotting both v–i characteristics on the same axes, as in Figure 3-36, and finding the intersection of the two graphs.

The v–i characteristic of the network N is shown as a solid line; that for the network within the dashed box, as given in (3.13), is shown as a dashed line. The solution of the circuit is the open circle. From the graphical solution, we see that the solution occurs in the linear portion of the v–i characteristic for N. We can use this fact to find numerical values for v and i if our application requires more accuracy than we get from reading the solution from the graph directly. In the linear region, we see that

$$v = \frac{3}{4}i + \frac{3}{2},$$

so that the equilibrium solution is

$$v = 15/7 \ V$$

$$i = 6/7 \ A.$$

  DEMO: *Graphical Analysis*

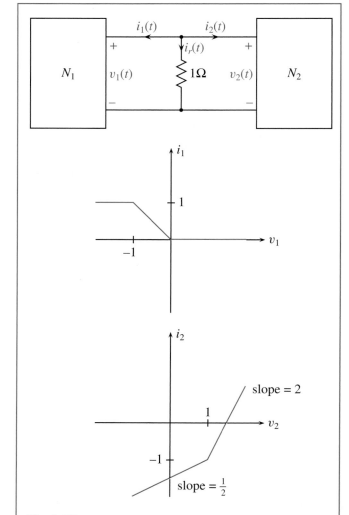

Fig. 3-37. Two connected two-terminal networks and their v–i characteristics.

Example 3-16   Circuit Solution by Graphical Analysis

As a considerably more complex example, consider the arrangement of two 2-terminal networks shown in Figure 3-37.

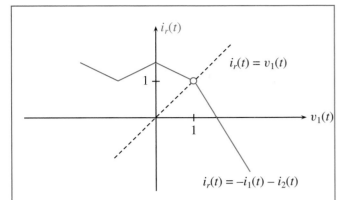

Fig. 3-38. Constraints on $v_1(t)$ and $i_r(t)$. The solid curve is the constraint $i_r(t) = -i_1(t) - i_2(t)$ imposed by N_1 and N_2. The dashed curve is the constraint imposed by the 1Ω resistor, $v_r(t) = i_r(t)$. The solution is given by the intersection point, marked by the open circle.

In this case, both networks are nonlinear. Notice that the v–i relations have been graphed with the voltage on the horizontal axis instead of one of the currents. Since the networks are connected in parallel and $v_1(t) = v_2(t)$, it is easier to work with the graphs when their common variable is plotted on the horizontal axis. This network has three components: the subnetwork N_1, the subnetwork N_2, and the 1Ω resistor. By KCL,

$$i_1(t) + i_2(t) + i_r(t) = 0.$$

Rewriting this relation yields

$$i_r(t) = -i_1(t) - i_2(t). \qquad (3.14)$$

From the graphs, we know that $i_1(t)$ and $i_2(t)$ can both be written as functions of $v_1(t)$, which means that we can express $i_r(t)$ as a function of $v_1(t)$. This is the constraint on $v_1(t)$ and $i_r(t)$ imposed by the two networks. The resistor imposes the additional constraint

$$v_1(t) = v_r(t) = i_r(t). \qquad (3.15)$$

These two constraints are plotted on the graph in Figure 3-38.

The solid curve is a plot of $i_r(t) = -i_1(t) - i_2(t)$ (from (3.14)), and the dashed curve is from (3.15). From these graphs, we see that the solution is given by $v_1(t) = v_2(t) = 1\,V, i_r(t) = 1\,A$. From the graphs in Figure 3-36, we also see that when $v_1(t) = v_2(t) = 1\,V$, then $i_1(t) = 0$ and $i_2(t) = -1\,A$. ∎

3-6   Chapter Summary

3-6-1   Important Points Introduced

- A subnetwork with two terminals behaves like an element. It has a voltage, a current, and a v–i relation.
- Two 2-terminal networks are network equivalent if they have the same v–i relation.
- Any two-terminal network containing only resistors and dependent sources is network equivalent to a single resistor.
- Any two-terminal network (with one exception) containing only resistors and sources is network equivalent to a Thévenin equivalent network consisting of a voltage source connected in series with a resistor.
- Any two-terminal network (with one exception) containing only resistors and sources is network equivalent to a Norton equivalent network consisting of a current source connected in parallel with a resistor.
- The Thévenin or Norton equivalent circuit can be constructed from measurements of the open-circuit voltage and the short-circuit current at the terminals of the two-terminal network.
- All of the solution methods that we have seen will produce the same solution. They differ only in the amount of effort that might be required.

3-6-2   New Abilities Acquired

You should now be able to do the following:

(1) Replace resistors connected in series by their equivalent resistance.

(2) Replace resistors connected in parallel by their equivalent resistance.

(3) Use a voltage divider for apportioning the voltages among several resistors connected in series.

(4) Use a current divider for apportioning the currents among several resistors connected in parallel.

(5) Find the v–i relation of a two-terminal network by attaching a generic current source to the terminals and solving for the voltage drop.

(6) Find the Thévenin equivalent of a two-terminal network.

(7) Find the Norton equivalent of a two-terminal network.

(8) Using graphical methods, find the solution of a circuit containing resistors, sources, and one or more nonlinear elements.

3-7   Problems

3-7-1   Drill Problems

All drill problems have solutions on the web site.

P3-1 This problem is similar to Example 3-1.

(a) Compute the equivalent resistance of the two-terminal network in Figure P3-1(a) as seen at the terminals a–a'.

(b) Repeat for the network in Figure P3-1(b) at the terminals b–b'.

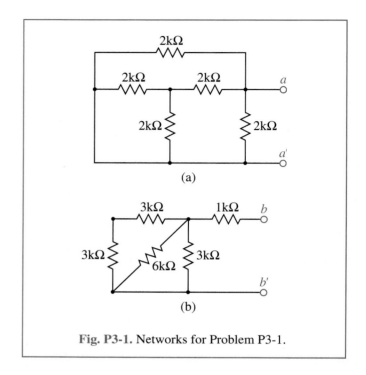

(a)

(b)

Fig. P3-1. Networks for Problem P3-1.

P3-2 Use superposition to find $v(t)$ in the circuit in Figure P3-2 in terms of $v_s(t)$, $i_{s_1}(t)$, and $i_{s_1}(t)$. This problem is similar to Example 3-2.

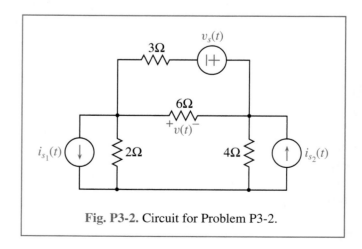

Fig. P3-2. Circuit for Problem P3-2.

P3-3 Find the v–i relation at the terminals of the two-terminal network in Figure P3-3 (i.e., express $v(t)$ as a function of $i(t)$). This problem is similar to Example 3-3. (Do not assume that the current $i(t)$ is zero.)

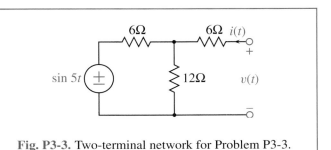

Fig. P3-3. Two-terminal network for Problem P3-3.

P3-4 Express $v(t)$ as a function of $i(t)$ and $i_s(t)$ for the circuit in Figure P3-4. Do this by attaching a current source with current $i(t)$ to the two terminals and then solving the resulting circuit, as in Example 3-4.

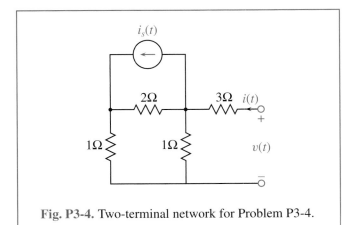

Fig. P3-4. Two-terminal network for Problem P3-4.

P3-5 Show that the two-terminal network in Figure P3-5 is equivalent to a single resistor, and calculate the value of the equivalent resistance, R_{eq}. This problem is similar to Example 3-5.

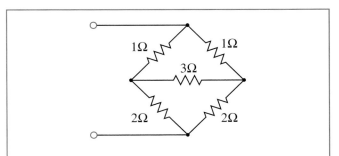

Fig. P3-5. Two-terminal network for Problem P3-5.

P3-6

(a) For the two-terminal network in Figure P3-6, find the open-circuit voltage $v_{oc}(t)$.

(b) Compute the short-circuit current, $i_{sc}(t)$.

(c) Create and sketch the Thévenin equivalent network.

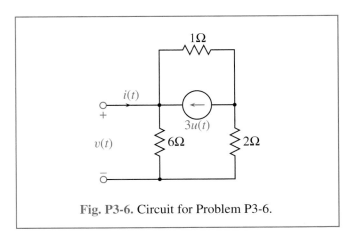

Fig. P3-6. Circuit for Problem P3-6.

P3-7 Find the Thévenin equivalent network that has the same v–i relation at its terminals as the two-terminal network in Figure P3-7. This problem is similar to Example 3-7.

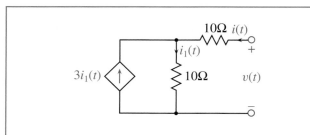

Fig. P3-7. Two-terminal network for Problem P3-7.

P3-8 Find the Norton equivalent network corresponding to the two-terminal network in Figure P3-8, using an approach similar to Example 3-8.

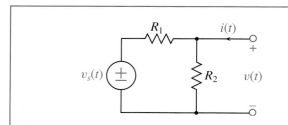

Fig. P3-8. Two-terminal network for Problem P3-8.

P3-9 Use source substitution to solve for $i(t)$ in the circuit of Figure P3-9. All of the resistors have a resistance of R ohms. This problem is similar to Example 3-9.

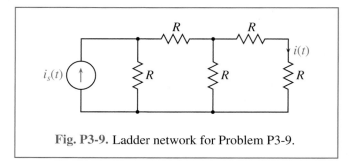

Fig. P3-9. Ladder network for Problem P3-9.

3-7-2   Basic Problems

P3-10 A switch is a device that acts like an open circuit when it is open and a short circuit when it is closed. Find the equivalent resistance R_{eq} at the indicated terminals in the circuit of Figure P3-10. All resistances are measured in ohms. *Hint:* Redraw the circuits for the two indicated cases.

(a) Find the equivalent resistance when the switch is open.

(b) Find the equivalent resistance when the switch is closed.

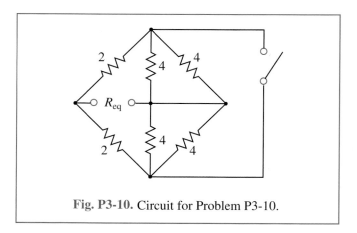

Fig. P3-10. Circuit for Problem P3-10.

P3-11 Find the equivalent resistance of the two-terminal network shown in Figure P3-11.

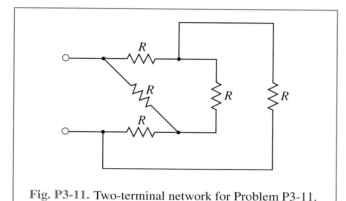

Fig. P3-11. Two-terminal network for Problem P3-11.

P3-12 The circuit of Figure P3-12 contains a potentiometer. Referring to the model of a potentiometer in Figure 3-3, assume that

$$R_1 = kR,$$

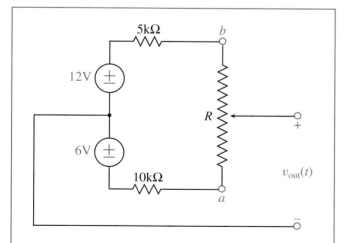

Fig. P3-12. Circuit containing a potentiometer that produces an adjustable constant voltage at its output.

where k varies over the range $0 \le k \le 1$ as the potentiometer shaft is turned. If $R = 40\,\mathrm{k\Omega}$ what are the minimum and maximum values of the output voltage, $v_{\mathrm{out}}(t)$? Notice that the labels on the terminals correspond to the labels in Figure 3-3.

P3-13 Figure P3-13 shows three potentiometers configured as variable resistors. Referring to the model of a potentiometer in Figure 3-3, assume that

$$R_1 = kR,$$

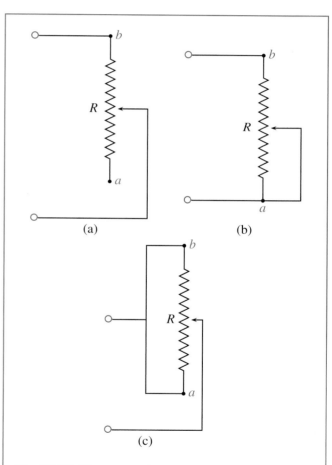

Fig. P3-13. Three potentiometers configured as variable resistors for Problem P3-13.

where k varies over the range $0 \leq k \leq 1$ as the potentiometer shaft is turned. Let $R_{eq}(k)$ denote the equivalent resistance measured at the terminals as a function of k. Notice that the labels on the terminals correspond to the labels in Figure 3-3.

(a) Plot R_{eq} as a function of k for the circuit in Figure P3-13(a).

(b) Repeat for the circuit in Figure P3-13(b). Notice that the terminals have been reversed from part (a).

(c) Repeat for the circuit in Figure P3-13(c).

P3-14

(a) Determine $i_1(t)$ (in terms of $v_s(t)$) for the circuit in Figure P3-14.

(b) Determine $i_2(t)$ (in terms of $v_s(t)$).

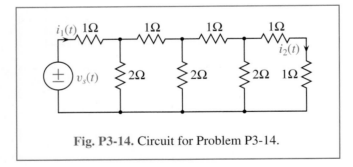

Fig. P3-14. Circuit for Problem P3-14.

P3-15   Each of the resistors in Figure P3-15 has a maximum power rating of 1 W.

(a) What is the maximum amount of power that the two-terminal network in Figure P3-15(a) can absorb without any of the resistors failing?

(b) Repeat for the two-terminal network in Figure P3-15(b).

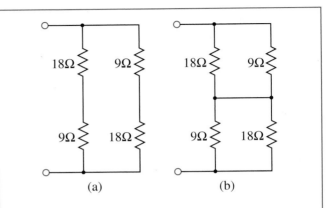

Fig. P3-15. Two-terminal networks for Problem P3-15. Each of the resistors can dissipate a maximum of 1W.

P3-16   Consider a two-terminal network consisting of two inductors with inductances L_1 and L_2 connected in series, as shown in Figure P3-16.

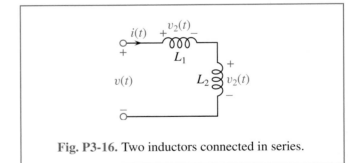

Fig. P3-16. Two inductors connected in series.

(a) Show that this network is equivalent to a single inductor—that is, that $v(t)$ is proportional to the first derivative of $i(t)$.

(b) Derive a formula for the equivalent inductance L_{eq} in terms of L_1 and L_2.

(c) Derive expressions for the voltage $v_1(t)$ measured across inductor L_1 and the voltage $v_2(t)$ measured across L_2 in terms of the voltage $v(t)$ appearing across the series connection.

P3-17 Consider a two-terminal network consisting of two inductors with inductances L_1 and L_2 connected in parallel, as in Figure P3-17.

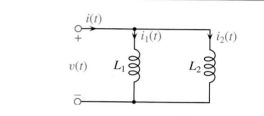

Fig. P3-17. Two inductors connected in parallel.

(a) Show that this subnetwork is network equivalent to a single inductor.

(b) Derive a formula for the equivalent inductance L_{eq} in terms of L_1 and L_2.

(c) Derive expressions for the current $i_1(t)$ that passes through inductor L_1 and the current $i_2(t)$ that passes through L_2 in terms of the current $i(t)$ entering the two-terminal network.

P3-18 Consider a two-terminal network consisting of two capacitors with capacitances C_1 and C_2 connected in series, as shown in Figure P3-18.

(a) Show that this network is equivalent to a single capacitor.

(b) Derive a formula for the equivalent capacitance C_{eq} in terms of C_1 and C_2.

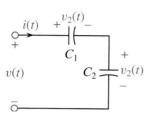

Fig. P3-18. Two capacitors connected in series.

(c) Derive expressions for the voltage $v_1(t)$ measured across capacitor C_1 and the voltage $v_2(t)$ measured across C_2 in terms of the voltage $v(t)$ appearing across the series connection.

P3-19 Consider a two-terminal network consisting of two capacitors with capacitances C_1 and C_2 connected in parallel, as shown in Figure P3-19.

(a) Show that this network is equivalent to a single capacitor.

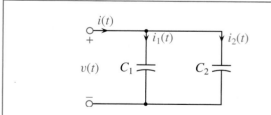

Fig. P3-19. Two capacitors connected in parallel.

(b) Derive a formula for the equivalent capacitance C_{eq} in terms of C_1 and C_2.

(c) Derive expressions for the current $i_1(t)$ that passes through capacitor C_1 and the current $i_2(t)$ that passes through C_2 in terms of the current $i(t)$ entering the network.

P3-20 Find the v–i relation of the two-terminal network in Figure P3-20.

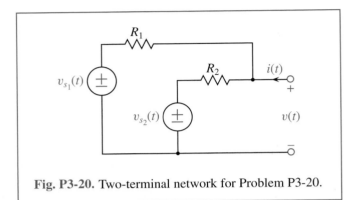

Fig. P3-20. Two-terminal network for Problem P3-20.

P3-21

Find the v–i relation of the two-terminal network shown in Figure P3-21.

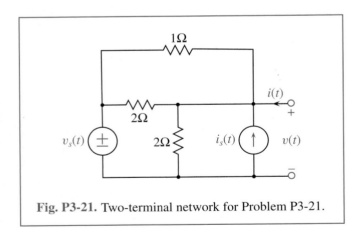

Fig. P3-21. Two-terminal network for Problem P3-21.

P3-22 Find the Thévenin equivalent of the two-terminal network "seen" by the R Ohm resistor in the network of Figure P3-22. In other words, all of the circuit *except* for the resistor R should be modeled as a two-terminal network with the terminals indicated.

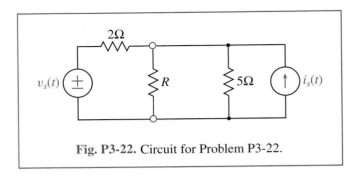

Fig. P3-22. Circuit for Problem P3-22.

P3-23 Find a Thévenin equivalent network that has the same v–i relation as the circuit in Figure P3-23.

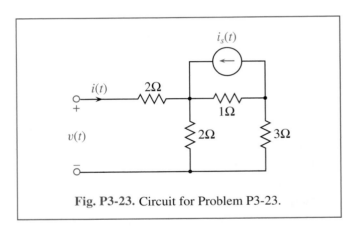

Fig. P3-23. Circuit for Problem P3-23.

P3-24 Real batteries do not behave like ideal voltage sources. The electrochemical reactions that take place within the battery can be better modeled by an ideal voltage source connected in series with a resistance as shown in Figure P3-24, where R_i is the internal resistance of the battery. In order to measure the value of R_i, a 1Ω resistor is connected across the terminals and the voltage across the resistor is measured to be 1.48V. What is the internal resistance of the battery?

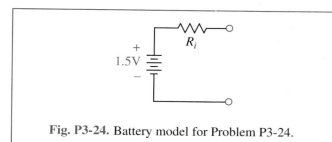

Fig. P3-24. Battery model for Problem P3-24.

P3-25 A network has the v–i characteristic shown graphically in Figure P3-25.

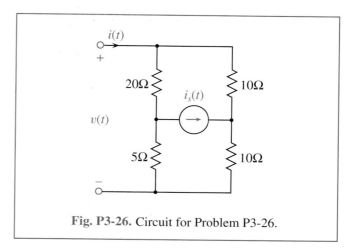

Fig. P3-26. Circuit for Problem P3-26.

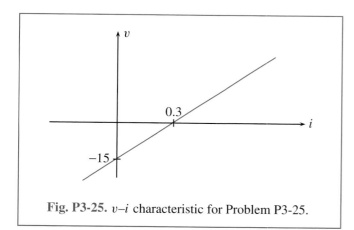

Fig. P3-25. v–i characteristic for Problem P3-25.

(a) Diagram the Thévenin equivalent model corresponding to the network.
(b) Diagram the Norton equivalent model corresponding to the network.

P3-26 Find the Norton equivalent of the two-terminal network shown in Figure P3-26.

P3-27 Sketch the Thévenin equivalent circuit corresponding to the two-terminal network in Figure P3-27.

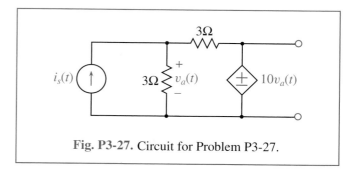

Fig. P3-27. Circuit for Problem P3-27.

P3-28

(a) Find the Thévenin equivalent network corresponding to the two-terminal circuit in Figure P3-28.
(b) Find the Norton equivalent network.

P3-29

(a) Find the v–i relation for the two-terminal network in Figure P3-29.
(b) Find and *sketch* the Norton equivalent network that has the same v–i relation.

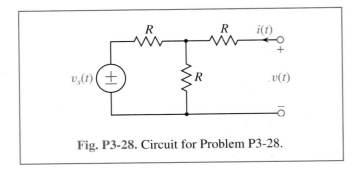

Fig. P3-28. Circuit for Problem P3-28.

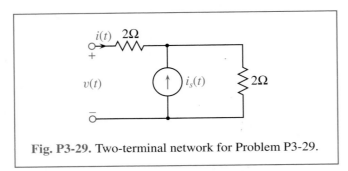

Fig. P3-29. Two-terminal network for Problem P3-29.

P3-30 Compute the power supplied by the battery in the circuit in Figure P3-30.

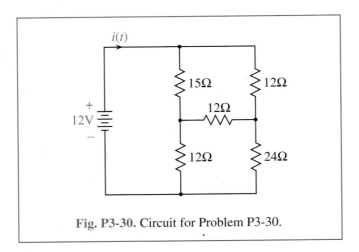

Fig. P3-30. Circuit for Problem P3-30.

3-7-3   Advanced Problems

P3-31 Find the equivalent resistance for the infinite ladder network shown in Figure P3-31. *Suggested approach:* Begin by replacing the subnetwork to the right of the resistor connecting nodes b and b' by its equivalent resistance.

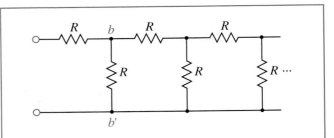

Fig. P3-31. An infinite ladder network for Problem P3-31.

P3-32 Two 2-terminal networks, N_1 and N_2, have the v–i relations

$$N_1 : v(t) = R_1 i(t) + v_{T1}$$
$$N_2 : v(t) = R_2 i(t) + v_{T2}.$$

(a) The two networks are connected in series if the same current passes through both, as illustrated in Figure P3-32(a). Find the v–i characteristic for the series connection of N_1 and N_2.

(b) The two networks are connected in parallel if the same voltage appears across the terminals of both networks, as illustrated in Figure P3-32(b). Find the v–i characteristic for the parallel connection of N_1 and N_2.

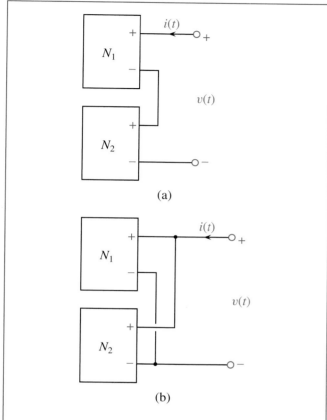

(a)

(b)

Fig. P3-32. (a) A series connection of two 2-terminal networks. (b) A parallel connection of two 2-terminal networks.

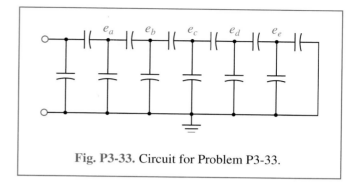

Fig. P3-33. Circuit for Problem P3-33.

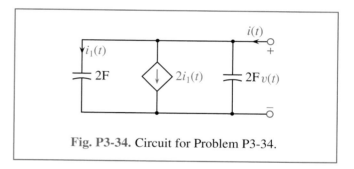

Fig. P3-34. Circuit for Problem P3-34.

P3-33 For the network in Figure P3-33, each of the capacitors has the same capacitance C.

(a) Find the equivalent capacitance measured at the terminals.

(b) If the voltage across the terminals is 72 V, compute each of the indicated node potentials.

P3-34 Show that the two-terminal network in Figure P3-34 is network equivalent to a single capacitor, and derive a formula for the equivalent capacitance.

P3-35 A two-terminal network N has the v–i characteristic

$$v_N(t) = 5i_N(t) - 3,$$

where $v_N(t)$ is the voltage across the terminals of the network and $i_N(t)$ is the current entering its + terminal. Determine the v–i characteristics of the four 2-terminal networks that are constructed from N and shown in Figure P3-35.

P3-36

(a) Solve for the voltage $v(t)$ in the circuit in Figure P3-36 in terms of $i_s(t)$ and $v_s(t)$. *Hint:* Setting up and solving a large system of linear equations is not the recommended approach.

(b) Solve for the current $i(t)$ in terms of $i_s(t)$ and $v_s(t)$.

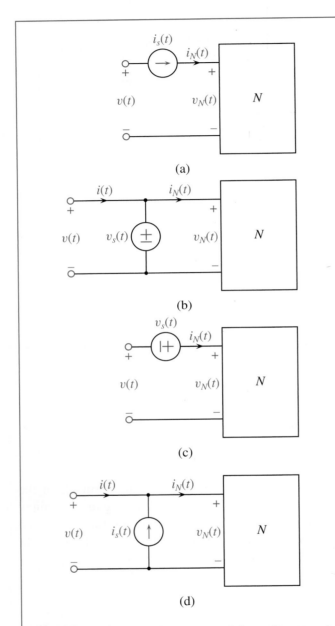

(a)

(b)

(c)

(d)

Fig. P3-35. Four circuits constructed from the network N of Problem P3-30 whose v–i relations need to be discovered.

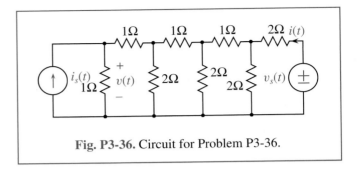

Fig. P3-36. Circuit for Problem P3-36.

P3-37 Consider the two 2-terminal networks N_1 and N_2 shown in Figure P3-37(a).

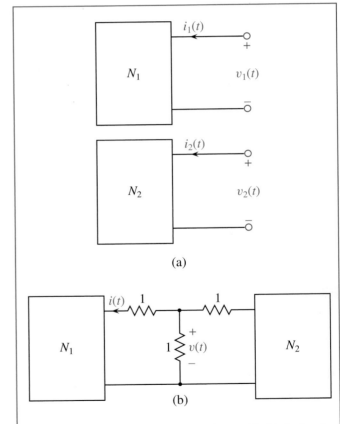

(a)

(b)

Fig. P3-37. (a) Networks for Problem P3-37. (b) A circuit constructed from the two networks in Figure P3-37(a).

The two networks have the v–i relations

$$N_1 : \quad v_1(t) = 4i_1(t) - 8$$

$$N_2 : \quad v_2(t) = 2i_2(t) + 3.$$

Figure out the waveforms $v(t)$ and $i(t)$ if the two networks are connected as shown in Figure P3-37(b).

P3-38 This circuit analyzes the model for a transistor amplifier shown encircled by the dashed box in Figure P3-38. A microphone, which is modelled as an independent voltage source connected in series with a resistance (Thévenin equivalent), is attached to the input terminals of the amplifier; an 8Ω resistor, which models a loudspeaker, is attached to the output terminals.

(a) Compute the voltage gain $G_v = v_{\text{out}}(t)/v_{\text{in}}(t)$ of the amplifier.

(b) Compute the power gain $G_p = P_{\text{out}}(t)/P_{\text{in}}(t)$.

(c) Generate a Thévenin equivalent model for the microphone and amplifier (everything to the left of the load resistor).

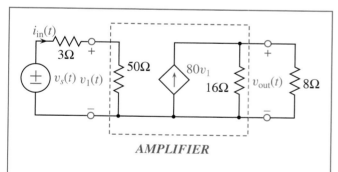

Fig. P3-38. A circuit model for an amplifier connected to a microphone and a loudspeaker.

P3-39 In an attempt to discover a Thévenin equivalent for a network containing only resistors and sources, two experiments are performed. First, a resistor with a value of R ohms is connected across the terminals, and the voltage $v_1(t)$ is measured, as shown in Figure P3-39. Then, a resistor with a value of $2R$ ohms is connected

across the terminals, and the voltage $v_2(t)$ is measured. Find the Thévenin equivalent model for the network in terms of the measured voltages $v_1(t)$ and $v_2(t)$—that is, express $v_T(t)$ and R_T in terms of the measured voltages.

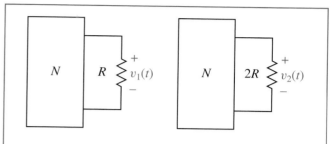

Fig. P3-39. Two experimental measurements on a circuit made to work out a Thévenin equivalent model.

P3-40 The circuit in Figure P3-40(a) contains a two-terminal subnetwork N_1 that has the v–i relation shown graphically in Figure P3-40(b).

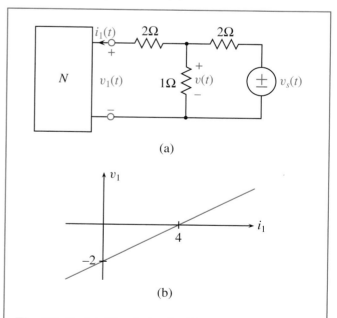

Fig. P3-40. (a) Circuit for Problem P3-40, containing a two-terminal subnetwork. (b) v–i relation for the subnetwork N in Problem P3-40.

(a) Write a mathematical expression (equation) for the v–i relation of the subnetwork.

(b) Solve for $v(t)$.

P3-41

(a) Find the v–i relation of the two-terminal network in Figure P3-41.

(b) Draw the Norton equivalent network.

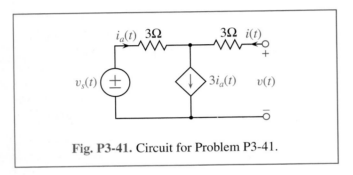

Fig. P3-41. Circuit for Problem P3-41.

P3-42 Find the Norton equivalent of the two-terminal network shown in Figure P3-42.

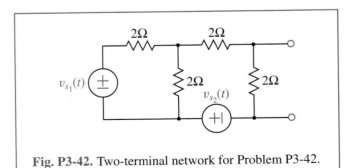

Fig. P3-42. Two-terminal network for Problem P3-42.

P3-43 The two-terminal network N_1 in Figure P3-43 has the v–i relation

$$v_1(t) = 2i_1(t) - 1.$$

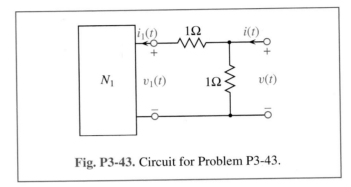

Fig. P3-43. Circuit for Problem P3-43.

(a) Sketch a two-terminal network that is the Norton equivalent of N_1.

(b) Express $v(t)$ as a function of $i(t)$.

P3-44 A load, modelled by a resistor with resistance R_L, is attached to a power source, which is modelled by the Thévenin equivalent network shown in Figure P3-44. For what value of R_L is the power delivered to the load maximized?

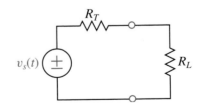

Fig. P3-44. A resistive load connected to a power source, modeled by its Thévenin equivalent network for Problem P3-44.

P3-45 Find the equivalent resistance of the two-terminal network in Figure P3-45. All resistances are in ohms.

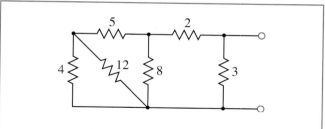

Fig. P3-45. Two-terminal network for Problem P3-45.

P3-46 The unusual-looking device in the circuit in Figure P3-46(a) is an *ideal diode*. Its v–i characteristic is shown in Figure P3-46(b). When the voltage v_d is negative, it behaves like an open circuit; otherwise it behaves like a short circuit. Find the v–i relation for the two-terminal network.

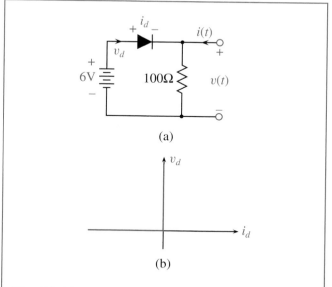

Fig. P3-46. (a) A two-terminal network containing an ideal diode. (b) The v–i characteristic of the ideal diode.

3-7-4 Design Problems

P3-47 You have a box containing an unlimited number of $10\text{k}\Omega$ resistors. Show how to connect some of these together to construct equivalent resistances with the following values:

(a) $20\,\text{k}\Omega$

(b) $25\,\text{k}\Omega$

(c) $6.667\,\text{k}\Omega$.

P3-48 You have a box containing an unlimited number of $5\,\text{k}\Omega$ resistors and another box containing an unlimited number of $2\,\text{k}\Omega$ resistors. Using resistors taken from these boxes, construct 10 two-terminal networks that have equivalent resistances of $1\,\text{k}\Omega$, $2\,\text{k}\Omega$, ..., $10\,\text{k}\Omega$. In each case, use as few resistors as possible.

P3-49 It is desired to construct a variable voltage reference that will vary over the range from -5V to 10V as the knob on the potentiometer in Figure P3-49 is turned. Select values for R_1 and R_2 that will make this possible. *Hint:* You might wish to refer to Problem P3-12 to get started.

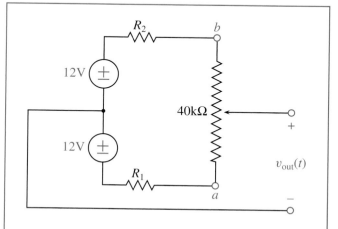

Fig. P3-49. Circuit containing a potentiometer that produces an adjustable constant voltage at its output.

CHAPTER 4

Operational Amplifiers

Objectives

By the end of this chapter, you should be able to do the following:

1. *Analyze a circuit containing one or more operational amplifiers, using the node method.*

2. *Construct circuits using operational amplifiers that will add or subtract signals, amplify signals, integrate signals, and differentiate them.*

Operational amplifiers (opamps) are extremely useful circuit elements, particularly for manipulating signal waveforms. As we shall see in this chapter, they can be used to add signals together or to multiply signals by gains. They can be configured as integrators, as differentiators, or to minimize loading effects, as buffers. Furthermore, because they can be used to isolate one part of a circuit from another, they can greatly simplify problems associated with circuit design. We shall look at

many useful operational-amplifier configurations later in the chapter, but first we need to formally introduce ideal operational amplifiers and develop some procedures that can be used to analyze circuits that contain them.

EXTRA INFORMATION: *Biomedical Amplifiers*

4-1 The Ideal Opamp

An operational amplifier is a high-gain differential amplifier that contains a number of active components (transistors) and has several terminals, as shown in Figure 4-1. At first glance the device looks complicated. It has two input signals, $v_+(t)$ and $v_-(t)$, and one output signal, $v_{\text{out}}(t)$.[1] In this respect it is different from the two-terminal elements that we have seen to this point. Because

---

[1] All three of these voltages are measured as node potentials relative to a ground somewhere in the circuit.

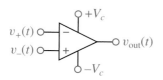

Fig. 4-1. Terminal connections for an operational amplifier.

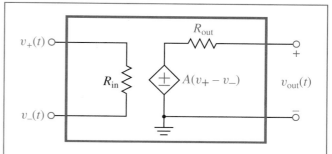

Fig. 4-2. A simple equivalent circuit model for an operational amplifier.

it is an active device that is capable of adding power to a signal, it also has connections to power sources—one to a positive supply voltage $+V_c$, one to a negative supply voltage $-V_c$. The supply voltages are constant as a function of time.

Under certain conditions, the output voltage can be written in terms of the input voltages as

$$v_{\text{out}}(t) = A[v_+(t) - v_-(t)], \qquad (4.1)$$

where the gain, A, is large, typically 10^5 or more. Thus, in terms of our network models, an operational amplifier behaves like the voltage-dependent voltage source shown in Figure 4-2. Opamps are designed so that the input resistance, R_{in}, is very high and the output resistance, R_{out}, is relatively low.

For a real operational amplifier to behave in a manner that resembles the idealized model, some constraints need to be imposed on the input and output voltages and currents. For example, electronics considerations dictate that the maximum and minimum values of the output voltage can never exceed the range defined by the power supply voltages, which are often in the range of ± 15 volts. This means that

$$|v_{\text{out}}(t)| < V_c,$$

which, in turn, implies that

$$|v_+(t) - v_-(t)| < \frac{V_c}{A}.$$

A is large, however, and V_c is finite, so a further consequence is that $|v_+(t) - v_-(t)|$ must be very small. This means that we will have

$$v_+(t) \approx v_-(t). \qquad (4.2)$$

Furthermore, since the voltage across the input terminals is small and the input resistance, R_{in} is large, the current passing through the input terminals will be extremely small.

$$i_+(t) = -i_-(t) = \frac{v_+(t) - v_-(t)}{R_{\text{in}}} \approx 0 \qquad (4.3)$$

Schematically, opamps are usually drawn without their attendant power supply connections, as in Figure 4-3. The input $v_-(t)$, which is applied to the terminal marked by the minus sign, is called the *inverting input*; the input $v_+(t)$, which is applied to the terminal marked by the plus sign, is called the *noninverting input*. The output terminal is the one connected to the vertex of the triangle.

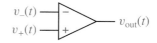

Fig. 4-3. An ideal operational amplifier.

Leaving out any specific consideration of the supply voltages simplifies our circuit drawings; nonetheless they are there. Indeed, their presence is critical to the operation of the device. This means that we need to be careful when we analyze circuits and we need to modify some of our procedures. We *cannot*, for example, draw a surface around the opamp in Figure 4-3, apply KCL, and infer that $i_{out}(t) + i_-(t) + i_+(t) = 0$, because the opamp also draws current from the power supplies and possibly from the ground, none of which are shown.

The circuit element that we shall call an *ideal operational amplifier* idealizes these characteristics. It is a three-terminal element for which:

1. The gain A is infinite;
2. The output voltage $v_{out}(t)$ and output current $i_{out}(t)$ are finite;
3. The input resistance R_{in} is infinite;
4. The output resistance R_{out} is zero;
5. $v_+(t) = v_-(t)$;
6. $i_+(t) = i_-(t) = 0$.
7. To limit saturation, the output voltage $v_{out}(t)$ must lie in the range $|v_{out}(t)| \le V_c$.

When operational amplifiers are configured as linear circuit elements, they are generally used with *negative feedback* to limit the degree of saturation. This means that there is a circuit element (not an open circuit) connecting the output of the amplifier back to the inverting input. This makes these devices quite different from other types of voltage-controlled voltage sources. With negative feedback, the performance of a circuit containing an operational amplifier is not critically dependent on the value of the gain A, as long as it is large. Operational amplifiers are also used in several nonlinear applications, such as threshold devices, where the saturation characteristic of the device is exploited directly. We will not explore these applications here, apart from a brief introduction in Problem P4-22. To

introduce the opamp, we begin with a simple example that represents a very common type of operational amplifier circuit called an *inverting amplifier*.

Figure 4-4(a) shows an operational amplifier configured as an inverting amplifier. For the moment, we shall not assume that the operational amplifier is ideal. The resistor R_2 is the feedback element that connects the output of the operational amplifier back to the inverting input. In Figure 4-4(b), we have inserted the nonideal model given in Figure 4-2 for the opamp; in part (c) of that figure, we have geometrically rearranged the elements so that the circuit has a more familiar look. The signs on the voltages might look confusing unless one remembers that the input voltage is applied to the inverting input of the opamp and the noninverting input is connected to the ground.

To solve for $v_{out}(t)$ we use the node method, writing KCL equations at the two circled nodes in Figure 4-4(c). Let the node potential at the left node be $e_a(t)$; at the right node, it is $v_{out}(t)$. We also notice that $v_1(t) = -e_a(t)$. The KCL equations are given by

$$\frac{e_a(t) - v_{in}(t)}{R_1} + \frac{e_a(t)}{R_{in}} + \frac{e_a(t) - v_{out}(t)}{R_2} = 0$$

$$\frac{v_{out}(t) - e_a(t)}{R_2} + \frac{v_{out}(t) + Ae_a(t)}{R_{out}} = 0.$$

These equations can then be put into a more useful form for solution by regrouping the terms:

$$\left[\frac{1}{R_1} + \frac{1}{R_{in}} + \frac{1}{R_2} \right] e_a(t) - \frac{1}{R_2} v_{out}(t) = \frac{1}{R_1} v_{in}(t)$$

$$\left[\frac{A}{R_{out}} - \frac{1}{R_2} \right] e_a(t) + \left[\frac{1}{R_2} + \frac{1}{R_{out}} \right] v_{out}(t) = 0.$$

To solve, we multiply the top equation by $\left[\frac{A}{R_{out}} - \frac{1}{R_2} \right]$ and the lower one by $\left[\frac{1}{R_1} + \frac{1}{R_{in}} + \frac{1}{R_2} \right]$ and then subtract

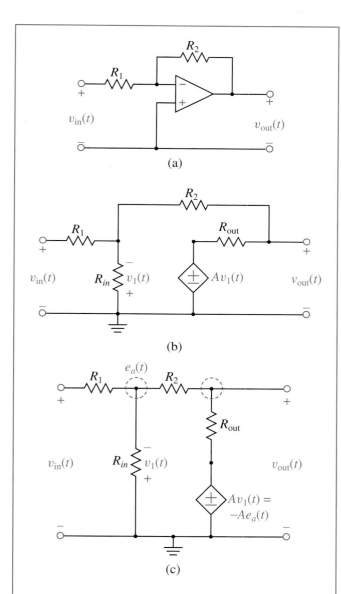

(a)

(b)

(c)

Fig. 4-4. An inverting amplifier implemented using an operational amplifier. (b) Circuit with a model for the opamp inserted. (c) Rearrangement of the circuit in (b).

the two equations to eliminate $e_a(t)$. Finally, solving the resulting equation for $v_{out}(t)$ we get

$$v_{out}(t) =$$

$$\frac{\left[\frac{1}{R_2} - \frac{A}{R_{out}}\right]\frac{1}{R_1}}{\left[\frac{1}{R_1} + \frac{1}{R_2} + \frac{1}{R_{in}}\right]\left[\frac{1}{R_2} + \frac{1}{R_{out}}\right] - \left[\frac{1}{R_2} - \frac{A}{R_{out}}\right]\frac{1}{R_2}} v_{in}(t).$$

(4.4)

Now let us make the assumption that A is large. Those terms containing A will dominate the terms to which they are added. This simplifies (4.4) considerably:

$$v_{out}(t) \approx -\frac{\dfrac{A}{R_{out}}\dfrac{1}{R_1}}{\dfrac{A}{R_{out}}\dfrac{1}{R_2}} v_{in}(t) = -\frac{R_2}{R_1} v_{in}(t) \qquad (4.5)$$

We observe that, *as long as the output voltage is not allowed to saturate,* this circuit is a voltage amplifier, because the output voltage is proportional to the input voltage. The magnitude of the gain of the amplifier is R_2/R_1 and the sign of the gain is negative, which is why this configuration is called an ***inverting amplifier.*** We also see that the size of the gain is controlled by the values of the two resistors—the input resistor R_1 and the feedback resistor R_2—that we placed in the circuit. If these two resistance values are precisely controlled, we have precise control over the gain of the amplifier. We should also notice what is missing in (4.5): The value of the voltage gain does not depend on the actual value of A, except that it should be large, and there is no dependence on the values of the internal parameters of the model, R_{in} and R_{out}. This means that the operational amplifier itself does not need to be a precision component.

The procedure that we used to derive (4.5) was an awkward one. It involved first replacing the amplifier by a model with multiple elements, finding an exact solution to the resulting circuit, and then approximating that solution by letting the gain of the amplifier become large. Effectively, we started with a simple problem, found the solution to a more complicated one, and then simplified the result by idealizing the opamp. It is much easier to incorporate the properties of the ideal element from the beginning. This is what we do in the next section.

 DEMO: *Operational Amplifier Circuits*

4-2   The Node Method for Opamp Circuits

In dealing with a circuit containing operational amplifiers, we need to be aware of a number of the properties of these devices that we enumerated earlier. First, since the amplifier gain is infinite,

$$v_a(t) = v_b(t);$$

that is, the voltages at the noninverting and inverting input terminals are virtually equal.[2] Second, the current flowing into (or out of) these terminals is negligible. Thus,

$$i_a(t) = 0; \qquad i_b(t) = 0.$$

Finally, there will usually be a component of the output current that comes from the power supplies and ground connections. Because these connections are ignored in our model, KCL equations that we write at these nodes will not be accurate, and we need to avoid writing them. Thus, *for circuits containing ideal opamps, we will not write KCL equations at the output terminal of the opamp or at the ground node.* Fortunately, these equations are not necessary.

Of the two systematic procedures for finding equilibrium solutions that we developed in Chapter 2, the node method and the mesh method, only the node method can easily be modified to deal with these extra

constraints. This is because the definitions of the mesh currents automatically satisfy KCL at all of the nodes of the circuit, including the output terminal of the opamp and the ground node; yet, as we have already mentioned, KCL will not be satisfied at these nodes unless the power supplies are taken into account. Without an explicit modelling of the power supplies, therefore, the mesh method will produce an incorrect solution. Thus, we limit our solution methods to the node method. We begin by reexamining the earlier example of an inverting amplifier.

Example 4-1   Inverting Amplifier

Let us look at the inverting amplifier again using the node method and an ideal model for the opamp. The circuit is redrawn in Figure 4-5 to identify the ground and highlight the node at which KCL needs to be applied. If we denote the node potential (measured relative to the ground) at the circled node as $e^-(t)$, then the KCL equation at that node becomes

$$\frac{1}{R_1}[e^-(t) - v_{\text{in}}(t)] + \frac{1}{R_2}[e^-(t) - v_{\text{out}}(t)] = 0. \quad (4.6)$$

Notice that we have not included the current entering the opamp, since, for an ideal opamp, it is zero; all of the current that flows through resistor R_1 also flows through R_2. Furthermore, if $e^+(t)$ and $e^-(t)$ are the potentials at the noninverting and inverting terminals, respectively,

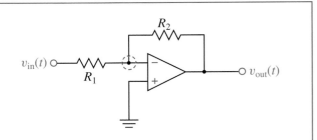

Fig. 4-5. The inverting amplifier redrawn to focus attention on the input node.

---

[2]With a nonideal opamp, there may be a difference between these voltages of a few microvolts, but this difference will be negligible in comparison with other voltages in the circuit.

A Confusing Issue: Writing KCL Equations at the Input Nodes

The model for an ideal operational amplifier specifies that the voltage across its two input terminals must be zero. This, in turn implies that the node potentials at those two terminals will be the same. Thus, if we know the value of one of them—say, because it is grounded—then we also know the value of the other. If we do not know either node potential, then we must assign the same variable to each. The fact that we assign the same node potential to the two different nodes, however, does not change the fact that **these are still two separate nodes and so we must write two independent KCL equations at them, not one.**

To illustrate this point, consider the circuit below, which contains two input signals and a single output.

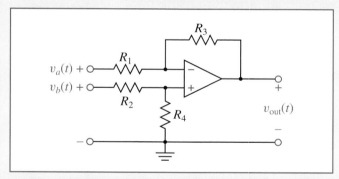

To solve this circuit via the node method, the first step is to identify a potential with each node. The ground node is already identified, and for the three remaining nodes associated with terminals, we can use the input and output variables $v_a(t)$, $v_b(t)$, and $v_{out}(t)$. That leaves only the two nodes connected to the inverting and noninverting input nodes of the operational amplifier. Since we know that these two nodes must have the same potential, but we do not yet know what that potential is, we treat it as an unknown variable, $e(t)$.

Having identified or defined all of the node potentials, the next step is to write KCL equations at all of the nodes, except for the output node of the operational amplifier and the ground node, in terms of the node potentials. In this case there will be two KCL equations—one at the inverting input of the opamp and one at the noninverting input. This is consistent with the fact that there are two unknowns—the node potential at each of the nodes $e(t)$ and the output voltage $v_{out}(t)$. Even though the two nodes where we are writing the equations have the same potential, they have independent KCL equations. If we were to combine these into a single equation, we would not have sufficient equations to solve the circuit.

For the inverting ($-$) node, we know that the current entering the node through R_1 must equal the current leaving through R_3. (Remember that the current entering the terminals of an ideal opamp is zero.) Thus,

$$\frac{1}{R_1}[v_a(t) - e(t)] + \frac{1}{R_3}[v_{out}(t) - e(t)] = 0.$$

At the noninverting node, we write a similar equation:

$$\frac{1}{R_2}[v_b(t) - e(t)] - \frac{1}{R_4}e(t) = 0.$$

Next we need to solve these equations. From the second equation, we can write $e(t)$ in terms of $v_b(t)$:

$$e(t) = \frac{R_4}{R_2 + R_4}v_b(t).$$

Notice that, since there is no current entering the opamp, we could have obtained this same result by using a voltage divider. We can now substitute for $e(t)$ into the first KCL equation. After algebraic manipulation, this gives the final result:

$$v_{out}(t) = \frac{R_3}{R_1}v_a(t) - \frac{R_4(R_1 + R_3)}{R_1(R_2 + R_4)}v_b(t).$$

then $e^+(t) = e^-(t)$; but, the positive terminal is grounded, so $e^-(t) = 0$. This reduces (4.6) to

$$-\frac{1}{R_1} v_{\text{in}}(t) - \frac{1}{R_2} v_{\text{out}}(t) = 0$$

or

$$v_{\text{out}}(t) = -\frac{R_2}{R_1} v_{\text{in}}(t),$$

as before. By incorporating the properties of the ideal opamp into the analysis early, the analysis via the node method becomes quite simple.

Before we leave this example, it is important to notice the role played by the negative feedback. Since all of the current that flows through R_1 also flows through R_2, we can write

$$\frac{v_{\text{in}} - e^-}{R_1} = \frac{e^- - v_{\text{out}}}{R_2}$$

or

$$e^- = \frac{R_2}{R_1 + R_2} v_{\text{in}} + \frac{R_1}{R_1 + R_2} v_{\text{out}}. \tag{4.7}$$

It also must be true, however, that

$$v_{\text{out}} = -Ae^-,$$

which follows from (4.1) with $e^+ = 0$. If e^- is perturbed so that it becomes slightly positive, then v_{out} will become more negative, and, through (4.7), this will reduce the value of e^- and effectively offset the perturbation. The feedback will similarly counteract a negative perturbation so that e^- is always driven back to zero. ∎

 To check your understanding, try Drill Problem P4-1.

 WORKED SOLUTION: *Node Method for Opamps*

The modification to the node method that was applied to this problem will always work. Recall that the node method required that, after we first identify a supernode encircling each voltage source, we define a node potential at each node of the network. Then we write a KCL equation at every node but one. If the network contains n nodes/supernodes, this results in $n - 1$ equations to be solved for $n - 1$ variables. For a circuit that contains m operational amplifiers, the number of KCL equations is reduced to $n - 1 - m$, because of the restriction that we not write KCL equations at the ground or at the m output nodes of the opamps, but the number of independent node potentials is correspondingly reduced, because, for each opamp, the potential at the node connected to the noninverting input must be the same as the potential at the node connected to the inverting input. This means that we will still get enough independent KCL equations to be able to find the solution. We shall see several examples of this methodology in the additional examples of the next section.

 FLUENCY EXERCISE: *Inverting Amplifier*

4-3   Additional Examples of Operational-Amplifier Circuits

In this section, we will give several examples of circuits incorporating operational amplifiers. We have two motivations. The first is the instructional goal that is achieved by looking at how the node method is applied in several examples. The second is that these examples are widely used for combining, amplifying, and filtering waveforms. The operational amplifier is a particularly useful circuit element, and many of these examples will reappear in later chapters.

4-3-1   Noninverting Amplifier

As a second example, we consider the noninverting amplifier of Figure 4-6. Notice, in this case, that the

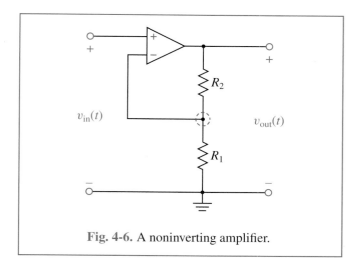

Fig. 4-6. A noninverting amplifier.

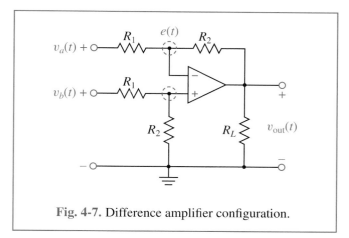

Fig. 4-7. Difference amplifier configuration.

input signal is applied to the noninverting terminal of the operational amplifier, but that there is still a feedback element connecting the output to the inverting input terminal. As before, we avoid writing KCL equations at the output terminal and the ground. This leaves only the node indicated by the dashed circle. The potential of that node is $v_{in}(t)$, since it is connected to the inverting terminal of the opamp, the voltage across the input terminals of the opamp is zero, and the noninverting terminal has the potential $v_{in}(t)$. No current flows into the opamp, so the single KCL equation will involve only two currents; all of the current flowing down through R_2 will also flow down through R_1. From KCL,

$$\frac{1}{R_2}[v_{in}(t) - v_{out}(t)] + \frac{1}{R_1}v_{in}(t) = 0.$$

This gives

$$v_{out}(t) = \left[1 + \frac{R_2}{R_1}\right]v_{in}(t).$$

This configuration also corresponds to an amplifier, but notice that the gain is positive, whereas with the inverting configuration, it was negative. With this (noninverting) configuration, however, the magnitude of the gain is

constrained to be greater than 1, whereas, with the inverting configuration it was unconstrained. As with the inverting amplifier, this relation will define the output voltage only if saturation is avoided. This means that the maximum value of the input waveform must be no greater than the saturation voltage of the opamp divided by the gain of the circuit, $1 + R_2/R_1$.

4-3-2   Difference Amplifier

The input signal was applied to the noninverting terminal of the opamp for the noninverting amplifier configuration and to the inverting terminal for the inverting configuration. If we apply input signals to both terminals indirectly, we can amplify their difference. This can be done by using the differential amplifier configuration shown in Figure 4-7.

For this circuit, there are two nodes where KCL needs to be applied, which are indicated by the dashed circles. The potentials at these two nodes must be equal, since each is the potential at one of the input terminals of the operational amplifier. Since that node potential is initially unknown, we denote it by $e(t)$. We can solve for $e(t)$ in one of the KCL equations and then substitute that value into the other to find the desired relationship between

$v_a(t)$, $v_b(t)$, and $v_{\text{out}}(t)$. At the inverting (upper) node, KCL says

$$\frac{1}{R_1}[e(t) - v_a(t)] + \frac{1}{R_2}[e(t) - v_{\text{out}}(t)] = 0.$$

Rearranging terms gives

$$e(t)\left[\frac{1}{R_1} + \frac{1}{R_2}\right] = \frac{1}{R_1}v_a(t) + \frac{1}{R_2}v_{\text{out}}(t). \qquad (4.8)$$

At the noninverting (lower) node, the KCL equation is

$$\frac{1}{R_1}[e(t) - v_b(t)] + \frac{1}{R_2}e(t) = 0.$$

A similar simplification yields

$$e(t)\left[\frac{1}{R_1} + \frac{1}{R_2}\right] = \frac{1}{R_1}v_b(t). \qquad (4.9)$$

Since the left-hand sides of (4.8) and (4.9) are the same, we can equate their right-hand sides to simplify the result and get the relation that we have been seeking:

$$v_{\text{out}}(t) = \frac{R_2}{R_1}[v_b(t) - v_a(t)].$$

The output of the operational amplifier is equal to the difference between the two input signals, amplified by a factor that can be controlled by the resistances R_1 and R_2. Furthermore, we notice that this output voltage is independent of the value of the load resistor R_L.

In this example, we have assumed that the two resistors with a value of R_1 and the two with a value of R_2 are *matched*. The case where each of these four resistors has a different value is worked out in the first "Confusing Issue" of this chapter.

   WORKED SOLUTION: *Difference Amplifier*

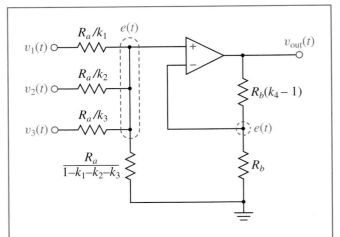

Fig. 4-8. An operational amplifier configured as a summing amplifier.

4-3-3   Summing Amplifier

Figure 4-8 shows an operational amplifier configured as a summing amplifier, with three inputs: $v_1(t)$, $v_2(t)$, and $v_3(t)$. Notice that the resistor values are defined relative to two basic resistances, R_a and R_b. The values of the constants k_1, k_2, k_3, and k_4 can be chosen by the user, but must be subject to the following constraints:

$$k_1 > 0; \quad k_2 > 0; \quad k_3 > 0;$$
$$k_1 + k_2 + k_3 < 1;$$
$$k_4 > 1.$$

These constants serve to weight the input voltages. The constraints are necessary to ensure that all of the resistance values are positive. Again, notice that a feedback element is present.

This circuit contains two nodes at which we must apply Kirchhoff's current law. These are the nodes enclosed by the dashed lines. Since each of these nodes is connected to one of the inputs of the ideal operational amplifier,

A Confusing Issue: Opamp Output Current

In our modification of the node method for analyzing a circuit containing one or more operational amplifiers, we proscribed writing a KCL equation at the node connected to the output of the opamp or at the ground. What do we do, however, when we want to learn the output current of the opamp? This is clearly a meaningful quantity for many problems. Since KCL is still valid at the output node of the amplifier, we can simply use that KCL equation to find the output current **after** we have solved for the node potentials in the circuit. Writing a KCL equation at the output node before we have solved the circuit is not helpful, because, along with that additional equation, we will have to introduce an additional variable (the output current). The situation is analogous to the one that we encounter when solving a general circuit with the node method when we are seeking the current flowing through a voltage source.

To illustrate the procedure, consider the inverting amplifier below. We would like to compute the total output current of the amplifier, $i(t)$.

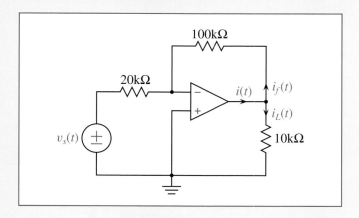

Clearly, it is not difficult to compute the output current **after** we have determined the node potentials, since we can use the node potentials to compute the currents through the 100 kΩ feedback resistor, $i_f(t)$, and the 10 kΩ load resistor, $i_L(t)$. Then KCL can be applied at the output node to find the output current:

$$i(t) = i_f(t) + i_L(t).$$

To determine the node potentials, we use our standard analysis approach. First we define the potential at the output node to be $v_{\text{out}}(t)$, and we observe that the current at the inverting input node is zero. The KCL equation at that node is

$$\frac{1}{20,000} v_s(t) + \frac{1}{100,000} v_{\text{out}}(t) = 0,$$

which gives $v_{\text{out}}(t) = 5 v_s(t)$. Given this value, the load current, $i_L(t)$, is

$$i_L(t) = \frac{1}{10,000} v_{\text{out}}(t) = 0.0005\, v_s(t).$$

The feedback current is

$$i_f(t) = \frac{1}{100,000} v_{\text{out}}(t) = 0.00005\, v_s(t)$$

and

$$i(t) = (0.0005 + 0.00005) v_s(t) = 0.00055\, v_s(t).$$

their node potentials are again seen to be equal; let this common node potential be $e(t)$. At the node on the left, we get

$$\frac{e(t) - v_1(t)}{\frac{R_a}{k_1}} + \frac{e(t) - v_2(t)}{\frac{R_a}{k_2}}$$

$$+ \frac{e(t) - v_3(t)}{\frac{R_a}{k_3}} + \frac{e(t)}{\frac{R_a}{1-k_1-k_2-k_3}} = 0.$$

If we multiply both sides of this expression by R_a, rewrite the fractions, and group terms, it simplifies to

$$e(t) = k_1 v_1(t) + k_2 v_2(t) + k_3 v_3(t). \qquad (4.10)$$

Applying KCL to the node on the right gives

$$\frac{e(t) - v_{\text{out}}(t)}{R_b(k_4 - 1)} + \frac{1}{R_b} e(t) = 0.$$

Now we multiply both sides by $R_b(k_4 - 1)$ and solve for $v_{\text{out}}(t)$:

$$v_{\text{out}}(t) = k_4 e(t).$$

Finally, substituting from (4.10), we obtain

$$v_{\text{out}}(t) = k_4[k_1 v_1(t) + k_2 v_2(t) + k_3 v_3(t)]. \qquad (4.11)$$

This shows that the output voltage is a weighted linear combination of the input voltages, the weights being controlled by the four constants. For this configuration, all of the weights are constrained to be positive, but an alternative configuration that borrows from this structure and that of the differential amplifier can be used in the more general case. (See Problem P4-40.)

Notice that the result is independent of the actual values of R_a and R_b. This is true only because the operational amplifier is assumed to be ideal. When a real opamp is used, its limitations could constrain the values of these resistors somewhat. In particular, the amount of current that the opamp can supply to its output will be limited, thereby putting a lower limit on the value of

R_b. In addition, if R_a is too small, the currents entering through the input terminals might load the input sources excessively. The issue of loading is discussed in the next subsection.

4-3-4   Buffer Amplifiers and Loading

The circuit in Figure 4-9 is called a *voltage follower* or *buffer amplifier*. It perhaps appears to do very little, since

$$v_{\text{out}}(t) = v_{\text{in}}(t), \qquad (4.12)$$

but, in fact, buffer amplifiers are frequently used to prevent loading, because the current entering the input terminal is minimal, while the current provided at the output can be much greater. Loading can be explained in terms of the circuits in Figure 4-10.

Referring to the circuit at the top of that figure, we see that, when there is no current entering the terminals, we can compute the output voltage, $v_a(t)$, by using a voltage divider:

$$v_a(t) = \frac{60\text{k}\Omega}{60\text{k}\Omega + 20\text{k}\Omega} v_{\text{in}}(t) = \frac{3}{4} v_{\text{in}}(t).$$

Now assume that these terminals are connected to a load, as shown in part (b), where we have modelled the load by a $30\,\text{k}\Omega$ resistor. Because current flows through the load resistor, the voltage across the load, $v_b(t)$, will be different from $v_a(t)$, which assumed that all of the current flowed through the $60\,\text{k}\Omega$ resistor. For this simple circuit,

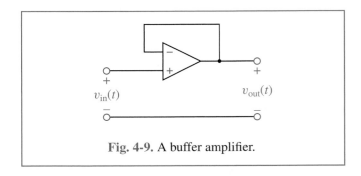

Fig. 4-9. A buffer amplifier.

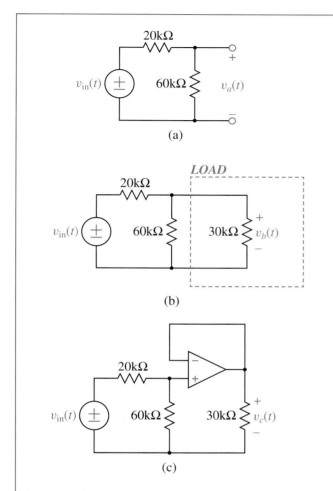

(a)

LOAD

(b)

(c)

Fig. 4-10. An illustration of the use of a buffer amplifier to prevent loading. (a) An unloaded circuit. (b) The circuit in (a) after the application of a resistive load. (c) The circuit in (a) in which the load is isolated from the remainder of the circuit through a buffer amplifier.

$v_b(t)$ is also simple to calculate. The $30 \, \text{k}\Omega$ resistor in parallel with the $60 \, \text{k}\Omega$ resistor is equivalent to a $20 \, \text{k}\Omega$ resistor. Applying the voltage divider in this case gives

$$v_b(t) = \frac{1}{2} \, v_{\text{in}}(t).$$

Because of the current draw through the load, the voltage that appears across it is reduced. This effect is known as *loading*.

One way to prevent loading is with the use of a buffer amplifier. This is the situation that is depicted in Figure 4-10(c). Since the current entering the operational amplifier is negligible, the voltage at the input terminal to the opamp is the same as $v_a(t)$, and, from (4.12),

$$v_c(t) = v_a(t) = \frac{3}{4} \, v_{\text{in}}(t).$$

The current that passes through the load is provided by the operational amplifier instead of by the voltage source. Of course, there are limits. As we commented at the beginning of this chapter, there is a maximum value of output current that can be supplied.

While loading is an important effect for resistive circuits, it is even more important in circuits involving capacitors and inductors because it can change the frequency response of a circuit. We shall see operational amplifiers used as buffers in later chapters.

    WORKED SOLUTION: *Buffer Amplifier*

4-3-5   Voltage-Controlled Current Source

Both voltage sources and current sources can be used to inject signals into circuits, but voltage sources are far more common. For those occasions when we would like to have a current as the input to a circuit rather than a voltage, we might use a voltage-controlled current source, such as the one shown in Figure 4-11. The output of interest here is the current $i_{\text{out}}(t)$, which flows into a load that is modelled here as a resistor with resistance R_L. We would like to show that the value of this current is independent of the value of R_L. We shall also show that it is proportional to the voltage $v_{\text{in}}(t)$. Thus, while we have chosen to call this device a voltage-controlled

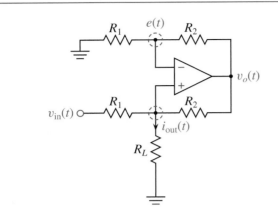

Fig. 4-11. A voltage-controlled current source implemented via an operational amplifier.

current source, we could have equally well chosen to call it a voltage-to-current converter.

We can write KCL equations at the two circled nodes. Because the potentials at these nodes are the potentials at the two input nodes of the opamps, they must be equal. Let $e(t)$ denote this common potential. From the upper node,

$$\frac{1}{R_1}e(t) + \frac{1}{R_2}[e(t) - v_o(t)] = 0, \qquad (4.13)$$

and from the lower one,

$$\frac{1}{R_1}[e(t) - v_{\text{in}}(t)] + \frac{1}{R_2}[e(t) - v_o(t)]$$
$$+ i_{\text{out}}(t) = 0. \quad (4.14)$$

Now, we can rewrite (4.13) as

$$e(t)\left[\frac{1}{R_1} + \frac{1}{R_2}\right] = \frac{1}{R_2}v_o(t),$$

from which we find that

$$v_o(t) = R_2\left[\frac{1}{R_1} + \frac{1}{R_2}\right]e(t).$$

Substituting this value into (4.14) gives

$$\frac{1}{R_1}[e(t) - v_{\text{in}}(t)] + i_{\text{out}}(t) + \frac{1}{R_2}e(t)$$
$$- \left[\frac{1}{R_1} + \frac{1}{R_2}\right]e(t) = 0.$$

We can cancel out the common terms to simplify this equation:

$$-\frac{1}{R_1}v_{\text{in}}(t) + i_{\text{out}}(t) = 0.$$

So, finally,

$$i_{\text{out}}(t) = \frac{1}{R_1}v_{\text{in}}(t),$$

which is independent of R_L and proportional to $v_{\text{in}}(t)$, as was desired.

Although we have not dwelled on the distinction between our ideal devices and real ones, we need to be aware of these distinctions. In particular, recall that the maximum voltage at the terminals of the opamp can be no higher than the supply voltages. This limits the maximum voltage that can be applied to the load. There is also a limit to the amount of current that this circuit can supply.

4-3-6   A "Negative Resistor"

The network in Figure 4-12 is a two-terminal network for which the voltage across the terminals is proportional to the current entering the network, but for which the proportionality constant is negative. In this respect, the circuit behaves like a resistor with a negative resistance. Normal resistors absorb power, as we have seen, but negative resistors supply it. Notice that the input signal is the current entering the terminals at the left and that the output signal is the voltage across those terminals.

The two nodes on which we can write KCL equations are circled. As before, we notice that the potentials at the two nodes are equal and that each is equal to $v_{\text{out}}(t)$. Let

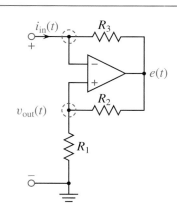

Fig. 4-12. A two-terminal network that looks like a resistor with a negative resistance.

the potential at the output of the operational amplifier be $e(t)$. Then, at the upper circled node, we have

$$\frac{1}{R_3}[v_{out}(t) - e(t)] = i_{in}(t)$$

$$\Longrightarrow e(t) = v_{out}(t) - R_3 i_{in}(t).$$

At the lower node,

$$\frac{1}{R_1}v_{out}(t) + \frac{1}{R_2}[v_{out}(t) - e(t)] = 0.$$

If we substitute for $e(t)$ into this equation and perform the obvious algebraic simplifications, it reduces to

$$v_{out}(t) = -\left(\frac{R_1 R_3}{R_2}\right) i_{in}(t),$$

which looks like the v–i relation of a resistor with a negative resistance, since $v_{out}(t)$ and $i_{in}(t)$ are measured at the same terminal pair, although we observe that one terminal of the negative resistor must be grounded.

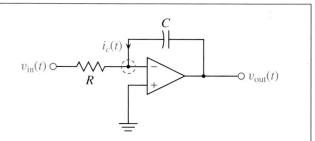

Fig. 4-13. An operational amplifier configured as an integrator.

4-3-7   Integrator

Figure 4-13 shows an operational amplifier configured like an inverting amplifier except that the feedback resistor has been replaced by a feedback capacitor. With this arrangement of elements, the opamp functions as an *integrator*. We have deferred discussing circuits containing capacitors and inductors to this point, because these introduce integrals and derivatives that change the circuit equations from algebraic equations into differential ones, but this particular circuit is simple enough to analyze at this time.

We again write a KCL equation at the input node. Observe that the potential at that node is zero because of the virtual ground. Let $i_c(t)$ temporarily denote the current entering that node through the capacitor. Then

$$\frac{1}{R}v_{in}(t) + i_c(t) = 0.$$

Recalling the element relation for a capacitor, we have

$$i_c(t) = C\frac{d\,v_c(t)}{dt} = C\frac{d\,v_{out}(t)}{dt},$$

and inserting this result into the KCL equation gives

$$\frac{1}{R}v_{in}(t) + C\frac{d\,v_{out}(t)}{dt} = 0.$$

We can solve this equation for $v_{out}(t)$. We have

$$v_{out}(t) = -\frac{1}{RC}\int_{t_0}^{t} v_{in}(\tau)\,d\tau + v_{out}(t_0),$$

for $t > t_0$, where $v_{out}(t_0)$ is the capacitor voltage at time $t = t_0$. For this circuit, the output voltage is proportional to the (negative of the) integral of the input voltage. The output voltage of a nonideal operational amplifier will saturate if it grows too large in magnitude, so this circuit will work as an integrator only if its input does not contain a constant component.

4-3-8   Differentiator

What happens if we interchange the resistor and the capacitor, as shown in Figure 4-14? The result is a circuit whose output voltage is proportional to the *derivative* of the input.

   We again evaluate a KCL equation at the input node, where the node potential is again zero. The current that leaves the node through the resistor is equal to the current that enters through the capacitor:

$$C\frac{d\,v_{in}(t)}{dt} = -\frac{1}{R}v_{out}(t)$$

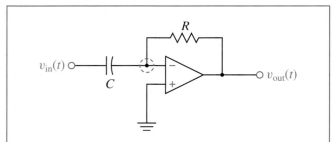

Fig. 4-14. An operational amplifier configured as a differentiator.

or

$$v_{out}(t) = -RC\frac{d\,v_{in}(t)}{dt}.$$

  FLUENCY EXERCISE: *Opamp Circuits*

4-4   Chapter Summary

4-4-1   Important Points Introduced

- An operational amplifier is a high-gain differential amplifier that must be connected to a power supply.

- An ideal operational amplifier contains three terminals—an inverting input terminal, a noninverting input terminal, and an output terminal.

- Operational amplifiers are frequently used with an element, voltage source, or short circuit connecting the output terminal to the inverting input terminal.

- The currents entering the two input terminals of an ideal operational amplifier are zero; the potentials of the two terminals are equal.

- In analyzing circuits containing operational amplifiers, we must avoid writing KCL equations at the output terminal of the amplifier and at the ground.

- Operational amplifier circuits are analyzed by using the node method. Special care must be taken to assign the node potentials correctly and to identify those node potentials which must be equal because they are attached to the two input terminals of the opamp.

- Operational amplifiers can be used as buffer amplifiers (isolating networks) to prevent loading.

4-4-2 New Abilities Acquired

You should now be able to do the following:

(1) Analyze a circuit containing one or more operational amplifiers, using the node method.

(2) Analyze and design inverting and noninverting amplifiers that use an operational amplifier as the central circuit element.

(3) Analyze and design circuits for adding weighting and for adding a number of signals together by using an operational amplifier.

(4) Analyze and design circuits for subtracting two signals by using an operational amplifier.

(5) Analyze circuits for integrating and differentiating a signal by using an operational amplifier.

4-5 Problems

4-5-1 Drill Problem

P4-1 Express the output voltage $v_{out}(t)$ in terms of the input voltage $v_{in}(t)$ for the circuit in Figure P4-1.

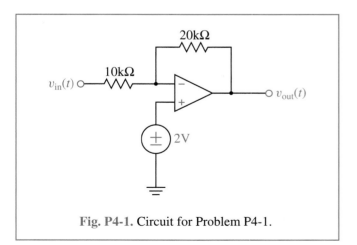

Fig. P4-1. Circuit for Problem P4-1.

4-5-2 Basic Problems

P4-2 Express $v_{out}(t)$ in terms of $v_{in}(t)$ for the circuit in Figure P4-2.

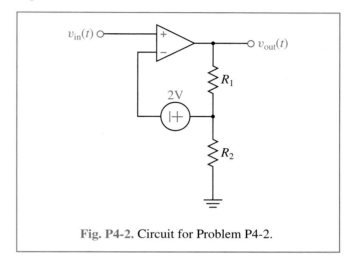

Fig. P4-2. Circuit for Problem P4-2.

P4-3 In the circuit in Figure P4-3,

(a) Circle the nodes at which you can write valid KCL equations.

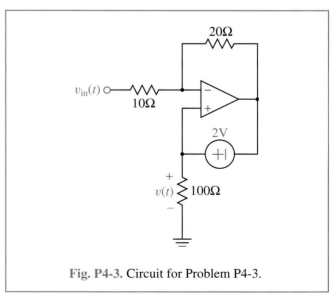

Fig. P4-3. Circuit for Problem P4-3.

(b) Label the potential of each node in the circuit.

(c) Solve for $v(t)$.

P4-4 Express $v_{out}(t)$ in terms of $i_s(t)$ and $v_s(t)$ for the circuit in Figure P4-4.

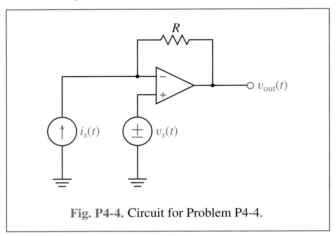

Fig. P4-4. Circuit for Problem P4-4.

P4-5 For the circuit in Figure P4-5, do the following:

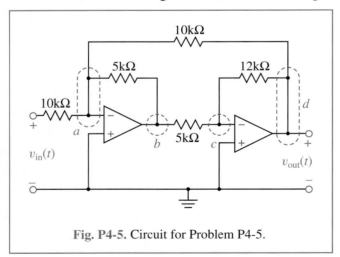

Fig. P4-5. Circuit for Problem P4-5.

(a) Indicate the electric potential at each of the four indicated nodes. For each potential that is unknown before the circuit is analyzed, give that potential a symbolic name, such as $e_a(t)$.

(b) Identify the nodes at which you can write a meaningful KCL equation.

(c) Express $v_{out}(t)$ in terms of $v_{in}(t)$.

P4-6 Find the output voltage $v_{out}(t)$ in terms of the input voltage $v_{in}(t)$ for the circuit in Figure P4-6.

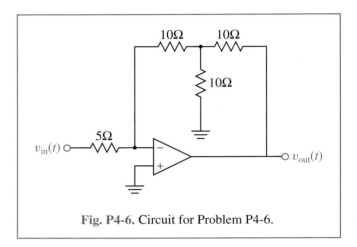

Fig. P4-6. Circuit for Problem P4-6.

P4-7 Express the output voltage $v_{out}(t)$ in terms of the input voltage $v_{in}(t)$ for the circuit in Figure P4-7.

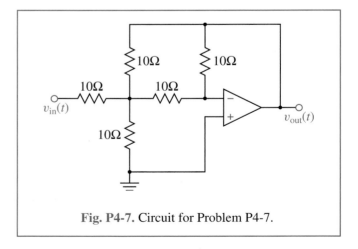

Fig. P4-7. Circuit for Problem P4-7.

P4-8 Compute the output voltage $v_{out}(t)$ for the circuit in Figure P4-8.

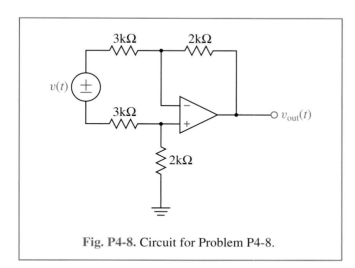

Fig. P4-8. Circuit for Problem P4-8.

P4-9 Express the output voltage $v_{out}(t)$ in terms of the input voltage $v_{in}(t)$ for the circuit in Figure P4-9.

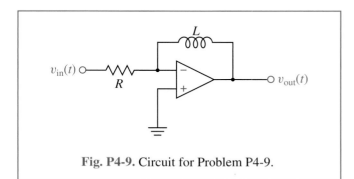

Fig. P4-9. Circuit for Problem P4-9.

P4-10 Express the output voltage $v_{out}(t)$ in terms of input voltage $v_{in}(t)$ for the circuit in Figure P4-10.

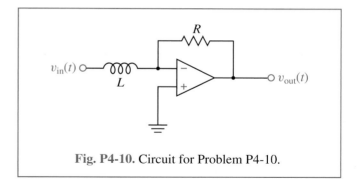

Fig. P4-10. Circuit for Problem P4-10.

P4-11 For the circuit in Figure P4-11, find the input–output relationship between $v_{out}(t)$ and $v_{in}(t)$.

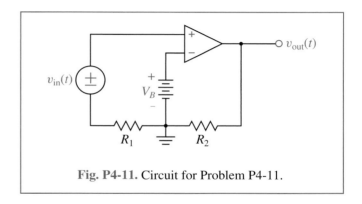

Fig. P4-11. Circuit for Problem P4-11.

P4-12 Express the output voltage $v_{out}(t)$ in terms of the input voltage $v_{in}(t)$ for the circuit in Figure P4-12.

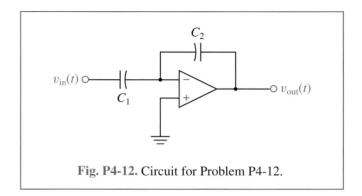

Fig. P4-12. Circuit for Problem P4-12.

P4-13 Express the output voltage $v_{out}(t)$ in terms of the input voltage $v_{in}(t)$ for the circuit in Figure P4-13.

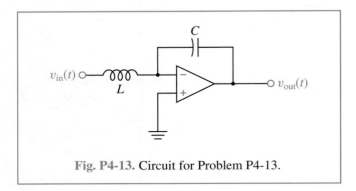

Fig. P4-13. Circuit for Problem P4-13.

P4-14 Find the output voltage $v_{out}(t)$ in terms of $v_{in}(t)$ for the circuit in Figure P4-14.

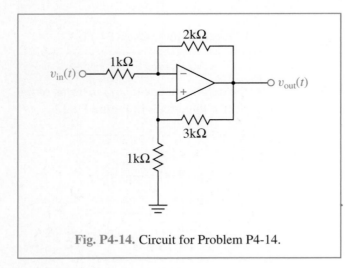

Fig. P4-14. Circuit for Problem P4-14.

P4-15 Find the voltage gain of the circuit in Figure P4-15. The voltage gain is defined as $G = v_{out}(t)/v_{in}(t)$.

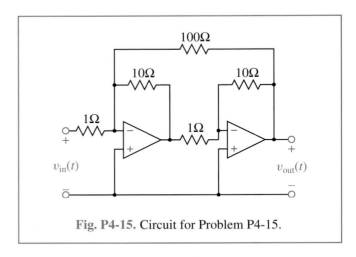

Fig. P4-15. Circuit for Problem P4-15.

P4-16 Express $v_{out}(t)$ in terms of $v_{in}(t)$ for the circuit in Figure P4-16.

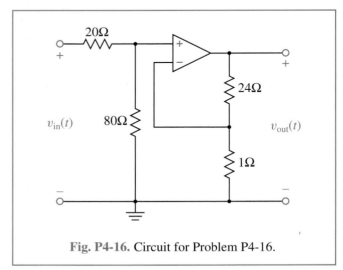

Fig. P4-16. Circuit for Problem P4-16.

P4-17

(a) Express $v(t)$ for the circuit in Figure P4-17 in terms of $v_s(t)$ and the resistor values.

(b) Express $i(t)$ in that same circuit in terms of $v_s(t)$ and the resistor values.

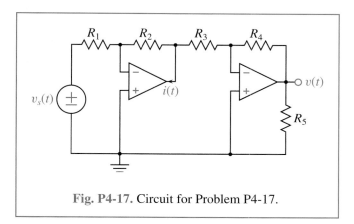

Fig. P4-17. Circuit for Problem P4-17.

P4-18 Express $v_{out}(t)$ in terms of $v_{in}(t)$ for the circuit in Figure P4-18.

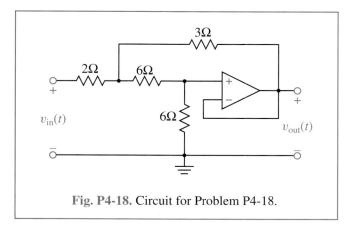

Fig. P4-18. Circuit for Problem P4-18.

P4-19 Express the output voltage $v_{out}(t)$ for the circuit in Figure P4-19 in terms of $v_{in}(t)$.

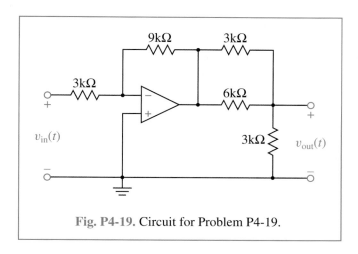

Fig. P4-19. Circuit for Problem P4-19.

P4-20 Express the current $i_{out}(t)$ as a function of the input voltage $v_{in}(t)$ for the circuit in Figure P4-20.

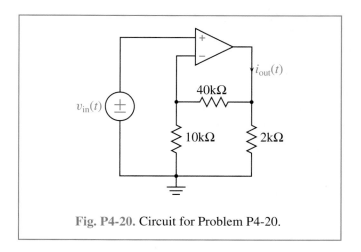

Fig. P4-20. Circuit for Problem P4-20.

P4-21 The amplifier in Figure P4-21 is to be used to amplify cardiac signals as part of an electrocardiograph (ECG). The differential gain is defined as

$$G = \frac{v_{\text{out}}(t)}{v_1(t) - v_2(t)}.$$

(a) If $R_1/R_2 = R_4/R_3$, find $v_{\text{out}}(t)$ when $v_1(t) = v_2(t)$.

(b) Find the value of the differential gain when $R_1/R_2 = R_4/R_3$.

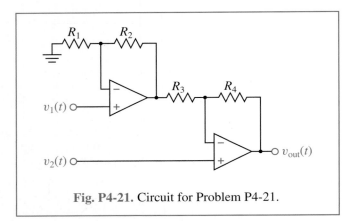

Fig. P4-21. Circuit for Problem P4-21.

4-5-3   Advanced Problems

P4-22 The voltage at the output terminal of an operational amplifier saturates at $\pm V_{\text{sat}}$, where V_{sat} is determined by the power-supply voltages. When an opamp saturates, it no longer performs as a linear device. The periodic input to each of the four opamp circuits shown in Figure P4-22(a)–(d) is the voltage waveform $v_{\text{in}}(t)$ drawn in Figure P4-22(e). For each circuit find and sketch the output waveform $v_{\text{out}}(t)$. In each circuit, the input is applied at the terminal on the left and the output is the voltage measured with respect to the ground at the terminal on the right.

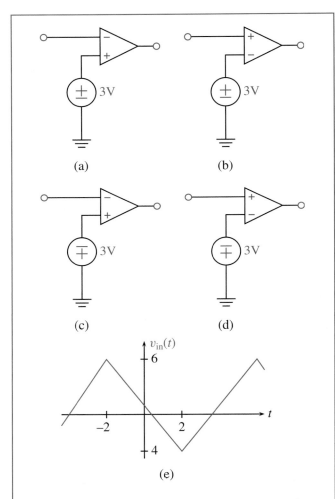

Fig. P4-22. (a)–(d) Circuits for Problem P4-22. (e) Input voltage waveform for Problem P4-22.

P4-23 Find $v_o(t)$ for the circuit in Figure P4-23 in terms of the input voltages $v_a(t)$ and $v_b(t)$. The potentials at all of the terminals are measured with respect to a ground that is not shown.

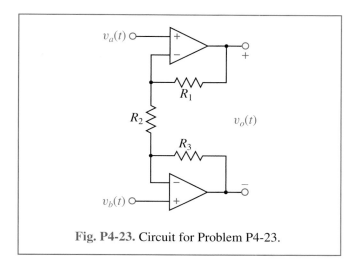

Fig. P4-23. Circuit for Problem P4-23.

P4-24 Express $v_{\text{out}}(t)$ as a function of $v_1(t)$ and $v_2(t)$ for the network in Figure P4-24.

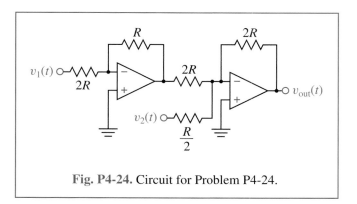

Fig. P4-24. Circuit for Problem P4-24.

P4-25 Express the output voltage $v_{\text{out}}(t)$ in terms of the input voltages $v_1(t)$ and $v_2(t)$ for the circuit containing three operational amplifiers in Figure P4-25. All resistances are measured in kilohms.

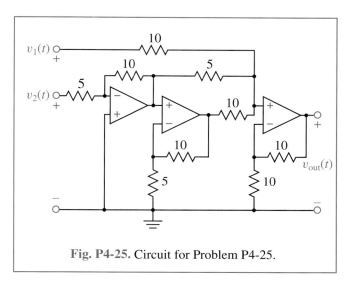

Fig. P4-25. Circuit for Problem P4-25.

P4-26 Express the output $v_{\text{out}}(t)$ of the circuit of Figure P4-26 in terms of the inputs $v_a(t)$ and $v_b(t)$.

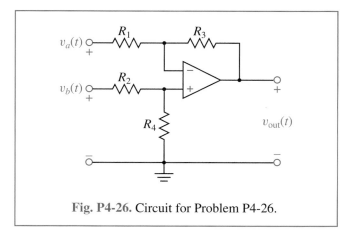

Fig. P4-26. Circuit for Problem P4-26.

P4-27 Compute the output voltage for the circuit in Figure P4-27.

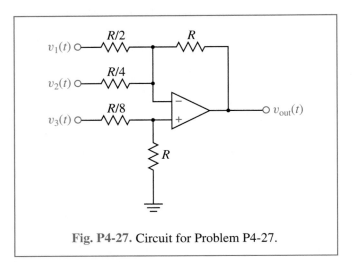

Fig. P4-27. Circuit for Problem P4-27.

P4-28 Use the node method to solve for the voltage gain of the circuit in Figure P4-28.

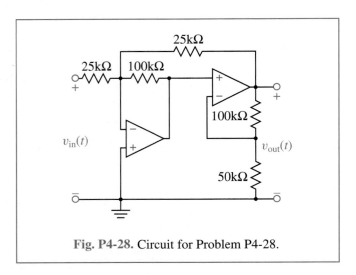

Fig. P4-28. Circuit for Problem P4-28.

P4-29 The two circuits in Figure P4-29(a) and (b) correspond to implementations of current sources. They differ in the direction of the current.

(a) Find the current $i_L(t)$ for the circuit in Figure P4-29(a) and show that it is independent of the load resistance R_L.

(b) Find the current $i_L(t)$ for the circuit in Figure P4-29(a) and show that it is also independent of the load resistance R_L.

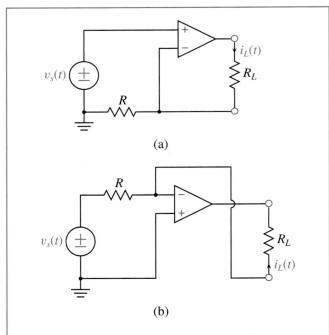

Fig. P4-29. Two implementations of current sources for Problem P4-29.

P4-30 Find the Thévenin equivalent network corresponding to the circuit in Figure P4-30 at the indicated terminals.

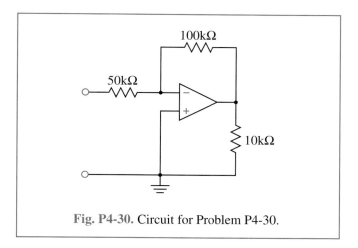

Fig. P4-30. Circuit for Problem P4-30.

P4-31 Find the Thévenin equivalent network corresponding to the circuit in Figure P4-31 at the indicated terminals.

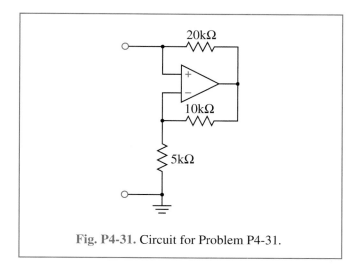

Fig. P4-31. Circuit for Problem P4-31.

P4-32 Find the Thévenin equivalent network corresponding to the circuit in Figure P4-32 at the indicated terminals.

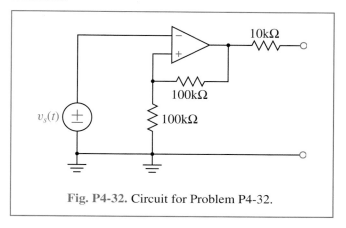

Fig. P4-32. Circuit for Problem P4-32.

P4-33 The circuit in Figure P4-33 is a current amplifier. Compute the value of the current gain,

$$G = \frac{i_L(t)}{i_s(t)}.$$

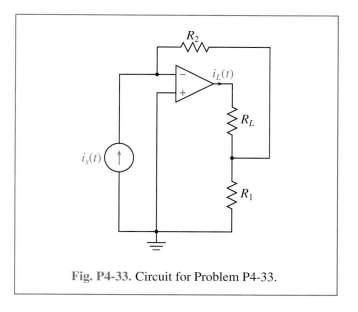

Fig. P4-33. Circuit for Problem P4-33.

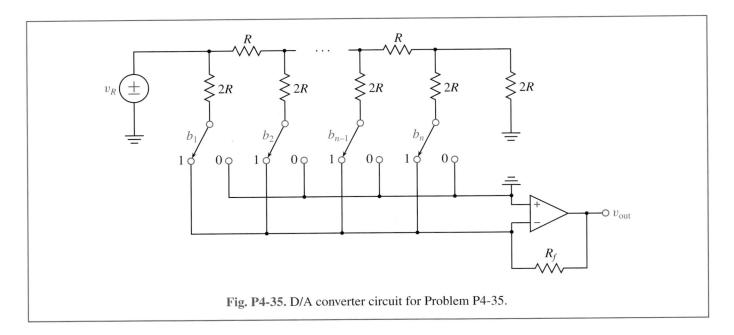

Fig. P4-35. D/A converter circuit for Problem P4-35.

P4-34   The circuit in Figure P4-34 contains a *Wheatstone bridge*, consisting of three equal resistors with resistance R and a fourth with resistance $R + \Delta R$.

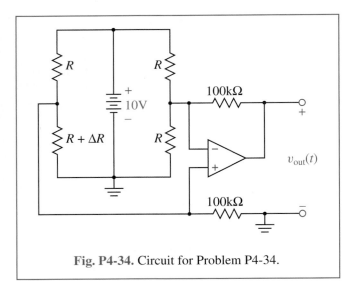

Fig. P4-34. Circuit for Problem P4-34.

The variable resistor is usually a transducer whose resistance varies in proportion to some outside variable, such as temperature or pressure. The bridge is connected to a battery and to an opamp. Show that, for $\Delta R << R$, the output voltage of the opamp is proportional to the deviation in the resistance, ΔR.

P4-35   The circuit shown Figure P4-35 is called an inverted R–$2R$ ladder digital-to-analog converter (DAC). The input to this circuit is a binary code represented by $b_1, b_2, \ldots, b_n$, where b_i is either 1 or 0. Each switch shown in the figure is controlled by one of the bits in the binary code. If $b_i = 1$, that switch will be in the '1' position; if $b_i = 0$, that switch will be in the '0' position. Depending on the position of the switch, each current i_k is diverted either to true ground (adding to the $+$ terminal of the opamp) or to the virtual ground (adding to the $-$ terminal.)

(a) If i is the current flowing out of the voltage source, show that $i = v_R/R$ regardless of the digital input code.

(b) Show that the output voltage can be expressed as

$$v_{out}(t) = -\frac{R_f}{R} v_R \left(b_1 2^{-1} + b_2 2^{-2} + \ldots \right.$$
$$\left. + b_{n-1} 2^{-n+1} + b_n 2^{-n} \right)$$

4-5-4 Design Problems

P4-36 Design a circuit, using a single operational amplifier, for which the output voltage $v_{out}(t)$ is equal to $3v_1(t) + 2v_2(t)$, where $v_1(t)$ and $v_2(t)$ are two input voltages to the circuit.

P4-37 Design a circuit, using a single operational amplifier, for which the output voltage $v_{out}(t)$ is equal to $3v_1(t) - 2v_2(t)$, where $v_1(t)$ and $v_2(t)$ are two input voltages to the circuit.

P4-38 Design circuits using resistors and operational amplifiers to implement each of the following input-output relationships:

(a) $v_{out}(t) = 4v_1(t) - 4v_2(t)$;

(b) $v_{out}(t) = v_1(t) - 2v_2(t)$;

(c) $v_{out}(t) = 2v_1(t) + 4v_2(t) - v_3(t)$.

P4-39 Design a circuit containing a single operational amplifier that will produce an output voltage $v_{out}(t)$ that is the integral of the difference of two input voltages, $v_a(t)$ and $v_b(t)$. Verify that your circuit works.

P4-40

(a) Design a circuit with four inputs $v_1(t)$, $v_2(t)$, $v_3(t)$, and $v_4(t)$ containing a single operational amplifier, such that the output voltage $v_{out}(t)$ satisfies

$$v_{out}(t) = k_5[k_1 v_1(t) + k_2 v_2(t) - k_3 v_3(t) - k_4 v_4(t)]$$

for positive values of the four constants. *Hint:* You might begin by designing a circuit that combines features of the differential-amplifier configuration and the summing-amplifier configuration.

(b) Prove that your design in (a) works correctly.

(c) State any additional constraints that need to be imposed on the five constants for the circuit to be buildable.

P4-41 Design a circuit that will produce an output voltage that is the average of several input voltages. For N inputs $v_1(t)$, $v_2(t)$, $\ldots$, $v_N(t)$, the average is

$$v_o(t) = \frac{1}{N}(v_1(t) + v_2(t) + \cdots + v_N(t)).$$

Use only 10Ω resistors in your design. One suggested approach would be to use a single operational amplifier which is constructed as a noninverting summing amplifier. The final value of the gain can be corrected by using a voltage divider. Verify that your design produces the proper output signal.

P4-42 An electrocardiograph (ECG) is used to measure the voltage generated by the human heart during the cardiac cycle. An ECG lead II signal is a differential signal measured between the right arm and left leg of a patient. Assuming that the ECG signal has a peak-to-peak amplitude of 20 mV and that the impedance of each skin electrode is 10kΩ, design a two-stage amplifier that will yield an output signal suitable for use as input to a 5 V peak-to-peak A/D converter.

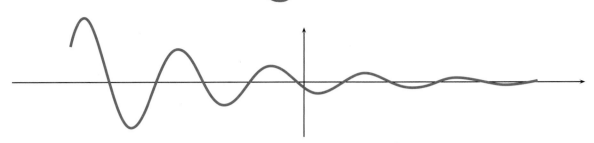

Laplace Transforms

Objectives

By the end of this chapter, you should be able to do the following:

1. *Express a sinusoidal signal as a sum of complex exponential functions.*

2. *Construct the phasor representation associated with a sinusoidal signal.*

3. *Find the Laplace transform of a signal by evaluating the Laplace transform integral.*

4. *Compute the Laplace transform of a signal by using a few basic known transforms and the properties of the Laplace transform.*

5. *Compute the signal associated with a particular rational Laplace transform.*

6. *Manipulate singularity functions, using their special properties.*

The treatment to this point has dealt almost exclusively with resistive circuits, because we wanted to focus on the distinct roles played by the element relations and by Kirchhoff's laws in discovering the complete solution of a circuit. For a resistive circuit, this task can be accomplished via familiar algebraic equations. Unfortunately, circuits that contain only resistors and sources are somewhat limited in what they can do, because the output signals (element variables) are always simple linear combinations of the input signals (independent sources). Circuits become much more powerful, and much more useful, when we add circuit elements that can store energy, such as inductors and capacitors (collectively called *reactive elements*). Unfortunately, adding inductors and capacitors to a circuit introduces derivatives and integrals into the element relations, as we have seen. These turn the algebraic equations that define the complete solution of the circuit into differential equations. To solve for the output signals then requires that we be able to solve these differential equations. The Laplace transform is a powerful tool that will help us to either solve those equations or avoid them completely.

Among its many uses, the Laplace transform is an extremely useful tool for solving linear differential equations with constant coefficients—the kind of equations that result from networks containing resistors, inductors, capacitors, operational amplifiers, and independent and dependent sources. That is why we want to look at Laplace transforms now. But the Laplace transform, like its discrete-time counterpart, the z-transform, has many more uses as well. By looking at a circuit in the Laplace domain we can understand how it will respond to both sinusoidal and transient source waveforms; we can learn how quickly the circuit will respond to sudden changes in the sources or to changes in the circuit; and we can predict how the behavior of the circuit will change in response to small perturbations in the element values.

The approach that we shall be using to analyze circuits with reactive elements is illustrated in Figure 5-1. Assume that a circuit contains a single independent

source whose underlying signal waveform is $x(t)$. Such a signal is usually referred to as its *input*. Let $y(t)$ be the element variable that is of primary interest to us in the circuit, which we denote as the *output*. The input and output can be either voltages or currents. Because both $x(t)$ and $y(t)$ are expressed as functions of time, we say that $x(t)$ and $y(t)$ are *time-domain descriptions* of these waveforms. In general, $x(t)$ and $y(t)$ are related by a differential equation with constant coefficients of the form

$$a_n \frac{d^n y(t)}{dt^n} + \cdots + a_1 \frac{dy(t)}{dt} + a_0 y(t)$$
$$= b_m \frac{d^m x(t)}{dt^m} + \cdots + b_1 \frac{dx(t)}{dt} + b_0 x(t). \quad (5.1)$$

If we are given $x(t)$, we can find $y(t)$ by solving the differential equation. This is the path indicated by the dashed line on the left. Our alternative will be to compute the Laplace transform of $x(t)$, denoted by $X(s)$, use $X(s)$ to compute the Laplace transform of the output signal $Y(s)$, and then compute the inverse Laplace transform of $Y(s)$ to get $y(t)$. The final result will be the same whichever path we follow, but if we follow the path indicated by solid lines, all of the operations can be performed by using straightforward algebra.

This chapter will introduce the Laplace transform as a mathematical tool. In order to understand this tool, we shall set circuits aside for the moment. We will come back to them in Chapter 6, when we use the Laplace transform to find the complete solution of circuits containing capacitors and inductors.

The Laplace transform is a complex function of a complex variable and we will make extensive use of complex numbers in the remainder of this book. We assume that you are familiar with manipulating complex numbers and with converting between their Cartesian (real–imaginary) and polar (magnitude–angle)

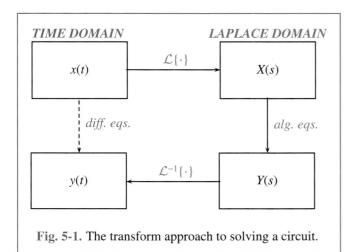

Fig. 5-1. The transform approach to solving a circuit.

representations. If you feel that you need to brush up on these topics, they are reviewed in Appendix A at the end of the book. We begin the chapter by introducing an extremely useful class of signals, the complex exponential time functions.

5-1   Some Basic Signals

5-1-1   Sinusoids

Sinusoids are among the most basic and most useful signals encountered in engineering. Mathematically, we can express a sinusoid as

$$x(t) = A \cos(\omega_0 t + \phi); \qquad (5.2)$$

graphically, it looks like the waveform in Figure 5-2.

A sinusoid is completely described by three parameters: its *amplitude* A, its *frequency* $\omega_0 = 2\pi f_0$, and its *phase* ϕ. Sinusoids are bounded in magnitude; the amplitude measures the height of the oscillations and the frequency measures how rapidly they occur. For the sinusoid in (5.2),

$$|x(t)| \leq A.$$

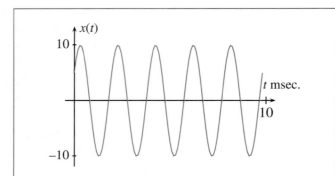

Fig. 5-2. The sinusoidal signal with $A = 10$, $\omega_0 = 2\pi[500]$, and $\phi = \pi/3$, plotted for $t \geq 0$.

A sinusoid is an example of a *periodic* waveform, because it repeats itself in time. More formally, a signal is periodic if it can be written as

$$x(t) = x(t + T), \qquad \text{for all } t. \qquad (5.3)$$

Any value of $T > 0$ for which (5.3) holds is called a *period* of the periodic signal $x(t)$. For the sinusoid in (5.2), the minimum period is

$$T = \frac{2\pi}{\omega_0} = \frac{1}{f_0},$$

although $2T$, $3T$, etc. are also periods of $x(t)$ (by our definition). The frequency $f_0 = 1/T$, which is measured in Hertz (Hz), is the reciprocal of the minimum period. It tells how fast the waveform oscillates. The quantity $\omega_0 = 2\pi f_0$, which is measured in radians per second, is called the *radian frequency*. Sometimes, however, we will be a little sloppy and also refer to this quantity simply as the frequency. There will be no sloppiness with the symbols, however. We shall use the Greek letter ω (possibly with subscripts) to denote radian frequencies and the Roman letter f, also possibly with subscripts, to denote frequencies measured in Hz; the two frequencies are related by the scaling factor 2π.

The final parameter in (5.2), ϕ, is called the *phase*. It corresponds to the angle of the cosine at $t = 0$. Changing the phase is equivalent to shifting the sinusoid on the time axis:

$$x(t) = A \cos(\omega_0 t + \phi) = A \cos(\omega_0[t + \phi/\omega_0]).$$

It will be important, for circuit applications, to notice that the derivative of a sinusoid is also a sinusoid at the same frequency, but with a different amplitude and phase, because

$$\frac{d}{dt}[A \cos(\omega_0 t + \phi)] = -A\omega_0 \sin(\omega_0 t + \phi)$$

$$= [A\omega_0] \cos(\omega_0 t + [\phi + \pi/2]).$$

Euler's relation allows us to express a sinusoid as a sum of two *complex exponential time functions*:

$$x(t) = A\cos(\omega_0 t + \phi) = \frac{A}{2}\left[e^{j(\omega_0 t + \phi)} + e^{-j(\omega_0 t + \phi)}\right]$$

$$= \frac{A}{2}e^{j\phi}e^{j\omega_0 t} + \frac{A}{2}e^{-j\phi}e^{-j\omega_0 t}$$

$$= \frac{1}{2}Xe^{j\omega_0 t} + \frac{1}{2}X^*e^{-j\omega_0 t} \qquad (5.4)$$

$$= \Re e(Xe^{j\omega_0 t}). \qquad (5.5)$$

The complex number $X = Ae^{j\phi}$ is called the *phasor*, or *complex amplitude*, associated with the sinusoid. If we consider the graph for the exponential

$$\hat{x}(t) = Ae^{j\phi}e^{j\omega_0 t},$$

we see that

$$|\hat{x}(t)| = A \quad \forall t$$

and that

$$\arg(\hat{x}(t)) = \omega_0 t + \phi.$$

The vector corresponding to $\hat{x}(t)$ traces out a circle of radius A in the complex plane as time increases. It rotates counterclockwise at a uniform rate of ω_0 radians per second. The phasor $Ae^{j\phi}$ is equal to the value of the complex exponential time function at $t = 0$. In Cartesian form,

$$\hat{x}(t) = Ae^{j\phi}e^{j\omega_0 t} = A\cos(\omega_0 t + \phi) + jA\sin(\omega_0 t + \phi).$$

The sum or difference of two sinusoids with the same frequency is another sinusoid at that frequency. Before we work out the general case, let us consider a specific example.

Example 5-1   Adding Sinusoids

Express $x(t) = \cos(5t) + 2\sin(5t)$ as a single sinusoid.

The first step is to find the phasor representation of each signal—that is, to write each in the form $\Re e(Xe^{j5t})$ for some X. For the cosine, this is easy, since

$$\cos(5t) = \Re e(1 \cdot e^{j5t}),$$

which follows from Euler's relation. For the other signal, because

$$\sin(5t) = \cos\left(5t - \frac{\pi}{2}\right),$$

we have

$$2\sin(5t) = \Re e(2e^{-j(\pi/2)}e^{j5t}).$$

Then

$$x(t) = \Re e(1 \cdot e^{j5t}) + \Re e(2e^{-j(\pi/2)}e^{j5t})$$

$$= \Re e([1 + 2e^{-j(\pi/2)}]e^{j5t})$$

$$= \Re e(\sqrt{5}e^{-j\tan^{-1}2}e^{j5t})$$

$$= \sqrt{5}\cos(5t - \tan^{-1}2)$$

$$= 2.236\cos(5t - 1.1071).$$

■

To check your understanding, try Drill Problem P5-1.

Now let us consider the general case. To compute the amplitude and phase of the sum sinusoid, which are functions of the magnitudes and phases of the sinusoids being added, it is easiest to work with the phasor representation. Let $x_1(t) = A_1\cos(\omega t + \phi_1)$ and $x_2(t) = A_2\cos(\omega t + \phi_2)$ be the sinusoids to be added. Notice that they have the same frequency. Then

$$x_1(t) + x_2(t) = \Re e(A_1 e^{j\phi_1}e^{j\omega t}) + \Re e(A_2 e^{j\phi_2}e^{j\omega t})$$

$$= \Re e([A_1 e^{j\phi_1} + A_2 e^{j\phi_2}]e^{j\omega t})$$

$$= \Re e(X e^{j\omega t}),$$

where

$$X = A_1 e^{j\phi_1} + A_2 e^{j\phi_2}$$
$$= A_1 \cos(\phi_1) + jA_1 \sin(\phi_1)$$
$$+ A_2 \cos(\phi_2) + jA_2 \sin(\phi_2)$$
$$= A e^{j\phi}.$$

The phasor associated with the sum sinusoid is thus seen to be equal to the sum of the phasors associated with the two sinusoids that are added. We can get the amplitude A and phase ϕ of the sum sinusoid from its phasor:

$$A = \sqrt{A_1^2 + 2A_1 A_2 \cos(\phi_1 - \phi_2) + A_2^2} \qquad (5.6)$$

and

$$\phi = \tan^{-1} \frac{A_1 \sin(\phi_1) + A_2 \sin(\phi_2)}{A_1 \cos(\phi_1) + A_2 \cos(\phi_2)}. \qquad (5.7)$$

Putting the pieces together, we get the final result

$$x_1(t) + x_2(t) = A \cos(\omega t + \phi). \qquad (5.8)$$

Rather than try to remember and use this result as a formula, it is recommended that one remember the process and go through that process whenever sinusoids need to be added: find the phasors associated with the sinusoids, then add these to get the phasor associated the sum sinusoid, as was done in the example.

 DEMO: *Sinusoids*

5-1-2   Exponentially Weighted Sinusoids

An *exponentially weighted sinusoid* is a signal of the form

$$x(t) = A e^{\sigma t} \cos(\omega t + \phi). \qquad (5.9)$$

These differ from ordinary sinusoids because they include the damping term $e^{\sigma t}$, where σ is a real number. When $\sigma = 0$, this term becomes a constant and the exponentially weighted sinusoid reduces to an ordinary sinusoid. Exponentially weighted sinusoids thus represent a generalization of the class of sinusoidal signals or, equivalently, ordinary sinusoids are a special case of exponentially weighted sinusoids. When σ is negative, the most common case, the amplitude of the signal decreases as t increases. When $\sigma > 0$, then the signal grows in amplitude with increasing t. Figure 5-3 shows a graph of an exponentially weighted sinusoidal waveform that has a negative value of σ. The term $A e^{\sigma t}$ can be interpreted as a time-varying amplitude; it is also a bound on the signal, because the signal values always lie in the range $\pm A e^{\sigma t}$ and the signal oscillates

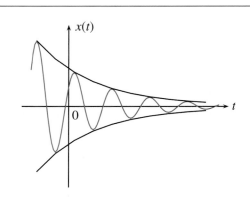

Fig. 5-3. An exponentially weighted sinusoid drawn for the case $\sigma < 0$. The waveform oscillates between the curves $x_u(t) = A e^{-\sigma t}$ and $x_\ell(t) = -A e^{-\sigma t}$, shown in black.

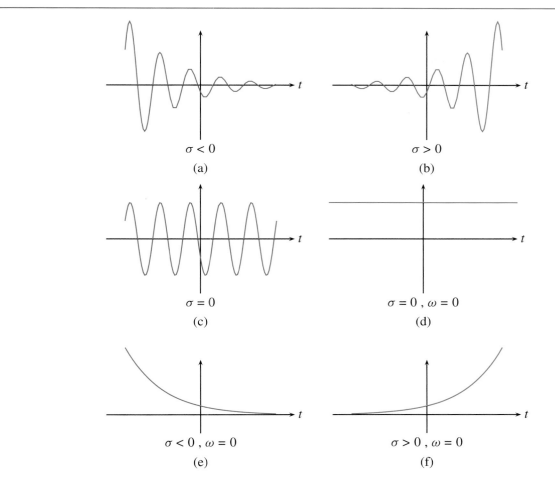

Fig. 5-4. Special cases of exponentially weighted sinusoids. (a) An exponentially damped sinusoid with $\sigma < 0$. (b) An exponentially growing sinusoid with $\sigma > 0$. (c) An ordinary sinusoid corresponding to $\sigma = 0$. (d) A constant signal: $\sigma = 0$, $\omega = 0$. (e) A decaying real exponential: $\sigma < 0$, $\omega = 0$. (f) A growing exponential: $\sigma > 0$, $\omega = 0$.

between these amplitude limits with a frequency of ω_0 radians/second. Notice that, for $\sigma < 0$, an exponentially weighted sinusoid decays to zero as $t \to \infty$ and becomes arbitrarily large as $t \to -\infty$.

The class of exponentially weighted sinusoidal signals includes signals other than the ordinary sinusoids as special cases. These include real exponential signals and constant signals. Several of the different appearances of these signals are illustrated in Figure 5-4.

Exponentially weighted sinusoids are closely related to complex exponential time functions. Let $s = \sigma + j\omega$ be a complex variable. Then the complex exponential time function $x(t) = e^{st}$ has exponentially weighted sinusoids for its real and imaginary parts, since we can write

$$x(t) = e^{st} = e^{(\sigma+j\omega)t}$$
$$= e^{\sigma t} e^{j\omega t}$$
$$= e^{\sigma t} \cos \omega t + je^{\sigma t} \sin \omega t.$$

We can also decompose real, exponentially weighted sinusoids into complex exponential time functions, since

$$y(t) = Ae^{\sigma t} \cos(\omega t + \phi)$$
$$= \frac{Ae^{\sigma t}}{2} \left[e^{j(\omega t + \phi)} + e^{-j(\omega t + \phi)} \right]$$
$$= \frac{1}{2} Ae^{j\phi} \cdot e^{(\sigma+j\omega)t} + \frac{1}{2} Ae^{-j\phi} \cdot e^{(\sigma-j\omega)t}$$
$$= \frac{1}{2} X \cdot e^{st} + \frac{1}{2} X^* \cdot e^{s^*t}$$
$$= \Re e\{X e^{st}\},$$

where $s = \sigma + j\omega$ and $X = Ae^{j\phi}$. The parameter $\sigma = \Re e\{s\}$ controls the envelope of the signal and $\omega = \Im m\{s\}$ controls its rate of oscillation.

5-1-3   Switched-Exponential Signals

Switched signals are signals that change abruptly at certain times. When there is only one such change, the time variable is often defined so that the change happens at $t = 0$. The simplest switched signal is the *unit step function*, defined as

$$u(t) = \begin{cases} 1, t > 0 \\ 0, t < 0 \end{cases}, \qquad (5.10)$$

which is shown in Figure 5-5. The unit step function undergoes an abrupt change from a constant value of zero to a constant value of unity at $t = 0$. Switched inputs are frequently used to model sources that change from a state corresponding to a logical value of "zero" to a logical "one," to model disturbances that are suddenly applied

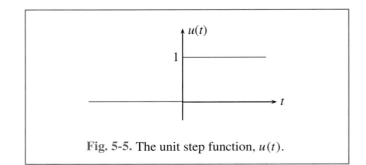

Fig. 5-5. The unit step function, $u(t)$.

or to model circuit behavior immediately after a circuit is "turned on."

We shall frequently encounter *switched exponential* and *switched sinusoidal signals*. A *switched exponential signal* is a signal of the form

$$x(t) = \begin{cases} e^{-at}, t \geq 0 \\ 0, \quad t < 0 \end{cases}. \qquad (5.11)$$

This is illustrated in Figure 5-6 for a real, positive value of a. This function can also be written as

$$x(t) = e^{-at} u(t), \qquad (5.12)$$

where $u(t)$ is the unit step function. The unit step function itself is a special case of a switched exponential with $a = 0$. Equations (5.11) and (5.12) are interchangeable, and we shall use both forms.

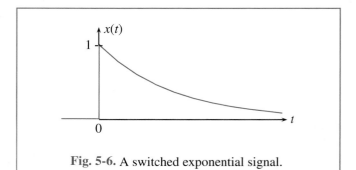

Fig. 5-6. A switched exponential signal.

We shall also encounter *switched sinusoids*, which are sinusoidal signals that are suddenly turned on and are given by

$$x(t) = A \cos(\omega_0 t + \phi)u(t),$$

switched exponentially weighted sinusoids with equation,

$$x(t) = Ae^{\sigma t} \cos(\omega_0 t + \phi)u(t),$$

and *switched complex exponentials*, given by,

$$x(t) = e^{st}u(t).$$

DEMO: *Complex Exponentials*

5-2 Definition of the Laplace Transform

The (unilateral) *Laplace transform*, $X(s)$, of a waveform $x(t)$ is defined as the integral

$$\mathcal{L}\{x(t)\} = X(s) = \int_0^\infty x(t)e^{-st} \, dt. \qquad (5.13)$$

The variable s is complex and, in general, $X(s)$ is complex, even when the time function $x(t)$ is real. The adjective *unilateral* serves to distinguish this Laplace transform from the *bilateral* Laplace transformindexbilateral Laplace transform, which has a similar definition except that the lower limit of its integral is $-\infty$. If $x(t) = 0$ for $t < 0$—that is, if $x(t)$ is a switched signal—then the unilateral and bilateral Laplace transforms are the same.

Notice, from the definition, that $X(s)$ ignores $x(t)$ for $t < 0$. Thus, if we have two waveforms $x_1(t)$ and $x_2(t)$ that are different for $t < 0$, but are identical for $t \geq 0$, their Laplace transforms will be the same. We can state this condition more formally by observing that if

$$x_1(t) = x_2(t), \ t \geq 0,$$

then

$$X_1(s) = X_2(s).$$

When we compute an inverse Laplace transform to $x(t)$ from $X(s)$, we will be able to find $x(t)$ only for $t > 0$. This is generally not a problem. Since we will usually use the Laplace transform to describe the behavior of a circuit in response to a switched waveform (a waveform that is zero for $t < 0$), we might be able to find the value of the output waveform for $t < 0$ trivially by other means.

The integral in (5.13) multiplies the function $x(t)$ by a complex exponential time function and then computes the area under the product between 0 and ∞. Changing the variable s changes the exponential, the product, and the integral. For some values of s, the area will be infinite; for others, it will be finite. When $X(s)$ is finite, we say that the Laplace transform integral *converges*; in general, it converges only for limited values of s. The issue of convergence is more important for the bilateral Laplace transform than it is for the unilateral one, but, for either transform, it is important that the defining integral converge for some values of s; if we are working with the Laplace transforms of multiple signals, it is important that there be a value of s at which all of them converge. For signals that are bounded in amplitude, the unilateral Laplace transform will converge for all values of s that have a positive real part; therefore, as long as we are restricted to such signals, convergence will not be an issue. To illustrate how a transform can converge for some values of s and not others, it is helpful to consider a specific example.

Example 5-2 **Laplace Transform of a Switched Exponential**

Let $x(t)$ correspond to the switched exponential signal

$$x(t) = \begin{cases} e^{-at}, t \geq 0 \\ 0, \quad t < 0 \end{cases}, \qquad (5.14)$$

which can also be written as

$$x(t) = e^{-at}u(t),$$

where $u(t)$ is the unit step function. To compute the Laplace transform of $x(t)$, we begin with its definition:

$$X(s) = \int_0^\infty x(t)e^{-st}\,dt = \int_0^\infty e^{-at}u(t)e^{-st}\,dt$$

$$= \int_0^\infty e^{-(s+a)t}\,dt = -\frac{1}{s+a}e^{-(s+a)t}\bigg|_{t=0}^{t=\infty}.$$

The behavior of this quantity at $t = \infty$ is critically dependent on the value of s. To see this, express s in terms of its real and imaginary parts as $s = \sigma + j\omega$. Then

$$X(s) = -\frac{1}{\sigma + j\omega + a}e^{-(\sigma+j\omega+a)t}\bigg|_{t=0}^{t=\infty}$$

$$= -\frac{1}{\sigma + j\omega + a}e^{-(\sigma+a)t}e^{-j\omega t}\bigg|_{t=0}^{t=\infty}.$$

If $\sigma + a > 0$, the exponential envelope goes to zero as $t \to \infty$, and $X(s)$ is well defined. Otherwise, the envelope becomes infinitely large as $t \to \infty$, and the upper limit does not exist. Therefore, we can state

$$X(s) = -\frac{1}{s+a}\{0 - 1\} = \frac{1}{s+a},$$
if and only if $\Re e(s) > -a$.

$X(s)$ is undefined otherwise. This is equivalent to saying that the transform integral converges only for values of s such that $\Re e(s) > -a$. Using a bidirectional arrow to denote a Laplace-transform pair yields

$$e^{-at},\ t > 0 \longleftrightarrow \frac{1}{s+a}, \quad \textit{if } \Re e(s) > -a. \quad (5.15)$$

This transform pair is also valid for complex values of a, except that the convergence condition changes to $\Re e(s) > -\Re e(a)$. ∎

 To check your understanding, try Drill Problem P5-2.

Example 5-3   **Laplace Transform of a Switched Sinusoid**

As a second example, let us evaluate the Laplace transform of the switched sinusoid signal $x(t) = \cos(\omega_0 t)\,u(t)$. To simplify the integral, we make use of the inverse Euler relation

$$\cos(\omega_0 t) = \frac{1}{2}e^{j\omega_0 t} + \frac{1}{2}e^{-j\omega_0 t}.$$

Then

$$X(s) = \int_0^\infty \left[\frac{1}{2}e^{j\omega_0 t} + \frac{1}{2}e^{-j\omega_0 t}\right]e^{-st}\,dt$$

$$= \frac{1}{2}\int_0^\infty e^{-(s-j\omega_0)t}\,dt + \frac{1}{2}\int_0^\infty e^{-(s+j\omega_0)t}\,dt.$$

We recognize these integrals to be essentially the same as the one that we evaluated for the exponential signal, except that a has been replaced by $\pm j\omega_0$. Therefore, making use of the earlier result, we see that

$$X(s) = \frac{\frac{1}{2}}{s - j\omega_0} + \frac{\frac{1}{2}}{s + j\omega_0} = \frac{s}{s^2 + \omega_0^2},$$

provided that $\Re e(s) > 0$. Thus,

$$\cos\omega_0 t,\ t > 0 \longleftrightarrow \frac{s}{s^2 + \omega_0^2}, \quad \textit{if } \Re e(s) > 0. \quad (5.16)$$

By using the complementary inverse Euler relation

$$\sin(\omega_0 t) = \frac{1}{2j}e^{j\omega_0 t} - \frac{1}{2j}e^{-j\omega_0 t}$$

and following a very similar derivation, it is also straightforward to show that

$$\sin \omega_0 t, \ t > 0 \longleftrightarrow \frac{\omega_0}{s^2 + \omega_0^2}, \quad \text{if } \Re e(s) > 0. \quad (5.17)$$

■

Example 5-4   Laplace Transform of a Unit Step or Constant

Another important signal is the step function, $x(t) = Ku(t)$. This is really a special case of the exponential that we saw earlier with $a = 0$. Therefore,

$$K, \ t > 0 \longleftrightarrow \frac{K}{s}, \quad \text{if } \Re e(s) > 0. \quad (5.18)$$

Recall that the unilateral Laplace transform depends upon only the portion of $x(t)$ for $t > 0$. Thus, while this is the Laplace transform of the step $x(t) = Ku(t)$, it is also the Laplace transform of the constant signal $x(t) = K$.   ■

In Table 5-1, we have summarized these examples of Laplace transforms. For the sake of completeness, you will notice that the table includes a few additional transform pairs as well. These additional transforms will be derived a little later. First, however, it is helpful to look at some properties of the Laplace transform.

 **FLUENCY EXERCISE:** *LTs of Exponentials*

Table 5-1. Common Laplace-transform pairs

$x(t),\ t > 0$	$X(s)$	Region of Convergence
1	$\dfrac{1}{s}$	$\Re e(s) > 0$
e^{-at}	$\dfrac{1}{s+a}$	$\Re e(s) > -\Re e(a)$
$\cos(\omega_0 t)$	$\dfrac{s}{s^2 + \omega_0^2}$	$\Re e(s) > 0$
$\sin(\omega_0 t)$	$\dfrac{\omega_0}{s^2 + \omega_0^2}$	$\Re e(s) > 0$
$e^{-at}\cos(\omega_0 t)$	$\dfrac{s+a}{(s+a)^2+\omega_0^2}$	$\Re e(s) > -a$
$e^{-at}\sin(\omega_0 t)$	$\dfrac{\omega_0}{(s+a)^2+\omega_0^2}$	$\Re e(s) > -a$
$\delta(t)$	1	all s
$k_\ell u_\ell(t)$	$k_\ell s^\ell$	all s
te^{-at}	$\dfrac{1}{(s+a)^2}$	$\Re e(s) > -\Re e(a)$

5-3   Some Properties of the Laplace Transform

Computing Laplace transforms by evaluating the integrals that define them is not done very often, except for the simple examples that we saw in the previous section. Instead, a number of properties of the transform are often used to extend existing transform pairs. These properties also play an important role in enabling us to evaluate inverse Laplace transforms and they help us to understand circuit behavior when we apply them to the waveforms that correspond to element variables. There are many more properties of the Laplace transform than those we have chosen to enumerate here. We have limited ourselves to those that will be of the greatest use in solving, understanding, and designing circuits.

Linearity

The Laplace transform of the sum of two signals is equal to the sum of their Laplace transforms, at least for those values of s that lie within both regions of convergence. Actually, the linearity property is a little more general than this. Let $x_1(t)$ and $x_2(t)$ be two signals that have Laplace transforms $X_1(s)$ and $X_2(s)$, respectively. Now create a third waveform $x(t)$ as a linear combination of $x_1(t)$ and $x_2(t)$—that is, for arbitrary complex constants, a and b, define

$$x(t) = ax_1(t) + bx_2(t).$$

Then

$$X(s) = \int_0^\infty [ax_1(t) + bx_2(t)] e^{-st} \, dt$$

$$= a \int_0^\infty x_1(t)e^{-st} \, dt + b \int_0^\infty x_2(t)e^{-st} \, dt$$

$$= aX_1(s) + bX_2(s).$$

This means that, if $x(t)$ is formed as a linear combination of two waveforms $x_1(t)$ and $x_2(t)$, then the Laplace transform of $x(t)$, $X(s)$, is the same linear combination of their Laplace transforms. That is,

$$ax_1(t) + bx_2(t) \longleftrightarrow aX_1(s) + bX_2(s). \qquad (5.19)$$

The derivation required that the two transform integrals be valid for the same values of s. Thus, it is valid only if the regions of convergence for the two transforms overlap.

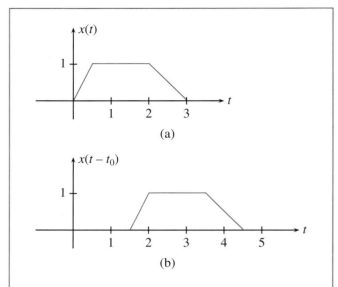

Fig. 5-7. (a) A signal $x(t)$ that is zero for $t < 0$. (b) The delayed signal $x(t - t_0)$, drawn for the special case $t_0 = 1.5$.

Delay

Let $x(t)$ be a signal that is zero for $t < 0$. Then the signal $x(t - t_0)$, where $t_0 > 0$, represents a *delayed* version of $x(t)$, as depicted in Figure 5-7.

The Laplace transform of $x(t - t_0)$ is closely related to the Laplace transform of $x(t)$. To understand the relationship, let $X(s)$ denote the Laplace transform of $x(t)$, and let $Y(s)$ denote the Laplace transform of $y(t) = x(t - t_0)$. From the definition of $Y(s)$, we have

$$Y(s) = \int_0^\infty x(t - t_0)e^{-st} \, dt.$$

Now let $v = t - t_0$ (or $t = v + t_0$) and make this substitution into the above integral. This gives

$$Y(s) = \int_{-t_0}^\infty x(v)e^{-s(v+t_0)} \, dv.$$

Since we require that $x(t) = 0$ for $t < 0$, we can change the lower limit of integration, to get

$$Y(s) = \int_0^\infty x(v) e^{-s(v+t_0)} \, dv$$

$$= e^{-st_0} \int_0^\infty x(v) e^{-sv} \, dv = e^{-st_0} X(s).$$

Thus, if $x(t) = 0$ for $t < 0$, then

$$x(t - t_0) \longleftrightarrow e^{-st_0} X(s). \qquad (5.20)$$

The regions of convergence for $Y(s)$ and $X(s)$ are the same.

Notice that this property is true only if $x(t) = 0$ for $t < 0$. If this condition is not met, then the values of $x(t)$ for $-t_0 < t < 0$ will affect the transform $Y(s)$, but will not affect $X(s)$. In this case we cannot express $Y(s)$ in terms of $X(s)$.

Example 5-5   Laplace Transform of a Pulse

As an example of how the properties can be used, consider evaluating the Laplace transform of the signal

$$x(t) = u(t) - u(t - T) = \begin{cases} 1, \, 0 < t < T \\ 0, \, \text{otherwise} \end{cases},$$

which is shown in Figure 5-8. This signal has the value 1 V for times between $t = 0$ and $t = T$ and is zero

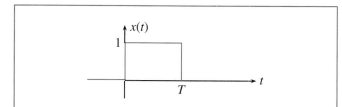

Fig. 5-8. A pulse-like signal, $x(t) = u(t) - u(t - T)$.

outside this range. Notice that it can also be expressed as the difference of two delayed step functions:

$$x(t) = u(t) - u(t - T).$$

Using both the linearity and delay properties, we can write

$$X(s) = \mathcal{L}\{u(t)\} - \mathcal{L}\{u(t - T)\}$$

$$= \mathcal{L}\{u(t)\} - e^{-sT} \mathcal{L}\{u(t)\}$$

$$= (1 - e^{-sT}) \mathcal{L}\{u(t)\} = (1 - e^{-sT}) \frac{1}{s}.$$

Here we used the symbol $\mathcal{L}$ to denote the Laplace transform operator. For this example, we could get the same result by evaluating the Laplace-transform integral. ∎

Transform-Domain Differentiation

Another property of the Laplace-transform is one that we call the *transform-domain differentiation property*. It states that

$$tx(t) \longleftrightarrow -\frac{dX(s)}{ds}. \qquad (5.21)$$

It is simple to derive. We begin with the definition of $X(s)$,

$$X(s) = \int_0^\infty x(t) e^{-st} \, dt,$$

and differentiate both sides with respect to s:

$$\frac{dX(s)}{ds} = \int_0^\infty -tx(t) e^{-st} \, dt = -\mathcal{L}\{tx(t)\}.$$

Our only need for this property will be to establish some additional Laplace-transform pairs.

If we let

$$x(t) = e^{-at}u(t) \longleftrightarrow \frac{1}{s+a},$$

then

$$\mathcal{L}\{tx(t)\} = \mathcal{L}\{te^{-at}u(t)\} = -\frac{d}{ds}\left(\frac{1}{s+a}\right)$$

$$= \frac{1}{(s+a)^2}.$$

This transform pair has been included in Table 5-1. By applying the property many times to this signal, it can be shown that

$$\frac{1}{(k-1)!}t^{k-1}e^{-at} \longleftrightarrow \frac{1}{(s+a)^k}. \qquad (5.22)$$

Time-Domain Differentiation

One other property of the Laplace-transform that will prove to be particularly useful for solving circuits with reactive elements is the time-domain differentiation property. It answers the following question: What is the Laplace-transform of $\frac{dx(t)}{dt}$? Once again, we begin with the definition

$$\mathcal{L}\left\{\frac{dx(t)}{dt}\right\} = \int_0^\infty \frac{dx(t)}{dt}e^{-st}\,dt.$$

Integrating by parts, we have

$$\mathcal{L}\left\{\frac{dx(t)}{dt}\right\} = s\int_0^\infty x(t)e^{-st}\,dt + x(t)e^{-st}\Big|_0^\infty.$$

If we assume that the term on the right vanishes at its upper limit, as must be true if s lies within the region of convergence, we get

$$\frac{dx(t)}{dt} \longleftrightarrow sX(s) - x(0). \qquad (5.23)$$

Thus, the Laplace-transform of the derivative of a function is s times the Laplace-transform of the function

Table 5-2. Some Properties of the Laplace Transform. ($X(s)$ is the Laplace-transform of $x(t)$.)

Property	Limitations
$ax(t) + b(y(t) \longleftrightarrow aX(s) + bX(s)$	
$x(t - t_d) \longleftrightarrow e^{-st_d}X(s)$	$x(t) = 0,\ t < 0$ $t_d \geq 0$
$tx(t) \longleftrightarrow -\frac{dX(s)}{ds}$	
$\frac{dx(t)}{dt} \longleftrightarrow sX(s) - x(0)$	
$\lim\limits_{s\to\infty} sX(s) = x(0)$	$\lim\limits_{t\to\infty} x(t)$ finite
$\lim\limits_{s\to 0} sX(s) = \lim\limits_{t\to\infty} x(t)$	$X(\sigma + j\omega)$ finite for all $\sigma \geq 0$

minus the initial value. We shall use this property in Chapter 6.

These properties of the Laplace-transform, plus the initial- and final-value properties (from Problem P5-27) are summarized in Table 5-2. Some of the properties are valid only under certain circumstances; those limitations are listed in the column on the right.

 FLUENCY EXERCISE: *LT Properties*

5-4　Inverse Laplace Transforms

In order to solve a circuit by using Laplace transforms in accordance with the approach indicated in Figure 5-1, we need to be able to perform an inverse transform. There are formal procedures for doing this, but they involve computing integrals with respect to complex variables, which is a fairly involved subject. Instead, we shall develop an alternative procedure that relies only on the linearity property of the Laplace-transform, the table of Laplace transforms that we have already derived, and the fact that the inverse Laplace-transform is unique for $t > 0$. Our procedure will be limited to Laplace-transforms of the form

$$X(s) = \frac{b_m s^m + \cdots + b_1 s + b_0}{a_n s^n + \cdots + a_1 s + a_0} = \frac{B(s)}{A(s)}, \quad (5.24)$$

but these are the only ones that we will need for analyzing linear circuits, so this is not a serious limitation. A Laplace transform $X(s)$ that can be written as a ratio of two polynomials, as in (5.24), is called a *rational function*. The roots of the numerator polynomial $B(s)$ are called the *zeros* of the rational function, and the roots of the denominator polynomial $A(s)$ are called its *poles*.

Our approach will be to build up to the most general case in stages. Initially, we consider the case where the poles are distinct (none of the roots of the denominator polynomial are repeated), and where there are more poles than zeros ($m < n$). This is the simplest case. Then we consider the case where $m \geq n$, but where the poles are still distinct. Finally, we remove this last restriction. Case 3 is the most general, Case 2 is a special case of it, and Case 1 is a special case of Case 2. Our reason for treating the special cases separately is that they are the situations that occur most often and they are simpler than the general case.

5-4-1 Case 1: More Poles than Zeros

Our approach to computing the inverse Laplace-transform of $X(s)$ is to manipulate $X(s)$ into a form in which we can recognize the inverse transform by making use of the properties and transforms that we have already seen. We can get a preview of this approach by computing the inverse Laplace-transform of $X(s) = \frac{10}{s+5}$, using the following reasoning:

$$x(t) = \mathcal{L}^{-1}\left\{\frac{10}{s+5}\right\} = 10\mathcal{L}^{-1}\left\{\frac{1}{s+5}\right\} = 10e^{-5t},$$

for $t > 0$. First, we used the linearity property to extract the constant factor 10, then we observed that the remaining transform to be inverted was of the form $\frac{1}{s+a}$ with $a = 5$. Therefore, from Table 5-1, the inverse transform was seen to be of the form e^{-at} for $t > 0$ and with $a = 5$.

The same approach will also work in some less obvious cases. Consider

$$X(s) = \frac{4s + 14}{s^2 + 6s + 8} = \frac{4s + 14}{(s+2)(s+4)}. \quad (5.25)$$

Equation (5.25) can be written as the sum of two rational functions with first-order denominators, each of which can be inverse transformed. To see this, let us try to find such a decomposition by writing

$$X(s) = \frac{4s + 14}{s^2 + 6s + 8} = \frac{A}{s+2} + \frac{B}{s+4} \quad (5.26)$$

and then trying to find appropriate values of the constants A and B. Recombining the two terms on the right side of (5.26) gives

$$X(s) = \frac{(s+4)A + (s+2)B}{(s+2)(s+4)}. \quad (5.27)$$

For the expansion in (5.26) to work, the numerators in (5.25) and (5.27) must be the same. Therefore, we write

$$(s+4)A + (s+2)B = 4s + 14$$

and thence,

$$(A + B)s + (4A + 2B) = 4s + 14,$$

which is satisfied only when A and B satisfy the following two simultaneous equations (obtained by equating like powers of s):

$$A + B = 4$$
$$4A + 2B = 14.$$

Their solution is $A = 3$, $B = 1$. Therefore, we have established that

$$X(s) = \frac{4s + 14}{s^2 + 6s + 8} = \frac{3}{s + 2} + \frac{1}{s + 4}.$$

When written in this form, the inverse transform is easy to perform by using the linearity property and the table of Laplace transforms. In this case,

$$x(t) = 3e^{-2t} + e^{-4t}, \text{ for } t > 0.$$

The procedure that we followed for this example is called a *partial-fraction expansion*. It enabled us to express a rational $X(s)$ as a sum of first-order rational functions. In a more general setting, the decomposition can be stated as

$$X(s) = \frac{B(s)}{A(s)} = \sum_{i=1}^{n} \frac{A_i}{s - s_i}. \tag{5.28}$$

Then the inverse transform (for $t > 0$) is

$$x(t) = \sum_{i=1}^{n} A_i e^{s_i t}, \text{ for } t > 0. \tag{5.29}$$

The coefficients A_i are called the *residues*. For a decomposition in the form of (5.28) to exist, we must have $m < n$ (the numerator polynomial must be of lower order than the denominator polynomial) and the poles must be distinct ($s_i \neq s_j$ for all $i \neq j$).

Before proceeding to more efficient methods for finding the residues, let us summarize the approach that was exemplified by this first example.

Approach #1 for Performing a Partial-Fraction Expansion

1. Find the roots of the denominator polynomial $\{s_i\}$.
2. Express $X(s)$ in terms of the unknown residues $\{A_i\}$, as in (5.28).
3. Algebraically recombine the sum of first-order rational functions into a single high-order rational function.

4. Write a series of n linear equations in the n unknown residues by equating the coefficients of the various powers of s in the numerator obtained in step 3 to their known values.
5. Solve the resulting set of n linear equations to obtain the n residues.

Example 5-6   Computing an Inverse Laplace Transform

Let us compute the inverse Laplace-transform of

$$X(s) = \frac{1}{s(s^2 + 3s + 2)} = \frac{1}{s(s + 1)(s + 2)}. \tag{5.30}$$

The first step is to perform a partial-fraction expansion and write $X(s)$ as

$$X(s) = \frac{A}{s} + \frac{B}{s + 1} + \frac{C}{s + 2}. \tag{5.31}$$

To find A, B, and C, we can recombine the three terms into a single rational function, obtaining

$$X(s) = \frac{A(s + 1)(s + 2) + Bs(s + 2) + Cs(s + 1)}{s(s + 1)(s + 2)}$$

$$= \frac{(A + B + C)s^2 + (3A + 2B + C)s + (2A)}{s(s + 1)(s + 2)}.$$

Equating the numerator of the latter equation to the numerator in (5.30) gives three equations that we can solve for A, B, and C:

$$A + B + C = 0,$$

$$3A + 2B + C = 0,$$

$$2A = 1.$$

A straightforward solution of these equations gives $A = \frac{1}{2}$, $B = -1$, $C = -\frac{1}{2}$. Thus,

$$X(s) = \frac{\frac{1}{2}}{s} - \frac{1}{s + 1} - \frac{\frac{1}{2}}{s + 2}.$$

Using the linearity property and Table 5-1, we can compute the inverse Laplace transform, term by term:

$$x(t) = \frac{1}{2} - e^{-t} - \frac{1}{2}e^{-2t}, \qquad t > 0. \qquad (5.32)$$

■

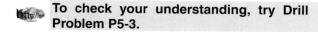 To check your understanding, try Drill Problem P5-3.

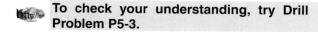 **MULTIMEDIA TUTORIAL:** *Inverse Laplace Trans.*

This procedure is computationally involved for n larger than two or three, if performed without the aid of a computer. The steps that are particularly demanding are steps 1, 3, and 5. Furthermore, since the poles $\{s_1\}$ are often complex, the equations that will need to be solved in Step 5 may have complex coefficients, and the final residue values may also be complex. Fortunately, there is another approach that we can use to find the residues that reduces the level of complexity.

We begin by writing out the partial-fraction expansion in terms of the unknown residues as before:

$$X(s) = \frac{A_1}{s - s_1} + \frac{A_2}{s - s_2} + \cdots + \frac{A_n}{s - s_n},$$

Here, we assume that the poles $\{s_i\}$ have been found by factoring the denominator of $X(s)$. Next, we multiply both sides of this equation by $(s - s_1)$, which results in

$$(s - s_1)X(s) = A_1 + A_2\frac{s - s_1}{s - s_2} + \cdots + A_n\frac{s - s_1}{s - s_n}.$$

Now take the limit as $s \to s_1$. Since the poles are distinct, we see that

$$\lim_{s \to s_1} \frac{s - s_1}{s - s_k} = 0 \quad \text{for } k \neq 1$$

and that

$$\lim_{s \to s_1} (s - s_1)X(s) = A_1.$$

By similar reasoning, we can show that

$$\lim_{s \to s_2} (s - s_2)X(s) = A_2$$

and, in general, that

$$\lim_{s \to s_k} (s - s_k)X(s) = A_k. \qquad (5.33)$$

This procedure allows us to find the residues without having to perform all of the algebraic manipulation of Step 3 of the earlier procedure and without having to solve a set of linear equations as required by Step 5.

Approach #2 for Performing a Partial-Fraction Expansion (PFE)

1. Find the roots of the denominator polynomial $\{s_i\}$.
2. For $i = 1, 2, \ldots, n$ find the residue A_i by using the formula

$$A_i = \lim_{s \to s_i} (s - s_i)X(s).$$

Example 5-7   Using Limits to Find Residues

Consider the example that we solved earlier:

$$X(s) = \frac{4s + 14}{(s + 2)(s + 4)}.$$

By letting $s_1 = -2$ and $s_2 = -4$, we get

$$A_1 = \lim_{s \to -2} (s + 2)X(s) = \lim_{s \to -2} \frac{4s + 14}{s + 4} = 3$$

$$A_2 = \lim_{s \to -4} (s + 4)X(s) = \lim_{s \to -4} \frac{4s + 14}{s + 2} = 1.$$

Thus,

$$X(s) = \frac{3}{s + 2} + \frac{1}{s + 4},$$

as before.

■

 To check your understanding, try Drill Problem P5-4.

Example 5-8   A PFE with Complex Poles

This procedure works equally well when the roots are complex. Let

$$X(s) = \frac{s+a}{(s+a)^2 + b^2} = \frac{s+a}{(s+a-jb)(s+a+jb)}.$$

If we let $s_1 = -a + jb$ and $s_2 = -a - jb$, then

$$A_1 = \lim_{s \to (-a+jb)} \frac{s+a}{s+a+jb} = \frac{jb}{j2b} = \frac{1}{2}$$

and

$$A_2 = \lim_{s \to (-a-jb)} \frac{s+a}{s+a-jb} = \frac{-jb}{-j2b} = \frac{1}{2}.$$

Therefore,

$$X(s) = \frac{\frac{1}{2}}{s + (a-jb)} + \frac{\frac{1}{2}}{s + (a+jb)}$$

and

$$x(t) = \frac{1}{2}e^{(-a+jb)t} + \frac{1}{2}e^{(-a-jb)t}, \quad \text{for } t > 0$$

$$= e^{-at}\cos(bt), \quad \text{for } t > 0.$$

We should save this result for the future:

$$e^{-at}\cos(bt), \quad for\ t > 0 \longleftrightarrow \frac{s+a}{(s+a)^2 + b^2}. \quad (5.34)$$

A similar derivation would show that

$$e^{-at}\sin(bt), \quad for\ t > 0 \longleftrightarrow \frac{b}{(s+a)^2 + b^2}. \quad (5.35)$$

These results are included in Table 5-1. ∎

> **To check your understanding, try Drill Problem P5-5.**

When $X(s) = \frac{B(s)}{A(s)}$ and the coefficients of $A(s)$ and $B(s)$ are real, as will be the case for circuits problems, the poles and zeros occur in complex-conjugate pairs. This means that, if s_i is a zero (pole), then s_i^* is also a

zero (pole). The residues associated with two complex-conjugate poles are complex conjugates of each other. This means that, if $s_j = s_i^*$, then $A_j = A_i^*$. This observation can provide further computational savings when $X(s)$ has complex poles.

Approach #2 is considerably more efficient than Approach #1 when the residues need to be computed analytically, but both approaches are quite mechanical and both can be readily programmed on a computer. In MATLAB, the pole and residue calculations can be performed by using the function `residue`. Its calling statement is

```
[Ai,si] = residue(B,A)
```

where B is the vector of coefficients of the numerator polynomial $B(s)$ and A is the vector of coefficients of the denominator polynomial $A(s)$. The function returns two arrays:[1] si, a vector containing the poles of the system (the roots of the denominator polynomial) and a vector Ai containing the residues. In general, both of the returned arrays are complex.

5-4-2   Case 2: Fewer Poles than Zeros ($m \geq n$)

Case #1 is the case that we encounter most often, but it makes two assumptions that are not always valid. The first of these is that the order of the numerator polynomial is lower than the order of the denominator polynomial. We remove this restriction in this section. The second is that the poles must be distinct; we remove that restriction in the following section.

When the order of the numerator polynomial is equal to or greater than the order of the denominator polynomial, the Laplace-transform cannot be decomposed as in (5.28). However, it can be decomposed by using a more general partial-fraction expansion, and once we develop this generalization, we shall need to make only minor alterations to our procedures for computing

---

[1] It also returns a third array that we shall discuss in the next section.

inverse Laplace-transforms. As before, assume that $X(s) = B(s)/A(s)$, where the degree of the numerator polynomial is m and the degree of the denominator polynomial is n. When $m > n$, we can write

$$X(s) = \frac{B(s)}{A(s)} = \sum_{i=0}^{m-n} k_i s^i + \sum_{i=1}^{n} \frac{A_i}{s - s_i}. \qquad (5.36)$$

This is a generalization of the earlier partial-fraction expansion by addition of the polynomial term embodied in the first summation.

Consider, as an example, the transform

$$X(s) = \frac{s^2 + 1}{s^2 + 5s + 6} = \frac{s^2 + 1}{(s + 2)(s + 3)}. \qquad (5.37)$$

This does not correspond to Case #1, because the degrees of the numerator and denominator polynomials are equal. According to the partial-fraction expansion in (5.36), however, it can be written in the form

$$X(s) = k_0 + \frac{A_1}{s + 2} + \frac{A_2}{s + 3}. \qquad (5.38)$$

To find the values of the unknown parameters, we can as before recombine the terms in (5.38) into a single rational function; doing so yields

$$X(s) = \frac{k_0(s + 2)(s + 3) + A_1(s + 3) + A_2(s + 2)}{(s + 2)(s + 3)}$$

$$= \frac{k_0 s^2 + (5k_0 + A_1 + A_2)s}{(s + 2)(s + 3)}$$

$$+ \frac{(6k_0 + 3A_1 + 2A_2)}{(s + 2)(s + 3)}. \qquad (5.39)$$

Now we can equate the numerators in (5.37) and (5.39), which gives the following three linear equations:

$$k_0 = 1,$$

$$5k_0 + A_1 + A_2 = 0,$$

$$6k_0 + 3A_1 + 2A_2 = 1:$$

We can solve these equations simultaneously, to arrive at the values

$$k_0 = 1; \quad A_1 = 5; \quad A_2 = -10.$$

Plugging these values into (5.38) completes the partial-fraction expansion. This approach illustrates the generalization of the first method that we derived in the previous section. We will not formally state it as a procedure, since the approach should be quite clear from the example. This method for computing the coefficients of the partial-fraction expansion is conceptually the most straightforward of the methods that we will consider, but it requires that we solve a set of linear equations; doing so might be tedious when the number of equations is large.

As an alternative to solving linear equations to find the coefficients of the expansion, we can again take limits to find the residues $\{A_i\}$; computationally this approach is more efficient. To compute the coefficients $\{k_i\}$, we begin by rewriting (5.36) as

$$X(s) = K(s) + \frac{B'(s)}{A(s)}, \qquad (5.40)$$

where

$$K(s) = \sum_{i=0}^{m-n} k_i s^i$$

is a polynomial of degree $m - n$ and

$$\frac{B'(s)}{A(s)} = X'(s) = \sum_{i=1}^{n} \frac{A_i}{s - s_i}.$$

We need to make two observations about $X'(s)$. First, its numerator polynomial, $B'(s)$, is of lower degree than its denominator polynomial, $A(s)$. Second, the denominator polynomials of $X(s)$ and $X'(s)$ are the same. From (5.36) and (5.40), we see that the original numerator polynomial can be written as

$$B(s) = K(s)A(s) + B'(s).$$

If we formally perform a long division of the polynomial $B(s)$ by $A(s)$, then $K(s)$ is the polynomial quotient and $B'(s)$ is the polynomial remainder. This suggests a two-step procedure: We find $K(s)$ by performing a long division, then find $\{A_i\}$ by performing a PFE on $X'(s)$, using any of the Case #1 methods that we discussed in the previous section. Let's attack the previous example, using this approach.

From the long division,

$$X(s) = \frac{s^2 + 1}{s^2 + 5s + 6} = 1 - \frac{5s + 5}{(s + 2)(s + 3)}.$$

Notice that the numerator of this fraction is of lower degree than the denominator polynomial. Since it is in the form of Case #1, we can now find the PFE of the resulting rational function by using limits. We obtain

$$A_1 = \lim_{s \to -2} \left(-\frac{5s + 5}{s + 3} \right) = 5$$

$$A_2 = \lim_{s \to -3} \left(-\frac{5s + 5}{s + 2} \right) = -10,$$

which together give

$$X(s) = 1 + \frac{5}{s + 2} - \frac{10}{s + 3}.$$

This is the same result that we got earlier by solving a series of linear equations.

In this example, we computed the residues by using the formula

$$A_i = \lim_{s \to s_1} (s - s_i) X'(s), \qquad (5.41)$$

but it is equally true that

$$A_i = \lim_{s \to s_1} (s - s_i) X(s). \qquad (5.42)$$

If we had used this formula, it would have led to the alternative computations:

$$A_1 = \lim_{s \to -2} \frac{s^2 + 1}{s + 3} = 5$$

$$A_2 = \lim_{s \to -3} \frac{s^2 + 1}{s + 2} = -10.$$

Using (5.42) spares us the necessity of finding the polynomial remainder (but we still need to compute the polynomial quotient.) Notice that the leading coefficient of the quotient polynomial is

$$k_{m-n} = \frac{b_m}{a_n},$$

which is the ratio of the leading coefficients of the numerator and denominator polynomials When the numerator and denominator polynomials are of the same degree, a case that occurs frequently, this is the complete quotient polynomial.

Since this is the approach that we will usually use when we are performing partial-fraction expansions analytically, it is appropriate to summarize the procedure.

Method for Performing a Partial-Fraction Expansion when $m \geq n$

1. Reduce $X(s)$ to the form

$$X(s) = K(s) + \frac{B'(s)}{A(s)}$$

by long division. The degree of $B'(s)$ must be lower than the degree of $A(s)$.

2. Find the roots of the denominator polynomial $A(s)$. Call these $\{s_i\}$.

3. For $i = 1, 2, \ldots, n$, find the residues A_i by using the formula

$$A_i = \lim_{s \to s_i} (s - s_i) X(s).$$

4. Then

$$X(s) = \frac{B(s)}{A(s)} = K(s) + \sum_{i=1}^{n} \frac{A_i}{s - s_i}.$$

The MATLAB function `residue` can handle this case as well as the earlier one. Its complete calling statement is

`[Ai,si,k] = residue(B,A)`

where the extra array on the left side, `k`, is the array that will be returned with the coefficients of the polynomial $K(s)$.

Unfortunately, generalizing the partial-fraction expansion does not provide the complete solution to the problem of finding the inverse Laplace-transform for Case #2, because we must still identify the inverse transforms of the terms that make up the polynomial $K(s)$. To do this requires that we introduce a class of *singularity functions*. These are not well-behaved, or even well-defined, functions in a mathematical sense, but they play an important role in describing circuits and systems. The singularity functions are related to derivatives of the step function, $u(t)$.

In order to motivate these functions, let us begin by considering the Laplace-transform of the signal $x(t) = u(t - T)$ for $T > 0$, which is a delayed step function. From the shift property and our known transforms,

$$X(s) = \frac{e^{-sT}}{s}.$$

Now, let us extend this result by computing the Laplace transform of the derivative of this signal. The delayed step

function is differentiable everywhere except at $t = T$. Glossing over this point for the moment and applying the derivative-in-the-time-domain property gives

$$\frac{du(t - T)}{dt} \longleftrightarrow s \cdot \frac{e^{-sT}}{s} - x(0) = e^{-sT}. \qquad (5.43)$$

Taking the limit[2] as $T \to 0$ gives the first of the results that we are seeking:

$$\delta(t) \triangleq \frac{du(t)}{dt} \longleftrightarrow 1 \qquad (5.44)$$

Here we have formally defined $\delta(t)$ as the derivative of the step function $u(t)$, even though the latter function is not differentiable everywhere. To do this rigorously, we would need to acknowledge that $\delta(t)$ is not a function but a distribution; we shall not do this here. $\delta(t)$ is called the *unit impulse* "function".

What does $\delta(t)$ look like? Consider Figure 5-9(a), where we have graphed the delayed step function $u(t - T)$. If we interpret the derivative as the slope of the curve, we see that, for $t < T$, the derivative is zero and that, for $t > T$, the derivative is also zero, but that the slope is infinite at $t = T$. This is illustrated symbolically by the graph in Figure 5-9(b). The impulse is drawn as a bold arrow. The location of the arrow places the impulse on the time axis and the number in parentheses next to it denotes the area underneath the curve—that is, the height of the discontinuity of its integral. An impulse is the mathematical equivalent of a point mass or a point charge in physics.

By extension, we can apply the derivative property to a delayed impulse and take a limit:

$$\frac{d\delta(t - T)}{dt} \longleftrightarrow s \cdot e^{-sT}$$

$$\frac{d\delta(t)}{dt} = \lim_{T \to 0} \frac{d\delta(t - T)}{dt} \longleftrightarrow \lim_{T \to 0} se^{-sT} = s.$$

---

[2]We need to work with delayed signals and take limits so that the discontinuity in the step function and the resulting singularity functions do not appear at the lower limit of integration in the definition of the Laplace transform.

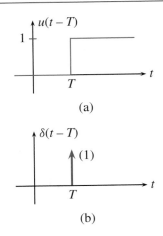

Fig. 5-9. (a) A delayed step function. (b) A graphical representation of a delayed unit impulse, which behaves like the derivative of the waveform in (a).

Thus, we show that

$$\frac{d\delta(t)}{dt} \triangleq u_1(t) \longleftrightarrow s.$$

By similar reasoning,

$$\frac{du_1(t)}{dt} \triangleq u_2(t) \longleftrightarrow s^2,$$

and so on for the whole family of related singularity functions. The function $u_1(t)$ is called the **unit doublet**, and $u_2(t)$ is called the **unit triplet**. The subscript associated with the singularity function is the power of s in its Laplace transform.[3]

The singularity functions can be shown to exhibit a number of properties, but we need only the few of those that are summarized in Table 5-3.

---

[3]In some texts, the symbol $u_0(t)$ is used in place of $\delta(t)$ for the unit impulse and $u_{-1}(t)$ is used in place of $u(t)$ for the unit step.

Table 5-3. Properties of the Singularity Functions. (For the sake of these properties, $\delta(t) = u_0(t)$ and $u(t) = u_{-1}(t)$.)

$$u_\ell(t) = \frac{d}{dt} u_{\ell-1}(t)$$

$$u_{\ell-1}(t) = \int_{-\infty}^{t} u_\ell(\beta)\, d\beta$$

$$f(t)\delta(t - T) = f(T)\delta(t - T)$$

$$k_\ell u_\ell(t) \overset{\mathcal{L}}{\longleftrightarrow} k_\ell s^\ell$$

Finally, we are able to complete our inverse-Laplace-transform example.

$$X(s) = \frac{s^2 + 1}{(s + 2)(s + 3)} = 1 + \frac{5}{s + 2} - \frac{10}{s + 3}$$

$$x(t) = \delta(t) + 5e^{-2t} - 10e^{-3t}, \quad \text{for } t \geq 0.$$

This function is plotted in Figure 5-10. Notice that the inverse Laplace-transforms of all rational functions whose numerator degree is equal to or greater than the denominator degree will contain singularity functions.

  WORKED SOLUTION: *The case $m \geq n$*

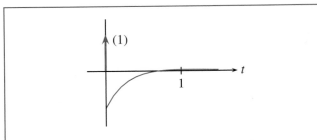

Fig. 5-10. The function $\delta(t) + 5e^{-2t} - 10e^{-3t}$ for $t \geq 0$.

A Confusing Issue: Using Limits to Compute Residues

Using limits to compute partial-fraction expansions is a very useful work-saving technique. The technique is not foolproof, however, and errors will result if the method is not applied properly. One common mistake can happen when the factors in the denominator polynomial are not monic. This is illustrated in the following example.

Suppose that we wish to compute the inverse Laplace-transform of the function

$$H(s) = \frac{1}{(s+1)(3s+2)}.$$

We know that this equation can be written in the form

$$H(s) = \frac{A}{s+1} + \frac{B}{s+\frac{2}{3}}$$

and that

$$A = \lim_{s \to -1} (s+1)H(s)$$

and

$$B = \lim_{s \to -2/3} (s+\frac{2}{3})H(s).$$

The cautious, and safest, way to approach these computations is to first write $H(s)$ with its denominator expressed as a product of monic first-order factors (i.e., with a leading coefficient of one for each denominator factor). This gives

$$H(s) = \frac{\frac{1}{3}}{(s+1)(s+\frac{2}{3})}.$$

Now the residue calculations take the following form:

$$A = \lim_{s \to -1} \left(\frac{\frac{1}{3}}{s+\frac{2}{3}} \right) = \frac{\frac{1}{3}}{-1+\frac{2}{3}} = -1$$

and

$$B = \lim_{s \to -2/3} \left(\frac{\frac{1}{3}}{s+1} \right) = \frac{\frac{1}{3}}{-\frac{2}{3}+1} = 1.$$

When the denominator has monic factors, the computation of a term such as $(s - s_i)H(s)$ involves simply cancelling one of the factors from the denominator of $H(s)$. This is not the case when the denominator factors are not monic and, if done in haste, will result in an incorrect answer!

$$B = \lim_{s \to -2/3} \left(s + \frac{2}{3} \right) \frac{1}{(s+1)(3s+2)}$$

$$\neq \lim_{s \to -2/3} \frac{1}{s+1} = 3.$$

The correct reasoning would yield

$$\left(s + \frac{2}{3} \right) \frac{1}{3s+2} = \frac{1}{3}.$$

Alternatively,

$$B = \lim_{s \to -2/3} \left(s + \frac{2}{3} \right) \frac{1}{(s+1)(3s+2)}$$

$$= \lim_{s \to -2/3} \frac{1}{3} \left(\frac{1}{s+1} \right) = 1,$$

although, for most people, the extra care involved means that this approach is more likely to cause an error.

To minimize the likelihood of an error, **it is strongly recommended that you express the denominator of a rational function as a product of monic first-order factors before computing a partial-fraction expansion.** This can be done by dividing the numerator and denominator polynomials by the leading coefficient of the denominator polynomial, a_n, prior to computing the inverse Laplace-transform.

5-4-3　Case 3: Repeated Roots

Neither of the forms for the partial-fraction expansion used for Case 1 or Case 2 will work when the poles are not distinct, as is the case for the transform

$$X(s) = \frac{5}{(s+1)^2(s+3)}, \tag{5.45}$$

which has two poles located at $s = -1$. In this case, our partial fraction expansion must also contain terms with double roots. It is straightforward to show that $X(s)$ can be decomposed as

$$\begin{aligned} X(s) &= \frac{5}{(s+1)^2(s+3)} \\ &= \frac{A}{s+1} + \frac{B}{(s+1)^2} + \frac{C}{s+3}. \end{aligned} \tag{5.46}$$

In this example, the order of the numerator polynomial is smaller than the order of the denominator. Were this not the case, it would be necessary to perform an initial long division, just as we did for Case #2.

How can we find the values of the constants A, B, and C? We could recombine the partial-fraction expansion, equate the numerator polynomials, and solve the three linear equations that result for these three unknowns. This approach works if we are willing to solve some linear equations. As an alternative, we can get some of the coefficients by taking limits. For example,

$$C = \lim_{s \to -3}(s+3)X(s) = \lim_{s \to -3}\frac{5}{(s+1)^2} = \frac{5}{4}. \tag{5.47}$$

Similarly, by looking at (5.46), we see that

$$B = \lim_{s \to -1}(s+1)^2 X(s) = \lim_{s \to -1}\frac{5}{s+3} = \frac{5}{2}. \tag{5.48}$$

Unfortunately, this approach will not work to compute A. (Of course, since we now know B and C, it is fairly straightforward to find A by recombining terms and

equating numerators, which results in only one equation in one unknown to be solved.) If we examine the quantity

$$(s+1)X(s) = A + \frac{B}{s+1} + C \cdot \frac{s+1}{s+3},$$

we see that the limit of this function as $s \to -1$ is infinite. As an alternative, consider

$$(s+1)^2 X(s) = A(s+1) + B + C\frac{(s+1)^2}{s+3}.$$

If we compute the derivative of this expression with respect to s, we get

$$\frac{d}{ds}(s+1)^2 X(s) = A + C\frac{2(s+3)(s+1) - (s+1)^2}{(s+3)^2}.$$

If we now take the limit as $s \to -1$, the second term goes to zero. Therefore,

$$A = \lim_{s \to -1}\frac{d}{ds}(s+1)^2 X(s).$$

For the transform in (5.45), this computation yields

$$\begin{aligned} A &= \lim_{s \to -1}\frac{d}{ds}(s+1)^2 X(s) = \lim_{s \to -1}\frac{d}{ds}\left(\frac{5}{s+3}\right) \\ &= \lim_{s \to -1}\frac{-5}{(s+3)^2} = -\frac{5}{4}. \end{aligned}$$

Therefore,

$$X(s) = \frac{5}{(s+1)^2(s+3)} = \frac{-\frac{5}{4}}{s+1} + \frac{\frac{5}{2}}{(s+1)^2} + \frac{\frac{5}{4}}{s+3},$$

and by using the table of Laplace transforms, we discover finally that

$$x(t) = -\frac{5}{4}e^{-t} + \frac{5}{2}te^{-t} + \frac{5}{4}e^{-3t}, \quad \text{for } t \geq 0.$$

It is rare to encounter roots with an even higher multiplicity, and we shall not treat them explicitly here, except to observe that, if we have an $X(s)$ of the form

$$X(s) = \frac{B(s)}{(s - s_i)^p(s - s_j)},$$

it will have a partial-fraction expansion involving terms of the form $A_1/(s - s_i)$, $A_2/(s - s_i)^2, \ldots, A_p/(s - s_i)^p$ and its inverse Laplace-transform will have terms of the form $e^{s_i t}, te^{s_i t}, \ldots, t^{p-1}e^{s_i t}$.

The MATLAB function `residue` can handle rational functions with repeated roots.

5-5 Chapter Summary

5-5-1 Important Points Introduced

- A real sinusoidal waveform is completely defined by three parameters: its amplitude, its frequency, and its phase.
- A real sinusoidal waveform can be expressed as the sum of two complex exponential time functions or as the real part of a complex exponential time function.
- Switched signals are zero for $t < t_0$, for some time t_0.
- The Laplace transform of any switched exponentially weighted sinusoid is a rational function of the Laplace variable s.
- The Laplace transform satisfies the linearity property.
- The inverse Laplace transform of a polynomial is a linear superposition of singularity functions.
- Singularity functions behave like "derivatives" of the unit step function.

5-5-2 New Abilities Acquired

You should now be able to do the following:

(1) Express a sinusoidal signal as a sum of complex exponential functions.

(2) Compute the phasor representation associated with a sinusoidal signal.

(3) Find the Laplace transform of a signal by evaluating the Laplace-transform integral.

(4) Find the Laplace transform of a signal by using the entries in Table 5-1 and the properties of the Laplace transform.

(5) Perform a partial-fraction expansion on a rational function (Laplace transform)
- by setting up and solving a set of linear equations;
- by evaluating limits;
- by using MATLAB.

(6) Compute the inverse Laplace transform of a rational function.
- when the numerator order is lower than the denominator order;
- when the numerator order is greater than or equal to the denominator order;
- when the poles are not distinct.

(7) Manipulate singularity functions, using their special properties.

5-6 Problems

5-6-1 Drill Problems

P5-1 Express each of the following signals as a single sinusoid of the form $A\cos(\omega_0 t + \phi)$.

(a) $x_a(t) = \cos(5\pi t) - 2\sin(5\pi t)$

(b) $x_b(t) = 3\cos(200t + 3\pi/4) + 3\cos(200t + \pi/2)$

(c) $x_c(t) = \cos(40\pi t + 3\pi) + \sqrt{2}\cos(40\pi t + \pi/4)$
$$+ \sqrt{2}\cos(40\pi t - \pi/4)$$

P5-2 Find the Laplace transforms of the following signals.

(a) $x_a(t) = u(t)$

(b) $x_b(t) = \begin{cases} e^{-3(t-3)}, & t > 3 \\ 0, & t < 3 \end{cases}$

(c) $x_c(t) = e^{2t}u(t)$

P5-3 Construct the inverse Laplace transform of

$$X(s) = \frac{s}{s^3 + 6s^2 + 11s + 6} = \frac{s}{(s+1)(s+2)(s+3)}$$

by setting up and solving a set of linear equations.

P5-4 Repeat Problem P5-3 and compute the inverse Laplace transform of

$$X(s) = \frac{s}{s^3 + 6s^2 + 11s + 6},$$

but use limits instead of linear equations to perform the partial-fraction expansion.

P5-5 Compute the inverse Laplace-transform of

$$X(s) = \frac{s}{s^2 + 2s + 2}.$$

5-6-2 Basic Problems

P5-6 Express each of the following as an exponentially weighted sinusoidal time function of the form $x(t) = Ae^{\sigma t}\cos(\omega t + \phi)$:

(a) $x_a(t) = \dfrac{1+j}{2}e^{(-2+j3)t} + \dfrac{1-j}{2}e^{(-2-j3)t}$

(b) $x_b(t) = \sqrt{3}e^{-t}\cos(2t) - e^{-t}\sin(2t)$

(c) $x_c(t) = \dfrac{j\omega_0}{1+j\omega_0}e^{j\omega_0 t} - \dfrac{j\omega_0}{1-j\omega_0}e^{-j\omega_0 t}$

P5-7 Express each of the following as an exponentially weighted sinusoidal time function of the form $x(t) = Ae^{\sigma t}\cos(\omega t + \phi)$:

(a) $x_a(t) = -e^{-2t}\cos t + 2e^{-2t}\sin t$

(b) $x_b(t) = (3 - j2)e^{(-5+j10)t} + (3 + j2)e^{(-5-j10)t}$

(c) $x_c(t) = \dfrac{1}{1+j5}e^{j\omega_0 t} + \dfrac{1}{1-j5}e^{-j\omega_0 t}$

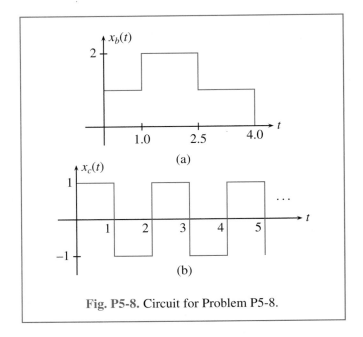

Fig. P5-8. Circuit for Problem P5-8.

P5-8 Many pulse-like signals can be expressed as a sum of delayed step functions. These representations are frequently very useful. Express each of the following signals as a sum of delayed step functions:

(a)

$$x_a(t) = \begin{cases} 1, & |t| < 2 \\ 0, & \text{otherwise} \end{cases}$$

(b) The signal $x_b(t)$ drawn in Figure P5-8(a)

(c) The signal $x_c(t)$ drawn in Figure P5-8(b)

P5-9 Find the Laplace transforms of the following time waveforms:

(a) $e^{-2t}[u(t) - u(t-1)]$

(b) $(t+1)u(t-1)$

(c) $e^{-2t}u(t-2)$

(d) $\frac{d}{dt}[e^{-t}\sin 2t]$

P5-10 Find the Laplace transforms of the following time waveforms:

(a) $x_a(t) = \begin{cases} 1, 0 \leq t \leq T \\ 0, \text{ otherwise} \end{cases}$

(b) $x_b(t) = t^2 e^{-3t}, \quad t > 0$

(c) $x_c(t) = e^{-4t} \sin 5t, \quad t > 0$

(d) $x_d(t) = t, \quad t > 0$

P5-11 Find the Laplace transforms of the following time waveforms:

(a) $x_a(t) = A(1 - e^{-bt})u(t)$

(b) $x_b(t) = \begin{cases} \sin \pi t, 0 \leq t \leq 1 \\ 0, \quad \text{ otherwise} \end{cases}$

(c) $x_c(t) = t^2, \quad t > 0$

(d) $x_d(t) = A\cos(\omega_0 t + \phi), \quad t > 0$

P5-12 Find the Laplace transforms of the following signals:

(a)

$$x_a(t) = \begin{cases} 3, t \geq 0 \\ 1, t < 0 \end{cases}$$

(b)

$$x_b(t) = \begin{cases} 10e^{-2t}, t \geq 3 \\ 0, \quad t < 3 \end{cases}$$

P5-13 Construct the inverse Laplace transform of

$$X(s) = \frac{s}{(s+a)^2 + b^2}.$$

P5-14 Construct the inverse Laplace transform of

$$X(s) = \frac{2s + 6}{s(s^2 + 3s + 2)}.$$

P5-15 Construct the inverse Laplace transform of

$$X(s) = \frac{2s^2 + 5s + 6}{s^3 + 5s^2 + 6s}.$$

P5-16 Construct the inverse Laplace transforms of the following functions:

(a) $X_a(s) = \frac{5s}{2s+1}$

(b) $X_b(s) = \frac{2s-1}{(s+3)^2+4}$

(c) $X_c(s) = \frac{7s^2}{s^3+3s^2+3s+1}$

P5-17 Find the signal $x(t)$ whose Laplace transform is

$$X(s) = \frac{s^2}{s^2 + 25}.$$

P5-18 Construct the inverse Laplace transform of

$$X(s) = \frac{s+3}{s^3 + 3s^2 + 6s + 4}.$$

P5-19 Construct the inverse Laplace transforms of the following functions:

(a) $Y_a(s) = \frac{3}{2s^2+1}$

(b) $Y_b(s) = \frac{s-3}{s^2+6s+9}$

(c) $Y_c(s) = \frac{6s}{(s^2+1)(s^2+4)}$

P5-20 Find the signals whose Laplace transforms are

(a) $X_a(s) = \frac{1}{s^2(s+a)}$

(b) $X_b(s) = \frac{s^3}{(s+1)^2(s+2)}$

P5-21 Construct the inverse Laplace transform of

$$X(s) = \frac{5s - 1}{s^3 - 3s - 2}.$$

P5-22 Find the signal $x(t)$ for $t \geq 0$ if its Laplace transform is

$$X(s) = \frac{s^3}{(s + 1)([s + 1]^2 + 4)}.$$

5-6-3 Advanced Problems

P5-23 If the Laplace transform of $x(t)$ is $X(s)$, derive the Laplace transform of the signal $e^{at}x(t)$.

P5-24 If two signals $x(t)$ and $h(t)$ that are both zero for $t < 0$ are convolved, the Laplace transform of their convolution is $X(s)H(s)$. More formally, if $x(t) = 0$, $t < 0$, $h(t) = 0$, $t < 0$, and

$$y(t) = \int_{-\infty}^{\infty} x(\tau)h(t - \tau)\,d\tau,$$

then

$$Y(s) = X(s)H(s).$$

Using this result, we can evaluate continuous-time convolutions by means of Laplace transforms. Determine the convolution of $h(t) = e^{-at}u(t)$ with each of the following signal waveforms:

(a) $x_a(t) = u(t)$

(b) $x_b(t) = e^{-bt}u(t)$

(c) $x_c(t) = \begin{cases} 1, & 0 < t < T \\ 0, & \text{otherwise} \end{cases}$

P5-25 The system function of a circuit, $H(s)$, is a Laplace-domain description of the circuit's behavior that we will explore extensively in Chapter 7. The inverse Laplace transform of the system function, $h(t)$, is called the *impulse response* of a circuit. We will frequently have occasion to compute the impulse response of a circuit from its system function; this simply involves computing an inverse Laplace transform.

(a) If the system function of a particular circuit is

$$H_{\text{lp}}(s) = \frac{10}{s^2 + 2s + 10},$$

what is its impulse response, $h_{\text{lp}}(t)$, for $t > 0$?

(b) If the system function of a circuit is

$$H_{\text{bp}}(s) = \frac{2s}{s^2 + 2s + 10},$$

what is its impulse response, $h_{\text{bp}}(t)$, for $t > 0$?

(c) If the system function of a circuit is

$$H_{\text{hp}}(s) = \frac{s^2}{s^2 + 2s + 10},$$

what is its impulse response, $h_{\text{hp}}(t)$, for $t > 0$?

P5-26 Compute the inverse Laplace transforms of the following functions:

(a) $X_a(s) = \frac{2s^2}{s^2 + 3s + 2}$

(b) $X_b(s) = \frac{5s + 1}{s^3 - s}$

(c) $X_c(s) = \frac{6s^4}{s^2 - s^3}$

(d) $X_d(s) = \frac{4(s^3 - s^2 + 3s - 15)}{(s^2 + 9)(s^2 + 4s + 13)}$

P5-27 Two interesting properties about the Laplace transform are the Initial and Final Value Theorems.

Initial Value Theorem:

> If both $x(t)$ and $\frac{dx}{dt}$ are Laplace-transformable and if $\lim\limits_{t \to \infty}$ exists, then
>
> $$\lim_{s \to \infty} sX(s) = x(0).$$

Final Value Theorem:

> If both $x(t)$ and $\frac{dx}{dt}$ are Laplace-transformable, and if $sX(s)$ has no poles on the $j\omega$-axis or in the right half plane, then
>
> $$\lim_{s \to 0} sX(s) = \lim_{t \to \infty} x(t).$$

(a) Apply these theorems to each of the following functions $X(s)$ to find (where possible) $x(0)$ and $\lim\limits_{t \to \infty} x(t)$:

> i. $\frac{1 - e^{-sT}}{s}$ iv. $\frac{1}{s^2(s+1)}$
>
> ii. $\frac{1}{s}e^{-sT}$ v. $\frac{2}{s^2+1}$
>
> iii. $\frac{1}{s(s+1)^2}$

(b) In each case, find $x(t)$ and show that the results of the theorems agree.

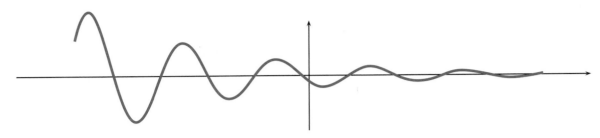

Circuits in the Laplace Domain

Objectives

By the end of this chapter, you should be able to do the following:

1. *Find and solve the differential equations that specify the element variables for first-order circuits.*

2. *Solve differential equations of the first order by using Laplace transforms.*

3. *Map a circuit to the Laplace domain.*

4. *Find the element variables of a circuit with reactive elements by mapping the circuit to the Laplace domain, setting up and solving the relevant equations, and computing inverse Laplace transforms.*

5. *Analyze first-order circuits with constant inputs by inspection.*

Up to now, we have focused our attention on circuits containing only resistors, amplifiers, and sources. These provide a nice vehicle for exploring the implications of Kirchhoff's laws and Ohm's law, but they have a very serious and unfortunate limitation: they can't do very much! Looking back at every example of a resistive circuit that has been presented reveals that the instantaneous value of every element variable is always a simple linear combination of the instantaneous values of the source variables. Adding inductors and capacitors (collectively called *reactive elements*) to a circuit allows it to do a lot more. Because these reactive elements are capable of storing energy in their magnetic and electric fields, they can be used to smooth noisy source waveforms and to decompose source waveforms into their components. This stored energy gives them a sort of memory. Their increased utility, however, comes at a price: reactive circuits are more difficult to analyze than resistive circuits. This is because reactive elements have v–i relations that are different from Ohm's law. As a result, the element variables and the source waveforms are related by differential equations, not algebraic ones. The Laplace transform will prove to be a very powerful tool for analyzing these circuits. Initially, we shall use it to solve the differential equations that describe them; later we shall show that it is also extremely useful to enable us to write those equations. Still later, in Chapter 7, we will see that the Laplace-transform domain analysis of a circuit provides additional insight into how it behaves.

The Laplace transform is much more than a tool, however; it is also an enabler. It allows us to define a different domain, the *Laplace domain*, in which Kirchhoff's laws and Ohm's law still apply, even for reactive elements. By creating a new domain in which these familiar rules are again applicable, all of the techniques that were developed in Chapters 1–4 can be extended to analyze reactive circuits. The three-step approach we shall be using for reactive circuits was summarized in Figure 5.1:

1. First, compute the Laplace transforms of the (input) source time functions.

2. Next, solve the circuit *in the Laplace domain* for the Laplace transforms of the desired element variables or other variables of interest.

3. Finally, use the inverse Laplace transform to discover the time behavior of the variables of interest.

Chapter 5 focused on Steps 1 and 3, the methods for determining forward and inverse Laplace transforms. This chapter develops the Laplace-domain analysis techniques necessary for Step 2 and then shows how the overall procedure works. First, however, we look at the time-domain behavior of a few simple reactive circuits. This will help us to understand how these circuits work and enable us to verify later that the Laplace-domain

solution and the differential-equation solution are, in fact, the same.

6-1 Circuits with One Reactive Element

Before jumping into the Laplace domain, it is important to look at the differential equations that describe the input–output behavior of some simple circuits and solve them. We have several reasons. First, by verifying that the solution of the differential equation agrees with the solution obtained via Laplace transforms, we hope to demystify the transform approach. In addition, solving circuits by using both methods will expose the relative simplicity of the transform approach vis-a-vis the direct solution and provide an appreciation of its power. Both of these motivations, however, are secondary. The most important reason for looking at the differential equations behind a circuit and their solution is to provide some important insight into how circuits behave. For example, it will allow us to identify the forced and transient response of the circuit with the particular and homogeneous solutions, respectively, of the differential equation and to understand how the initial values of certain voltages and currents in the circuit affect its behavior. It will also make it possible for us to understand noncircuit systems whose input–output behavior is described by a differential equation and to solve those equations via transforms.

6-1-1 Differential-Equation Descriptions

To analyze a circuit containing one or more inductors and capacitors, we begin as we have always begun, by writing KVL equations, KCL equations, and element relations and then solving them. The element relations for capacitors and inductors, however, involve derivatives, which means that we cannot solve for the element variables without solving a differential equation. This section focuses on the problem of finding the differential equation that relates an element variable of interest and its derivatives to the source variables and possibly their derivatives as well; we defer the solution methods until the following section. Finding the differential equation that defines an element variable for a complex circuit without using Laplace transforms can be algebraically tedious. For this reason, we confine our initial attention to the case of circuits with one inductor or one capacitor, for which the problem is manageable. For the more general case, we shall use transform techniques exclusively. (Problems P6-34 and P6-37 give you an opportunity to set up differential equations for some circuits with two reactive elements.)

Every element variable in a circuit can be figured out from the source variables by solving a differential equation. All of these equations will be of the same order, which is called the *order of the circuit*.

> *The order of a circuit is never greater than the number of reactive elements, although it can be less.*

Circuits containing one reactive element are of first order. For the solution of a differential equation to be unique, we must specify some auxiliary conditions, which are related to the internal energy stored in the inductors and/or capacitors at some time instant.

> *The number of auxiliary conditions required is equal to the order of the circuit. These auxiliary conditions will normally be the values of the capacitor voltages and the inductor currents at a specific time, usually $t = 0$.*

The procedure for finding the differential equations for a first-order circuit is demonstrated through the following examples.

Example 6-1   A Parallel RC Circuit

Consider the circuit in Figure 6-1, which consists of a parallel connection of a resistor, a capacitor, and a current source for which it is known that the capacitor voltage at time $t = 0$ is $v_c(0) = v_0$. The reference directions for the element currents, $i_r(t)$ and $i_c(t)$, are given by the default sign conventions. Since the network contains only a single mesh that does not include an exterior current source and two nodes, the values for the four element variables can be found by setting up and solving four equations (one KVL equation, one KCL equation, and two element relations):

$$-v_r(t) + v_c(t) = 0 \qquad (6.1)$$

$$i_r(t) + i_c(t) = i_s(t) \qquad (6.2)$$

$$i_r(t) = \frac{1}{R}v_r(t) \qquad (6.3)$$

$$i_c(t) = C\frac{dv_c(t)}{dt}. \qquad (6.4)$$

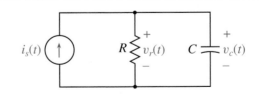

Fig. 6-1. A first-order circuit containing a single capacitor and resistor connected in parallel with a current source. The initial value of the capacitor voltage is v_0.

Using (6.1) to equate $v_r(t)$ and $v_c(t)$ and substituting (6.3) and (6.4) into (6.2) gives the following differential equation involving only $v_c(t)$ and $i_s(t)$:

$$\frac{dv_c(t)}{dt} + \frac{1}{RC}v_c(t) = \frac{1}{C}i_s(t). \qquad (6.5)$$

The other element variables in the circuit can also be found as the solutions of first-order differential equations through algebraic manipulation of (6.1), (6.3), (6.4), and (6.5):

$$\frac{dv_r(t)}{dt} + \frac{1}{RC}v_r(t) = \frac{1}{C}i_s(t)$$

$$\frac{di_r(t)}{dt} + \frac{1}{RC}i_r(t) = \frac{1}{RC}i_s(t)$$

$$\frac{di_c(t)}{dt} + \frac{1}{RC}i_c(t) = \frac{d\,i_s(t)}{dt}.$$

The auxiliary conditions corresponding to these are

$$v_r(0) = v_0$$

$$i_r(0) = v_0/R$$

$$i_c(0) = i_s(0) - \frac{1}{R}v_0.$$

When we want all four element variables (as is rarely the case), instead of solving all four differential equations, a better approach is to solve (6.5) first and then to substitute the result for $v_c(t)$ into (6.1), (6.3), and (6.4) for the other variables. ■

 DEMO: *Parallel RC Circuit*

Example 6-2   A Series RL Circuit

Figure 6-2 shows another first-order circuit, this one consisting of a resistor and an inductor connected in series with a voltage source. The only variable of interest is the

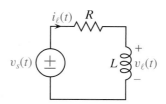

Fig. 6-2. A series RL circuit. The variable of interest is the voltage $v_\ell(t)$.

voltage across the terminals of the inductor, $v_\ell(t)$, and we are given the auxiliary condition that $i_\ell(0) = 0$. Because of the series connection, $i_r(t) = i_\ell(t)$. This leaves one KVL equation and the two element relations:

$$v_r(t) + v_\ell(t) = v_s(t) \qquad (6.6)$$

$$v_r(t) = Ri_\ell(t) \qquad (6.7)$$

$$v_\ell(t) = L\frac{di_\ell(t)}{dt}. \qquad (6.8)$$

From (6.6) and (6.7),

$$i_\ell(t) = \frac{1}{R}\left[v_s(t) - v_\ell(t)\right],$$

which can be differentiated and substituted into (6.8). The result is a differential equation involving only $v_\ell(t)$:

$$\frac{dv_\ell(t)}{dt} + \frac{R}{L}v_\ell(t) = \frac{dv_s(t)}{dt}.$$

The auxiliary condition is

$$v_\ell(0) = v_s(0) - Ri_\ell(0) = v_s(0).$$

■

 To check your understanding, try Drill Problem P6-1.

 DEMO: *Series RL Circuit*

6-1-2   Solving Differential Equations with Exponential Inputs

A simple example of a linear, first-order differential equation with constant coefficients is

$$\frac{dy(t)}{dt} + \frac{1}{\tau}y(t) = Xe^{s_p t}, \qquad t > 0, \qquad (6.9)$$

with the auxiliary condition $y(0) = y_0$. In this equation, $y(t)$ is the variable of interest. It might be an element variable, a node potential, a mesh current, etc. The parameter τ is called the *time constant* of the circuit for reasons that will become apparent later. In this equation, the expression on the right-hand side, sometimes called the *forcing function* or simply the *input*, is a complex exponential. X is its complex amplitude and s_p is its complex frequency. This differential equation is similar in form to the one derived in the first example of the previous section, except that we have assumed that the forcing function (source variable) is a complex exponential time function. Although placing such a restriction on the input might appear to limit our results, this is not really the case, because all of the source waveforms that we shall treat analytically can be constructed from complex exponential time functions. Notice that, when $s_p = 0$, the right-hand side is constant (for $t > 0$); when s_p is real, it is a real exponential; and when s_p is purely imaginary, it is a complex sinusoid. Our approach will find $y(t)$ for $t > 0$.

The solution of a differential equation with constant coefficients and appropriate initial conditions is guaranteed to be unique. Exploiting this fact, our approach will be to guess at the form of the solution and then adjust its free parameters until the result satisfies both the equation and the auxiliary (initial) condition.

Since the right-hand side of the equation is a complex exponential time function with complex frequency s_p, the left-hand side—the sum $\frac{dy(t)}{dt} + \frac{1}{\tau}y(t)$—must also be a complex exponential time function. Complex exponential time functions have the property that they are proportional to their derivatives. Thus, if $y(t)$ is a complex exponential time function with complex frequency s_p, then $\frac{dy(t)}{dt} + \frac{1}{\tau}y(t)$ will also be such a function with the same complex frequency. This suggests beginning with a trial solution of the form

$$y_f(t) = Ye^{s_p t}, \qquad t > 0,$$

where the complex amplitude Y remains to be computed. If we substitute this trial solution into (6.9), we get

$$(s_p + \frac{1}{\tau})Ye^{s_p t} = Xe^{s_p t}, \qquad t > 0,$$

which is satisfied if

$$Y = \frac{X}{s_p + \frac{1}{\tau}}.$$

Thus, one solution of this equation is

$$y_f(t) = \frac{X}{s_p + \frac{1}{\tau}}e^{s_p t} \qquad t > 0. \qquad (6.10)$$

We call this solution the *forced solution* of the differential equation. The forced solution has the same form as the source waveform and it satisfies the differential equation. It is not the final solution of that equation, however, because usually it does *not* satisfy the auxiliary condition.

Another solution can be found by adding to the forced solution a solution of the differential equation with the right-hand side set to zero, that is, a solution of the *homogeneous equation*

$$\frac{dy(t)}{dt} + \frac{1}{\tau}y(t) = 0. \qquad (6.11)$$

A solution of the homogeneous equation is called a *homogeneous solution*. If a homogeneous solution $y_h(t)$ is added to the forced solution, the result still satisfies the original differential equation. To see this, we first write the separate equations

$$\frac{dy_f(t)}{dt} + \frac{1}{\tau}y_f(t) = Xe^{s_p t}, \qquad t > 0,$$

$$\frac{dy_h(t)}{dt} + \frac{1}{\tau}y_h(t) = 0,$$

and then add them together:

$$\frac{d}{dt}[y_f(t) + y_h(t)] + \frac{1}{\tau}[y_f(t) + y_h(t)] = Xe^{s_p t}, \qquad t > 0.$$

This equation implies that the sum of the forced solution and a homogeneous solution is also a solution of the original equation.

To find a homogeneous solution, we observe that the homogeneous equation implies

$$\frac{dy_h(t)}{dt} = -\frac{1}{\tau}y_h(t). \qquad (6.12)$$

Recognizing again that an exponential function is proportional to its derivative, we try the solution

$$y_h(t) = Ae^{st}, \qquad (6.13)$$

for some value of s and a constant value of A. Substituting (6.13) into (6.11) gives

$$Ase^{st} + \frac{1}{\tau}Ae^{st} = 0,$$

or

$$(s + \frac{1}{\tau})Ae^{st} = 0,$$

which is satisfied if

$$s = -\frac{1}{\tau}$$

for all values of A. The general solution of the equation is the sum of the forced and homogeneous solutions:

$$y(t) = Ae^{-t/\tau} + \frac{X}{s_p + \frac{1}{\tau}}e^{s_p t}, \qquad t > 0. \quad (6.14)$$

The solution in (6.14) satisfies the original differential equation for all values of the constant A. Only one value of A, however, will satisfy the initial condition. To find that value, we take the limit as $t \rightarrow 0$ in (6.14). This gives

$$A + \frac{X}{s_p + \frac{1}{\tau}} = y(0) = y_0,$$

which is satisfied if

$$A = y_0 - \frac{X}{s_p + \frac{1}{\tau}}.$$

Putting all of the pieces together, we finally have

$$y(t) = \frac{X}{s_p + \frac{1}{\tau}}(e^{s_p t} - e^{t/\tau}) + y_0 e^{t/\tau}, \qquad t > 0,$$

$$(6.15)$$

which satisfies both the differential equation and the initial condition.

Example 6-3   Finding the Response to a Constant Input

To illustrate the procedure that we have just derived, let us find the solution for $t > 0$ to the circuit in Figure 6-3 when $i_s(t) = 1$ and $v_c(0) = 0$. Notice that this is the same circuit as Example 6-1, except that we have assigned numerical values for R and C and set $i_s(t) = 1$.

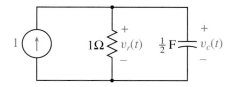

Fig. 6-3. The first-order circuit of Figure 6-1 with specific values specified. The current source is 1 for $t > 0$, and the initial value of the capacitor voltage is zero.

Borrowing from the solution to that example, we know that $v_c(t)$ is the solution of the differential equation

$$\frac{dv_c(t)}{dt} + 2v_c(t) = 2. \qquad (6.16)$$

Since the right-hand side is a constant (i.e., a complex exponential time function with $s_p = 0$), the forced solution, $v_{cf}(t)$, will be constant. In fact, it is straightforward to verify that

$$v_{cf}(t) = 1.$$

The homogeneous equation is

$$\frac{dv_{ch}(t)}{dt} + 2v_{ch}(t) = 0,$$

and the solution of that equation (the homogeneous solution) is

$$v_{ch}(t) = Ae^{-2t}.$$

(You should verify that this is indeed a solution of the homogeneous equation.) This gives as a total solution

$$v_c(t) = 1 + Ae^{-2t}, \qquad t > 0.$$

Finally, for $v_c(0)$ to be zero, we set $A = -1$. This results in the final answer:

$$v_c(t) = 1 - e^{-2t}, \qquad t > 0, \qquad (6.17)$$

which is sketched in Figure 6-4. The voltage on the capacitor begins at its initial value of zero and asymptotically approaches a constant value of unity as t increases. For this problem, the forced solution is the solution for large values of t. The homogeneous solution contributes to the transitional behavior. The time constant τ ($\tau = RC = \frac{1}{2}$ for this problem) is a measure of the duration of the transition, because, when $t = \tau$, $v_c(t) = 1 - e^{-1} \approx 0.63$.

■

WORKED SOLUTION: *Constant Input*

Example 6-4   **Finding the Response to an Exponential Input**

As another example, consider the series RL circuit in Figure 6-5, which has an exponential voltage source waveform and an initial current flowing through the inductor, $i_\ell(0) = 1$. The variable of interest is the voltage across the terminals of the resistor.

To find the differential equation, we can begin with a simple statement of KVL:

$$v_\ell(t) + v_r(t) = e^{-t}. \qquad (6.18)$$

Next, we observe that

$$i_\ell(t) = i_r(t) = v_r(t)$$

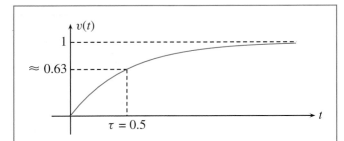

Fig. 6-4. The output waveform for Example 6-3.

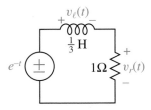

Fig. 6-5. A series RL circuit with an exponential source waveform.

and that

$$v_\ell(t) = \frac{1}{3}\frac{di_\ell(t)}{dt} = \frac{1}{3}\frac{dv_r(t)}{dt},$$

so that (6.18) becomes

$$\frac{dv_r(t)}{dt} + 3v_r(t) = 3e^{-t}, \qquad t > 0, \qquad (6.19)$$

with the initial condition

$$v_r(0) = i_\ell(0) = 1.$$

To solve this equation for $t > 0$, we try a forced solution of the form $v_{rf}(t) = Ve^{-t}$ and substitute into (6.19) to compute V. This gives

$$-Ve^{-t} + 3Ve^{-t} = 3e^{-t},$$

which is satisfied if

$$V = \frac{3}{2}.$$

Therefore, the forced solution is $v_{rf}(t) = \frac{3}{2}e^{-t}$. To find the homogeneous solution, substitute a trial solution $v_{rh}(t) = Ae^{st}$ into the homogeneous equation

$$\frac{dv_{rh}}{dt} + 3v_{rh}(t) = 0.$$

This results in

$$(s + 3)Ae^{st} = 0,$$

which is satisfied when $s = -3$. Therefore, the total solution for $t > 0$ is

$$v_r(t) = \frac{3}{2}e^{-t} + Ae^{-3t}.$$

We choose the free parameter A so that the initial condition is satisfied:

$$A = v_r(0) - \frac{3}{2} = -\frac{1}{2}.$$

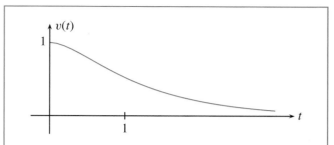

Fig. 6-6. The computed resistor voltage $v_r(t)$ for the circuit in Figure 6-5.

Then the final solution is

$$v_r(t) = \frac{3}{2}e^{-t} - \frac{1}{2}e^{-3t}, \qquad t > 0,$$

which is sketched in Figure 6-6. ∎

To check your understanding, try Drill Problem P6-2.

This approach to solving differential equations with constant coefficients and complex exponential inputs can be extended to equations of higher order. The approach is indicated in Problem P6-35.

6-1-3   Solving Differential Equations by Using Laplace Transforms

Laplace transforms are a very natural tool for solving differential equations of any order; however, in this section we shall continue to limit ourselves to first-order differential equations corresponding to first-order circuits.

To illustrate the method, let's again solve the differential equation that we saw in the previous section (6.9), which is recopied below.

$$\frac{dy(t)}{dt} + \frac{1}{\tau}y(t) = Xe^{s_pt}, \qquad t > 0 \qquad (6.20)$$

It is helpful to interpret this equation as an implicit equation that relates two time waveforms: one, on the left-hand side, that is derived from $y(t)$; the other a complex exponential time function on the right-hand side. Since these two time waveforms are equal, their Laplace transforms must be equal as well. For computing these transforms, the following transform pairs are relevant :

$$Xe^{s_p t}, \quad t > 0 \longleftrightarrow \frac{X}{s - s_p} \qquad (6.21)$$

$$\frac{1}{\tau}y(t), \quad t > 0 \longleftrightarrow \frac{1}{\tau}Y(s) \qquad (6.22)$$

$$\frac{dy(t)}{dt}, \quad t > 0 \longleftrightarrow sY(s) - y_0. \qquad (6.23)$$

Equation (6.21) comes from Table 5-1 of Laplace transform pairs, with $a = -s_p$; (6.22) follows from the linearity property; and (6.23) follows from the differentiation-in-time property of the Laplace transform. (Note that $y(0) = y_0$.)

In the Laplace-transform domain, (6.20) becomes

$$sY(s) - y_0 + \frac{1}{\tau}Y(s) = \frac{X}{s - s_p}, \qquad (6.24)$$

which we can solve for $Y(s)$:

$$Y(s) = \frac{X}{(s - s_p)(s + \frac{1}{\tau})} + \frac{y_0}{s + \frac{1}{\tau}}. \qquad (6.25)$$

We can now use familiar methods from Chapter 5 to compute the inverse Laplace transform. We begin with a partial-fraction expansion,

$$Y(s) = \frac{\frac{X}{s_p + \frac{1}{\tau}}}{s - s_p} - \frac{\frac{X}{s_p + \frac{1}{\tau}}}{s - \frac{1}{\tau}} + \frac{y_0}{s - \frac{1}{\tau}}.$$

By taking the inverse Laplace transforms of the individual terms and using the linearity property, we arrive at the result

$$y(t) = \frac{X}{s_p + \frac{1}{\tau}}(e^{s_p t} - e^{t/\tau}) + y_0 e^{t/\tau}, \qquad t > 0. \qquad (6.26)$$

Notice that this is the same as (6.15) and that it satisfies both the differential equation and the initial condition. When we use the Laplace transform, we do not need to address the forced solution and the homogeneous solutions separately, since it takes care of both at the same time. Similarly, the initial conditions are also incorporated into the solution directly.

Example 6-5   Example 6-3 Revisited

In Example 6-3, we found the voltage that would be measured across a capacitor that was connected in parallel to a resistor and a current source when the input was constant for $t > 0$ and the initial value of the voltage was zero at $t = 0$. The differential equation that was satisfied by that voltage was

$$\frac{dv_c(t)}{dt} + 2v_c(t) = 2.$$

Taking the Laplace transforms of the functions on the two sides of the equation yields

$$sV_c(s) + 2V_c(s) = \frac{2}{s},$$

which we can solve for $V_c(s)$:

$$V_c(s) = \frac{2}{s(s + 2)} = \frac{1}{s} - \frac{1}{s + 2}.$$

Thus,

$$v_c(t) = 1 - e^{-2t}, \quad t > 0,$$

as before.   ∎

Example 6-6   Laplace-Transform Solution to Example 6-4

In Example 6-4, the variable of interest satisfied the differential equation

$$\frac{dv_r(t)}{dt} + 3v_r(t) = 3e^{-t}, \qquad t > 0, \qquad (6.27)$$

with an initial condition of $v_r(0) = 1$. Computing the Laplace transform of both sides of (6.27) gives

$$s V_r(s) - 1 + 3 V_r(s) = \frac{3}{s+1}.$$

Notice the presence of the -1 on the left-hand side of the equation. This is introduced because of the initial condition. Solving for $V_r(s)$, we obtain

$$V_r(s) = \frac{3}{(s+1)(s+3)} + \frac{1}{s+3} = \frac{\frac{3}{2}}{s+1} - \frac{\frac{1}{2}}{s+3}$$

from which

$$v_r(t) = \frac{3}{2}e^{-t} - \frac{1}{2}e^{-3t}, \quad t > 0.$$

■

To check your understanding, try Drill Problem P6-3.

6-2   Circuits in the Laplace Domain

In Section 6-1, we found the solutions to first-order circuits containing a reactive element via a two-step procedure:

1. First, we found the differential equation that was satisfied by the voltage or current of interest and its initial values.

2. Then, we solved that differential equation, either directly or by using Laplace transforms.

The Laplace transform simplified the labor in the second of these steps, but finding the differential equation itself in the first step proved to be somewhat tricky. It becomes even more difficult when there are multiple reactive elements in the circuit.

The key to simplifying this task is to incorporate the Laplace transform at an earlier point in the analysis, to avoid the necessity of constructing the differential equation at all. In fact, it is helpful to move to the Laplace-transform domain at the earliest possible time—before even writing the KCL and KVL equations and the element relations.

Figure 6-7(a) shows a fairly simple circuit containing a single voltage source and three elements. It looks much like all of the other circuits that we have seen; the elements and source are connected together, and we

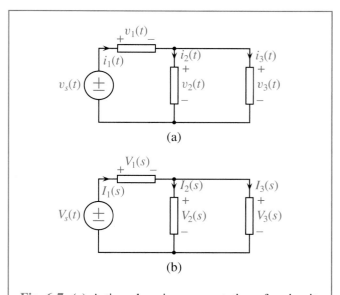

(a)

(b)

Fig. 6-7. (a) A time-domain representation of a circuit. (b) The Laplace-domain representation of the same circuit.

have identified the currents and the voltage associated with all the elements, which vary as functions of time. At any time, we can measure the instantaneous values of all these variables, using a measurement device, such as a meter or an oscilloscope. This familiar representation of the circuit, in which all of the voltages and currents are expressed as functions of time, is called the *time-domain representation of the circuit*.

Figure 6-7(b) shows the *Laplace-domain represen-tation* of the same circuit. It also contains a number of sources and elements connected together, and we associate a current and a voltage with each element, but here the voltages and currents are labelled as functions of the Laplace-domain variable s instead of the time-domain variable t. The Laplace-domain element and source variables are the Laplace transforms of the corresponding time-domain variables—for example, $V_s(s)$ is the Laplace transform of $v_s(t)$ and $I_1(s)$ is the Laplace transform of $i_1(t)$. To try to minimize confusion, we will continue to use lowercase letters to denote time-domain element variables and uppercase letters to denote the corresponding Laplace-domain element variables.

In the Laplace domain, the currents at a node, which are functions of s, still satisfy Kirchhoff's current law, and voltages summed over any closed path still satisfy Kirchhoff's voltage law. Furthermore, there are element relations that tie the voltages across each element to the currents flowing through them, but the element relations in the time and Laplace domains are different for some elements. Unfortunately, we cannot measure the Laplace-domain voltages and currents with oscilloscopes or meters. If we connect actual devices to build the circuit and make measurements, we will always measure the time-domain variables. This means that, whenever we perform an analysis in the Laplace domain, the final step must always be an inverse Laplace transformation to convert the signals of interest back to their time-domain representation. Fortunately, we already know how to do this.

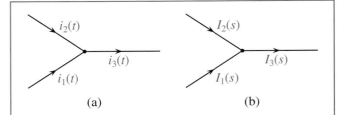

Fig. 6-8. Kirchhoff's current law. (a) Time domain. (b) Laplace domain.

6-2-1   KCL in the Laplace Domain

Consider a node in a circuit, such as the one shown in Figure 6-8(a) that connects three branches. KCL, applied to this node, tells us that

$$i_1(t) + i_2(t) - i_3(t) = 0, \qquad (6.28)$$

given the reference directions of the currents. However, from the linearity property of the Laplace transform, we know that the Laplace transform of the sum of three signals is the sum of their Laplace transforms. Therefore, if we take the Laplace transforms of the quantities on both sides of (6.28), we get a Laplace-domain statement of KCL:

$$I_1(s) + I_2(s) - I_3(s) = 0. \qquad (6.29)$$

This states that

> *The sum of the Laplace transforms of the currents entering a node is zero.*

The implication is that KCL applies in the Laplace domain just as it did in the time domain. Equation (6.29) corresponds to KCL at the Laplace-domain node shown in Figure 6-8(b). Notice that the reference directions of the currents in the Laplace domain are the same as the corresponding reference directions of the variables in the time domain. The conclusion is that

> *KCL is the same for both the time-domain and the Laplace-domain representations of a circuit.*

The simple derivation of this result depends only on the time-domain statement of KCL and on the linearity property of the Laplace transform. It should be clear that the Laplace-domain statement of KCL is valid whenever the time-domain statement is valid. This means that it is true when the number of currents entering a node is different from three or when KCL is applied to the currents crossing an arbitrary closed surface that is not a node. Furthermore, when writing KCL equations in the Laplace domain, the number of independent KCL equations that we can write is the same as it is when we write them in the time-domain: one fewer than the number of nodes in the circuit, since there is a one-to-one relationship between the time-domain currents and their Laplace transforms.

6-2-2   KVL in the Laplace Domain

Now consider an arbitrary closed path in a circuit, such as the one in Figure 6-9(a) that contains four elements. For this path, the familiar time-domain statement of Kirchhoff's voltage law tells us that

$$v_1(t) + v_2(t) - v_4(t) - v_3(t) = 0, \qquad (6.30)$$

but once again we can take the Laplace transform of the quantities on both sides of the equation by exploiting the linearity property of the Laplace transform. This provides a Laplace domain statement of KVL. For this path, it states that

$$V_1(s) + V_2(s) - V_4(s) - V_3(s) = 0. \qquad (6.31)$$

> *The sum of the Laplace transforms of the voltages along any closed path is zero.*

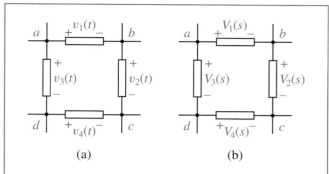

Fig. 6-9. A closed path from a network illustrating KVL. (a) Time-domain representation. (b) Laplace-domain representation.

The Laplace-domain representation of this equation is shown in Figure 6-9(b). The conclusion is similar to the one that we drew for KCL:

> *KVL is the same for either the time-domain or the Laplace-domain representation of a circuit.*

The Laplace-domain statement of KVL is virtually identical to the time-domain statement, and its implications are also the same: KVL equations written over the paths corresponding to meshes in a circuit are independent of each other and sufficient to guarantee that KVL will be satisfied over *all* closed paths.

6-2-3   Element Relations in the Laplace Domain

We have just seen that Kirchhoff's current and voltage laws are the same in both the time and the Laplace domains. This is not the case for the element relations for reactive elements (inductors and capacitors). Let's begin, however, with resistors and dependent sources, for which the element relations are similar in the two domains.

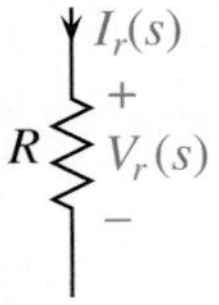

$I_r(s)$

$+$

$R \lessgtr V_r(s)$ $V_r(s) = RI_r(s)$

$I_r(s) = \frac{1}{R} V_r(s) = GV_r(s)$

$-$

Fig. 6-10. An ideal resistor in the Laplace domain.

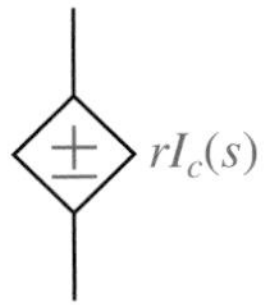

$rI_c(s)$

Fig. 6-11. A current-controlled voltage source (CCVS) in the Laplace domain.

Resistors: Resistors obey Ohm's law; the instantaneous value of the voltage across their terminals is proportional to the instantaneous value of the current flowing between them. If $v_r(t)$ and $i_r(t)$ denote the voltage and current, respectively, then

$$v_r(t) = Ri_r(t).$$

We again use the linearity property of the Laplace transform to evaluate Laplace transforms of the functions on both sides of this expression to get

$$V_r(s) = RI_r(s),$$

or

$$I_r(s) = \frac{1}{R} V_r(s) = GI_r(s).$$

Thus, we see that Ohm's law applies to resistors in both the time and the Laplace domains without modification. The Laplace-domain behavior of a resistor is summarized in Figure 6-10.

Dependent Sources: We have included dependent sources before inductors and capacitors because, like resistors, their *v–i* relations are similar in the time and Laplace domains. If we use a current-controlled voltage source as an example, the voltage of the source is given by

$$v_s(t) = ri_c(t),$$

where the current $i_c(t)$ is a current elsewhere in the circuit. Taking Laplace transforms of the quantities on the two sides shows that

$$V_s(s) = rI_c(s).$$

As long as the controlling relation for a dependent source is linear, that same relation is valid in both the time and Laplace domains. The Laplace-domain rendition of a current-controlled voltage source is shown in Figure 6-11. The other types of dependent sources (current-controlled current sources, etc.) extend to the Laplace domain in a similar fashion.

Inductors: Figure 6-12 shows the familiar time-domain representation of an inductor with an inductance L, current $i_\ell(t)$, and voltage $v_\ell(t)$. Its *v–i* relation is

$$v_\ell(t) = L\frac{di_\ell(t)}{dt}. \tag{6.32}$$

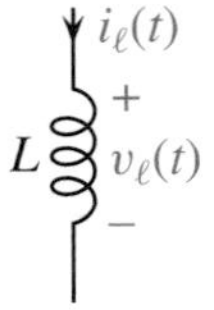

$i_\ell(t)$

$+$

$L \lessgtr v_\ell(t)$

$-$

$v_\ell(t) = L\frac{di_\ell(t)}{dt}$

$i_\ell(t) = \frac{1}{L} \int_{t_0}^{t} v_\ell(\beta)d\beta + i_\ell(t_0)$

Fig. 6-12. The time-domain representation of an ideal inductor.

We can use the derivative property of the Laplace transform to evaluate the Laplace transform of both sides. The result will be an equation involving $I_\ell(s)$ and $V_\ell(s)$, which we will call the *Laplace-domain v–i characteristic of an inductor*:

$$V_\ell(s) = L \cdot \mathcal{L}\left\{\frac{di_\ell(t)}{dt}\right\}$$

$$= L\{sI_\ell(s) - i_\ell(0)\}$$

$$= LsI_\ell(s) - Li_\ell(0). \qquad (6.33)$$

Notice that this *v–i* relation is different from others in two significant ways. First, it contains two terms, one of which involves the initial value of the current flowing through the inductor. Second, the coefficient of proportionality depends on the variable s.

Based on our earlier discussions of differential-equation representations for circuits, it is very important that the *v–i* relations for the reactive elements contain a dependence on the initial value of some variable. This provides a means for including the initial conditions, which are needed to solve the differential equation, in the analysis of the circuit when the differential equation is not written explicitly. To manipulate (6.33) into a more familiar form, we use it to solve for $I_\ell(s)$:

$$I_\ell(s) = \frac{1}{Ls}V_\ell(s) + \frac{1}{s}i_\ell(0). \qquad (6.34)$$

This is similar in form to the *v–i* relation of a Norton equivalent network constructed from a current source connected in parallel with a resistor, which leads to the equivalent model shown in Figure 6-13(b). Notice that both the current flowing through the current source, $\frac{1}{s}i_\ell(0)$, and the "resistance" of the other element are functions Laplace variable s. This is not a problem (indeed, it will prove to be extremely useful), but it is different from what we have seen with resistive circuits in the time domain, and it will take some getting used to.

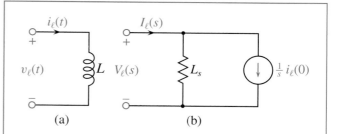

Fig. 6-13. Two representations for an inductor with an initial current. (a) Time-domain representation. (b) Laplace-domain representation.

Before we go any further, it is useful to comment on the two elements that make up the model in Figure 6-13(b). The first element has the *v–i* relation

$$I(s) = \frac{1}{Ls}V(s),$$

or

$$V(s) = (Ls) \cdot I(s).$$

This is similar to Ohm's law, with the quantity Ls playing a role similar to a resistance and $\frac{1}{Ls}$ playing a role similar to a conductance. This is why we choose, for the time being, to draw the element with the same symbol that we use for a resistor. The quantity Ls is formally called the *impedance* of the inductor, and the quantity $\frac{1}{Ls}$ is called its *admittance*. Impedances are measured in ohms (Ω) and admittances in siemens (S), just as are resistances and conductances, respectively. *This Ohm's law–like statement of the v–i relation is valid only in the Laplace domain;* in the time domain, the inductor voltage and current still have their familiar *v–i* relation involving a derivative that is given in (6.32).

Now consider the current source in Figure 6-13(b). The Laplace-domain source value is $\frac{1}{s}i_\ell(0)$. This is the Laplace transform of a constant. When the initial current through the inductor $i_\ell(0)$ is zero, this source reduces to

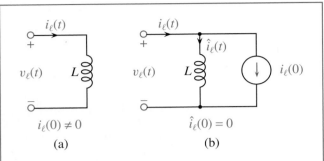

Fig. 6-14. (a) An inductor with an initial current $i_\ell(0)$. (b) An equivalent time-domain model for that inductor, one that models the initial current by using a constant current source.

an open circuit; it will not be present at all! In this case, the Laplace-domain model for the inductor reduces to a simple impedance. When the initial current is not zero, the current source will generate a Laplace-domain current $\frac{1}{s}i_\ell(0)$, corresponding to a constant *time-domain* current $i_\ell(0)$. This provides our first piece of useful insight from looking at circuit behavior in the Laplace domain:

> *An inductor with an initial current $i_\ell(0)$ can be modelled in the time domain as an inductor with an initial current of zero connected in parallel with a constant current source $i_\ell(0)$ as shown in Figure 6-14.*

Before we look at the capacitor, we should summarize the main result of this section:

> *An inductor with an inductance L can be modelled in the Laplace domain by an impedance of value Ls connected in parallel with a current source of value $\frac{1}{s}i_\ell(0)$.*

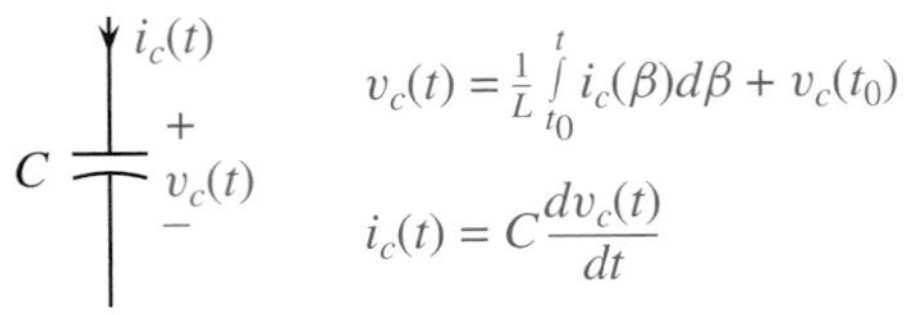

Fig. 6-15. The time-domain representation of an ideal capacitor.

Capacitors: Capacitors are also elements that contain derivatives in their time-domain v–i relations. Like inductors, their time-domain and Laplace-domain descriptions are quite different. Figure 6-15 repeats the time-domain representation of a capacitor, which has the v–i relation

$$i_c(t) = C\frac{dv_c(t)}{dt}. \qquad (6.35)$$

Once again, we compute the Laplace-domain v–i characteristic by transforming the quantities on both sides of (6.35), using the derivative property of the Laplace transform:

$$\begin{aligned} I_c(s) &= C \cdot \mathcal{L}\left\{\frac{dv_c(t)}{dt}\right\} \\ &= C\{sV_c(s) - v_c(0)\} \\ &= CsV_c(s) - Cv_c(0). \qquad (6.36) \end{aligned}$$

To get a Laplace-domain model involving more familiar circuit elements, as we did for the inductor, we solve (6.36) for $V_c(s)$:

$$V_c(s) = \frac{1}{Cs}I_c(s) + \frac{1}{s}v_c(0). \qquad (6.37)$$

This is the same as the potential difference that we would measure across two elements that are connected in series. Since the first term contributes a voltage that is proportional to the current flowing through both elements, we model it as an impedance with a value of $\frac{1}{Cs}$;

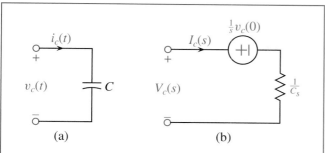

Fig. 6-16. Two representations for a capacitor with an initial voltage. (a) Time-domain representation. (b) Laplace-domain representation.

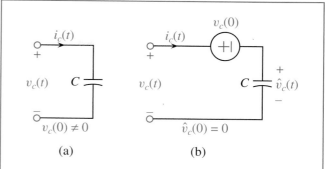

Fig. 6-17. A capacitor with an initial voltage $v_c(0)$. (b) An equivalent time-domain model for that capacitor, one that models the initial voltage by using a constant voltage source.

and since the voltage from the second term is independent of the current, we model it as a voltage source. The resulting Laplace-domain model for a capacitor with an initial voltage of $v_c(0)$ is shown in Figure 6-16.

When the initial capacitor voltage, $v_c(0)$, is zero, the voltage source is replaced by a short circuit, and the Laplace-domain representation of the capacitor reduces to a single impedance with a value of $\frac{1}{Cs}$. The admittance of the capacitor is Cs. Notice that, for both the inductor and the capacitor, the impedance varies with the variable s, but for the inductor it is directly proportional to s, whereas for the capacitor it is inversely proportional to s.

The voltage source that is connected in series with the impedance in the Laplace-domain model for the capacitor has a value of $\frac{1}{s}v_c(0)$. This source also corresponds to a constant signal in the time domain.

A capacitor with an initial voltage $v_c(0)$ can be modelled in the time domain as a capacitor with an initial voltage of zero connected in series with a constant voltage source $v_c(0)$, as shown in Figure 6-17.

To reiterate the major point of this section:

A capacitor with a capacitance C can be modelled in the Laplace domain by an impedance of value $\frac{1}{Cs}$ connected in series with a voltage source of value $\frac{1}{s}v_c(0)$.

The complete procedure for analyzing a circuit in the Laplace domain is elucidated in some detail in the examples that follow.

Example 6-7   A Series RC Circuit at Initial Rest

We begin with the simple circuit in Figure 6-18. The voltage source provides the input signal, which has a constant value of 1A after $t = 0$, and the voltage on the capacitor is zero at that time (i.e., $v_c(0) = 0$, a condition known as *initial rest*). The goal is to compute $v_r(t)$ for all $t \geq 0$.

Our approach is a three-step procedure. First, we map the circuit into the Laplace domain; next, we solve the Laplace-domain circuit for $V_r(s)$; finally, we compute the inverse Laplace transform of $V_r(s)$, to find $v_r(t)$ for $t \geq 0$.

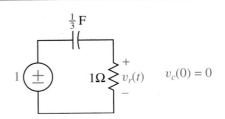

Fig. 6-18. A simple series RC circuit excited by a voltage source. We wish to compute $v_r(t)$ for $t \geq 0$ given that $v_c(0) = 0$.

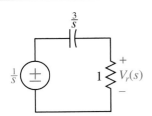

Fig. 6-19. The circuit of Figure 6-18 mapped to the Laplace domain.

Mapping the circuit to the Laplace domain is fairly straightforward in this case, because the initial voltage on the capacitor is zero, so we do not need to introduce an additional source to model it. The Laplace-domain circuit is drawn in Figure 6-19. Instead of replacing the capacitor symbol with that of a resistor to denote an impedance, we retain the capacitor symbol and simply replace the capacitance written next to it by the impedance value. This helps us to keep track of the identity of the various elements and limits the amount of redrawing that we have to do. It is important to remember, however, that in the Laplace domain the V-I relation for this element is $I(s) = CsV(s)$, whereas in the time domain the v-i relationship is $i(t) = C\frac{dv(t)}{dt}$. Notice also that each element variable in the Laplace domain is the Laplace transform of the corresponding variable in the time domain. Thus, the variable that is labelled $V_r(s)$ is the Laplace transform of $v_r(t)$, and the source variable $V_s(s)$ is the Laplace transform of $v_s(t) = 1$. Recall from the table of Laplace transforms (Table 5-1) that

$$1 \overset{\mathcal{L}}{\longleftrightarrow} \frac{1}{s}.$$

Since, in the Laplace domain, a resistor with a value of 1Ω still behaves as a resistor with a value of 1Ω, there is no change for this element, and we still draw it like a resistor. The capacitor has been replaced by an impedance with

a value of $\frac{1}{Cs} = \frac{3}{s}$. If the initial voltage on the capacitor were nonzero, we would have needed to include a voltage source to account for it (as in the next example), but this provision is not necessary here.

We have already shown that, in the Laplace domain,

(a) KCL is satisfied at every node,

(b) KVL is satisfied over every closed path,

(c) all of the impedances satisfy Ohm's law, and

(d) all RLC circuits can be modelled by using only connected impedances and sources.

This means that *in the Laplace domain* we have created a domain in which KCL, KVL, and Ohm's law again apply. We can set up and solve a series of linear equations to find (the Laplace transforms of) all the element variables, using familiar methods from our earlier study of resistive circuits. Some of the element values in this new domain are functions of s, but this is the only change.

For this particular circuit, we begin by writing a KVL equation around the only closed path:

$$V_r(s) + V_c(s) = V_s(s) = \frac{1}{s}. \qquad (6.38)$$

Since the two elements are connected in series, the same current passes through each. From Ohm's law, it follows that

$$V_c(s) = \frac{3}{s}I_c(s) = \frac{3}{s}I_r(s) = \frac{3}{s}V_r(s)/1 = \frac{3}{s}V_r(s).$$

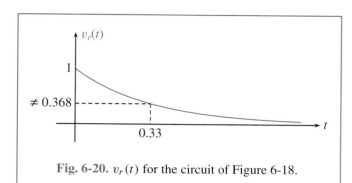

Fig. 6-20. $v_r(t)$ for the circuit of Figure 6-18.

Substituting for $V_c(s)$ in (6.38) gives

$$V_r(s) + \frac{3}{s}V_r(s) = \frac{1}{s},$$

which we can solve for $V_r(s)$:

$$V_r(s) = \frac{1}{s+3} \qquad (6.39)$$

Notice that $V(s)$ is indeed a rational function of s.

The final step is to compute the inverse Laplace transform of $V_r(s)$, to determine $v_r(t)$. This is straightforward by using the Laplace-transform pairs in Table 5-1. Specifically,

$$v_r(t) = e^{-3t}, \quad t > 0, \qquad (6.40)$$

which is plotted in Figure 6-20. The resistor voltage begins at a value of 1 V and decays to zero, with a time constant of 0.333 second. This is the time it takes to drop to $1/e \approx 0.368$ of its original value. The voltage across the capacitor begins at zero and asymptotically approaches 1 V with the same time constant, since the voltages across the resistor and capacitor must always sum to 1 V for $t > 0$.

$\blacksquare$

 To check your understanding, try Drill Problem P6-4.

Example 6-8   A Circuit with an Initial Capacitor Voltage

As a second example, consider the circuit shown in Figure 6-21, for which the initial value of the voltage on the capacitor is nonzero. This complicates the analysis considerably. In this case, we are told that the initial value of $v_c(t)$ is -1 V.

For all of these problems with reactive elements, the first step is to map the circuit to the Laplace domain. Here this step involves replacing the capacitor by an impedance connected in series with a voltage source, as shown in Figure 6-22.

In that figure, each of the time-domain element variables has been replaced by its Laplace transform. Because the initial voltage on the capacitor is nonzero, it has been replaced by an impedance with a value $\frac{1}{Cs}$,

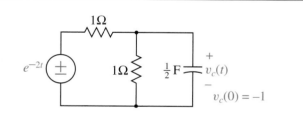

Fig. 6-21. A more general RC circuit with a nonzero initial value of the voltage on the capacitor.

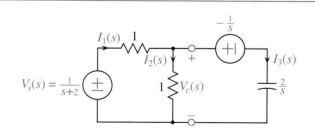

Fig. 6-22. The circuit of Figure 6-21 mapped to the Laplace domain.

connected in series with a voltage source with a voltage of $v_c(0)/s$. We need to be very careful here, however, to correctly locate the voltage $V_c(s)$. It is the Laplace transform of the voltage across the terminals of the original capacitor. Since the time-domain capacitor was replaced by a series connection in the Laplace domain, $V_c(s)$ is the voltage across the series connection and not simply the voltage across the impedance. This is indicated by the two indicated terminals in the circuit. Since the original capacitor was connected in parallel with a resistor, in both cases, the voltage of interest is also the voltage across the resistor.

Now that the circuit has been completely mapped to the Laplace domain, we are on familiar ground. Systematic methods, such as the node or mesh methods can be used here, but since the circuit is so simple, *ad hoc* methods will also work. For example, we could start with a simple statement of KCL, namely,

$$I_1(s) - I_2(s) - I_3(s) = 0, \qquad (6.41)$$

and then express each of the currents in terms of $V_c(s)$ and the source voltages:

$$I_1(s) = [V_s(s) - V_c(s)]/1 = \frac{1}{s+2} - V_c(s),$$

$$I_2(s) = V_c(s)/1 = V_c(s),$$

$$I_3(s) = \frac{s}{2}\left[V_c(s) + \frac{1}{s}\right] = \frac{s}{2}V_c(s) + \frac{1}{2}.$$

Substituting for the currents in (6.41) results in

$$\frac{1}{s+2} - V_c(s) - V_c(s) - \frac{s}{2}V_c(s) - \frac{1}{2} = 0,$$

which can be solved for $V_c(s)$ to give

$$V_c(s) = \frac{-s}{(s+2)(s+4)} = \frac{1}{s+2} - \frac{2}{s+4}. \qquad (6.42)$$

The time-domain waveform is

$$v_c(t) = e^{-2t} - 2e^{-4t}, \quad t > 0, \qquad (6.43)$$

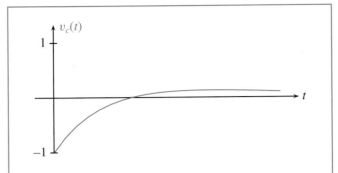

Fig. 6-23. The output waveform for the circuit of Figure 6-21.

which is what we would measure across the capacitor terminals, using an oscilloscope. This waveform appears in Figure 6-23. The capacitor voltage begins at its initial voltage, reaches a maximum value and then decays to zero as the forced solution goes to zero.

■

To check your understanding, try Drill Problem P6-5.

Example 6-9   A Second-Order Circuit at Initial Rest

We have confined our attention to circuits of first order, but the approach works as well for higher-order circuits. To demonstrate this fact, we will compute the voltage across the capacitor in the second-order circuit in Figure 6-24 when the circuit is at initial rest and the input is a constant for $t > 0$. Initial rest in this case means that both the voltage across the capacitor $v_c(t)$ and the current flowing through the inductor are zero at $t = 0$. Both the time-domain and the Laplace-domain versions of the circuit are included in the figure. Notice that the *initial rest* condition means that no auxiliary sources need to be introduced in the Laplace-domain version of the circuit.

A Confusing Issue: Modelling Initial Capacitor Voltages and Inductor Currents

Mapping a circuit to the Laplace domain is fairly straightforward *unless* there are initial voltages on the capacitors and/or initial currents on the inductors. In those cases, there are several opportunities to make errors. The most common of these are the following:

- improperly drawing the circuit after the substitution.
- forgetting that the inserted source variable is $i_\ell(0)/s$ or $v_c(0)/s$ (i.e., leaving off the s in the denominator).
- improperly labelling the capacitor voltage or the inductor current. Remember that the original element is replaced by an impedance and a source. If you are solving for an inductor current, that is the current flowing into the parallel combination, not merely the current flowing into the impedance. Similarly, if you are solving for a capacitor voltage, that is the voltage across the series connection of an impedance and a voltage source, not merely the voltage across the impedance.
- misstating the polarity of the voltage source or the direction of the current source, which should be the same as the reference direction of the capacitor voltage or inductor current, respectively.

Consider this time-domain circuit:

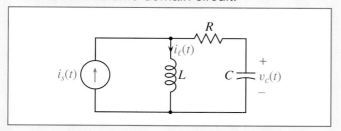

If there is a nonzero initial current in the inductor and a nonzero initial voltage across the plates of the capacitor, the Laplace-domain version of the circuit must accommodate these by using additional sources. A drawing of the Laplace-domain model for the circuit is shown here:

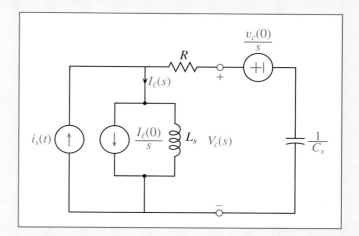

Notice that the Laplace-domain model for the inductor contains an impedance connected in parallel with a current source. Because the inductor current has a reference direction that points downward, the current source points downward also. Also notice that $I_\ell(s)$, the Laplace transform of the inductor current, is the current that flows into the parallel combination; the current that flows into the impedance is $I_\ell(s) - \frac{i_\ell(0)}{s}$.

The Laplace-domain model for a capacitor is a series connection of an impedance and a voltage source. Notice that the reference direction for the voltage source is in the same direction as the reference direction for the capacitor voltage. Furthermore, $V_c(s)$, the Laplace transform of the capacitor voltage, is the voltage across the terminals of the model. Thus, it is measured across the terminals of the series connection, as indicated on the Laplace-domain drawing.

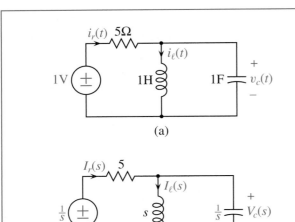

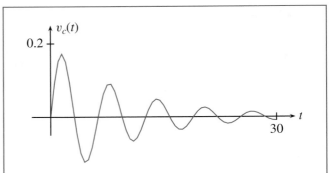

Fig. 6-25. The capacitor voltage for the second-order circuit of Figure 6-24.

Fig. 6-24. A second-order RLC circuit at initial rest with a constant input for $t > 0$. (a) Time-domain circuit. (b) Laplace-domain circuit.

can be used to eliminate $I_c(s)$, (6.47) can be used to eliminate $I_\ell(s)$, (6.46) can be used to eliminate $V_r(s)$, and (6.45) can be used to eliminate $I_r(s)$. This leaves

$$5s\,V_c(s) + V_c(s) + \frac{5}{s}V_c(s) = \frac{1}{s}$$

from which we see that

We can solve for $V_c(s)$ in the Laplace-domain circuit by writing Laplace-domain V–I relations for the three elements, writing a KCL equation at one of the nodes where the three elements are connected together, and writing a KVL equation around the left mesh.

$$V_c(s) = \frac{0.2}{s^2 + 0.2s + 1}$$
$$= \frac{-j(0.1005)}{s + 0.1 - j(0.995)} + \frac{j(0.1005)}{s + 0.1 + j(0.995)}. \quad (6.49)$$

The partial-fraction expansion reveals a pair of complex roots and complex values for the residues, but the familiar Laplace-transform pairs still apply in this case. The inverse Laplace transform is

$$V_r(s) + V_c(s) = \frac{1}{s} \quad (6.44)$$

$$I_r(s) - I_\ell(s) - I_c(s) = 0 \quad (6.45)$$

$$V_r(s) = 5I_r(s) \quad (6.46)$$

$$V_c(s) = sI_\ell(s) \quad (6.47)$$

$$I_c(s) = sV_c(s) \quad (6.48)$$

$$v_c(t) = 0.201e^{-0.1t}\left[\frac{1}{2j}e^{j0.995t} - \frac{1}{2j}e^{-j0.995t}\right]$$

$$= 0.201e^{-0.1t}\sin(0.995t). \quad (6.50)$$

The capacitor voltage is the damped sinusoid pictured in Figure 6-25.

Even though these equations involve the variable s, which precludes their numerical solution, solving them algebraically is nonetheless straightforward. Equation (6.48)

■

 To check your understanding, try Drill Problem P6-6.

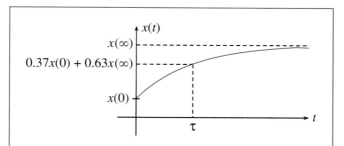

Fig. 6-26. A generic element variable waveform for a first-order circuit with a step input.

6-3   Inspection Methods for First-Order Circuits with Constant Inputs

A circuit is of first order when it contains a single reactive element (i.e., one inductor or one capacitor)[1] and at least one resistor. When the input to a first-order circuit is a constant for $t > 0$, we can write down the output waveform by inspection. While this approach is only valid for a limited class of circuits and for a certain family of input signals, these occur frequently enough that these inspection methods can be useful. This section presents these techniques.

All of the element variables in a first-order circuit whose input comes from a voltage or current source that is constant for $t > 0$ have waveforms that resemble the one shown in Figure 6-26, where we have denoted the generic element variable by $x(t)$. It begins at an initial value $x(0)$ and asymptotically approaches a limiting value $x(\infty)$. Between these points, the curve has an exponential rate parameter, or time constant, τ, that determines the rate of transition. This waveform is described by the equation

$$x(t) = x(\infty) + [x(0) - x(\infty)] e^{-t/\tau} \quad t \geq 0. \quad (6.51)$$

---

[1]More properly, a circuit is of first order if the input signal and output signal are related by a first-order differential equation. In special cases, a first-order circuit can contain more than one capacitor or one inductor (e.g., two capacitors connected in series or in parallel), but we shall limit ourselves to circuits with a single reactive element.

Notice that the first example in the previous section had an output that was of this form. It is completely determined by three numbers: the initial value $x(0)$, the final value $x(\infty)$, and the time constant τ. Using inspection techniques, we can find the values of these three parameters by simple measurements on the circuit and substitute the values into (6.51) to write down the time evolution of the element variable.

6-3-1   Circuit Behavior at $t = \infty$

The circuit behavior for large t is perhaps the easiest to understand, so let us begin here. We make our arguments in the time domain, because we can gain more insight here, but the same result could be shown by reasoning in the Laplace domain.

Consider any stable[2] circuit containing (constant) sources, resistors, and a single *capacitor*. As $t \rightarrow \infty$, all of the element values approach a constant value, including the capacitor voltage, because this is the forced response of the differential equation. However, the current flowing through the capacitor is proportional to the derivative of the capacitor voltage. If the capacitor voltage becomes constant, its current must become zero. In the limit ($t = \infty$), the capacitor current must actually be equal to zero. This means that, *at $t = \infty$, the capacitor looks like an open circuit.* Exploiting this line of reasoning, we can find the value of $x(\infty)$ by *replacing* the capacitor by an open circuit and solving the resulting resistive circuit for the element variable of interest. The result is not a waveform, but a number, $x(\infty)$. Notice that, once the single capacitor has been replaced by an open circuit, the circuit that remains contains only resistors and constant sources. It can usually by analyzed by exploiting series and parallel connections and voltage and current dividers.

---

[2]Any RLC circuit that does not contain a dependent source is stable. We will have more to say about stability in Chapter 7.

Now consider a circuit containing only constant sources, resistors, and a single *inductor*. The reasoning here is similar. The inductor *current* must approach a constant value, which means that its derivative, which is proportional to the inductor voltage, must approach zero. Thus, *as $t \to \infty$, the inductor looks like a short circuit.* We can find the value of $x(\infty)$ by replacing the inductor by a short circuit and solving the resulting circuit for this variable. Remember that $x(t)$ can be any variable in the circuit.

Summarizing these results:

> *For a first-order circuit with a constant input:*
>
> - *At $t = \infty$, an inductor looks like a short circuit.*
> - *At $t = \infty$, a capacitor looks like an open circuit.*

6-3-2   Circuit Behavior at $t = 0$

The behavior of the circuit at $t = 0$ is limited by the fact that the current flowing through an inductor and the voltage across a capacitor must be continuous when the source signals are piecewise constant (discontinuities in the source signals are allowed). We can show this by contradiction. Consider the simple *RC* circuit in Figure 6-27. This represents a more general circuit in which we have modelled the resistive circuit connected to the capacitor by its Thévenin equivalent and we have inserted a voltage source to accommodate the initial voltage on the capacitor, as we did in Figure 6-17. We also specifically assume that the constant voltage source is "turned on" at $t = 0$. Now consider what would happen if the capacitor voltage were discontinuous at $t = 0$

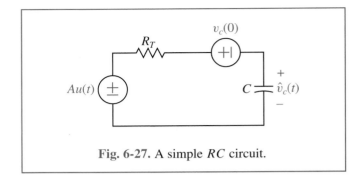

Fig. 6-27. A simple *RC* circuit.

(i.e., $\hat{v}_c(0^+) - \hat{v}_c(0^-) = \Delta)$[3]. Since the capacitor current is proportional to the derivative of the capacitor voltage, this means that the capacitor current, which is the same as the resistor current, will contain an impulse term of the form $C\Delta\delta(t)$. Now consider a KVL equation around the single closed path:

$$-A + [RC\hat{i}_r(t) + RC\Delta\delta(t)] + v_c(0) + \hat{v}_c(t) = 0.$$

The term in brackets is the voltage across the resistor, in which the impulse term has been indicated explicitly and $\hat{i}_r(t)$ denotes the rest of the resistor current. Since neither A, $\hat{i}_r(t)$, nor $\hat{v}_c(t)$, contains an impulse, if KVL is to be valid, we must have $\Delta = 0$, a contradiction. Therefore, the capacitor voltage must be continuous even if the source voltage is discontinuous (i.e., $v_c(0^+) = v_c(0^-)$). Since the capacitor voltage at $t = 0$ is continuous independent of any discontinuities in its current, it behaves like a short circuit connected in series with a voltage source with voltage $v_c(0)$.

Similar reasoning can be used to show that the inductor current must be continuous at $t = 0$ and that it behaves like an open circuit connected in parallel with a current source representing the initial current (i.e., it looks like a current source with current $i_\ell(0)$).

---

[3]The notation $\hat{v}_c(0^+)$ refers to the limiting value of $\hat{v}_c(t)$ as $t \to 0$ along the positive time axis (i.e. from the right); $\hat{v}_c(0^-)$ is a similar limit along the negative time axis (i.e., from the left).

Summarizing these results:

> *For a first-order circuit with a constant input:*
>
> - *At $t = 0$, an inductor looks like a current source with current $i_\ell(0)$.*
> - *At $t = 0$, a capacitor looks like a voltage source with voltage $v_c(0)$.*
>
> *If the initial conditions are zero, the current source becomes an open circuit and the voltage source becomes a short circuit.*

6-3-3   Calculating the Time Constant τ

The remaining parameter is the time constant τ. If we assume that the resistive two-terminal network that is connected to the reactive element has been replaced by its Thévenin equivalent circuit as in Figure 6-27, then, for a circuit with a single capacitor, the time constant is

$$\tau = R_T C,$$

and for one with a single inductor, it is

$$\tau = L/R_T = G_T L.$$

Recall that, if there are no dependent sources, we can find the Thévenin equivalent resistance of a two-terminal network by turning off all of the independent sources, replacing the voltage sources by short circuits and the current sources by open circuits. This includes any sources introduced to accommodate initial conditions. The whole inspection procedure is summarized in Table 6-1.

Table 6-1. Substitutions for the inspection method for a first-order circuit.

	$t = 0$	$t = \infty$	τ
capacitor	voltage source	open circuit	$R_T C$
inductor	current source	short circuit	$G_T L$

DEMO: *Inspection Method*

Example 6-10   **Applying the Inspection Method**

Consider the first-order circuit shown in Figure 6-28(a), in which the current flowing through the inductor at $t = 0$ is zero. The variable of interest is the voltage across the terminals of the inductor, $v_\ell(t)$.

To find $v_\ell(0)$, we replace the inductor by a current source providing a current of zero (i.e., we replace it by an open circuit). This reduces the circuit to the one shown in Figure 6-28(b). The voltage of interest, $v_\ell(0)$, is equal to the voltage across resistor R_2, since this resistor is connected in parallel with the inductor. Using a voltage divider, we see that

$$v_\ell(0) = \frac{R_2}{R_1 + R_2} \cdot 1 = \frac{R_2}{R_1 + R_2}.$$

At $t = \infty$, the inductor is replaced by a short circuit, as shown in Figure 6-28(c). Inasmuch as we are looking for the voltage across the inductor when it is replaced by a short circuit, that voltage is clearly zero. Thus,

$$v_\ell(\infty) = 0.$$

To get the time constant, τ, we turn off the independent sources and measure the equivalent resistance "seen" by the inductor, R_T. Since this consists of R_1 connected in parallel with R_2, we have

$$R_T = \frac{R_1 R_2}{R_1 + R_2},$$

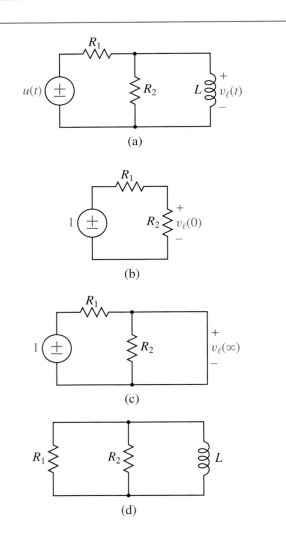

Fig. 6-28. (a) A first-order circuit to illustrate the inspection method. (b) Equivalent circuit at $t = 0$. (c) Equivalent circuit at $t = \infty$. (d) Circuit with independent sources turned off, for finding the time constant.

This gives, as the total solution for the inductor voltage,

$$v_\ell(t) = \frac{R_2}{R_1 + R_2} e^{-\frac{R_1 R_2}{L(R_1 + R_2)}t}, \quad t \geq 0. \qquad (6.52)$$

∎

To check your understanding, try Drill Problem P6-7.

Example 6-11   Inspection Method with an Initial Condition

As a second example, consider the circuit in Figure 6-29(a), which is an RC circuit with a nonzero initial voltage of $v_c(0)$. The variable of interest, $v(t)$, is the voltage across the resistor R_1. At $t = 0$, the capacitor looks like a voltage source with a voltage equal to the initial capacitor voltage, as shown in Figure 6-29(b). By source superposition, the voltage across R_1 is

$$v(0) = \frac{R_1 R_2}{R_1 + R_2} \cdot 1 + \frac{R_1}{R_1 + R_2} v_c(0)$$

$$= \frac{R_1}{R_1 + R_2} [R_2 + v_c(0)].$$

At $t = \infty$, the capacitor looks like an open circuit, resulting in the equivalent circuit shown in Figure 6-29(c). All of the current from the current source flows through R_1. Therefore,

$$v(\infty) = R_1 \cdot 1 = R_1.$$

Finally, to figure out the time constant, we turn off the independent current source, to produce the equivalent circuit shown in Figure 6-29(d). The equivalent resistance "seen" by the capacitor is $R_1 + R_2$. Therefore,

$$\tau = (R_1 + R_2)C.$$

from which it follows that

$$\tau = \frac{L}{R_T} = \frac{L(R_1 + R_2)}{R_1 R_2}.$$

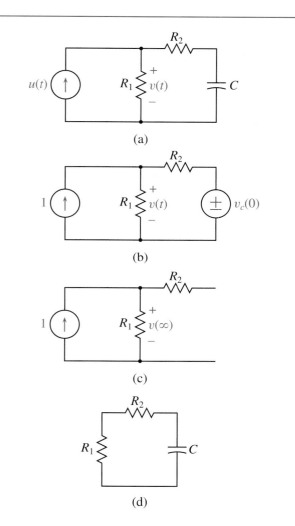

Fig. 6-29. (a) A first-order RC circuit. (b) Equivalent circuit at $t = 0$. (c) Equivalent circuit at $t = \infty$. (d) Circuit with independent sources turned off.

This gives, for the total solution,

$$v(t) = R_1 - \left[R_1 - \frac{R_1}{R_1 + R_2} (R_2 + v_c(0)) \right] e^{-j\frac{t}{(R_1+R_2)C}}$$

$$= R_1 \left\{ 1 - \frac{R_2 + v_c(0)}{R_1 + R_2} e^{-j\frac{t}{(R_1+R_2)C}} \right\}, \quad t \geq 0.$$

 To check your understanding, try Drill Problem P6-8.

6-4 Impedances and Admittances

Let N be any two-terminal network containing resistors, inductors, capacitors, and dependent sources, but no independent sources. Denote the Laplace transform of the voltage across its terminals as $V(s)$ and the transform of the current entering through the $+$ terminal as $I(s)$, as shown in Figure 6-30. Then $V(s)$ will be proportional to $I(s)$, or

$$V(s) = Z(s)I(s). \tag{6.53}$$

The quantity

$$Z(s) = \frac{V(s)}{I(s)}$$

is called the *impedance* of the network, and its reciprocal,

$$Y(s) = \frac{1}{Z(s)} = \frac{I(s)}{V(s)},$$

is called its *admittance*. In general, both the impedance and the admittance are rational functions of s.

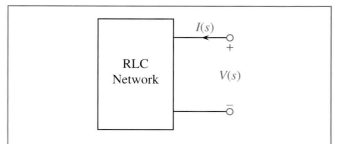

Fig. 6-30. A two-terminal network in the Laplace domain.

Table 6-2. Impedances and admittances of the elements.

	$Z(s)$	$Y(s)$
Resistor	R	$\dfrac{1}{R}$
Capacitor	$\dfrac{1}{Cs}$	Cs
Inductor	Ls	$\dfrac{1}{Ls}$

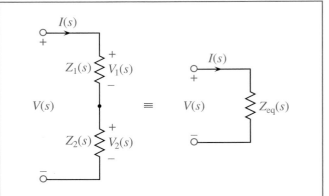

Fig. 6-31. Two impedances connected in series are equivalent to a single impedance: $Z_{\text{eq}}(s) = Z_1(s) + Z_2(s)$.

When the two-terminal network consists of a single isolated resistor, inductor (no initial current), or capacitor (no initial voltage), we already know its impedance (and its admittance). These expressions are summarized in Table 6-2. We can also create two-terminal networks by combining elements and characterize those by their admittances and impedances, or, when independent sources are present, by their Laplace-domain Thévenin and Norton equivalent circuits. These combinations are the subject of this section.

6-4-1   Impedances Connected in Series and Parallel

Two impedances are connected in series if all of the current flowing through the first also flows through the second, and vice versa. In the Laplace domain, this situation is depicted in Figure 6-31. Two impedances connected in series are network equivalent *in the Laplace domain* to a single impedance equal in value to their sum. We can show this by beginning with the Laplace-domain statement of KVL:

$$V(s) = V_1(s) + V_2(s).$$

From the impedances of the two elements and the fact that both have the same current, these two voltages are

equal to $V_1(s) = Z_1(s)I(s)$ and $V_2(s) = Z_2(s)I(s)$. Therefore,

$$V(s) = Z_1(s)I(s) + Z_2(s)I(s) = [Z_1(s) + Z_2(s)]I(s)$$

and

$$\frac{V(s)}{I(s)} = Z_{\text{eq}}(s) = Z_1(s) + Z_2(s).$$

We can also verify the Laplace-domain version of the voltage-divider relation. Since

$$I(s) = \frac{V(s)}{Z_{\text{eq}}(s)} = \frac{V(s)}{Z_1(s) + Z_2(s)},$$

it follows that

$$V_1(s) = Z_1(s)I(s) = \frac{Z_1(s)}{Z_1(s) + Z_2(s)}V(s) \quad (6.54)$$

$$V_2(s) = Z_2(s)I(s) = \frac{Z_2(s)}{Z_1(s) + Z_2(s)}V(s). \quad (6.55)$$

These results are similar to the ones that we observed for resistors connected in series, but $Z_{\text{eq}}(s)$ is a function of s, not simply a number. Furthermore, as a function, it can be more complicated than either $Z_1(s)$ or $Z_2(s)$, particularly if different types of elements make up the series connection. It is also very important to remember that *these impedance formulas are valid only in the Laplace domain!*

 DEMO: *Series and Parallel Impedances*

Example 6-12   Impedance of a Series Connection

Consider a resistor connected in series with a capacitor, as in Figure 6-32. The equivalent impedance seen at the terminals is

$$Z_{eq}(s) = Z_r(s) + Z_c(s) = R + \frac{1}{Cs} = \frac{1 + RCs}{Cs},$$

which is a rational function of s with a first-order numerator and a first-order denominator. If we use the voltage divider to express $V_c(s)$ in terms of $V(s)$, we get

$$V_c(s) = \frac{Z_c(s)}{Z_r(s) + Z_c(s)} V(s)$$

$$= \frac{\frac{1}{Cs}}{R + \frac{1}{Cs}} V(s) = \frac{1}{1 + RCs} V(s).$$

■

 To check your understanding, try Drill Problem P6-9.

Two elements are connected in parallel if the same voltage appears across both of them, as in the two-terminal network in Figure 6-33. Alternatively, they are connected in parallel if they connect the same pair of nodes. Two impedances in parallel are also equivalent to a single impedance. To show this, we begin with a statement of KCL, namely,

$$I(s) = I_1(s) + I_2(s),$$

and substitute the Laplace domain $V–I$ relations for the two elements:

$$I_1(s) = Y_1(s)V(s)$$
$$I_2(s) = Y_2(s)V(s).$$

Substituting these into the KCL statement gives

$$I(s) = Y_1(s)V(s) + Y_2(s)V(s) = [Y_1(s) + Y_2(s)]V(s),$$

which implies that

$$Y_{eq}(s) = Y_1(s) + Y_2(s), \qquad (6.56)$$

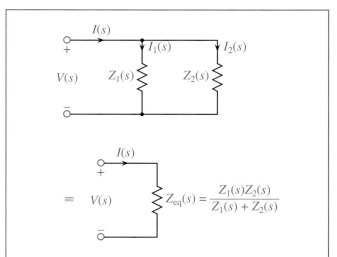

Fig. 6-33. Two impedances connected in parallel are equivalent to a single impedance.

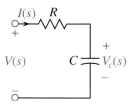

Fig. 6-32. A two-terminal network consisting of a resistor connected in series with a capacitor.

or

$$\frac{1}{Z_{eq}(s)} = \frac{1}{Z_1(s)} + \frac{1}{Z_2(s)}. \qquad (6.57)$$

There is also a Laplace-domain version of the current-divider relations:

$$I_1(s) = \frac{Y_1(s)}{Y_1(s) + Y_2(s)} I(s)$$

$$I_2(s) = \frac{Y_2(s)}{Y_1(s) + Y_2(s)} I(s).$$

Example 6-13   Impedance of a Parallel Connection

As a simple example, consider the impedance of the parallel connection of an inductor and a capacitor in Figure 6-34. The equivalent admittance is

$$Y_{eq}(s) = \frac{1}{Ls} + Cs = \frac{LCs^2 + 1}{Ls},$$

and the equivalent impedance is its reciprocal,

$$Z_{eq}(s) = \frac{Ls}{LCs^2 + 1}.$$

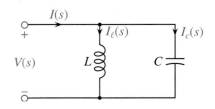

Fig. 6-34. A parallel connection of an inductor and a capacitor.

We can use the current divider to find the Laplace transforms of the currents flowing through the two elements:

$$I_\ell(s) = \frac{\frac{1}{Ls}}{\frac{1}{Ls} + Cs} I(s) = \frac{1}{LCs^2 + 1} I(s) \quad (6.58)$$

$$I_c(s) = \frac{Cs}{\frac{1}{Ls} + Cs} I(s) = \frac{LCs^2}{LCs^2 + 1} I(s). \quad (6.59)$$

■

> **To check your understanding, try Drill Problem P6-10.**

6-4-2   Thévenin and Norton Equivalent Circuits*

In Chapter 3, we showed that any two-terminal network containing only sources and resistors had a v–i relation of the form

$$v(t) = R_T i(t) + v_T(t). \qquad (6.60)$$

R_T was called the Thévenin equivalent resistance and $v_T(t)$ was the Thévenin voltage, or open-circuit voltage, of the network. Equation (6.60) is also the v–i relation of a simple network consisting of a voltage source connected in series with a resistor. This means that any resistive network with sources is network equivalent to a voltage source $v_T(t)$ connected in series with a resistor R_T, a circuit called the Thévenin equivalent network. Equation (6.60) is also the v–i relation of a current source connected in parallel with a resistor. The latter circuit is called a Norton equivalent network. The fact that the v–i relation of the two-terminal network has the form in (6.60) depends on only three things: (1) the currents in the network satisfy KCL at all nodes, (2) the voltages in the network satisfy KVL on all closed paths, and (3) the v–i relations for the resistors obey Ohm's law.

When we analyze an *RLC* network *in the Laplace domain*, the same three conditions apply. This means that the Laplace-domain V–I relation of any two-terminal network containing resistors, inductors, capacitors, and sources (called an RLC network) has the form

$$V(s) = Z_T(s)I(s) + V_T(s).$$

By extension, any two-terminal RLC network in the Laplace domain is network equivalent to a Thévenin equivalent network consisting of a voltage source connected in series with an impedance, or to a Norton equivalent network consisting of a current source connected in parallel with an impedance. We can find these network equivalent circuits by using methods similar to the ones that we used for resistive networks in the time domain. The approach is illustrated in the following example.

 WORKED SOLUTION: *Thévenin Equivalents*

Example 6-14   **Laplace-Domain Thevénin and Norton Equivalent Circuits**

Consider the RL circuit shown in Figure 6-35(a). To find the Laplace-domain Thévenin equivalent network, we first map the circuit to the Laplace domain by replacing all of the time-domain waveforms by their Laplace transforms and replacing the inductor by an impedance connected in parallel with a current source to handle the initial current. The result of this transformation is shown in Figure 6-35(b).

We begin by computing the Thévenin-equivalent impedance. To do this, we turn off both of the independent sources and then measure the equivalent impedance of the simplified circuit. Since turning off the current source replaces it by an open circuit and turning off the voltage source replaces it by a short circuit, the two-terminal network reduces to the two impedances connected in parallel. Their equivalent impedance is

$$Z_T(s) = \frac{2s}{s+2}.$$

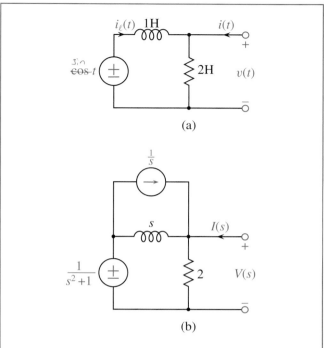

Fig. 6-35. A two-terminal RL circuit. The initial current in the inductor is 1 A. (a) Time-domain representation. (b) Laplace-domain representation. Notice the added current source to account for the initial value of the current.

We can construct the Laplace transform of the open-circuit voltage by using superposition of the two sources—that is, by turning off the current source to find the component of the open-circuit voltage created by the voltage source and then turning off the voltage source to find the component caused by the current source. Doing so, we get

$$V_T(s) = \frac{2}{s+2} \cdot \frac{1}{s^2+1} + \frac{2s}{s+2} \cdot \frac{1}{s}$$

$$= \frac{2s^2+3}{(s+2)(s^2+1)}.$$

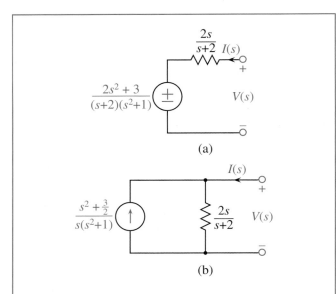

Fig. 6-36. Laplace-domain equivalents to the network in Figure 6-35. (a) Thévenin equivalent. (b) Norton equivalent.

6-5   Systematic Analysis Methods

The early examples in this chapter have demonstrated an approach that can be used to analyze circuits containing reactive elements. We map the circuit to the Laplace domain, solve for the Laplace transforms of the waveforms of interest, and identify the waveforms themselves by computing a final inverse Laplace transform. For circuits such as the ones we have considered as examples, which are quite simple, the analysis in the Laplace domain is easy to do using *ad hoc* techniques. For larger circuits, it is often appropriate to use Laplace-domain generalizations of some of the systematic procedures that we developed for resistive circuits in Chapter 2, including the node and mesh methods.

In this section, we show that our approach will always work by demonstrating two results: (1) we can use our familiar methods to determine a sufficient set of KCL and KVL equations to use with the element relations to find the complete solution for the element variables in the Laplace domain, just as we did earlier in the time domain; and (2) the Laplace transforms of the element variables are rational functions if the Laplace transforms of the source waveforms are rational functions. Then we show that the number of algebraic equations that must be solved can be reduced by using Laplace-domain versions of the node method and the mesh method, provided that we are willing to use node potentials or mesh currents as the variables.

This gives the Laplace-domain Thévenin equivalent network shown in Figure 6-36.

For the Norton equivalent network, we need the Laplace transform of the short-circuit current:

$$I_{\text{sc}}(s) = -\frac{V_T(s)}{Z_T(s)} = -\frac{s^2 + \frac{3}{2}}{s(s^2 + 1)}.$$

The Norton equivalent network is also shown in Figure 6-36. Notice that the Thévenin impedance is a rational function that does not correspond to a single resistor, inductor, or capacitor. The Thévenin and Norton sources also correspond to waveforms that are more complicated than the waveforms of the original circuit. ■

To check your understanding, try Drill Problem P6-11.

6-5-1   A Set of Sufficient Equations for Analyzing a Circuit*

Assume, as we did earlier, that our goal is to calculate one or more of the element variables in a circuit for $t \geq 0$, where the values of all of the inductor currents and all of the capacitor voltages are known at $t = 0$. The first step is to map the circuit from the time domain to the Laplace domain. This involves the following, now familiar, steps:

- The time waveform corresponding to each element variable is replaced by its Laplace transform.
- Each independent source is replaced by a Laplace-domain independent source in which the source variable is the Laplace transform of the time-domain source variable.
- Each resistor having a resistance R is replaced by an impedance R.
- Each inductor having inductance L is replaced by an impedance of value Ls connected in parallel with a current source with a current $\frac{i_\ell(0)}{s}$.
- Each capacitor having capacitance C is replaced by an impedance of value $\frac{1}{Cs}$ connected in series with a voltage source with a voltage $\frac{v_c(0)}{s}$.
- Each dependent source is replaced by a Laplace-domain dependent source with the appropriate controlling relation.

The number of variables in the Laplace-domain circuit is the same as in the time-domain circuit. Observe that, even when additional sources are introduced to account for initial conditions, we do not need any additional variables. The voltage drops across the current sources introduced to account for the initial currents in any inductors are the same as the inductor voltages themselves, and the currents flowing through the voltage sources introduced to account for the initial voltages on the capacitors are the same as the capacitor currents. No need is created for any additional KVL or KCL equations either, since the only additional meshes introduced are the trivial ones that state that the inductor voltages also appear across their associated current sources and the only nodes introduced are the trivial ones that state that the capacitor currents also flow through the associated voltage sources. Thus, we can write our KVL and KCL equations on the same nodes, supernodes, meshes, and supermeshes that we would use in the time domain. Furthermore, in the Laplace domain, all of the elements have impedances for which $V_k(s) = Z_k(s)I_k(s)$.

If the time-domain circuit is planar, the Laplace-domain circuit will also be planar. Furthermore, if the Laplace-domain circuit contains b branches, ℓ meshes, and n nodes, we must have

$$b = \ell + n - 1,$$

and the exhaustive method is guaranteed to produce a solution. The methodologies of the simplified exhaustive method can also be extended and are also guaranteed to produce a solution. In addition, by the reasoning outlined in the previous paragraph, the simplified exhaustive method will produce the same number of equations in the Laplace domain (without derivatives) as in the time domain (with derivatives).

How do we know that the Laplace transforms of the element variables will be rational functions, so that we will be able to evaluate the necessary inverse Laplace transforms? To show this, assume that we have systematically written all of the KCL, KVL, and $V\text{--}I$ relations with all terms involving the element variables on the left-hand sides and all source terms (including the sources representing initial values) on the right-hand sides. Initially the coefficients of the element variables in the KCL and KVL equations will all be constants. If the element relations are written in the form

$$V_r(s) - RI_r(s) = 0$$
$$V_\ell(s) - LsI_\ell(s) = 0$$
$$-CsV_c(s) + I_c(s) = 0,$$

then the coefficients of the element variables in the $V\text{--}I$ relations will be either constants or polynomials in s. Further assume that the Laplace transforms of the independent sources are all rational functions. Notice that those added to account for initial values are guaranteed to be rational. Now let us think about how we go about solving such a set of equations by using Gaussian elimination to work out a variable of interest. We multiply certain equations by either constants or polynomials in s and add the resulting equations together to eliminate an unwanted variable and reduce the number of remaining equations by one. We continue this process until we are left with a single equation involving only the variable of interest. If $V_1(s)$ is the variable of interest, we will be left with a single equation of the form

$$P(s)V_1(s) = \frac{A(s)}{B(s)},$$

where $P(s)$ is a polynomial. The quantity on the right-hand side must be a rational function, because we began with rational functions on the right-hand sides of all the equations, and adding rational functions together or multiplying rational functions by polynomials must produce only rational functions. Solving for $V_1(s)$, we get

$$V_1(s) = \frac{A(s)}{B(s)P(s)},$$

which is rational.

The implications of these results are very important— and also very comforting. They imply that the procedure we have been using is guaranteed to work, provided only that the source waveforms have rational Laplace transforms. We can map the circuit to the Laplace domain, use familiar methods for finding the equilibrium solution in the Laplace domain, and be assured that we will get a result that we will be able to inverse-transform to identify the time-domain waveform.

Example 6-15   Verification of the Method

To verify the method, consider the second-order circuit shown in Figure 6-37. We would like to find the voltage across the capacitor, $v_c(0)$, for $t > 0$, given that the initial current through the inductor $i_\ell(0)$ is 1 A and the initial voltage on the capacitor $v_c(0)$ is 2 V.

The first step is to map the circuit to the Laplace domain, as is done in Figure 6-38. There are two auxiliary

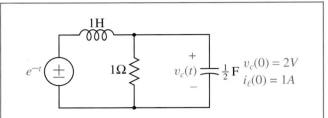

Fig. 6-37. A circuit for which the voltage $v_c(t)$ is desired for $t > 0$.

sources, because the initial values of the inductor current and the capacitor voltage are both nonzero. The voltage of interest is measured across the time-domain capacitor, which in the Laplace domain is the voltage across the series connection of the voltage source and impedance that were substituted for the capacitor. It is also the voltage across the resistor. Denote the currents through the elements as $I_\ell(s)$, $I_r(s)$, and $I_c(s)$ and the associated voltages by $V_\ell(s)$, $V_r(s)$, and $\hat{V}_c(s)$. We shall solve for $V_r(s)$.

After the voltage sources are incorporated into supernodes, the network has two supernodes and two meshes that do not pass through current sources. To

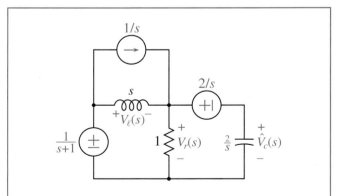

Fig. 6-38. The circuit of Figure 6-37 mapped to the Laplace domain.

specify the complete solution of the circuit, which contains six element variables, requires six independent equations—one KCL equation (at one of the supernodes),

$$I_\ell(s) - I_r(s) - I_c(s) = -\frac{1}{s};$$

two KVL equations,

$$V_r(s) - \hat{V}_c(s) = \frac{2}{s}$$

$$V_\ell(s) + V_r(s) = \frac{1}{s+1};$$

and three element relations,

$$s I_\ell(s) - V_\ell(s) = 0$$

$$V_r(s) - I_r(s) = 0$$

$$\frac{s}{2}\hat{V}_c(s) - I_c(s) = 0.$$

We can reduce these six equations to two by using the element relations to eliminate $V_\ell(s)$, $I_r(s)$, and $I_c(s)$ and by using the second KVL equation to eliminate $I_\ell(s)$. This leaves

$$V_r(s)[s+1] + \frac{s^2}{2}\hat{V}_c(s) = \frac{s+2}{s+1}$$

$$V_r(s) - \hat{V}_c(s) = \frac{2}{s}.$$

If we now multiply the second equation by $s^2/2$ and add it to the first, we can eliminate $\hat{V}_c(s)$. The result is

$$\left[\frac{s^2}{2} + s + 1\right] V_r(s) = \frac{s+2}{s+1} + s$$

or

$$V_r(s) = \frac{2s(s+2)}{(s+1)(s^2+2s+2)}.$$

The partial-fraction expansion involves three terms:

$$V_r(s) = \frac{-2}{s+1} + \frac{2}{s+1-j} + \frac{2}{s+1+j}.$$

Taking the term-by-term inverse Laplace transform and algebraically simplifying the complex exponentials results in

$$v_c(t) = -2e^{-t} + 2e^{-t}e^{jt} + 2e^{-t}e^{-jt}$$

$$= e^{-t}[-2 + 4\cos t], \quad t \geq 0.$$

∎

> **To check your understanding, try Drill Problem P6-12.**

6-5-2   Node and Mesh Methods in the Laplace Domain

The method presented in the previous section allows us to find the equilibrium solution for a circuit by solving a set of $2e$ equations in $2e$ unknowns, where e is the number of elements. This is not as bad as it sounds, because each of the equations typically contains only a few terms, but this can still be a very large number of equations. In Chapter 2, when we encountered a similar situation, we showed that, by replacing the element variables as unknowns by either the node potentials or the mesh currents, we could reduce the number of equations to be solved considerably. The node and mesh methods both generalize straightforwardly to the Laplace-domain analysis of circuits with inductors and capacitors. Again, we get a reduced number of linear equations, but the unknowns now are the Laplace transforms of either the node potentials or the mesh currents. We illustrate the approach with a simple example.

Example 6-16   The Laplace-Domain Node and Mesh Methods

Consider the first-order RL circuit shown in Figure 6-39(a). This is a circuit that we could most easily solve by inspection, but instead we shall use first the node method and then the mesh method to illustrate these

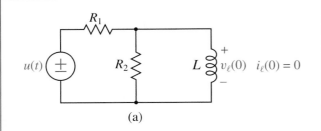

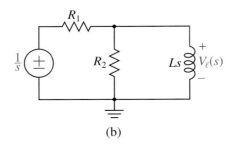

Fig. 6-39. A simple circuit to illustrate the node method in the Laplace domain. (a) Time-domain representation. (b) Laplace-domain representation.

approaches. Figure 6-39(b) shows the circuit redrawn in the Laplace domain.

Notice that the circuit contains only two nodes. We therefore denote the lower one as the ground and let the potential at the upper node be $V_\ell(s)$. The formalism of the node method requires that we write a KCL equation at the upper node, where the currents are expressed in terms of the node potentials. This gives the expression

$$\frac{1}{R_1}\left[V_\ell(s) - \frac{1}{s}\right] + \frac{1}{R_2}V_\ell(s) + \frac{1}{Ls}V_\ell(s) = 0.$$

This is an easy equation to solve for $V_\ell(s)$:

$$V_\ell(s) = \frac{\frac{\frac{1}{R_1}}{\frac{1}{R_1}+\frac{1}{R_2}}}{s + \frac{1}{L\left[\frac{1}{R_1}+\frac{1}{R_2}\right]}} = \frac{\frac{R_2}{R_1+R_2}}{s + \frac{R_1 R_2}{L(R_1+R_2)}}.$$

Now we compute the inverse Laplace transform to arrive at the time waveform:

$$v_\ell(t) = \frac{R_2}{R_1+R_2}e^{-\left(\frac{R_1 R_2 t}{L[R_1+R_2]}\right)}, \quad t \ge 0.$$

This answer is the same as (6.52), which we derived when we analyzed this circuit earlier.

We can also solve this circuit by using the mesh method. Since the circuit contains two meshes, this means that we need to set up and solve two KVL equations. Let $I_\alpha(s)$ and $I_\beta(s)$ be the Laplace transforms of the left and right mesh currents, respectively. Then

$$-\frac{1}{s} + R_1 I_\alpha(s) + R_2(I_\alpha(s) - I_\beta(s)) = 0$$
$$R_2(I_\beta(s) - I_\alpha(s)) + Ls I_\beta(s) = 0.$$

The next step is to rewrite these two equations:

$$[R_1 + R_2]I_\alpha(s) - R_2 I_\beta(s) = \frac{1}{s}$$
$$-R_2 I_\alpha(s) + [Ls + R_2]I_\beta(s) = 0.$$

We can eliminate $I_\alpha(s)$ by multiplying the top equation by R_2 and the bottom one by $[R_1 + R_2]$ and adding the results:

$$[LR_1 s + LR_2 s + R_1 R_2]I_\beta(s) = \frac{R_2}{s}.$$

This, in turn, can be rewritten as

$$I_\beta(s) = \frac{\frac{R_2}{s}}{s[L(R_1+R_2)] + R_1 R_2} = \frac{\frac{R_2}{Ls(R_1+R_2)}}{s + \frac{R_1 R_2}{L(R_1+R_2)}}.$$

Finally,

$$V_\ell(s) = Ls I_\beta(s) = \frac{\frac{R_2}{R_1+R_2}}{s + \frac{R_1 R_2}{L(R_1+R_2)}},$$

which is the same (Laplace-domain) answer that we obtained before. ■

To check your understanding, try Drill Problem P6-13.

 WORKED SOLUTION: *Lapl.-Dom. Node Meth.*

6-5-3 Operational Amplifiers in the Laplace Domain

In Chapter 4, we defined an ideal operational amplifier as a device with two input terminals with potentials $v^+(t)$ and $v^-(t)$ and one output terminal with potential $v_o(t)$. It had the following properties

- $v_o(t) = A[v^+(t) - v^-(t)]$, where A is virtually infinite.
- $v^+(t) \approx v^-(t)$.
- $i^+(t) = i^-(t) = 0$.

If we take the Laplace transforms of the signals involved in these equations, it is straightforward to derive their Laplace-domain equivalents

- $V_o(s) = A[V^+(s) - V^-(s)]$;
- $V^+(s) \approx V^-(s)$;
- $I^+(s) = I^-(s) = 0$.

From these observations, we see that an ideal operational amplifier in the Laplace domain behaves exactly as it does in the time domain.

Operational amplifiers are analyzed by using a slight modification of the node method. We derived the time-domain version in Chapter 4. It involved the following steps:

1. Set $v^+(t) = v^-(t)$, where $v^+(t)$ is the potential at the noninverting node of the opamp and $v^-(t)$ is the potential at the inverting node. Do this for each opamp in the circuit.

2. Label the potentials of all of the nodes in the circuit.

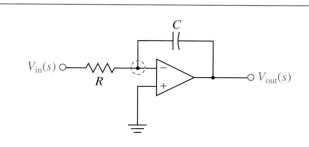

Fig. 6-40. A circuit with an operational amplifier and a reactive element.

3. Write a KCL equation at each node except the ground and the output node(s) of the operational amplifier(s). The currents entering the input terminals of the operational amplifiers are zero.

4. Solve the resulting equations, and use the results to compute the variables of interest.

We use the same procedure in the Laplace domain. The only differences are that the node potentials will be functions of the Laplace variable s, not of time, and that we need to perform a final inverse Laplace transform to get back to the time domain.

 WORKED SOLUTION: *Laplace Domain Opamps*

Example 6-17   Using Laplace Transforms for Reactive Opamp Circuits

We illustrate the procedure, using the circuit shown in Figure 6-40.

Assume that the input signal is an exponential for $t > 0$:

$$v_{in}(t) = e^{-at}, \quad t > 0$$

$$V_{in}(s) = \frac{1}{s+a}$$

Assume also that the voltage on the capacitor at $t = 0$ is zero. This means that, in the Laplace domain, the capacitor is replaced by a simple impedance with a value $\frac{1}{Cs}$. The potential at the circled node is zero, because it is connected to one input of the opamp and the other input is grounded. This is the only node at which we can write a KCL equation. This equation reveals that

$$-\frac{1}{R}V_{\text{in}}(s) - CsV_{\text{out}}(s) = 0.$$

We solve for $V_{\text{out}}(s)$:

$$V_{\text{out}}(s) = -\frac{1}{RCs}V_{\text{in}}(s) = -\frac{1}{RCs(s+a)}$$

$$= \frac{-\frac{1}{RCa}}{s} + \frac{\frac{1}{RCa}}{s+a}.$$

A final inverse Laplace transform yields

$$v_{\text{out}}(t) = -\frac{1}{RCa}\left[1 - e^{-at}\right], \quad t \geq 0. \qquad (6.61)$$

∎

To check your understanding, try Drill Problem P6-14.

6-6   Chapter Summary

6-6-1   Important Points Introduced

- The order of a circuit is the order of the differential equation that defines the element variables.
- The order of a circuit is never greater than the number of reactive elements that it contains.
- The number of auxiliary conditions required is equal to the order of the circuit.

- Auxiliary conditions are normally specified as the values of the inductor currents and capacitor voltages at a single time instant, typically $t = 0$.
- The solution of a differential equation contains two components: a forced solution and a homogeneous solution (which is a solution of the homogeneous equation).
- Laplace-domain circuits consist of sources and impedances. All variables are functions of the Laplace variable s. All currents satisfy KCL, all voltages satisfy KVL, and all impedances satisfy Ohm's law.
- An inductor with inductance L with an initial current $i_\ell(0)$ maps to the Laplace domain as an impedance Ls connected in parallel with a current source of value $i_\ell(0)/s$.
- A capacitor with a capacitance C with an initial voltage $v_c(0)$ maps to the Laplace domain as an impedance $1/Cs$ connected in series with a voltage source of value $v_c(0)/s$.
- For a first-order circuit with a constant input, an inductor looks like a current source at $t = 0$ and a short circuit at $t = \infty$. A capacitor looks like a voltage source at $t = 0$ and an open circuit at $t = \infty$.
- Two impedances connected in series are network equivalent in the Laplace domain to a single impedance whose value is equal to the sum of their impedances.
- Two impedances connected in parallel are network equivalent in the Laplace domain to a single impedance whose admittance is equal to the sum of their admittances.
- Voltage dividers and current dividers can be applied in the Laplace domain.

6-6-2 New Abilities Acquired

You should now be able to do the following:

(1) Find the differential equations that specify the element variables for first-order circuits.

(2) Solve first-order differential equations for exponential excitation signals.

(3) Solve differential equations of the first order, using Laplace transforms.

(4) Map a circuit to the Laplace domain.

(5) Find the element variables of a circuit with reactive elements by mapping the circuit to the Laplace domain, setting up and solving the relevant equations, and computing inverse Laplace transforms.

(6) Analyze first-order circuits with constant inputs by inspection.

(7) Find the Laplace domain Thévenin and Norton equivalent circuits with the same (Laplace domain) V–I relation as a two-terminal network containing reactive elements.

(8) Apply the node method and the mesh method in the Laplace domain to analyze a circuit.

6-7 Problems

6-7-1 Drill Problems

P6-1

(a) Write the differential equation that relates the voltage $v_c(t)$ to the source voltage $v_s(t)$ for the circuit of Figure P6-1.

(b) Write the differential equation whose solution is $v_r(t)$.

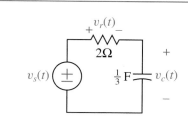

Fig. P6-1. Circuit for Problem P6-1.

(c) If $v_s(t) = e^{-2t}$ for $t > 0$ and $v_c(0) = 1$ V, find the value of $v_r(0)$ (i.e., the value of $v_r(t)$ immediately after $t = 0$).

P6-2   Find the solution of each of the following differential equations for $t > 0$.

(a) $\frac{dy(t)}{dt} + 2y(t) = 3, \qquad y(0) = 0$

(b) $2\frac{dy(t)}{dt} + y(t) = e^{-t}, \qquad y(0) = 0$

(c) $\frac{dy(t)}{dt} + \frac{1}{5}y(t) = e^{-t}e^{jt}, \qquad y(0) = 1$

P6-3   The input $x(t)$ and the output $y(t)$ of a circuit satisfy the following differential equation:

$$3\frac{dy(t)}{dt} + 2y(t) = 5x(t).$$

The input is the signal

$$x(t) = u(t) = \begin{cases} 1, t > 0 \\ 0, t \le 0 \end{cases},$$

and the output is known to satisfy the initial condition $y(0) = 1$.

(a) Construct $Y(s)$, the Laplace transform of $y(t)$.

(b) Compute $y(t)$ for $t > 0$.

P6-4 Find $v_{out}(t)$ for $t > 0$ for the circuit in Figure P6-4, if $v_{out}(0) = 0$.

(a) Redraw the circuit in the Laplace domain.

(b) Construct $V(s)$, the Laplace transform of the resistor voltage.

(c) Compute $v(t)$ for $t > 0$.

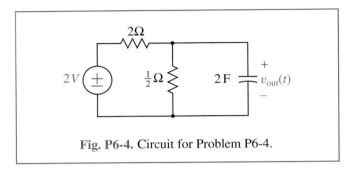

Fig. P6-4. Circuit for Problem P6-4.

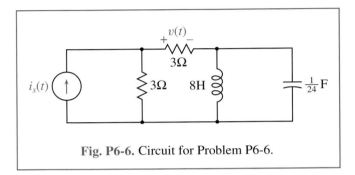

Fig. P6-6. Circuit for Problem P6-6.

P6-5 For the circuit shown in Figure P6-5.

(a) Find $v_r(t)$ for $t > 0$ if $i_\ell(0) = 0$.

(b) Find $v_r(t)$ for $t > 0$ if $i_\ell(0) = 5$.

P6-7 The circuit in Figure P6-7 is at initial rest—that is, the capacitor voltage is zero at $t = 0$, and $v_s(t) = 1$ for $t > 0$.

(a) Find $v_{out}(0)$.

(b) Determine $v_{out}(\infty)$.

(c) Determine $v_{out}(t)$ for all t.

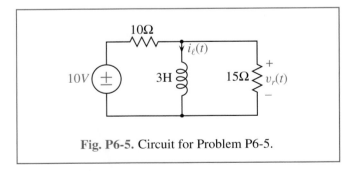

Fig. P6-5. Circuit for Problem P6-5.

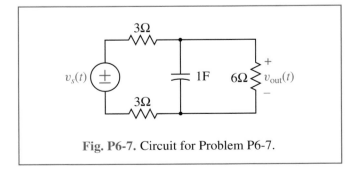

Fig. P6-7. Circuit for Problem P6-7.

P6-6 For the circuit in Figure P6-6, let $i_s(t) = 1$ for $t > 0$ and assume that, at $t = 0$, the current through the inductor is zero and that the voltage drop across the capacitor terminals is 1 volt.

P6-8 The voltage, $v_c(t)$, across the terminals of the capacitor in Figure P6-8 at $t = 0$ is 4 V. Determine $v_c(t)$ for all t if the source voltage is 3 for $t > 0$.

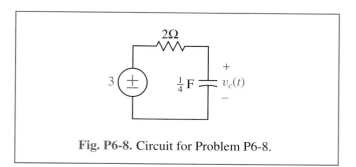

Fig. P6-8. Circuit for Problem P6-8.

P6-9 Two inductors with inductances L_1 and L_2, respectively, are connected in series. Show that the series connection behaves like a single inductor and compute the impedance of the series connection in terms of L_1 and L_2.

P6-10 Two capacitors with capacitances C_1 and C_2, respectively, are connected in parallel. Show that the parallel connection behaves like a single capacitor and compute the impedance of the parallel connection in terms of C_1 and C_2.

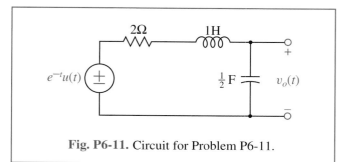

Fig. P6-11. Circuit for Problem P6-11.

P6-11 Find the Laplace-domain Thevenin–equivalent network that corresponds to the two-terminal network in Figure P6-11 at initial rest.

P6-12 In the circuit of Figure P6-12, the source waveform is $i_s(t) = 1$, for $t > 0$. The initial value of the current through the inductor is $i_\ell(0) = 1$ A, and the initial value of the potential difference across the terminals of the capacitor, defined under the default sign convention, is $v_c(0) = 1$ V. Compute $i_c(t)$, the capacitor current, for $t > 0$.

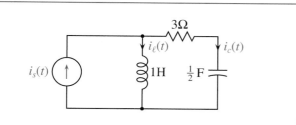

Fig. P6-12. A second-order circuit with initial conditions for Problem P6-12.

P6-13 Use the mesh method to find first $I_c(s)$, then $i_c(t)$ for the circuit in Figure P6-13. The initial value of the voltage drop on the capacitor is zero. The source current is 1 A for $t > 0$.

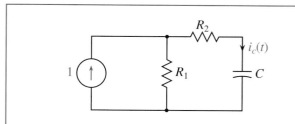

Fig. P6-13. A simple circuit for Problem P6-13.

P6-14 Find $v_{out}(t)$ in the circuit in Figure P6-14 if $v_{out}(0) = 0$ and $v_{in} = u(t)$.

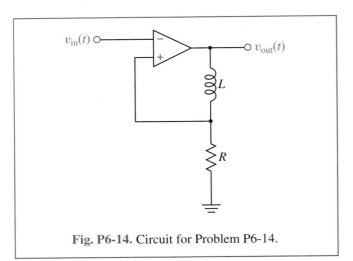

Fig. P6-14. Circuit for Problem P6-14.

6-7-2   Basic Problems

P6-15 Write the differential equation whose solution is the voltage $v(t)$ in the circuit in Figure P6-15. Also, find the value of $v(0)$.

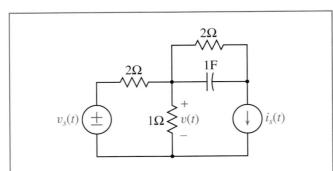

Fig. P6-15. A first-order circuit with two sources for Problem P6-15.

P6-16 Find the differential equation and initial condition that relate $v_{out}(t)$ to $v_{in}(t)$ in the circuit in Figure P6-16. Assume that the current flowing through the inductor is zero at $t = 0$.

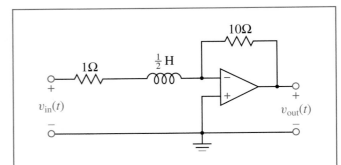

Fig. P6-16. A first-order circuit at initial rest for Problem P6-16.

P6-17

(a) For the circuit in Figure P6-17, compute $v_c(t)$ if $i_s(t) = 1$ for $t > 0$ under the assumption that $v_c(0) = 0$.

(b) Repeat for $v_c(0) = 1$.

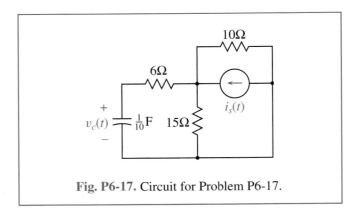

Fig. P6-17. Circuit for Problem P6-17.

P6-18 The circuit in Figure P6-18 is at initial rest. This means that, at $t = 0$, there is no current flowing through the inductor and no voltage across the terminals of the capacitor. Determine $v(t)$ for $t \geq 0$ if $i_s(t) = 2e^{-3t}u(t)$.

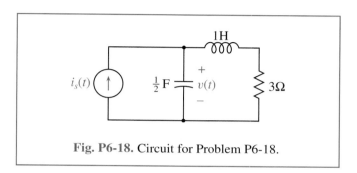

Fig. P6-18. Circuit for Problem P6-18.

P6-19 In the circuit in Figure P6-19, solve for $v(t)$ for $t > 0$. Assume that the circuit is at initial rest.

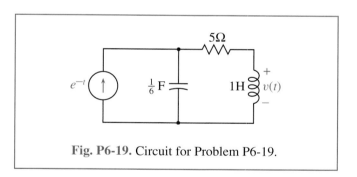

Fig. P6-19. Circuit for Problem P6-19.

P6-20 The second-order circuit in Figure P6-20 is at initial rest (i.e., the current flowing through the inductor is zero at $t = 0$ and the voltage across the capacitor is zero at $t = 0$). Solve for $i(t)$ for $t > 0$ by analyzing the circuit in the Laplace domain.

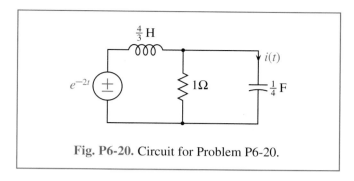

Fig. P6-20. Circuit for Problem P6-20.

P6-21 In the circuit in Figure P6-21, the initial current flowing through the inductor (top to bottom) is 2 amperes and the initial voltage on the capacitor in the indicated reference direction is 1 volt. The source current is

$$i_s(t) = u(t).$$

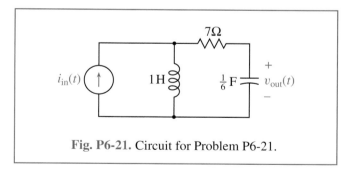

Fig. P6-21. Circuit for Problem P6-21.

(a) Draw the Laplace-domain representation of the circuit.

(b) Construct $V_{\text{out}}(s)$.

(c) Find $v_{\text{out}}(t)$, for $t > 0$.

P6-22 In the circuit of Figure P6-22, the current flowing through the inductor at $t = 0$ is 1 A and the voltage across the capacitor is 2 V; that is,

$$i_\ell(0) = 1, \quad v_c(0) = 2.$$

If $v_s(t) = e^{-t}u(t)$, compute $v_c(t)$ for $t \geq 0$.

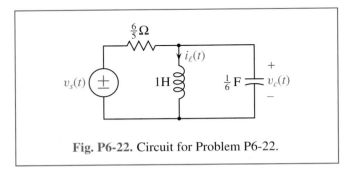

Fig. P6-22. Circuit for Problem P6-22.

P6-23 Work out the complete solution for $v_c(t)$, $t \geq 0$ in Figure P6-23, when $v_s(t) = e^{-2t}u(t)$ and the circuit is initially at rest.

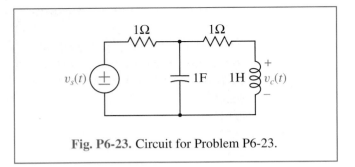

Fig. P6-23. Circuit for Problem P6-23.

P6-24 The circuit in Figure P6-24 is of first order and its input is constant for $t > 0$. Find the current $i(t)$ for $t \geq 0$, using the inspection technique.

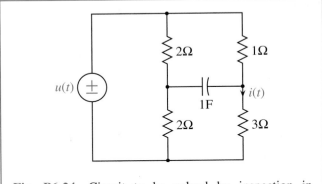

Fig. P6-24. Circuit to be solved by inspection in Problem P6-24.

P6-25 The first-order circuit in Figure P6-25, which is at initial rest, has a step input. We would like to find the signal $v(t)$ by inspection.

(a) Find $v(0)$.

(b) Find $v(\infty)$.

(c) Find $v(t)$ for $t > 0$.

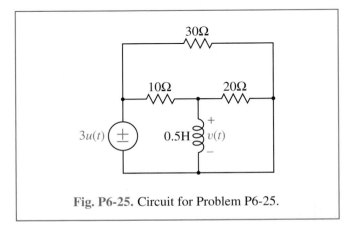

Fig. P6-25. Circuit for Problem P6-25.

P6-26 The circuit in Figure P6-26 is a first-order circuit with a constant excitation for $t > 0$. Compute $v(t)$ for $t > 0$, using the inspection method.

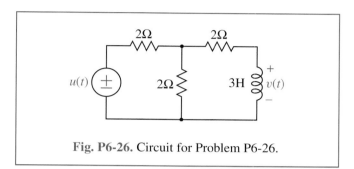

Fig. P6-26. Circuit for Problem P6-26.

P6-27 The first-order circuit in Figure P6-27 has a constant input for $t > 0$. The circuit is at initial rest—that is, $v_c(0) = 0$.

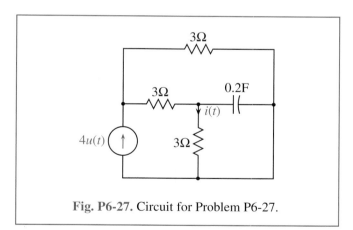

Fig. P6-27. Circuit for Problem P6-27.

(a) Find $i(0)$.

(b) Find $i(\infty)$.

(c) Find $i(t)$ for $t > 0$.

P6-28

(a) Two inductors with inductances L_1 and L_2, respectively, and no current flowing through them at $t = 0$ are connected in parallel. Show that the parallel connection behaves like a single inductor, and express the impedance of the parallel connection in terms of L_1 and L_2.

(b) Now assume that an inductor with inductance L_1 and initial current $i_1(0)$ is connected in parallel with a second inductor with inductance L_2 and initial current $i_2(0)$. Derive a simple Laplace-domain equivalent model for the connection, consisting of a single inductor and a single current source.

P6-29

(a) Two capacitors with capacitances C_1 and C_2, respectively, and no initial voltage drop across their terminals at $t = 0$ are connected in series. Show that the series connection behaves like a single capacitor, and express the impedance of the series connection in terms of C_1 and C_2.

(b) Now assume that a capacitor with capacitance C_1 and initial voltage drop $v_1(0)$ is connected in series with a second capacitor with capacitance C_2 and initial voltage drop $v_2(0)$. Derive a simple Laplace-domain equivalent model for the connection consisting of a single capacitor and a single voltage source.

P6-30 For each of the networks in Figure P6-30, write the equivalent impedance. Express your answers as ratios of polynomials in s. All of the elements are at initial rest.

P6-31 For each of the networks in Figure P6-31, write the Laplace-domain formula that relates the indicated output variable to the source variable. Your answers should be expressed in terms of the Laplace transforms of the source variables and the Laplace-domain variable s.

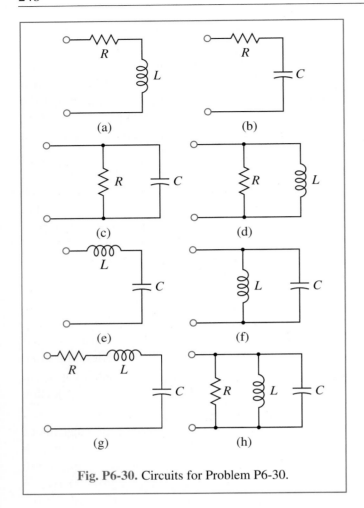

Fig. P6-30. Circuits for Problem P6-30.

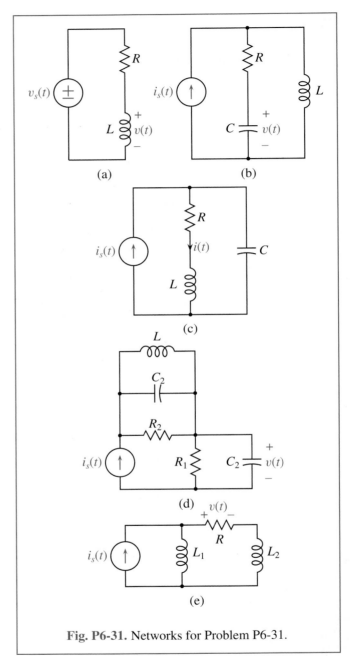

Fig. P6-31. Networks for Problem P6-31.

P6-32 Find the Laplace-domain Norton equivalent network corresponding to the two-terminal network in Figure P6-32, which is at initial rest.

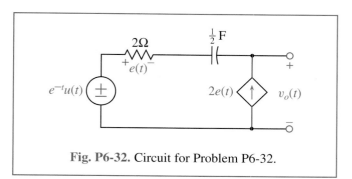

Fig. P6-32. Circuit for Problem P6-32.

P6-33 Find $v_{out}(t)$ for $t > 0$ when $v_{in}(t) = \cos(1000t)$ and $v_{out}(0) = 0$ for the circuit drawn in Figure P6-33.

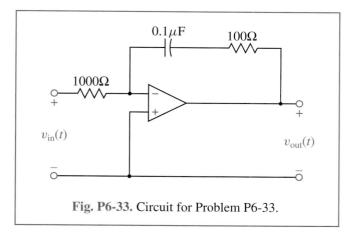

Fig. P6-33. Circuit for Problem P6-33.

6-7-3 Advanced Problems

P6-34 The network in Figure P6-34 is a second-order circuit.

(a) Write a differential equation that relates $v_c(t)$ to $v_s(t)$. *Hint:* Begin with a statement of KVL. Next

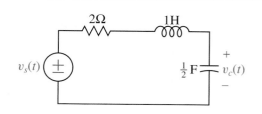

Fig. P6-34. A second-order circuit to be solved in Problem P6-34.

relate the current to $v_c(t)$. Use this current and the element relations for the other elements to relate their voltages to $v_c(t)$.

(b) To solve this differential equation, you will need to know the values of $v_c(0)$ and $\frac{dv_c(0)}{dt}$. Compute these initial conditions if we know that $i_\ell(0) = 0$ and $v_c(0) = 0$.

P6-35 This problem outlines a procedure for solving higher-order, linear, constant-coefficient differential equations with exponential input signals. The following example will serve to illustrate the method:

$$\frac{d^2 y(t)}{dt^2} + 3\frac{dy(t)}{dt} + 2y(t) = 2, \qquad t > 0,$$

$$y(0) = y_0; \qquad \frac{dy(0)}{dt} = z_0.$$

(a) Find the forced response of the differential equation. For an input of the form $x(t) = Xe^{s_p t}$, this will have the form $y_f(t) = Ye^{s_p t}$. Remember that a constant input is a special case of an exponential.

(b) To the forced response, we must add a solution of the homogeneous equation. Usually, a solution of the homogeneous equation has the form

$$y_h(t) = \sum_{i=1}^{d} A_i e^{s_i t},$$

in which d is the order of the differential equation, the A_i are constants, and the s_i are complex numbers chosen so that the homogeneous equation is satisfied. To find the s_i for our example, substitute a trial solution of the form $y_h(t) = Ae^{st}$ and find the values of s for which the equation should have a solution. (*Hint:* There should be two of them.) Write down the homogeneous solution to this particular equation.

(c) Write down the total solution, and use the initial conditions to determine the free parameters A_1 and A_2 from the homogeneous solution.

(d) Verify that your total solution satisfies both the original differential equation and the initial conditions.

(e) Solve the same differential equation by using the Laplace transform, and verify that the solution is the same that you got in part (d).

P6-36 Use Laplace transforms to solve the following higher-order differential equations for $t > 0$.

(a) $\frac{d^2y(t)}{dt^2} + 4y(t) = 2$,

$y(0) = 0; \quad \left.\frac{dy(t)}{dt}\right|_{t=0} = 0$

(b) $\frac{d^3y(t)}{dt^3} + 6\frac{d^2y(t)}{dt^2} + 11\frac{dy(t)}{dt} + 6y(t) = 6$,

$y(0) = 0; \quad \left.\frac{dy(t)}{dt}\right|_{t=0} = 0; \quad \left.\frac{d^2y(t)}{dt^2}\right|_{t=0} = 1$

P6-37 The circuit in Figure P6-37 is of second order.

(a) Find a second-order differential equation that relates the voltage $v(t)$ to the source current $i_s(t)$.

(b) At $t = 0$, it is known that the current through the inductor is zero and the voltage across the capacitor is 1 V. Find the auxiliary conditions on $v(t)$—that is, the values of $v(0)$ and $\frac{dv(0)}{dt}$ if $i_s(t) = 3$.

(c) Compute $v(t)$ for $t > 0$ if $i_s(t) = 3$.

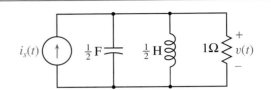

Fig. P6-37. A second-order circuit to be solved in Problem P6-37.

P6-38 Consider a first-order differential equation with an excitation function, $x(t)$ that is a complex exponential time function; that is,

$$\frac{dy(t)}{dt} + \frac{1}{\tau}y(t) = x(t),$$

with

$$y(0) = y_0,$$

for a real value of τ.

(a) Let $y(t)$ be the solution when $x(t) = Xe^{s_pt}$ and $y_0 = 0$. Show that $y^*(t)$ is the solution when the input is $x^*(t)$. (* denotes the complex conjugate.) Is it necessary that the initial condition be zero for this property to be true? Explain.

(b) Let $y_0 = 0$, let $y_1(t)$ be the solution when $x(t) = x_1(t)$, and let $y_2(t)$ be the solution when $x(t) = x_2(t)$. Show that $y_1(t) + y_2(t)$ is the solution when the excitation is $x_1(t) + x_2(t)$. Is this condition true if $y_0 \neq 0$? Explain.

(c) If $y_0 = 0$, show that if $x(t)$ is replaced by $\Re\{x(t)\}$, then the solution is $\Re\{y(t)\}$. (*Hint:* Use parts (a) and (b).)

(d) If $y_0 = 0$, show that, if $x(t)$ is replaced by $\Im\{x(t)\}$, then the solution is $\Im\{y(t)\}$.

P6-39 Solve the following differential equations, using Laplace transforms, for the case $x(t) = u(t)$.

(a)

$$\frac{d^2 y(t)}{dt^2} + 6\frac{dy(t)}{dt} + 10y(t) = \frac{dx(t)}{dt}$$

$$y(0) = 0; \quad \left. \frac{dy(t)}{dt} \right|_{t=0} = 0$$

(b)

$$\frac{d^2 y(t)}{dt^2} + 6\frac{dy(t)}{dt} + 5y(t) = x(t)$$

$$y(0) = 2; \quad \left. \frac{dy(t)}{dt} \right|_{t=0} = -1$$

(a) Verify that your answer to part (b) satisfies both the differential equation and the auxiliary conditions.

P6-40 The circuit of Figure P6-40 is initially at rest— that is, $v(0) = 0$. Compute the voltage $v(t)$ for $t \geq 0$ for each of these excitations:

(a) $i_s(t) = u(t)$.
(b) $i_s(t) = \sin t \, u(t)$.
(c) $i_s(t) = t u(t)$.
(d) $i_s(t) = \delta(t)$.

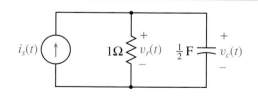

Fig. P6-40. Circuit for Problem P6-40.

P6-41 At $t = 0$, there is no stored energy in the circuit in Figure P6-41. Assume that $v_s(t) = 2e^{-t}u(t)$.

(a) Draw the Laplace-domain representation of the circuit.
(b) Determine $v(t)$ for $t > 0$.

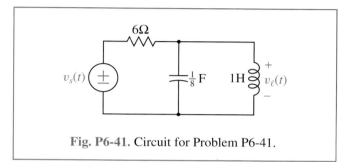

Fig. P6-41. Circuit for Problem P6-41.

P6-42 In the circuit in Figure P6-42, at $t = 0$, the inductor has a current of 1 A and the capacitor has a voltage of 6 V.

(a) Draw the Laplace-domain representation of the circuit.
(b) Compute $i_\ell(t)$ for $t > 0$, if $i_s(t) = \delta(t)$.

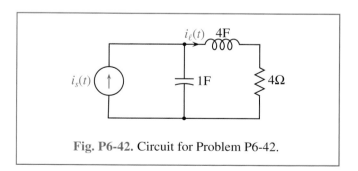

Fig. P6-42. Circuit for Problem P6-42.

P6-43 Compute $v(t)$ in Figure P6-43. The voltage on both capacitors at $t = 0$ is zero.

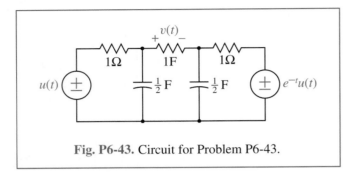

Fig. P6-43. Circuit for Problem P6-43.

P6-44 Construct the differential equation relating $v(t)$ and $i(t)$ at the terminals of the network shown in Figure P6-44.

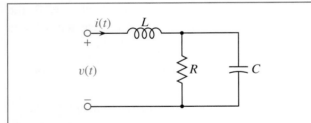

Fig. P6-44. A second-order circuit to be solved in Problem P6-44.

P6-45 Find $v_{\text{out}}(t)$ in the circuit in Figure P6-45 if $v_{\text{out}}(0) = 0$ and $v_{\text{in}}(t) = \cos(2t)$.

6-7-4 Design Problems

P6-46 When $i_s(t) = 2e^{-2t}$ in the circuit in Figure P6-46, the voltage across the inductor is observed to be $v_\ell(t) = 2e^{-4t} - e^{-2t}$ for $t > 0$. Calculate the values of R and L.

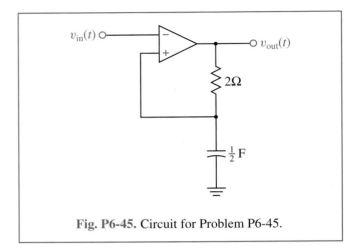

Fig. P6-45. Circuit for Problem P6-45.

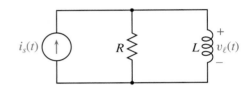

Fig. P6-46. A circuit with unknown values of R and L for Problem P6-46.

P6-47 Find a two-terminal RC network whose equivalent impedance, measured at the terminals, is

$$Z_{\text{eq}}(s) = \frac{5s + 12}{(s + 2)(s + 3)}.$$

Hint: Perform a partial-fraction expansion on Z_{eq}. This allows you to decompose the system into a series connection of two circuits with simpler equivalent impedances.

P6-48 Find a two-terminal RL network whose equivalent *admittance*, measured at the terminals, is

$$Y_{\text{eq}}(s) = \frac{5s + 12}{(s + 2)(s + 3)}.$$

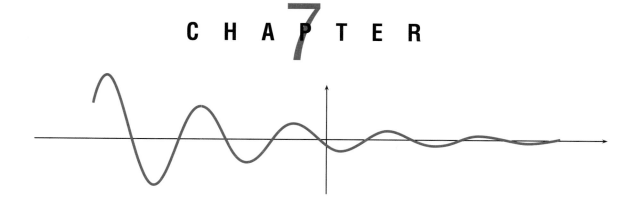

C H A P T E R 7

System Functions

Objectives

By the end of this chapter, you should be able to do the following:

1. *Compute the system function, $H(s)$, of a circuit having a single input and a designated output variable.*

2. *Compute the impulse response of a circuit at initial rest from the system function.*

3. *Find the poles and zeros of a linear circuit.*

4. *Ascertain the output response of a circuit to a pulse-type input by exploiting linearity and time invariance.*

5. *Relate the key features of the impulse and step responses of a first- or second-order circuit to the locations of its poles and zeros.*

Up to this point, our goal has been to develop systematic procedures for analyzing circuits—that is, methods for computing one or more of the voltages or currents in a circuit when a specific input-source waveform is given. As valuable as these methods are, they have a number of limitations. The most obvious limitation is that when the circuit contains a lot of elements, we need to solve a lot of equations and so often require computer assistance.[1] Unfortunately, most of this intellectual or computational energy is often wasted, because knowing every element voltage and current in a circuit is usually overkill. These methods have an even more serious limitation, however: Although they allow us to analyze circuits, they provide very little insight into how circuits behave and how they should be designed.

---

[1] Fortunately, there are some excellent circuit simulators for this task that can be used to help. PSPICE is the best known among them.

Typically, large circuits (or systems) are designed via a *top-down* strategy: The overall circuit is broken down into a series of smaller connected subsystems, each of which performs a specific task. If our intention is to design a stereo FM receiver, for example, we might first identify as subsystems such components as the FM detectors, lowpass filters, bandpass filters, oscillators, mixers, and amplifiers. These subsystems, in turn, might themselves be decomposed into component systems still smaller and still simpler. The circuit-design task then need be performed only at the lowest level, for simple circuits that contain only a few elements and have well-defined functions. This top-down design strategy introduces some questions of its own. How do we isolate the subnetworks, both conceptually and physically, so that they can be designed independently? And how can we specify the behavior that we want these component circuits to have in a language that relates to circuit behavior? The goal of this chapter is to begin to answer such questions as these.

7-1   Circuits as Systems

7-1-1   The Input-Output Point of View

    DEMO: *System Functions*

    EXTRA INFORMATION: *Systems*

In common English usage, the word *system* refers to a group of interrelated elements that form a complex whole. Electrical circuits certainly meet this definition, but circuits are not arbitrary, random connections of elements. Instead, the elements are selected and connected in a particular way for a purpose: to generate useful voltage and/or current waveforms. Therefore, for our purposes, we will define a *system* as a group of

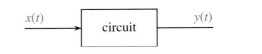

Fig. 7-1. A "black box" view of a system. The signal $x(t)$ is the system input, and $y(t)$ is the system output.

connected components that transforms one or more input signals (source waveforms) into one or more output signals. The components might be circuit elements or sources or they might be smaller systems. This systemic, or "black box," view of a circuit is captured in Figure 7-1. The source, or input, waveform, $x(t)$, enters the system on the left, and the output waveform, $y(t)$, leaves it on the right. The systems concept is prevalent throughout engineering. When the system is an electrical circuit, the input signals are the waveforms associated with independent sources, and the output signal is one of the currents or voltages in the circuit. Clearly, the output signal has two parents; it will depend both upon the specific input to the circuit and upon the circuit itself. We would like to separate and understand these two influences.

By way of illustration, consider the simple RC circuit in Figure 7-2. Here, the voltage source waveform, $v_s(t)$, is the input, and the voltage across the capacitor, $v_c(t)$,

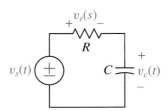

Fig. 7-2. A simple RC circuit. The signal $v_s(t)$ is the input; the voltage across the capacitor terminals, $v_c(t)$, is the output.

is the output. This circuit is at initial rest: $v_c(0) = 0$. The goal is to derive a formula that expresses the output signal as a function of the input. This turns out to be fairly difficult to do in the time domain,[2] but it is reasonably straightforward in the Laplace domain, which is where we shall work. Thus, instead of looking for a time-domain relation that relates $v_c(t)$ to $v_s(t)$, we seek a formula that expresses $V_c(s)$ in terms of $V_s(s)$.

We begin by writing a (Laplace-domain) KVL equation around the only closed path in the circuit. This gives

$$V_r(s) + V_c(s) = V_s(s), \qquad (7.1)$$

in which $V_r(s)$ is the Laplace transform of the voltage across the resistor. Since the resistor and the capacitor are connected in series, we know that $I_c(s) = I_r(s)$. Next, we relate the current variables to the voltage variables by using the Laplace-domain element relations for the resistor and the capacitor:

$$I_c(s) = Cs V_c(s)$$

$$V_r(s) = RI_r(s) = RI_c(s) = RCs V_c(s).$$

Now that we have $V_r(s)$ expressed in terms of $V_c(s)$, we can substitute for it in (7.1), which becomes

$$RCs V_c(s) + V_c(s) = V_s(s).$$

Finally, we solve this equation for $V_c(s)$. We obtain

$$V_c(s) = \left(\frac{1}{RCs+1}\right) V_s(s) = H(s)V_s(s), \qquad (7.2)$$

where

$$H(s) \triangleq \frac{1}{RCs+1}.$$

Equation (7.2), the result that we are looking for, has a particularly nice form. It states that the Laplace transform of the output signal $V_c(s)$ is the product of two functions:

_____
[2] See Section 7-4-2.

One, $V_s(s)$, comes from the input; the other, $H(s)$, called the *system function*, depends only on the circuit. Notice that *this multiplicative relationship occurs only in the Laplace domain.* The time-domain relationship between $v_c(t)$ and $v_s(t)$ is different and more complicated. (See Section 7-4-2.)

How general is this result? Can we always express the Laplace transform of the output as a product of a system function and an input term? The answer here is no, unless we impose some conditions on the circuit. To understand them, we need to look at the general case. This is the topic of the following section.

7-1-2   The Complete Solution of a Circuit

To help answer the questions we have just posed, we consider a completely general RLC circuit with multiple independent voltage and current sources. Because we want to consider the most general case, we allow the circuit to contain dependent sources and we also allow nonzero initial voltages on the capacitors and nonzero initial currents flowing through the inductors. We assume, without loss of generality, that the output signal of interest is one of the element variables, so that we can solve for it by using the procedure that we developed in Chapter 6. That procedure is restated here:

1. First, we map the circuit to the Laplace domain. This mapping may generate additional independent sources to accommodate the initial conditions.

2. Next, we write a Laplace-domain KCL equation at all but one of the closed surfaces corresponding to nodes in the basic network.

3. Then, we write a Laplace-domain KVL equation on each closed path that corresponds to a mesh in the basic network.

4. Next, we write a Laplace-domain element relation for each element.

5. Finally, we solve the resulting equations for the Laplace transform of the output variable.

We can stop here, because we are looking for a *Laplace-domain* relationship between the input and output signals; we do not need to compute a final inverse Laplace transform to get back to the time domain.

Let's consider each step of this procedure, beginning with Step 1. The mapping of the independent sources is straightforward. Independent sources in the time domain become independent sources in the Laplace domain; the time-domain source waveforms are simply replaced by their Laplace transforms. The mappings of any resistors and dependent sources are also straightforward, since these simply become constant impedances and dependent sources in the new domain. Inductors and capacitors are only slightly more complicated. As we saw in Chapter 6, when we map the kth inductor from the time domain to the Laplace domain, it is replaced by an impedance connected in parallel with a current source whose current is $i_{\ell_k}(0)/s$. The ℓth capacitor mapped to the Laplace domain is replaced by an impedance connected in series with a voltage source whose value is $v_{c_\ell}(0)/s$. These additional sources that must be added to accommodate the initial conditions are treated just like any other independent sources. Notice that these extra sources do not create the need for any additional equations, since they can be incorporated into supernodes and supermeshes.

Now consider Step 2 of the procedure. At each of the $n - 1$ surfaces that corresponds to a node or supernode network, we write a KCL equation. After all of the source terms have been moved to the right-hand sides of these equations, a typical KCL equation might resemble the following:

$$I_1(s) + I_2(s) = I_{s_1}(s) + \frac{1}{s} i_{\ell_1}(0).$$

Here, $I_1(s)$ and $I_2(s)$ are two element currents, and $I_{s_1}(s)$ is the current from an independent source. Notice that any source terms created by initial conditions have also been placed on the right-hand side. In this equation, we have included such a term caused by an initial current flowing through one of the inductors. Most of the terms in the KCL equations will be currents, but there might also be a voltage term introduced if the node is connected to a voltage-controlled current source.

In Step 3 of the procedure, we write a KVL equation on each of the ℓ closed paths that corresponds to a mesh or supermesh in the network that does not include an exterior current source. A typical one of these KVL equations might look like the following:

$$V_1(s) - V_2(s) + V_4(s) = V_{s_1}(s) + \frac{1}{s} v_{c_2}(0).$$

Again, we write it with the element variables on the left-hand side and any source terms on the right-hand side, including those introduced by the initial conditions.

Step 4 requires that we write an element relation for each of the impedances. For the kth resistor, the relation looks like

$$V_{r_k}(s) - R_k I_{r_k}(s) = 0;$$

for the mth capacitor, it looks like

$$Cs V_{c_m}(s) - I_{c_m}(s) = 0;$$

and for the nth inductor, we get an equation resembling

$$V_{\ell_n}(s) - Ls I_{\ell_n}(s) = 0.$$

Notice, once again, that we have been careful to write these equations with all of the element variables on the left-hand sides.

Finally we solve this set of $2b$ (b is the number of elements) equations for the output variable. This can be done in a number of ways. For example, we could use Cramer's rule. Instead, for this illustration, assume that we use a systematic reduction procedure similar to Gaussian elimination. This involves selecting two of the equations, multiplying each by an appropriate polynomial, and adding the two resulting equations so that one of the variables is eliminated. This equation then replaces one of the two original equations. If the selection of the equations is done carefully and systematically, we can get to a point where there is a single equation involving only the element variable of interest and the source terms. If $V_1(s)$ is the variable of interest, this remaining equation will have the form

$$A(s)V_1(s) = \sum_k B_k(s)I_{s_k}(s) + \sum_p C_p(s)V_{s_p}(s)$$
$$+ \frac{1}{s}\sum_n D_n(s)i_{\ell_n}(0) + \frac{1}{s}\sum_m E_m(s)v_{c_m}(0),$$

where $A(s)$, and all of the $B_k(s)$, $C_p(s)$, $D_n(s)$, and $E_m(s)$ are polynomials. After a final division of both sides by $A(s)$, we get the result that we have been looking for:

$$V_1(s) = \sum_k \frac{B_k(s)}{A(s)}I_{s_k}(s) + \sum_p \frac{C_p(s)}{A(s)}V_{s_p}(s)$$
$$+ \sum_n \frac{D_n(s)}{sA(s)}i_{\ell_n}(0) + \sum_m \frac{E_m(s)}{sA(s)}v_{c_m}(0)$$
$$= \sum_k H_k(s)X_k(s) + \sum_n H_{\ell_n}(s)i_{\ell_n}(0)$$
$$+ \sum_m H_{c_m}(s)v_{c_m}(0). \quad (7.3)$$

In this final expression, we have lumped the source terms together; some of the $X_k(s)$ correspond to voltage sources and some to current sources.

Several aspects of (7.3) are very important. First, notice that the Laplace transform of the output signal is always a superposition of the Laplace transforms of the sources. Next, observe that the coefficients of that superposition, the H's, are rational functions of the Laplace transform variable s. These facts mean that if the source waveforms have rational Laplace-transforms, a final inverse Laplace transform can be evaluated by using the methods that were developed in Chapter 5. Furthermore, all of the

denominators of those rational terms share the same denominator component, $A(s)$, which can be derived from the homogeneous equation.

What restrictions do we need to place on the circuit so that (7.3) is a simple product of a system function with the Laplace transform of the input signal, as in (7.2)? Clearly, the first requirement is that all of the $i_{\ell_n}(0)$ and $v_{c_m}(0)$ must be zero, so that the last two summations in (7.3) will disappear. In addition, there must be only a single independent source, which can be either a voltage source or a current source, so that the first summation reduces to a single term of the proper form: a product of the Laplace transform of the input signal and a system function. This important special case is called a *single-input, single-output system at initial rest*.

For a single-input, single-output system at initial rest, the Laplace transform of every variable can be written in the form

$$Y_i(s) = H_i(s)X(s),$$

where $X(s)$ is the Laplace transform of the source waveform and $H_i(s)$ depends only on the circuit.

For a single-input, single-output system at initial rest (no initial voltages or currents), we define the *system function* as the ratio of the Laplace transform of the output variable to the Laplace transform of the input variable:

$$H(s) \triangleq \frac{Y(s)}{X(s)}. \quad (7.4)$$

It is a function of the circuit only and does not depend upon the particular input that is applied. In many texts, it is also called the *transfer function* of the circuit.

For simple circuits, finding the system function often requires little more than replacing time-domain circuit elements by their Laplace-domain impedances and using familiar rules for circuit simplification, such as the series and parallel combination rules and voltage and current dividers. The fact that the circuit is at initial rest means that we do not need to insert any auxiliary sources to accommodate initial values. This procedure is illustrated in the following example.

 WORKED SOLUTION: *System Function*

Example 7-1	Determining the System Function of a Circuit

Consider the circuit in Figure 7-3, which is at initial rest. The input is the voltage-source waveform, $v_i(t)$, and the output is the voltage across the inductor, $v_o(t)$. We wish to identify the system function,

$$H_1(s) = \frac{V_o(s)}{V_i(s)}.$$

For a circuit at initial rest, such as this one, the mapping to the Laplace domain can be accomplished without redrawing the circuit, since the Laplace-domain and time-domain circuits are topologically identical. Even though we are not redrawing the circuit, however, it is important to recognize that we are operating in the Laplace domain.

The system function can be derived by using a voltage-divider. Let $Z_{r\ell}(s)$ denote the equivalent impedance of the resistor and the inductor connected in parallel. $V_o(s)$

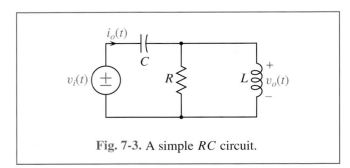

Fig. 7-3. A simple RC circuit.

is the (Laplace transform of the) voltage across this equivalent element, which, in turn, is connected in series with the capacitor. From the voltage-divider relation, we have

$$V_o(s) = \frac{Z_{r\ell}(s)}{Z_{r\ell}(s) + Z_c(s)} V_i(s), \qquad (7.5)$$

where $Z_c(s)$ is the impedance of the capacitor. $Z_{r\ell}(s)$ itself can be expressed in terms of the impedances of the resistor and inductor as

$$Z_{r\ell}(s) = \frac{Z_r(s)Z_\ell(s)}{Z_r(s) + Z_\ell(s)} = \frac{RLs}{R + Ls}.$$

Substituting this result into (7.5) gives

$$V_o(s) = V_i(s)\frac{\frac{RLs}{R+Ls}}{\frac{RLs}{R+Ls} + \frac{1}{Cs}}$$

$$= V_i(s)\frac{RLCs^2}{RLCs^2 + Ls + R}.$$

Therefore,

$$H_1(s) = \frac{RLCs^2}{RLCs^2 + Ls + R}$$

$$= \frac{s^2}{s^2 + \frac{1}{RC}s + \frac{1}{LC}}. \qquad (7.6)$$

■

 To check your understanding, try Drill Problem P7-1.

When we change which element variable is designated as the output variable, we change the behavior of the system and we change the system function. We illustrate this in the next example.

Example 7-2   Changing the Output Variable

Consider the same circuit as in the previous example, except that now we will consider the current flowing into the capacitor, $i_o(t)$, to be the output signal. For this new situation the system function is

$$H_2(s) \triangleq \frac{I_o(s)}{V_i(s)},$$

which is the equivalent admittance $Y_{eq}(s)$ of the two-terminal network "seen" by the voltage source. That network consists of the capacitor connected in series with the parallel RL. Therefore,

$$H_2(s) = \frac{1}{\frac{1}{Cs} + \frac{RLs}{R+Ls}} V_i(s)$$

$$= \frac{1}{R} \cdot \frac{s^2 + \frac{R}{L}s}{s^2 + \frac{1}{RC}s + \frac{1}{LC}}.$$

Notice that $H_1(s) \neq H_2(s)$; changing the output variable changes the system function. These two system functions, however, are not completely unrelated. Both have the same denominator polynomial, for example.   ■

 To check your understanding, try Drill Problem P7-2.

7-1-3   Circuits at Initial Rest

When we use the Laplace transform to analyze a circuit with reactive elements whose initial values are specified at $t = 0$, our methods are able to compute the output waveform only for $t \geq 0$. This may not be a serious limitation in most cases, but there are occasions when it is important to know the output waveform for all t. If the circuit is at initial rest, we can learn the complete response by appealing to the underlying differential equations.

First, however, we should confess that we have been specifying initial conditions at $t = 0$ largely for our own convenience. Our methodology could be modified to treat any "initial" time. In the more general case, when all of the initial conditions are specified at $t = t_0$, we can evaluate the element variables for $t \geq t_0$ by defining a new variable $\tau = t - t_0$ and using the Laplace transform to solve for the output variables as functions of τ. Since the initial conditions are specified at $\tau = 0$, this situation resembles the one we have already studied. Once we have its solution, we simply replace τ by $t - t_0$. Thus, it does not matter at what time the initial conditions are specified, although when there are multiple reactive elements, all of the initial values must be specified at the same time, and our methods find the solution only for times after the time at which the initial conditions are specified.

Consider an RLC circuit that is at initial rest, one for which there are no initial voltages on the capacitors or currents flowing through the inductors at $t = 0$ and for which all of the independent sources are zero for $t < 0$. For such a circuit, *all of the voltages and currents must be zero for $t < 0$*. This follows from the uniqueness of the solution of the system of equations that define the circuit. When all of the independent sources are zero, the all-zero solution for all of the element variables must satisfy KCL at all nodes and KVL over all closed paths, and it must also be consistent with the element relations for all of the resistors, inductors, capacitors, and dependent sources. In short, for a circuit at initial rest, nothing can happen until the independent sources are turned on.[3]. This fact specifies the time behavior for all of the element variables for $t < 0$.

 WORKED SOLUTION: *Initial Rest*

---

[3]In linear-systems terminology, this is a statement that circuits at initial rest are *causal* systems

Example 7-3   **Finding a Solution for all** t

The input signal to the circuit at initial rest in Figure 7-4 is the switched exponential $i_s(t) = e^{-t}u(t)$. Find the output $i_\ell(t)$ for all t.

To find the solution for $t > 0$, we begin by finding the system function

$$H(s) = \frac{I_\ell(s)}{I_s(s)} = \frac{4}{s+4},$$

which can be derived by inspection by using a current divider. Then

$$I_\ell(s) = H(s)I_s(s) = \frac{4}{s+4} \cdot \frac{1}{s+1}$$

$$= \frac{\frac{4}{3}}{s+1} - \frac{\frac{4}{3}}{s+4}.$$

An inverse Laplace transform will now provide the response for $t > 0$:

$$i_\ell(t) = \frac{4}{3}(e^{-t} - e^{-4t}), \quad t > 0.$$

Since the circuit is at initial rest, however, and the input signal is zero for $t < 0$, we know that $i_\ell(t) = 0$, for $t < 0$. Therefore, for the complete response, we can write

$$i_\ell(t) = \begin{cases} \frac{4}{3}(e^{-t} - e^{-4t}), & t > 0 \\ 0, & t < 0 \end{cases}$$

$$= \frac{4}{3}(e^{-t} - e^{-4t})u(t). \tag{7.7}$$

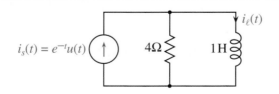

Fig. 7-4. A simple RL circuit with a switched-exponential excitation.

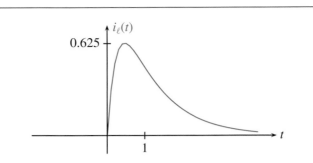

Fig. 7-5. The output waveform, $i_\ell(t)$, for the circuit of Figure 7-4.

The final expression uses the step function to capture both the positive-t and negative-t behavior in a single expression. This output waveform is sketched in Figure 7-5. It is the difference of two decaying exponentials for $t > 0$. At $t = 0$ the difference is zero, but since the term e^{-4t} decays faster than the e^{-t} term, for large t, $i_\ell(t) \approx \frac{4}{3}e^{-t}$.

■

 To check your understanding, try Drill Problem P7-3.

7-1-4   Impulse Responses, Poles, and Zeros

The inverse Laplace transform of the system function is called the *impulse response* of the circuit, $h(t)$. Formally,

$$h(t) \triangleq \mathcal{L}^{-1}\{H(s)\} \tag{7.8}$$

or

$$h(t) \longleftrightarrow H(s). \tag{7.9}$$

For a single-input RLC circuit at initial rest, the Laplace transform of the output signal, $Y(s)$, can be expressed in the form

$$Y(s) = H(s)X(s),$$

where $X(s)$ is the Laplace transform of the input. We have already commented on the useful fact that this relation effectively decouples the contributions to the output from the input signal and from the circuit. The system function captures the circuit-only component of the output in the Laplace domain. The impulse response similarly characterizes the behavior of the circuit in an input-independent fashion, but in the time domain.

Before proceeding further, we should first comment on the name *impulse response*. When the input signal is a unit impulse, $x(t) = \delta(t)$ and $X(s) = 1$. This means that $Y(s) = H(s)$ and $y(t) = h(t)$. Thus, $h(t)$ is the output of the circuit (system) when the input is a unit impulse. If we know the impulse response and the input to the system $x(t)$, then we can compute the system output in the time domain, although this involves evaluating an integral, called the *convolution integral*. We shall have more to say about this in Section 7-4-2. In general, it is easier to compute the system output by working in the Laplace domain.

In Section 7-1-2, we showed that the system function of an *RLC* circuit is a rational function in the variable s—that is, it can be written as a ratio of two polynomials:

$$\begin{aligned}
H(s) &= \frac{B(s)}{A(s)} \\
&= \frac{b_m s^m + b_{m-1} s^{m-1} + \cdots + b_1 s + b_0}{a_n s^n + a_{n-1} s^{n-1} + \cdots + a_1 s + a_0} \\
&= K \frac{(s - z_1)(s - z_2) \cdots (s - z_m)}{(s - p_1)(s - p_2) \cdots (s - p_n)}. \quad (7.10)
\end{aligned}$$

The *zeros of the system*, z_i, are those values of s for which $H(s) = 0$. These include the roots of the numerator polynomial, but there might also be one or more zeros at $s = \infty$. The *poles of the system*, p_i, are those values of s for which the system function is infinite. These are the roots of the denominator polynomial (also possibly augmented by poles at $s = \infty$). In general, the zeros and the poles can be (and usually are) complex numbers. The

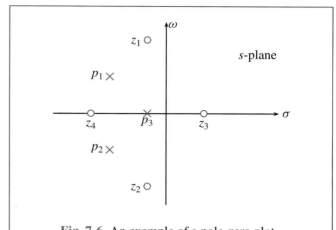

Fig. 7-6. An example of a pole-zero plot.

order of a circuit is equal to the number of finite poles, n, that it contains; this count of poles excludes any poles at infinity. The gain constant, K, is equal to b_m/a_n.

Each pole or zero can be represented by a point in the complex plane (or *s-plane*). A plot of all the pole and zero locations in the s-plane is called a *pole-zero plot*, an example of which is shown in Figure 7-6. The zeros are represented by circles and the poles by crosses. The coordinate axes are $\sigma = \Re(s)$ and $\omega = \Im(s)$. The coefficients of $B(s)$ and $A(s)$ are real,[4] so for every pole or zero that is complex, its complex conjugate will also be a pole or zero.

To compute $h(t)$, we first compute the partial-fraction expansion of $H(s)$. If there are more poles than zeros ($n > m$) and if all of the poles are distinct, this will be of the form

$$H(s) = \frac{A_1}{s - p_1} + \frac{A_2}{s - p_2} + \cdots + \frac{A_n}{s - p_n}$$

---

[4] They are products of element values, which are real.

for some (possibly complex) residues A_i. Notice that each pole contributes one term to the partial-fraction expansion. The inverse Laplace transform of $H(s)$ is

$$h(t) = A_1 e^{p_1 t} + A_2 e^{p_2 t} + \cdots + A_n e^{p_n t}, \quad t > 0.$$

The system function and the impulse response are, however, meaningful only for circuits at initial rest. For a circuit at initial rest, the response to an input that is zero for $t < 0$ must be zero for $t < 0$. Since the impulse is such an input, we can write the complete impulse response as

$$h(t) = (A_1 e^{p_1 t} + A_2 e^{p_2 t} + \cdots + A_n e^{p_n t})u(t), \quad (7.11)$$

which is a sum of switched-exponential time functions. Each pole of the system contributes one of the exponentials. When $m \geq n$, the impulse response can also contain some singularity functions located at the origin. When the poles are not distinct, some of the exponentials might be multiplied by polynomials in t, as we observed when performing inverse Laplace transforms in Chapter 5. Apart from these variations, however, the impulse responses of all RLC circuits will include terms of the form indicated in (7.11). The next example illustrates some impulse responses associated with a second-order circuit.

 DEMO: *Impulse Response*

Example 7-4   Finding the Impulse Response of a Circuit

To illustrate the procedure for finding the impulse response of a circuit, consider the network in Figure 7-7. Because there are six element variables, there are six possible system functions and thus six possible impulse responses, depending on which element variable is designated as the output. Only four of these, however, are independent, since the current flowing through the inductor is the same as the current flowing through the resistor and the resistor current is proportional to its

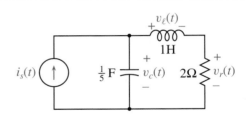

Fig. 7-7. A second-order RLC circuit. For this circuit, we can define four different system functions and impulse responses.

voltage. Therefore, let's find the four distinct impulse responses

$$h_1(t) \longleftrightarrow H_1(s) \triangleq \frac{V_c(s)}{I_s(s)} \qquad (7.12)$$

$$h_2(t) \longleftrightarrow H_2(s) \triangleq \frac{I_c(s)}{I_s(s)} \qquad (7.13)$$

$$h_3(t) \longleftrightarrow H_3(s) \triangleq \frac{V_\ell(s)}{I_s(s)} \qquad (7.14)$$

$$h_4(t) \longleftrightarrow H_4(s) \triangleq \frac{V_r(s)}{I_s(s)}. \qquad (7.15)$$

To find these, we first find the four system functions and then compute their inverse Laplace transforms.

$H_1(s)$ is same as the equivalent impedance of the three elements as seen by the current source. If we compute this function first, we can use it to calculate all of the others. We can find this impedance by using series and parallel connection simplifications:

$$H_1(s) = \frac{V_c(s)}{I_s(s)} = \frac{(s+2)\frac{5}{s}}{s+2+\frac{5}{s}} = \frac{5(s+2)}{s^2+2s+5}. \quad (7.16)$$

Now, we can use $H_1(s)$ to compute $H_2(s)$:

$$H_2(s) = \frac{I_c(s)}{I_s(s)} = \frac{V_c(s)}{I_s(s)} \cdot \frac{I_c(s)}{V_c(s)}$$

$$= H_1(s) \cdot \frac{s}{5} = \frac{s(s+2)}{s^2+2s+5}. \quad (7.17)$$

$H_3(s)$ and $H_4(s)$ can be derived from $H_1(s)$ by using voltage dividers:

$$H_3(s) = \frac{V_\ell(s)}{I_s(s)} = \frac{V_\ell(s)}{V_c(s)} H_1(s)$$

$$= \frac{s}{s+2} \cdot \frac{5(s+2)}{s^2+2s+5} = \frac{5s}{s^2+2s+5};$$

$$H_4(s) = \frac{V_r(s)}{I_s(s)} = \frac{V_r(s)}{V_c(s)} H_1(s)$$

$$= \frac{2}{s+2} \cdot \frac{5(s+2)}{s^2+2s+5} = \frac{10}{s^2+2s+5}.$$

The four impulse responses are computed by evaluating the inverse Laplace transforms of these four system functions. The calculations, using the methods developed in Chapter 5, are straightforward, but tedious. The details are left for the reader; the results are

$$h_1(t) = \frac{5\sqrt{5}}{2} e^{-t} \cos\left(2t - \left[\tan^{-1}\frac{1}{2}\right]\right) u(t),$$

$$h_2(t) = \delta(t) - \frac{5}{2} e^{-t} \sin(2t) u(t),$$

$$h_3(t) = \frac{5\sqrt{5}}{2} e^{-t} \cos\left(2t + \left[\tan^{-1}\frac{1}{2}\right]\right) u(t),$$

$$h_4(t) = 5 e^{-t} \sin(2t) u(t).$$

These four waveforms are plotted in Figure 7-8. Notice that we can specify the impulse responses for all t, since the system functions are measured with the circuit at initial rest.

There are some similarities and differences among these waveforms that are immediately apparent. Among the differences, we note that one of the signals contains an impulse and that the other three do not. Furthermore, one of the waveforms is continuous and the other three are not. In addition, all four of the signals contain multiple local maxima and minima, but they all have these extrema at different times. On the other hand, all are damped sinusoids for $t > 0$; all have the same exponential

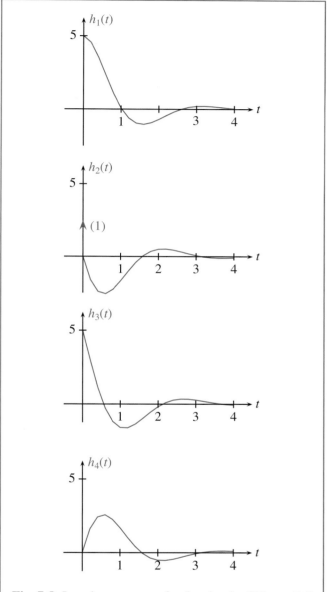

Fig. 7-8. Impulse responses for the circuit of Figure 7-7 with different signals considered as the output.

envelope and all oscillate at the same frequency. This is because these similar attributes of the signal are caused by the locations of the poles and all four of the system functions have the *same poles*. Conversely, those features that we have highlighted as differences are all related to both the poles *and* the zeros, and all four of the system functions have different configurations of zeros. ∎

 To check your understanding, try Drill Problem P7-4.

 WORKED SOLUTION: *Impulse Response*

7-1-5   The Unit-Step Response

Just as the impulse response is the output of a circuit or system when the input is a unit impulse function, so the step response is the output of a circuit or system when the input waveform is a unit step, $u(t)$:

$$u(t) = \begin{cases} 1, t > 0 \\ 0, t < 0 \end{cases}.$$

Step functions are often used to represent an abrupt change from one level to another, as might occur in digital switching circuits or when a device is turned on. Such events typically generate transient components in the element variables. When it is important to control the transient behavior of a circuit, the step response of the circuit could become an important facet of its design.

We shall denote the step response of a single-input, single-output circuit as $g(t)$ and its Laplace transform as $G(s)$. Since $g(t)$ is the output of a linear, time-invariant circuit when the input is $u(t)$, we know that

$$G(s) = \frac{1}{s} H(s). \tag{7.18}$$

Multiplying both sides of this equation by s, we see that

$$H(s) = sG(s).$$

Taking inverse Laplace transforms of both sides yields

$$h(t) = \frac{dg(t)}{dt}.$$

(Remember that a linear, time-invariant circuit must be at initial rest, so both $g(t)$ and $h(t)$ are zero for $t < 0$.) Thus, the impulse response is the derivative of the step response. Conversely, the step response is the integral of the impulse response:

$$g(t) = \int_{-\infty}^{t} h(\beta) \, d\beta. \tag{7.19}$$

Example 7-5   **Step and Impulse Responses**

In order to compare the unit-step response and the unit impulse response of a circuit, let us consider the circuit in Figure 7-9, in which $i_s(t)$ is the input and $i_\ell(t)$ is the output.

We can obtain the system function by applying a current divider in the Laplace domain:

$$H(s) = \frac{I_\ell(s)}{I_s(s)} = \frac{R}{Ls + R} = \frac{\frac{R}{L}}{s + \frac{R}{L}}.$$

Then we compute the impulse response from an inverse Laplace transform:

$$h(t) = \frac{R}{L} e^{-(R/L)t} u(t).$$

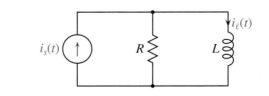

Fig. 7-9. A first-order parallel RL circuit.

To calculate the unit-step response, we let the input to the circuit be a unit step and compute the output. In the Laplace domain,

$$G(s) = \frac{1}{s}H(s) = \frac{\frac{R}{L}}{s(s + \frac{R}{L})}$$

$$= \frac{1}{s} - \frac{1}{s + \frac{R}{L}}.$$

Taking an inverse Laplace transform and exploiting the initial rest condition gives, for the step response,

$$g(t) = \left(1 - e^{-(R/L)t}\right)u(t).$$

The impulse response is indeed equal to the first derivative of the unit-step response. To verify this, we take the derivative of $g(t)$, using the product rule, and then use the fact that $f(t)\delta(t) = f(0)\delta(t)$:

$$\frac{dg(t)}{dt} = \left(1 - e^{-(R/L)t}\right)\delta(t) + \frac{R}{L}e^{-(R/L)t}u(t)$$

$$= 0 \cdot \delta(t) + \frac{R}{L}e^{-(R/L)t}u(t)$$

$$= \frac{R}{L}e^{-(R/L)t}u(t) = h(t).$$

∎

Example 7-6   **More Practice with Computing Unit-Step Responses**

In Example 7-4, we looked at a second-order circuit and computed several system functions and impulse responses for that circuit by varying the element variable that was designated as the output. Here we revisit that example and examine the corresponding four step

responses. The circuit is shown in Figure 7-7. The Laplace transforms of the four step responses are

$$G_1(s) = \frac{1}{s}H_1(s) = \frac{5(s+2)}{s(s^2 + 2s + 5)} \qquad (7.20)$$

$$G_2(s) = \frac{1}{s}H_2(s) = \frac{s+2}{s^2 + 2s + 5} \qquad (7.21)$$

$$G_3(s) = \frac{1}{s}H_3(s) = \frac{5}{s^2 + 2s + 5} \qquad (7.22)$$

$$G_4(s) = \frac{1}{s}H_4(s) = \frac{10}{s(s^2 + 2s + 5)}. \qquad (7.23)$$

Computing inverse Laplace transforms gives the four step responses. After all of the associated computation, the step responses are

$$g_1(t) = \left(2 + \frac{5}{2}e^{-t}\cos\left(2t + \tan^{-1}\frac{3}{4} - \pi\right)\right)u(t)$$

$$g_2(t) = \frac{\sqrt{5}}{2}e^{-t}\cos\left(2t - \tan^{-1}\frac{1}{2}\right)u(t)$$

$$g_3(t) = \frac{5}{2}e^{-t}\sin(2t)u(t)$$

$$g_4(t) = \left(2 + \sqrt{5}e^{-t}\cos\left(2t + \pi - \tan^{-1}\frac{1}{2}\right)\right)u(t).$$

These waveforms are plotted in Figure 7-10.

Observe that $g_2(t)$ and $g_3(t)$ both decay to zero, whereas $g_1(t)$ and $g_4(t)$ each have a step component. This is because $H_2(s)$ and $H_3(s)$ each have a zero at $s = 0$. The duration of the transient portion of the waveforms are the same for all four signals and are the same as the effective durations of the impulse responses. This duration is caused by the real parts of the pole locations that all eight Laplace transforms share.   ∎

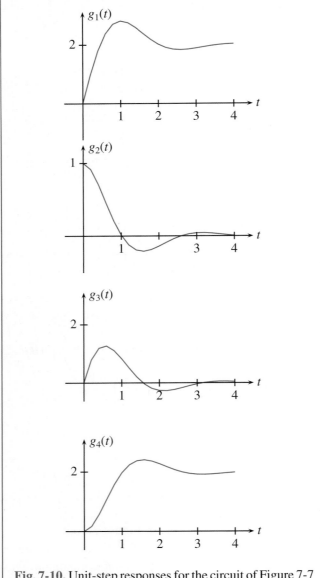

Fig. 7-10. Unit-step responses for the circuit of Figure 7-7 with different signals considered as the output.

7-2   Linearity and Time Invariance

RLC circuits at initial rest are linear, time-invariant systems. These two attributes tell us a great deal about how these circuits respond to a wide variety of input signals. This section looks at the properties of linearity and time invariance and explores a few of their implications.

7-2-1   Circuits at Initial Rest as Linear Systems

Figure 7-11 is meant to represent an arbitrary circuit with a single input and a single output variable of interest. The input signal is the source waveform, $x(t)$, which could correspond to either a voltage source or a current source, and the output signal is $y(t)$, which is shown here as a current. The circuit is at initial rest, and its system function is $H(s) = Y(s)/X(s)$.

Linearity is an attribute of a limited set of systems that tells how they respond to superpositions of inputs. Circuits at initial rest containing linear resistors, inductors, capacitors, and dependent sources are members of this set, but circuits that contain nonlinear elements or that are not at initial rest are not. To define the property, assume that we observe three outputs of the circuit. First, we select an arbitrary input waveform,

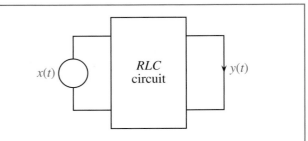

Fig. 7-11. A single-input, single-output circuit. $x(t)$ is the input (source) waveform and $y(t)$ is the output waveform, here represented as a current.

$x_1(t)$—for example, a sinusoid—and we measure the output waveform when that input is applied. Call the observed output waveform $y_1(t)$. Next, we select a different waveform $x_2(t)$—possibly a step, or maybe a sinusoid at a different frequency—apply it to the circuit in place of $x_1(t)$, and measure its output, $y_2(t)$. Finally, we form a third input signal, $x(t)$, by taking some linear superposition of $x_1(t)$ and $x_2(t)$, namely,

$$x(t) = ax_1(t) + bx_2(t),$$

where a and b are arbitrary complex constants. If the circuit is linear, then what we will observe when the third input is applied is the same linear superposition of $y_1(t)$ and $y_2(t)$:

$$y(t) = ay_1(t) + by_2(t). \tag{7.24}$$

This will be true no matter what original inputs $x_1(t)$ and $x_2(t)$ were selected and no matter what constants a and b were used to combine them. An example of a linear system is one whose output waveform is the first derivative of the input waveform. A system whose output is the square of the input waveform is an example of one that is not linear.

We do not provide a formal proof here that circuits at initial rest have this property, but we can easily demonstrate its plausibility by using the system function. When we apply the different input signals $x_1(t)$ and $x_2(t)$, we change only the system inputs, not the circuit itself. Therefore, $Y_1(s) = H(s)X_1(s)$ and $Y_2(s) = H(s)X_2(s)$. Since $Y(s)$ is the response to $X(s)$,

$$Y(s) = H(s)X(s)$$
$$= H(s)[aX_1(s) + bX_2(s)]$$
$$= aH(s)X_1(s) + bH(s)X_2(s) = aY_1(s) + bY_2(s).$$

By the linearity property of the Laplace transform, (7.24) follows.

> An RLC circuit at initial rest is a linear system.

Our real interest here is with systems that are not only linear, but also time invariant. So, before we look at the implications of linearity, let us discuss what it means for a circuit to be time invariant.

7-2-2   Circuits at Initial Rest as Time-Invariant Systems

Like linearity, time invariance can be used to define a class of systems. Time invariance and linearity are independent properties; neither property implies the other: Systems can be linear without being time invariant, and they can be time invariant without being linear.

As we did with linearity, we define the class of *time-invariant systems* by prescribing how a time-invariant system should behave when it sees related inputs. In this case, we are interested in the response of the system to input signals that are time-shifted (delayed) replicas of one another.

Let $x(t)$ be an arbitrary input signal to a system, and let the observed response to that signal be $y(t)$. Notationally, we write

$$x(t) \longrightarrow y(t).$$

Now let $z(t)$ be the response to the time-shifted signal $x(t - t_0)$ for an arbitrary time shift t_0:

$$x(t - t_0) \longrightarrow z(t).$$

The system is time invariant if and only if

$$z(t) = y(t - t_0).$$

In words, if a system is time invariant, then delaying the input signal will delay the output signal by the same amount, but otherwise the output will be unchanged.

> An *RLC* circuit at initial rest is a time-invariant system.

Again, we can demonstrate plausibility via the Laplace transform. Assume that $x(t) = 0$ for $t < 0$. Since $y(t)$ is the response to $x(t)$,

$$Y(s) = H(s)X(s).$$

If $t_0 > 0$, then

$$\mathcal{L}\{x(t - t_0)\} = e^{-st_0}X(s).$$

Therefore,

$$Z(s) = H(s)e^{-st_0}X(s) = e^{-st_0}[H(s)X(s)] = e^{-st_0}Y(s)$$

and

$$z(t) = y(t - t_0).$$

The requirements that $x(t) = 0$ for $t < 0$ and that t_0 be positive were necessary only for this derivation, which was based on the unilateral Laplace transform. The property can be shown to be valid without these restrictions, by using other methods.

7-2-3   Exploiting Linearity and Time Invariance

RLC circuits at initial rest are simultaneously linear and time invariant.[5] These properties can be extremely powerful. They, imply that when the input to a linear, time-invariant circuit can be decomposed into a sum of components, the output is similarly decomposable into the sum of the component outputs. We exploit this property in the following example.

 DEMO: *LTI Systems*

---

[5]By way of contrast, *RLC* circuits that are not at initial rest are neither linear nor time invariant.

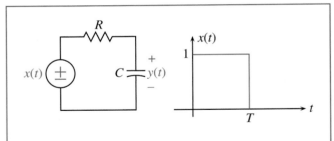

Fig. 7-12. A simple *RC* circuit with a pulse excitation.

Example 7-7   **Finding the Response to a Pulse Input**

Consider the simple *RC* circuit at initial rest shown on the left in Figure 7-12 for which the input signal is the T-second pulse shown graphically on the right. The circuit here is familiar, but the input signal is not. Nonetheless, we can find its output by exploiting the linearity and time-invariance properties of the circuit. The key here is to recognize that the pulse can be represented as the difference of two unit step functions:

$$x(t) = u(t) - u(t - T). \qquad (7.25)$$

This construction is illustrated in Figure 7-13.

Since the circuit is linear and time invariant and the input is the difference between two step functions, $y(t)$ must be equal to the difference between two *step responses*:

$$u(t) \longrightarrow g(t) \qquad (7.26)$$

$$u(t - T) \longrightarrow g(t - T) \qquad (7.27)$$

$$x(t) = u(t) - u(t - T)$$

$$\longrightarrow y(t) = g(t) - g(t - T) \qquad (7.28)$$

(7.26) defines $g(t)$ as the *step response*, the output of the circuit when the input is a unit-step function. (7.27) follows because the circuit is time invariant. Finally, (7.28) must be true because the circuit is linear. We can use this result to find $y(t)$ once we have $g(t)$. In one sense,

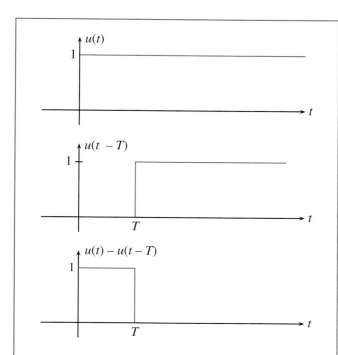

Fig. 7-13. The construction of a T-second pulse as a difference of a unit step and a delayed unit step.

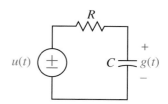

Fig. 7-14. The circuit of Figure 7-12 with a simpler input signal. The solution to this problem is used to find the solution to the circuit in Figure 7-12.

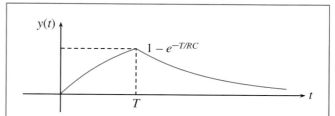

Fig. 7-15. The output waveform for the circuit in Figure 7-12.

we have merely replaced one problem with another, but the step input is a familiar one and we already know how to find $g(t)$.

We can find $g(t)$ by solving the circuit shown in Figure 7-14. This is the same as the original circuit, apart from the change in the input signal. The circuit is at initial rest. Using the inspection method, we know that

$$g(t) = \left(1 - e^{-t/RC}\right) u(t). \qquad (7.29)$$

Therefore,

$$y(t) = g(t) - g(t - T)$$
$$= \left(1 - e^{-t/RC}\right) u(t) - \left(1 - e^{-(t-T)/RC}\right) u(t - T)$$

$$= \begin{cases} 0, & t < 0 \\ 1 - e^{-t/RC}, & 0 \leq t \leq T \\ \left(e^{T/RC} - 1\right) e^{-t/RC}, & T < t. \end{cases}$$

This signal is sketched in Figure 7-15.

In using this approach to solve this problem, it is important to notice that we needed to know that the step response was zero for $t < 0$. This meant that the response to the delayed step did not begin until $t = T$. When the source voltage is nonzero, the capacitor is charging through the resistor; it discharges through the resistor when the source voltage returns to zero. Also notice that the voltage across the capacitor is continuous. ∎

 To check your understanding, try Drill Problem P7-5.

Linearity and time invariance also constrain how circuits respond to different input waveforms that are integrals or derivatives of each other. The resulting properties are frequently useful. Let $x(t)$ be the source waveform for a linear time-invariant (LTI) RLC circuit at initial rest, and let $y(t)$ be the output variable of interest— or, in shorthand notation,

$$x(t) \longrightarrow y(t).$$

Then

$$\frac{dx(t)}{dt} \longrightarrow \frac{dy(t)}{dt},$$

and

$$\int_{-\infty}^{t} x(\tau)\, d\tau \longrightarrow \int_{-\infty}^{t} y(\tau)\, d\tau.$$

The unit-impulse function is the first derivative of the unit step, and we have already implicitly used the first of the properties above to relate the impulse response to the step response.

$$h(t) = \frac{ds(t)}{dt}$$

This is a convenient method for finding the impulse response of a circuit when we have constructed the step response by using the inspection method.

| Example 7-8 | **Finding the Impulse Response by Differentiation** |

The goal of this example is to find the impulse response of the circuit in Figure 7-16. The output variable is the voltage across the capacitor. Since we would like

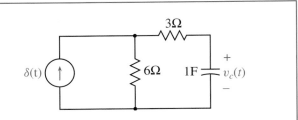

Fig. 7-16. A circuit at initial rest with a unit impulse input signal.

to find the impulse response of the circuit, a unit impulse is used as the current source waveform, which is the input.

Our approach is based on the fact that, since the circuit is at initial rest, it is a linear, time-invariant system. Therefore, its impulse response is the derivative of the step response. For a first-order circuit, we can find the step response, $g(t)$, by inspection. This means that we temporarily replace the unit impulse input by a step input to find the values of $g(0)$, $g(\infty)$, and the time constant τ.

To find $g(0) = v_c(0)$, we replace the capacitor by a short circuit. Since the voltage across a short circuit is zero, $g(0) = 0$. At $t = \infty$, the capacitor looks like an open circuit. All of the current from the current source flows through the 6-Ω resistor, and $g(\infty)$ is the voltage across that resistor. Therefore, $g(\infty) = 6$ V. The time constant $\tau = R_{eq}C = 9$ (sec). Putting all of this information together gives the complete step response:

$$g(t) = 6\left(1 - e^{t/9}\right) u(t). \qquad (7.30)$$

Recall that $g(t) = 0$ for $t < 0$, because of the initial-rest condition.

The impulse response is the derivative of the step response; that is,

$$h(t) = \frac{dg(t)}{dt},$$

which we can evaluate via the product rule:

$$h(t) = \frac{d}{dt}\left[6(1 - e^{-t/9})u(t)\right]$$

$$= 6(1 - e^{-t/9})\delta(t) + \frac{6}{9}e^{-t/9}u(t)$$

$$= 0 \cdot \delta(t) + \frac{2}{3}e^{-t/9}u(t)$$

$$= \frac{2}{3}e^{-t/9}u(t). \tag{7.31}$$

In simplifying this expression, we have made use of some of the properties of singularity functions listed in Table 5-2. The impulse of current places a finite charge on the capacitor at $t = 0$, which dissipates through the resistors for $t > 0$.

This is the same result for the impulse response that we get by using the system function. We can derive the system function by using a current divider to find the capacitor current and then using its Laplace domain V–I relation to find the voltage:

$$H(s) = \frac{1}{s} \cdot \frac{6}{6 + 3 + \frac{1}{s}} = \frac{\frac{2}{3}}{s + \frac{1}{9}}. \tag{7.32}$$

An inverse Laplace transform produces the impulse response

$$h(t) = \frac{2}{3}e^{-t/9}u(t), \tag{7.33}$$

as before.   ∎

 To check your understanding, try Drill Problem P7-6.

Example 7-9   Working with Other Pulse Shapes

The approach outlined in the Example 7-8 can be generalized to more general pulse-like input signals. For example, we can use it to find the response of the circuit

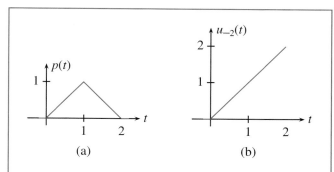

Fig. 7-17. (a) A triangular pulse input waveform. (b) The unit ramp signal, $u_{-2}(t)$.

in Figure 7-12 to the triangular pulse, $p(t)$, shown in Figure 7-17(a).

Although $p(t)$ cannot be written as a superposition of delayed step functions, it can be written as a superposition of delayed *unit ramp functions,* $u_{-2}(t) = tu(t)$. A unit ramp is shown in Figure 7-17(b). Specifically,

$$p(t) = u_{-2}(t) - 2u_{-2}(t - 1) + u_{-2}(t - 2). \tag{7.34}$$

Since the unit ramp is the integral of the unit step, or

$$u_{-2}(t) = \int_{-\infty}^{t} u(\tau)\, d\tau,$$

the response of the circuit to a unit ramp, which we shall denote as $r(t)$, is the integral of the step response:

$$r(t) = \int_{-\infty}^{t} g(\tau)\, d\tau. \tag{7.35}$$

(The step response, in turn, is the integral of the impulse response.) Let $z(t)$ denote the response of the circuit to $p(t)$; linearity, time invariance, and the representation of the pulse in (7.34) allow us to observe that $z(t)$ is related to $r(t)$ by

$$z(t) = r(t) - 2r(t - 1) + r(t - 2). \tag{7.36}$$

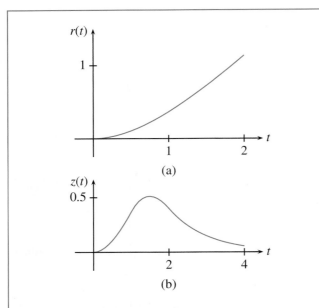

(a)

(b)

Fig. 7-18. The output waveform for the circuit in Figure 7-12 with (a) a ramp input, and (b) the triangular pulse input $p(t)$ shown in Figure 7-17.

We saw in Example 7-5 that

$$g(t) = \begin{cases} 1 - e^{t/RC}, & t \geq 0 \\ 0, & t < 0 \end{cases}.$$

Using this result in (7.35) gives

$$r(t) = \begin{cases} 0, & t < 0 \\ t - RC(1 - e^{-t/RC}), & 0 \leq t \end{cases}$$

for the ramp response This is plotted in Fig 7-18(a) for $RC = 1$. We get the response to the triangular pulse by using (7.36). This is shown in Fig 7-18(b).   ■

7-3   Responses to Switched-Exponential Inputs

Switched-exponential signals are very common. They include step functions and switched-sinusoidal signals

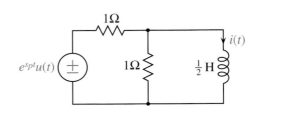

Fig. 7-19. A first-order circuit with an exponential input signal.

as special cases, which are particularly important in the analysis of digital circuits and communication circuits, respectively. We have also seen that these signals occur as components of the impulse responses of RLC circuits. This section looks specifically at the behavior of circuits when the input is a switched exponential.

To begin, consider the simple first-order circuit at initial rest in Figure 7-19; it has a generic switched-exponential input with a complex exponential parameter s_p. We calculate $i(t)$ by using the approach that has worked well for us in the past: we first map the circuit to the Laplace domain, then use a voltage divider and the Laplace-domain $V-I$ relation for the inductor to compute $I(s)$, and finally compute an inverse Laplace transform:

$$I(s) = V_s(s) \cdot \frac{V_\ell(s)}{V_s(s)} \cdot \frac{I(s)}{V_\ell(s)} = \frac{1}{s - s_p} \cdot \frac{\frac{s}{2}}{1 + \frac{\frac{s}{2}}{1 + \frac{s}{2}}} \cdot \frac{2}{s}$$

$$= \frac{1}{(s+1)(s-s_p)}.$$

To get $i(t)$, we perform the usual partial-fraction expansion and identify the inverse Laplace transforms of the individual terms. This produces the result

$$I(s) = \frac{\frac{1}{s_p+1}}{s - s_p} - \frac{\frac{1}{s_p+1}}{s + 1},$$

$$i(t) = \frac{1}{s_p + 1} \left[e^{s_p t} - e^{-t} \right] u(t).$$

This result is valid for any value of the parameter s_p. Notice that the output signal consists of a sum of two terms, one of which is an exponential with the same exponential behavior as the input, although with a different gain, and one of which is created by the system. The first of these terms is the *forced response* of the circuit, which we encountered in Section 6-1 when we were solving first-order differential equations; the second is called the *transient response of the circuit*. The transient response comes from the homogeneous solution of the differential equation. Observe that the transient response is indeed transitory; it decays to zero after some time, while the forced response does not when $\mathfrak{Re}(s_p) = 0$. Thus, after a short time, the output signal will be the same as the forced response, which resembles the input except for a gain term that depends upon the particular value of s_p.

7-3-1   The General Case

Now we would like to generalize this result and show that the response of any linear RLC circuit to a switched-exponential input signal also consists of a sum of a forced response and a transient response. In performing this derivation, we will also learn more about the nature of these two terms.

Consider an *arbitrary* linear RLC circuit at initial rest and having a single input that is again a switched-exponential time function, $x_{in}(t) = e^{s_p t}u(t)$. This input could be either an independent voltage source or an independent current source. Let the output of the circuit be $y_{out}(t)$. Assume that the output signal is one of the element voltages or currents in the circuit and that $H(s)$ is the system function that relates this input and output. Then

$$Y_{out}(s) = X_{in}(s)H(s) = \frac{1}{s - s_p}H(s).$$

We know that $H(s)$ is rational, which means that it can be described by its gain, poles, and zeros. Let the pole locations be denoted by p_i, $i = 1, 2, \ldots, n$. For simplicity, we assume that the poles are distinct and that the number of zeros is less than the number of poles, so that the inverse Laplace transform evaluation is of Class 1. Should either of these assumptions not be valid, we would need to modify some of the expressions below, but our basic conclusions would not change.

In order to compute $y_{out}(t)$ we first perform a partial-fraction expansion of $Y_{out}(s)$. This results in an expression of the form

$$Y_{out}(s) = \frac{A}{s - s_p} + \sum_{i=1}^{n} \frac{B_i}{s - p_i}.$$

The exact values of the residues B_i are not particularly important. These can be obtained from the partial-fraction expansion once s_p is known and the system function has been specified. It is important to notice, however, that there will be a term in the partial-fraction expansion that is due to the exponential input and also one from each of the poles. The coefficient A can be computed by evaluating the limit

$$A = \lim_{s \to s_p} (s - s_p)Y_{out}(s) = \lim_{s \to s_p} H(s) = H(s_p).$$

Thus, $y_{out}(t)$ for $t > 0$ will have the form

$$y_{out}(t) = \underbrace{H(s_p)e^{s_p t}}_{\substack{\text{forced} \\ \text{response}}} + \underbrace{\sum_{i} B_i e^{p_i t}}_{\substack{\text{transient} \\ \text{response}}}. \tag{7.37}$$

This is a very useful result and tells us a great deal about how a linear circuit behaves. As with the earlier first-order example, the output signal contains two components. The first of these, the forced solution, is an exponential with the same exponential dependence as the input. The gain of that exponential is $H(s_p)$, which provides an alternative interpretation for the system function. It also means that, by proper design, the circuit can be made to have a large forced response for certain values of s_p and a small response for others. This possibility suggests that a linear circuit can function as a *filter*.

The transient response consists of a superposition of exponential time functions, one for each pole of the system. These exponentials all decay with increasing time if

$$\Re e(p_i) < 0, \quad \forall i. \tag{7.38}$$

The rate at which the transient terms decay depends on the locations of the poles. The smaller the magnitude of the $\Re e(p_i)$, the slower the decay of the transients. Furthermore, whenever there is a sudden change in the input (which can be modelled as an added shifted step component in the input), transients will be introduced.

If the circuit is not at initial rest, the output will still have the form given by (7.37), because, as we recall from Chapter 6, the nonzero initial conditions have no effect on the forced response for $t > 0$. They will change the values of the coefficients B_i that appear in the transient response, but not the locations of the poles. This means that the transient behavior will be modified by the initial conditions, but it will still be similar in its gross aspects.

Whenever the condition in (7.38) holds, the circuit is said to be *stable*. The transient signals in stable circuits die out. On a few occasions, such as when we wish to design a circuit whose output signal is periodic (called an *oscillator*), we will deliberately design circuits that are not stable, by using dependent sources to force some of the poles to be purely imaginary. For an oscillator, the transient component of the response does not decay. (See Problem P7-26 at the end of the chapter.) On the pole-zero plot associated with a circuit, stability requires that all of the poles must lie to the left of the $\omega = \Im m(s)$ axis.

7-3-2   Impulse Responses of First- and Second-Order Systems

Looking at the impulse response of a system tells us a great deal about the transient behavior that it will exhibit. This is not surprising, since the impulse response is itself the transient response to a particular input, the unit impulse. Furthermore, both the transient response of a circuit to a suddenly applied input and the impulse response consist of a sum of complex exponential time functions whose exponents are tied to the poles of the circuit; only the coefficients of the linear combinations vary. In this section, we present an album of typical impulse responses that result from circuits of first or second order. We relate the pole (and zero) locations of the system functions to the transient behavior of the impulse response.

As a first example, consider the simple first-order system function

$$H_a(s) = \frac{1}{s + a} \longleftrightarrow h(t) = e^{-at}u(t), \tag{7.39}$$

which has a single real pole at $s = -a$ and no finite zeros. An impulse response of this form might result from the circuit shown in Figure 7-20(a). The impulse response, which is shown in that figure, decays monotonically, with a rate of decay that increases as the magnitude of a increases. For this impulse response, the time constant is $\tau = 1/a$, which is tied to the location of the pole. The location of the pole, in turn, depends on the component values.

Figure 7-20(b) shows the impulse response corresponding to the system function

$$H_b(s) = \frac{s}{s + a} \longleftrightarrow \delta(t) - ae^{-at}u(t),$$

along with a circuit that can have this response. This system function retains the single pole at $s = -a$, but adds a single zero at $s = 0$. Because the degrees of the numerator and denominator polynomials in the

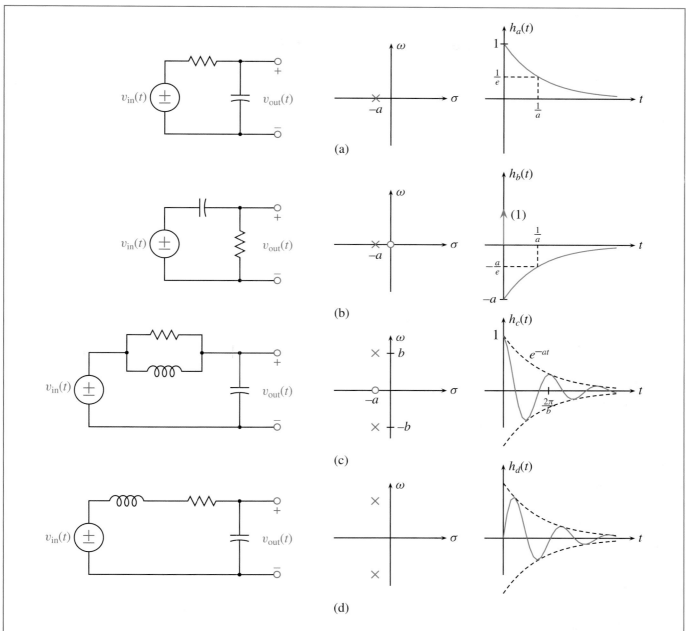

Fig. 7-20. Pole-zero plots and impulse responses of some first- and second-order circuits. (a) First-order system with a real pole and no zeros. (b) First-order system with a single real pole and a zero at s=0. (c) Second-order system with two complex poles and one zero. (d) Second-order system with two complex poles and no zeros.

system function are equal (Case 2 of inverse Laplace transform), there is an impulse term in $h_b(t)$ in addition to the exponential term, but the rate at which the impulse response decays to zero is the same as for the earlier example, and the time constant is again $1/a$. This example illustrates that the zero affects some aspects of the impulse response, but not its rate of decay. For this example, notice that $h_b(t)$ is the derivative of $h_a(t)$.

When a circuit has two poles and both an inductor and a capacitor, there are more possibilities. The system can have two real distinct poles, two real poles at the same location, or two complex poles. Figures 7-20(c,d) show the impulse responses of two second-order systems with two complex-conjugate poles. The pole locations in the two cases are the same, but the zeros are different. The two system functions are

$$H_c(s) = \frac{s+a}{(s+a)^2 + b^2} \longleftrightarrow h_c(t) = e^{-at} \cos(bt)\, u(t)$$

and

$$H_d(s) = \frac{b}{(s+a)^2 + b^2} \longleftrightarrow h_d(t) = e^{-at} \sin(bt)\, u(t).$$

Both systems have complex poles at $s = -a \pm jb$. $H_c(s)$ has a single real zero at $s = -a$, but $H_d(s)$ has no zeros. In both cases, the impulse response is a decaying switched sinusoid with an exponential decay that is controlled by a and a rate of oscillation that is controlled by the imaginary part of the pole location, b. The location of the single zero affects the initial phase of the sinusoid. For both of these systems, the time constant is $\tau = 1/a$, which must be positive if the system is to be stable.

Even when there are no zeros, the behavior of the impulse and step responses of a second-order circuit will depend upon where the poles are located. Consider a general second-order system function with no zeros:

$$H(s) = \frac{\omega_0^2}{s^2 + 2\zeta\omega_0 s + \omega_0^2}. \qquad (7.40)$$

The parameter ζ, which must be positive if the filter is to be stable, is called the *damping ratio*. When $\zeta > 1$, the circuit has two real poles and is said to be *overdamped*. When $\zeta = 1$, the circuit has a double pole and is said to be *critically damped*. When $0 < \zeta < 1$, there are two complex poles that are complex conjugates of each other, and the circuit is said to be *underdamped*. Figure 7-21 shows the impulse and step responses corresponding to the three system functions

$$H_1(s) = \frac{1}{s^2 + 3s + 1}$$

$$H_2(s) = \frac{1}{s^2 + 2s + 1}$$

$$H_3(s) = \frac{1}{s^2 + s + 1}.$$

The first of these is overdamped, the second is critically damped, and the third is underdamped. The impulse response of an overdamped second-order system is the difference of two real exponentials; it is positive for all $t > 0$ and decays with a time constant that is the longer of the time constants of the two exponentials. Its step response increases monotonically, approaching the value of one in the limit as $t \rightarrow \infty$. The critically damped impulse response is of the form $h_2(t) = te^{-t}u(t)$. It is also positive for all $t > 0$ and decays exponentially for large t. Its step response is also monotonic, but rises faster than that when the system is overdamped. The impulse response of an overdamped second-order system is a switched sinusoid; it alternates between regions that are positive and regions that are negative. Its step response is not monotonic; it alternately overshoots and undershoots the final limiting value. Notice that all three of these systems have step responses that approach 1 for large t.

 DEMO: *First- and Second-order Systems*

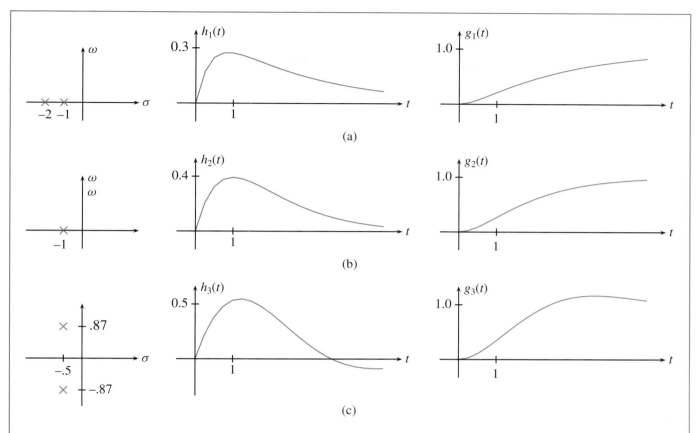

Fig. 7-21. Pole-zero plots, impulse responses, and step responses for some two-pole circuits. (a) System with two distinct real poles (overdamped). (b) System with two identical real poles (critically damped). (c) Second-order system with two complex poles (underdamped).

7-4 Two Additional Circuit Descriptions*

7-4-1 Differential-Equation Characterization at Initial Rest

Our motivation for introducing the Laplace transform to analyze circuits containing reactive elements was to avoid the necessity of setting up and solving differential equations for circuits of higher than first order. Now that we understand that procedure and know what the solutions of those circuits look like, it might be informative to go back to the differential-equation representation and see how the same solution can be derived from it. This will not only provide additional assurance that the Laplace transform approach that we have been using really works, it will also provide some additional insight.

To simplify the discussion, let's work with a specific circuit rather than trying to work out the general case. We can do this with the series RLC circuit with a unit-

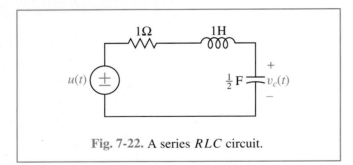

Fig. 7-22. A series RLC circuit.

step input shown in Figure 7-22. We first find the solution by using the familiar Laplace transform approach for the case when the circuit is at initial rest.

The system function is

$$H(s) = \frac{2}{s^2 + s + 2}.$$

Using this, we can construct the Laplace transform of the output signal:

$$V_c(s) = \frac{1}{s}H(s) = \frac{2}{s(s^2 + s + 2)} = \frac{2}{s([s + \frac{1}{2}]^2 + \frac{7}{4})}.$$

This system function has complex poles, which means that the evaluation of the inverse Laplace transform is tedious, although nevertheless straightforward, by using the Case 1 methodology of Chapter 5. We obtain

$$v_c(t) = \left(1 - e^{-t/2}\cos\frac{\sqrt{7}}{2}t - \frac{1}{\sqrt{7}}e^{-t/2}\sin\frac{\sqrt{7}}{2}t\right)u(t).$$
$$(7.41)$$

Notice that this expression evaluates to zero at $t = 0$, which is consistent with the known initial value of the capacitor voltage. As an aside, we notice that the first derivative of the capacitor voltage is

$$\frac{dv_c(t)}{dt} = \left[\left(\frac{\sqrt{7}}{2} + \frac{1}{2\sqrt{7}}\right)e^{-t/2}\sin\frac{\sqrt{7}}{2}t\right]u(t),$$
$$(7.42)$$

which also is equal to zero at $t = 0$. Since the derivative of the capacitor voltage is proportional to the capacitor current, and this is equal to the inductor current for this circuit, this observation is also seen to be consistent with the known initial condition. The constant term in (7.41) is the forced response for this example, and the two damped-sinusoidal terms represent the transient response.

As an alternative, we can find a single differential equation that relates the input signal to the output. We shall see shortly that it is easy to derive this equation from the system function; but, for this circuit, instead, we can get the equation directly. Let $i(t)$ represent the current flowing through all of the elements. Then, a KVL equation around the only mesh gives

$$\frac{di(t)}{dt} + i(t) + v_c(t) = u(t).$$
$$(7.43)$$

The three terms on the left side correspond to the voltages across the inductor, resistor, and capacitor, respectively. Since $i(t)$ is also the current through the capacitor, we know from its element relation that

$$i(t) = C\frac{dv_c(t)}{dt} = \frac{1}{2}\frac{dv_c(t)}{dt},$$

which we can substitute into (7.43). After multiplying both sides of the equation by two, the result is

$$\frac{d^2v_c(t)}{dt^2} + \frac{dv_c(t)}{dt} + 2v_c(t) = 2u(t).$$
$$(7.44)$$

We can solve this equation for $t > 0$ by using Laplace transforms. Let $V_c(s)$ be the Laplace transform of $v_c(t)$. Then

$$\frac{dv_c(t)}{dt} \longleftrightarrow sV_c(s) - v_c(0)$$
$$(7.45)$$

$$\frac{d^2v_c(t)}{dt^2} \longleftrightarrow s[sV_c(s) - v_c(0)] - \dot{v}_c(0)$$

$$= s^2V_c(s) - sv_c(0) - \dot{v}_c(0). \quad (7.46)$$

The quantity $\dot{v}(0)$ is the value of $\frac{dv_c(t)}{dt}$ at $t = 0$. We know that this value must be zero, because it is proportional to the current through the capacitor, which is also the current flowing through the inductor. We also know that $v_c(0) = 0$, since this is an initial condition. Therefore, if we substitute (7.45) and (7.46) into (7.44), we get

$$s^2 V_c(s) + s V_c(s) + 2V_c(s) = \frac{2}{s},$$

or

$$V_c(s) = \frac{2}{s(s^2 + s + 2)},$$

the same result that we obtained before.

The process of deriving a single differential equation that relates the input signal to the output signal in the time domain was deceptively easy for this example. In general, the necessary substitutions are not always obvious. Fortunately, there is a straightforward relationship between the differential equation and the system function that provides a much simpler method for finding the differential equation.

Let the input signal be $x(t)$, and the output signal $y(t)$, and let the circuit be at initial rest. The initial-rest condition means that not only are the values of the input signal and output signal zero for $t < 0$, but their derivatives of all degrees are zero for $t < 0$ also. Let the differential equation relating the input and output signals be

$$\frac{d^n y(t)}{dt^n} + a_{n-1}\frac{d^{n-1}y(t)}{dt^{n-1}} + \cdots + a_1\frac{dy(t)}{dt} + a_0 y(t)$$
$$= b_m\frac{d^m x(t)}{dx^m} + \cdots + b_1\frac{dx(t)}{dt} + b_0 x(t).$$

Next, we compute the Laplace transforms of the quantities on both sides of the equation. These transforms are simplified by the initial-rest condition:

$$s^n Y(s) + a_{n-1}s^{n-1}Y(s) + \cdots + a_1 s Y(s) + a_0 Y(s)$$
$$= b_m s^n X(s) + \cdots + b_1 s X(s) + b_0 X(s)$$

Since the system function is the ratio of the Laplace transform of the output signal to the Laplace transform of the input signal, we have

$$H(s) = \frac{Y(s)}{X(s)}$$
$$= \frac{b_m s^m + b_{m-1}s^{m-1} + \cdots + b_1 s + b_0}{s^n + a_{n-1}s^{n-1} + \cdots + a_1 s + a_0}. \quad (7.47)$$

The numerator polynomial contains the coefficients from the input side of the differential equation, and the powers of s identify the orders of the associated derivatives. Similarly, the denominator polynomial is derived from the output side of the differential equation. We also notice that each step in the derivation of (7.47) is reversible. Given a rational system function, we can reverse the steps to arrive at a differential equation that relates the input and output signals. The whole process is illustrated in the following example.

Example 7-10   Using $H(s)$ to Find the Differential Equation

Here, we wish to find the differential equation that relates the current $i(t)$ to the input signal $v_s(t)$ for the circuit in Figure 7-23. In this case, the system function is equal to the equivalent admittance seen by the voltage source:

$$H(s) = Y_{eq}(s) = \frac{1}{Z_{eq}(s)}$$
$$= \frac{1}{s + \frac{3}{s+1.5}} = \frac{s + 1.5}{s^2 + 1.5s + 3}. \quad (7.48)$$

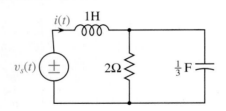

Fig. 7-23. A circuit for which the differential equation relating $i(t)$ and $v_s(t)$ is desired.

We can now use (7.48) to identify the differential equation relating $i(t)$ to $v_s(t)$:

$$\frac{d^2i(t)}{dt^2} + 1.5\frac{di(t)}{dt} + 3i(t) = \frac{dv_s(t)}{dt} + 1.5v_s(t).$$
(7.49)

∎

 To check your understanding, try Drill Problem P7-7.

For any circuit containing more than one independent reactive element, there are multiple differential-equation representations. These are equivalent to one another when the circuit is at initial rest, but they require different information for their initial conditions. To illustrate these differences, consider once again the circuit in Figure 7-23, but now let us *not* assume that the circuit is at initial rest. We can use the result of the previous example (7.49) to write a single second-order differential equation that implicitly relates the output signal to the input signal. Let $i(0)$ denote the value of $i(t)$ at $t = 0$ and $\hat{i}(0)$ denote the value of $\frac{di(t)}{dt}$ at $t = 0$. Using (7.45) and (7.46), we can equate the Laplace transforms of the two sides of (7.49). Doing so gives

$$s^2I(s) - si(0) - \hat{i}(0) + 1.5sI(s) - 1.5i(0) + 3I(s)$$
$$= sV_s(s) - v_s(0) + 1.5V_s(s).$$

We can solve this algebraic relation for $I(s)$:

$$I(s) = \frac{s + 1.5}{s^2 + 1.5s + 3}V_s(s)$$
$$+ \frac{[s + 1.5]i(0) + \hat{i}(0) - v_s(0)}{s^2 + 1.5s + 3}.$$
(7.50)

The first term is the initial-rest solution; the second term contains the contributions to $I(s)$ from the initial conditions (including $v_s(0)$).

Instead of writing a single second-order differential equation, we could have written two coupled first-order ones in the variables $i(t)$ and $v_c(t)$, the voltage across the capacitor. To derive these equations, we go back to basics and write one KCL equation, two KVL equations, and three element relations. This gives the six equations

$$i_r(t) = i(t) - i_c(t)$$
$$v_\ell(t) + v_r(t) = v_s(t)$$
$$v_r(t) = v_c(t)$$
$$v_r(t) = 2i_r(t)$$
$$\frac{di_\ell(t)}{dt} = v_\ell(t)$$
$$\frac{1}{3}\frac{dv_c(t)}{st} = i_c(t).$$

Now we use the first four equations to eliminate the variables $v_r(t)$, $i_r(t)$, $v_\ell(t)$, and $i_c(t)$. This leaves two coupled first-order equations:

$$\frac{di(t)}{dt} = v_s(t) - v_c(t)$$
$$\frac{dv_c(t)}{dt} = 3i(t) - \frac{3}{2}v_c(t).$$
(7.51)

We can use Laplace transforms to solve these. Taking the Laplace transform of each equation gives

$$sI(s) - i(0) = V_s(s) - V_c(s)$$
$$sV_c(s) - v_c(0) = 3I(s) - \frac{3}{2}V_c(s),$$

which reduce to

$$I(s) = \frac{s + 1.5}{s^2 + 1.5s + 3} V_s(s) + \frac{[s + 1.5]i(0) - v_c(0)}{s^2 + 1.5s + 3}.$$

$$(7.52)$$

The results in (7.50) and (7.52) look different because the initial conditions are specified differently, but in fact, they are the same. We can see this by noting that a KVL equation written at $t = 0$ would show that

$$v_\ell(0) = 1 \cdot \hat{i}(0) = v_s(0) - v_c(0). \qquad (7.53)$$

Using (7.53) to solve for $v_c(0)$ and substituting this result into (7.50) reveals that (7.50) and (7.52) are, in fact, the same.

When we write the time-domain equations as a series of coupled first-order differential equations, the initial conditions are the initial values of the inductor currents and the capacitor voltages. When we write the time-domain relation as a single high-order differential equation relating the input and output signals, the initial conditions are initial values of the input and output variables and a number of their derivatives. In this respect, the representation with coupled first-order equations is more natural and is generally preferred, since the initial conditions correspond to measurable quantities within the circuit. When the circuit is at initial rest, either set of initial conditions will be identically zero.

7-4-2   Impulse-Response Characterization

What happens when an arbitrary time waveform is the input to a circuit? If the circuit is at initial rest, so that it behaves like a linear, time-invariant system, we can derive a time-domain formula that relates the input to the output. The key is to represent the arbitrary input signal as a superposition of delayed step functions, for which we know how to compute the response.

Figure 7-24 illustrates how a time waveform, $x(t)$, can be approximated by a piecewise constant function. That

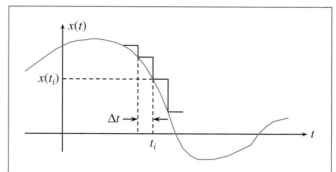

Fig. 7-24. Approximation of a curve by a sequence of small steps.

stairstep function, in turn, can be written as a sum of delayed step functions. The delays of the individual steps are multiples of a basic delay Δt. The superposition of delayed steps becomes a better and better approximation to $x(t)$ as Δt becomes smaller. The height of the jth step is

$$[x(t_j) - x(t_j - \Delta t)] \approx x'(t_j)\Delta t,$$

where $x'(t)$ is the first derivative of $x(t)$. This approximation becomes exact in the limit as $\Delta t \to 0$. Using this expression, we can write a formula for the approximation of $x(t)$ by the piecewise constant approximation, $x_\Delta(t)$

$$x(t) \approx x_\Delta(t) = \sum_j u(t - t_j)x'(t_j)\Delta t. \qquad (7.54)$$

Equation (7.54) approximates the arbitrary input signal by a superposition of delayed step functions.

Now we need to ascertain the circuit output when this signal is its input. Define $g(t)$ to be the step response of the circuit—in other words, $g(t)$ is the output signal when the input source signal is $u(t)$:

$$u(t) \longrightarrow g(t).$$

Then, because the circuit is time invariant, delaying the step input will simply delay the step response. Thus,

$$u(t - t_j) \longrightarrow g(t - t_j).$$

Next, we know that, because the circuit is linear, scaling the delayed step by a constant will simply scale the step response. Therefore,

$$x'(t_j)\Delta t \, u(t - t_j) \longrightarrow x'(t_j)\Delta t \, g(t - t_j).$$

The linearity property also ensures that, when we create an input signal by adding a number of components together, the output will be the sum of the outputs due to each individual component. In this case, the components are scaled and delayed step functions:

$$x_\Delta(t) = \sum_j u(t - t_j)x'(t_j)\Delta t$$

$$\longrightarrow \sum_j g(t - t_j)x'(t_j)\Delta t = y_\Delta(t). \quad (7.55)$$

Notice that $u(t - t_j)$ in the representation of the input signal is replaced by $g(t - t_j)$ in the representation of the output signal. Finally, we take the limit as $\Delta t \to 0$:

$$x(t) = \lim_{\Delta t \to 0} x_\Delta(t) = \lim_{\Delta t \to 0} \sum_j u(t - t_j)x'(t_j)\Delta t$$

$$= \int_{-\infty}^{\infty} u(t - \tau)x'(\tau)\, d\tau.$$

This is an esoteric relation that expresses $x(t)$ as the integral of its derivative. A much more useful relation occurs when we look at the effect of the limiting operation on $y_\Delta(t)$:

$$y(t) = \lim_{\Delta t \to 0} y_\Delta(t) = \int_{-\infty}^{\infty} g(t - \tau)x'(\tau)\, d\tau. \quad (7.56)$$

This integral is a formula for computing the output of the circuit from the input signal and the step response.

In order to remove the necessity for performing a derivative on $x(t)$, we can integrate by parts to derive a more useful relation:

$$y(t) = g(t - \tau)x(\tau)|_{-\infty}^{\infty} + \int_{-\infty}^{\infty} x(\tau)g'(t - \tau)\, d\tau.$$

Assuming that $g(-\infty) = 0$ and that $x(-\infty) = 0$ (from the initial-rest condition), this reduces to

$$y(t) = \int_{-\infty}^{\infty} x(\tau)h(t - \tau)\, d\tau, \quad (7.57)$$

where $h(t) = g'(t)$. For a linear, time-invariant circuit, the first time derivative of the step response is the familiar impulse response. The formula above is the time-domain relation that we have been seeking. It is a formula for computing the time-domain output waveform from the time-domain input waveform. (7.57) is called the *convolution integral*. To paraphrase an equation (7.57): The output of a linear, time-invariant circuit can be computed as the *convolution* of the excitation (input) of the circuit with the impulse response, $h(t)$. The impulse response is also the inverse Laplace transform of the system function. Thus,

$$h(t) \longleftrightarrow H(s). \quad (7.58)$$

Equation (7.57) is the time-domain relation that relates the input signal to the circuit. In the Laplace domain, this relationship is given by

$$Y(s) = X(s)H(s), \quad (7.59)$$

as we have already seen. The simplicity of (7.59) when compared to (7.57) provides further justification for working in the Laplace domain.

To illustrate how all of these various methods for computing the output of a circuit are related and to add evidence that they give the same result when applied properly, consider the following example.

🌐   **WORKED SOLUTION:** *Exponential Input*

Example 7-11   A Comparison of Methods

Consider the simple *RC* circuit in Figure 7-25, which is at initial rest. We shall evaluate its output several ways. First, however, we evaluate the system function and impulse response of the circuit, using a voltage divider to compute $H(s)$:

$$H(s) = \frac{\frac{2}{s}}{1 + \frac{2}{s}} = \frac{2}{s + 2}; \qquad (7.60)$$

and then we evaluate an inverse Laplace transform to compute $h(t)$:

$$h(t) = 2e^{-2t}u(t). \qquad (7.61)$$

Approach 1: We can use the system function to compute the Laplace transform of the capacitor voltage, $Y(s)$.

$$Y(s) = H(s)X(s) = \frac{2}{s + 2} \cdot \frac{1}{s + 3} = \frac{2}{s + 2} - \frac{2}{s + 3}.$$

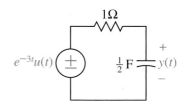

Fig. 7-25. A simple *RC* circuit with an exponential input signal.

Computing an inverse transform gives

$$y(t) = 2(e^{-2t} - e^{-3t})$$

for $t > 0$. Because of the initial-rest condition, however, we know that $v_c(t) = 0$ for $t < 0$. Therefore,

$$y(t) = 2(e^{-2t} - e^{-3t})u(t). \qquad (7.62)$$

Approach 2: Alternatively, we can set up and solve a differential equation for $y(t)$. We can derive the differential equation from the system function via the methods of the previous section:

$$\frac{dy(t)}{dt} + 2y(t) = 2v_s(t).$$

We can solve this equation by using Laplace transforms again. We get

$$sY(s) + 2Y(s) = \frac{2}{s + 3}$$

or

$$Y(s) = \frac{2}{(s + 2)(s + 3)}.$$

The inverse transform is

$$y(t) = 2(e^{-2t} - e^{-3t}), \qquad t > 0,$$

as before. Since the circuit is at initial rest, we also know that $y(t) = 0$ for $t < 0$.

Approach 3: Finally, we can compute $y(t)$, using the convolution integral

$$y(t) = \int_{-\infty}^{\infty} x(\tau)h(t - \tau)\, d\tau.$$

As a first step, we substitute for x and h.

$$y(t) = \int_{-\infty}^{\infty} 2e^{-3\tau}u(\tau)e^{-2(t-\tau)}u(t - \tau)\, d\tau \qquad (7.63)$$

The step functions in the integrand play a key role in evaluating the convolution integral and we must handle them very carefully. Their role is to limit the range of integration. The term $u(\tau)$ forces the integrand to be zero whenever $\tau < 0$. Similarly, the term $u(t - \tau)$ will be zero whenever $t - \tau < 0$ (or $\tau > t$). Taken together, these facts mean that the only values of τ that will contribute to the integral are those that lie in the range $0 \leq \tau \leq t$. If $t < 0$, then the integrand will be zero for all values of τ, which means that the integral will be zero also. For values of τ in the range $0 < \tau < t$, both step functions have a value of one. Incorporating all of these facts allows us to rewrite (7.63) as

$$y(t) = \begin{cases} \int_0^t 2e^{-3\tau} e^{-2(t-\tau)} \, d\tau, & t \geq 0 \\ 0, & t < 0 \end{cases}.$$

The integral for $t > 0$ is now in such a form that we can evaluate it:

$$y(t) = \int_0^t 2e^{-3\tau} e^{-2(t-\tau)} \, d\tau$$

$$= 2e^{-2t} \int_0^t e^{-\tau} \, d\tau$$

$$= 2e^{-2t} \left[-e^{-t} + 1 \right] = 2 \left[-e^{-3t} + e^{-2t} \right].$$

Now, if we put all of the pieces together, we get

$$y(t) = 2 \left[e^{-2t} - e^{-3t} \right] u(t),$$

as before.   ∎

 To check your understanding, try Drill Problem P7-8.

7-5   Chapter Summary

7-5-1   Important Points Introduced

- The system function allows us to capture the input/output behavior of a linear circuit.

- The system function of a circuit does not depend on the input (source) signals.

- If a circuit is at initial rest and the source waveform is zero for $t < t_0$, then all of the voltages and currents in the circuit must also be zero for $t < t_0$.

- The impulse response of a circuit is the inverse Laplace transform of the system function.

- If a linear circuit is stable, its transient response (and its impulse response) must decay to zero.

- A circuit is stable if and only if all of its poles have negative real parts,

- Circuits at initial rest are linear and time-invariant systems.

- Differentiating (integrating) the input to a linear circuit at initial rest differentiates (integrates) its output.

- The impulse response of a linear circuit is the derivative of its step response.

- The forced response of a linear circuit to an input of the form $e^{s_p t}$ is $H(s_p)e^{s_p t}$.

- The differential equation that related the input and output of a circuit can be derived from the system function, and vice versa.

- In the time domain, the output of a linear circuit at initial rest can be computed by convolving the input signal with the impulse response.

7-5-2 New Abilities Acquired

You should now be able to do the following:

(1) Compute the system function, $H(s)$ of a circuit with a single input and a designated output variable.

(2) Find the response of a circuit at initial rest, for all t.

(3) Derive the impulse response of a circuit at initial rest from the system function.

(4) Identify the poles and zeros of a linear circuit.

(5) Compute the output of a circuit from a pulse-type input by exploiting linearity and time invariance.

(6) Compute the impulse response of a circuit by differentiation.

(7) Relate the key features of the impulse response of a first- or second-order circuit to the locations of its poles and zeros.

(8) Construct the differential equation that relates the input and output of a circuit.

(9) Solve a circuit that is specified by a series of coupled-form differential equations.

(10) Calculate the output of a circuit by evaluating a convolution integral.

7-6 Problems

7-6-1 Drill Problems

P7-1 Write the system function $H(s)$ that relates the output $i_o(t)$ to the input $i_s(t)$ in the circuit of Figure P7-1.

P7-2 Write the system function $H(s)$ for the circuit in Figure P7-1 if the output variable is changed to $v_o(t)$.

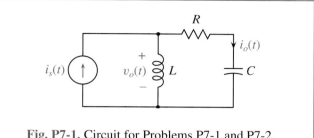

Fig. P7-1. Circuit for Problems P7-1 and P7-2.

P7-3 Find $v_c(t)$ for all t for the circuit of Figure P7-3, if the circuit is at initial rest and the source voltage is $v_s(t) = u(t)$.

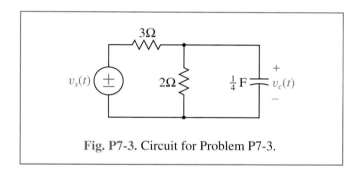

Fig. P7-3. Circuit for Problem P7-3.

P7-4

(a) For the circuit in Figure P7-4, write the system function that relates $V_c(s)$ and $V_s(s)$.

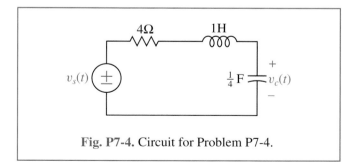

Fig. P7-4. Circuit for Problem P7-4.

(b) Find $v_c(t)$ for all time if the circuit is at initial rest and $v_s(t) = u(t)$.

(c) Find $v_c(t)$ for all time if the circuit is at initial rest and $v_s(t) = 2\delta(t)$.

P7-5 Find the voltage $v(t)$ in the circuit of Figure P7-5, if $i_s(t) = \delta(t) - \delta(t-2)$.

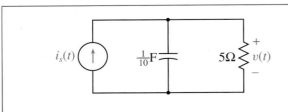

Fig. P7-5. Circuit for Problem P7-5.

P7-6 The step response of a circuit is observed to be

$$g(t) = [e^{-2t} - e^{-3t}]u(t).$$

(a) What is the system function, $H(s)$?
(b) What is the impulse response of the circuit, $h(t)$?

P7-7 Find the differential equation that relates $v(t)$, the output to $i_s(t)$, the input, for the circuit in Figure P7-7, by first finding the system function $H(s)$ and then using it to find the differential equation.

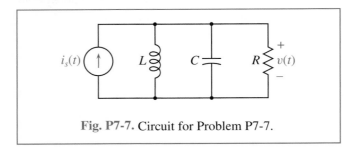

Fig. P7-7. Circuit for Problem P7-7.

P7-8 A circuit has the impulse response $h(t) = u(t)$. Using the convolution integral, calculate the response, $y(t)$, when the input signal is $u(t)$.

7-6-2   Basic Problems

P7-9

(a) For the circuit in Figure P7-9, state the system function $H(s)$ that relates the output $i(t)$ to the input $i_s(t)$, if $i_s(t)$ is the current in the current source. In other words, find

$$H(s) = \frac{I(s)}{I_s(s)}.$$

In this part of the problem, do not assume that $i_s(t) = u(t)$.

(b) Find $i(t)$ for all values of t if $i_s(t) = u(t)$. Assume that the circuit is at initial rest for $t < 0$.

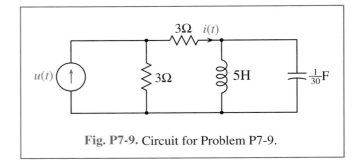

Fig. P7-9. Circuit for Problem P7-9.

P7-10 Find the system function $H(s) = V_{out}(s)/V_{in}(s)$ for the circuit in Figure P7-10.

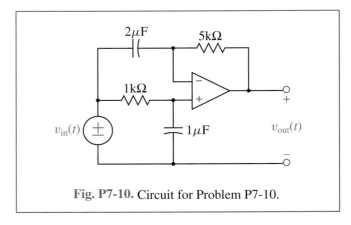

Fig. P7-10. Circuit for Problem P7-10.

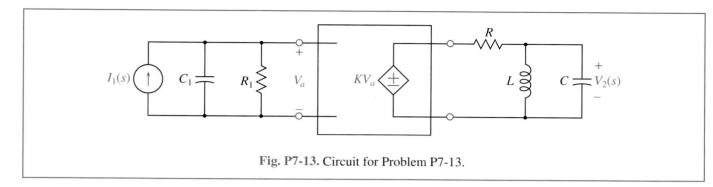

Fig. P7-13. Circuit for Problem P7-13.

P7-11 Find the system function, $H(s)$, for the circuit in Figure P7-11.

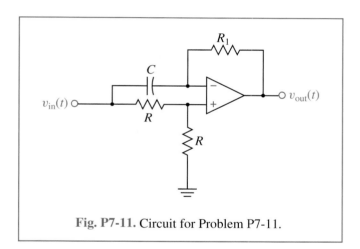

Fig. P7-11. Circuit for Problem P7-11.

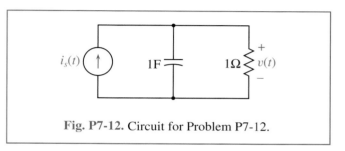

Fig. P7-12. Circuit for Problem P7-12.

P7-12 The network of Figure P7-12 is initially at rest. Compute the voltage $v(t)$ for each of the following inputs:

(a) $i_s(t) = u(t)$;

(b) $i_s(t) = (\sin t)u(t)$;

(c) $i_s(t) = tu(t)$.

P7-13 State the system function $H(s) = V_2(s)/I_1(s)$ of the circuit in Figure P7-13.

P7-14

(a) The system function of the circuit in Figure P7-14, which is at initial rest, has the form

$$H(s) = \frac{V_{\text{out}}(s)}{V_{\text{in}}(s)} = \frac{as + b}{s + c}.$$

Calculate the values of a, b, and c.

(b) Determine $v_{\text{out}}(t)$ for all t, if $v_{\text{in}}(t) = e^{-2t}u(t)$.

P7-15 This problem asks you to analyze the circuit in Figure P7-15.

(a) Find the system function of the circuit if $v_{\text{in}}(t)$ is the input signal and $v_{\text{out}}(t)$ is the output signal.

(b) Compute the impulse response of the circuit.

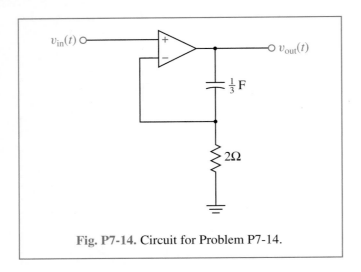

Fig. P7-14. Circuit for Problem P7-14.

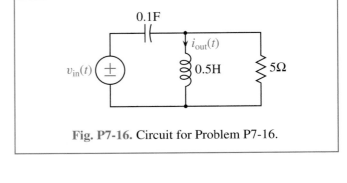

Fig. P7-16. Circuit for Problem P7-16.

P7-17   The circuit in Figure P7-17 is at initial rest. Compute $i(t)$ for all t.

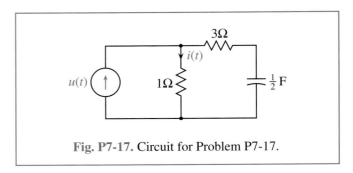

Fig. P7-17. Circuit for Problem P7-17.

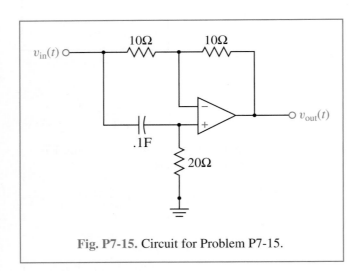

Fig. P7-15. Circuit for Problem P7-15.

P7-18   The circuit in Figure P7-18 is at initial rest. Compute $i(t)$.

P7-16   The circuit in Figure P7-16 is at initial rest.

(a) Find the system function $H(s)$ that relates the output, $I_{out}(s)$, to the input, $V_{in}(s)$.

(b) Find the impulse response of the system, $h(t)$.

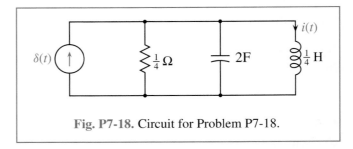

Fig. P7-18. Circuit for Problem P7-18.

P7-19

(a) Find the system function $H(s)$ of the circuit in Figure P7-19(a), which is at initial rest.

(b) Find the impulse response of that circuit.

(c) Find the response of the circuit to the input shown in Figure P7-19(b).

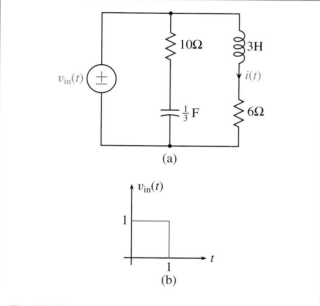

(a)

(b)

Fig. P7-19. (a) Circuit for Problem P7-19. (b) Waveform for Problem P7-19.

P7-20 Find the impulse response of each of the circuits in Figure P7-20. The output variables are the indicated voltages. All of the reactive elements are at initial rest.

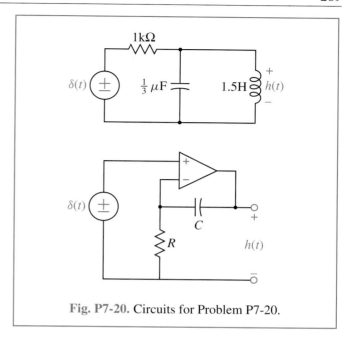

Fig. P7-20. Circuits for Problem P7-20.

P7-21

(a) Find the system function of the circuit in Figure P7-21(a). The circuit is at initial rest.

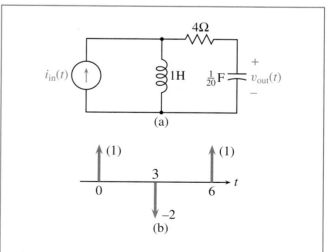

(a)

(b)

Fig. P7-21. (a) Circuit for Problem P7-21. (b) Input waveform, $v_{in}(t)$, for Problem P7-21.

(b) Find and sketch the impulse response.

(c) Sketch the response of the circuit to the input represented in Figure P7-21(b).

P7-22

(a) Express each of the waveforms in Figure P7-22 as a superposition of shifted unit-step functions $u(t)$ and/or unit-ramp function, $u_{-2}(t)$.

(b) Sketch each of the following signals.

    i. $y_i(t) = u(t) - 2u(t-4) + 2u(t-6) - u(t-7)$

    ii.

$$y_{ii}(t) = -u_{-2}(t) + 2u_{-2}(t-1) - 2u_{-2}(t-3)$$
$$+ 2u_{-2}(t-5) - u_{-2}(t-6)$$

    iii.

$$y_{iii}(t) = u_{-2}(t) - 2u_{-2}(t-5)$$
$$+ 2u_{-2}(t-10) - 2u_{-2}(t-15) + \dots$$

P7-23 For each of the circuits in Figure 7-20 in the text, let the inductors have an inductance of L, the capacitors a capacitance of C, and the resistors a resistance of R. Verify that each circuit produces the pole-zero plot indicated and express the location of its poles and zeros in terms of L, C, and R.

P7-24 Approximations to ideal delay elements can be achieved in a number of ways, but each has its limitation(s). Consider the circuit in Figure P7-24, which approximates an ideal delay of 2 seconds.

(a) Show that the system function for this circuit is $H(s) = \frac{s-1}{s+1}$.

(b) By inverse transforming, find the impulse response, $h(t)$.

(c) Compute the step response of the circuit—that is, the response to the input $v_{in}(t) = u(t)$.

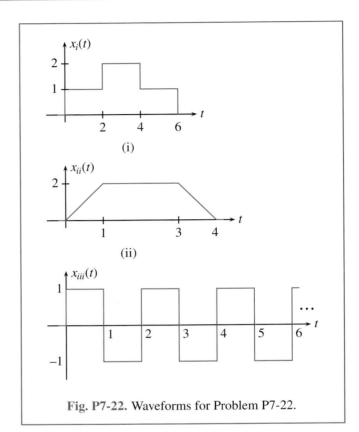

Fig. P7-22. Waveforms for Problem P7-22.

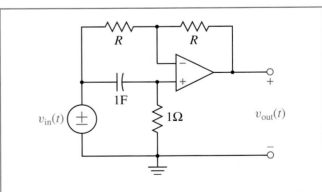

Fig. P7-24. Approximation to a two-second delay for Problem P7-24.

(d) Determine the response of the circuit to a unit ramp input $v_{in}(t) = tu(t)$.

(e) To test the extent to which this system approximates an ideal delay of two seconds, sketch carefully and approximately to scale on separate axes the impulse response, the step response, and the ramp response. Superimpose on each plot the ideal responses.

(f) What is the system function of an ideal delay of two seconds? Show that the system function of the circuit approximates that of the ideal delay for small values of s. (*Hint:* Expand both system functions in power series.)

(g) Use the result of part (f) to explain why the system is a better approximation to an ideal delay when the input signal is a ramp than when it is a step or impulse.

P7-25   A circuit has the impulse response $h(t) = e^{-2t}u(t)$. Using the convolution integral, compute the response, $y(t)$, when the input signal is $u(t)$.

7-6-3   Advanced Problems

P7-26

(a) For the circuit in Figure P7-26(a), write a formula that expresses $I(s)$ in terms of the source variables $V_s(s)$ and $I_s(s)$.

(b) If we are interested only in solving for the current flowing through the inductor, we can replace the whole circuit that is external to the inductor by a Thévenin equivalent network, as shown in Figure P7-26(b). Compute R_T and $V_T(s)$ for this equivalent network in terms of R_1, R_2, $V_s(s)$, and $I_s(s)$. Using this network, verify that the $I(s)$ that you get is the same as in part (a).

(c) Find $i(t)$ for all t if $v_s(t) = Au(t)$ and $i_s(t) = Bu(t)$.

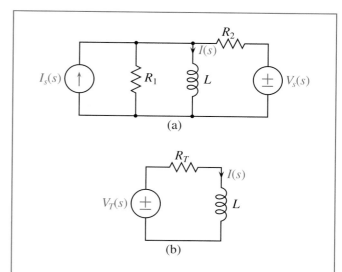

Fig. P7-26. (a) Circuit for Problem P7-26. (b) Thévenin equivalent circuit seen by the inductor in Figure P7-26(a).

P7-27   Show that the circuit in Figure P7-27 functions as a double integrator, in that the output is proportional to the double integral of the input.

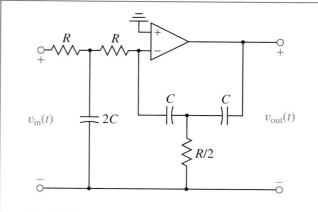

Fig. P7-27. A double integrator for Problem P7-27.

P7-28 An *oscillator* is a circuit whose system function has a pair of complex-conjugate poles located $s = \pm\omega_0$, where ω_0 is the frequency of oscillation. Once started it will continue to produce a sinusoidal output $v_{out}(t)$. In the network in Figure P7-28, find the value of K for which the system will oscillate and compute the frequency of oscillation.

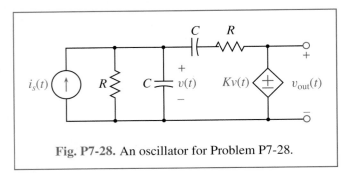

Fig. P7-28. An oscillator for Problem P7-28.

P7-29 Find the system function that relates $V_{out}(s)$ to $V_{in}(s)$ for the circuit in Figure P7-29.

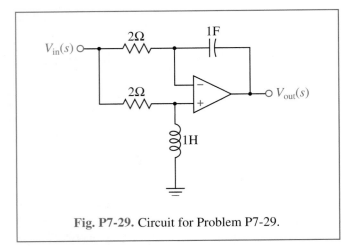

Fig. P7-29. Circuit for Problem P7-29.

P7-30 Find the system function that relates $V_{out}(s)$ to $V_{in}(s)$ for the circuit in Figure P7-30.

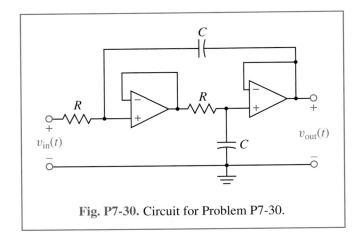

Fig. P7-30. Circuit for Problem P7-30.

P7-31 Switches are devices that can change the topology of a circuit. When they are closed, they behave like short circuits but when they are open, they behave like open circuits. For the circuit in Figure P7-31 the source has been 2 Amperes and the switch has been closed for a very long time prior to $t = 0$. At $t = 0$ the switch is opened and then it is closed again at $t = 1$.

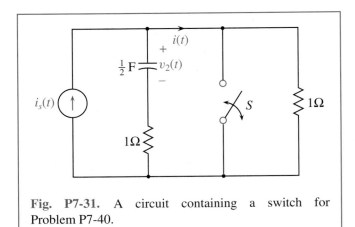

Fig. P7-31. A circuit containing a switch for Problem P7-40.

(a) Compute the voltage across the capacitor at $t = 0$.

(b) Using your result from part (a) as the initial condition for the capacitor voltage, determine the voltage $v(t)$ for $0 < t < 1$.

(c) Sketch $v(t)$ for $-1 < t < 4$.

(d) Compute and sketch $i(t)$ for $-1 < t < 4$.

P7-32 Find the impulse response for the circuit in Figure P7-32.

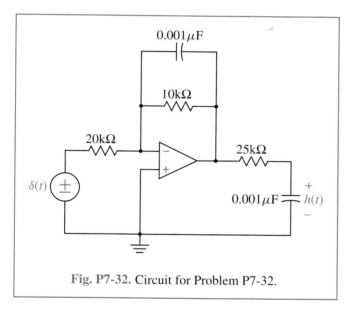

Fig. P7-32. Circuit for Problem P7-32.

P7-33 This problem is about cascading circuits.

(a) Suppose that we have two RLC circuits at initial rest (possibly containing dependent sources) with a voltage input and a voltage output, as illustrated in Figure P7-33(a). These systems are connected in cascade using a buffer amplifier as illustrated in Figure P7-33(b). Show that this cascade connection is equivalent to a single LTI (linear, time-invariant) system with a system function $H_1(s) H_2(s)$—that is, that we can use a buffer amplifier to cascade certain types of LTI circuits.

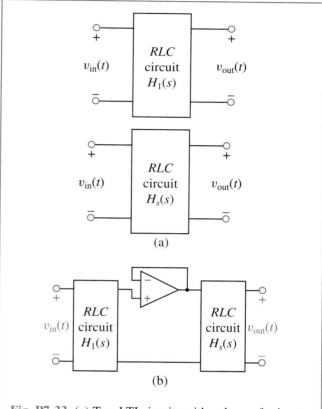

Fig. P7-33. (a) Two LTI circuits with voltages for inputs and outputs. (b) Cascade connection of two LTI circuits.

(b) Show that if an LTI circuit is at initial rest and

$$x(t) \longrightarrow y(t),$$

then

$$\int_{-\infty}^{t} x(\tau) \, d\tau \longrightarrow \int_{-\infty}^{t} y(\tau) \, d\tau.$$

Hint: Model the integration operation as a system. (What is its system function?) Then treat the combined operations of the integration and the system as a cascade.

P7-34 Compute the current $i(t)$ flowing through the resistor in the circuit in Figure P7-34 when the input is the indicated waveform. *Hint:* Use the property that you derived in Problem P7-33(b).

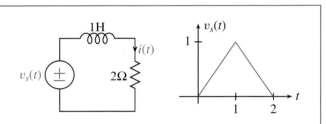

Fig. P7-34. Circuit and source waveform for Problem P7-34.

P7-35 Find the response of the circuit in Figure P7-35(a) to the input signal sketched in Figure P7-35(b).

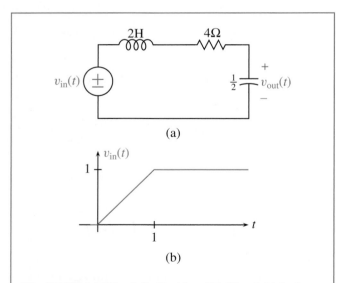

(a)

(b)

Fig. P7-35. (a) Circuit for Problem P7-35 and (b) Its input source waveform.

P7-36

(a) Find the system function $H(s)$ for the circuit in Figure P7-36 with the switch in position a.

(b) Assume that the switch has been in position a for a very long time prior to $t = 0$. At $t = 0$ the switch is moved to position b. Find $v(t)$ and $\frac{dv}{dt}$ just after the switch is moved.

(c) Compute $v(t)$ for $t > 0$.

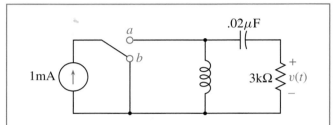

Fig. P7-36. Circuit containing a switch for Problem P7-36.

7-6-4   Design Problems

P7-37 For the circuit with two pairs of terminals shown in Figure P7-37, the relation between $v(t)$ and $i(t)$ is

$$v(t) = i(t) + 3\frac{di(t)}{dt}$$

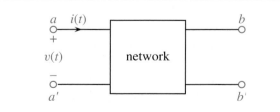

Fig. P7-37. The two-terminal network to be designed in Problem P7-37.

when the terminals b and b' are open-circuited, but

$$v(t) = \frac{di(t)}{dt}$$

when the terminals b and b' are short-circuited. Sketch a possible circuit having these properties.

P7-38 In Figure 7-20 in the text, we show four pole-zero patterns and four circuits that can produce them when the inputs and outputs are the voltages shown.

(a) Find four different circuits that will have the same pole-zero patterns, but for which the input is applied as a current source and the output variable is an element current.

(b) Repeat part (a), but with an output that is an element voltage.

(c) Repeat part (a), but with an input that is applied through a voltage source and where the output is an element current.

P7-39 In Chapter 4, we derived an operational-amplifier implementation of an integrator. Using integrators as building blocks, it is possible to construct a system that will have an arbitrary rational system function. This problem explores that approach to design for a second-order example.

(a) Figure P7-39(a) shows a signal-level implementation of a circuit. The input signal $x_{in}(t)$ (or $X_{in}(s)$) is shown on the left; the output signal $y_{out}(t)$ (or $Y_{out}(s)$) is shown on the right. The boxes labelled

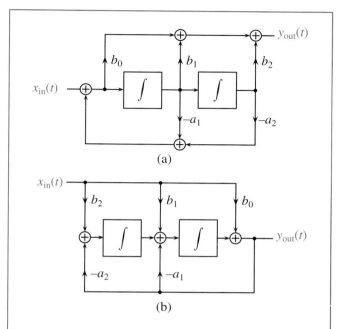

(a)

(b)

Fig. P7-39. Signal-level flow for Problem P7-39. The input signal is $x(t)$, and the output signal is $y(t)$. The implementation incorporates integrators, signal adders, and gains. (a) Canonic-form implementation. (b) Transpose-form implementation.

with integral signs are integrators. (The output signal is the integral of the input signal.) The circles containing +-signs are signal adders, and the coefficients drawn next to the arrows on the branches correspond to gains. Compute the system function

$$H(s) = \frac{Y_{out}(s)}{X_{in}(s)}$$

for this circuit. Assume that there is no initial value on the output of the integrators (initial-rest condition).

(b) Repeat part (a) for the implementation shown in Figure P7-39(b).

P7-40 A parallel RC circuit is connected in series with an unknown circuit N excited by a current source, as shown in Figure P7-40(a). The current-source waveform is shown in Figure P7-40(b). The goal of the problem is to design the network N so that the voltage $v(t)$ will be that shown in Figure P7-40(c).

(a) Use Laplace-transform methods to compute the voltage $v_2(t)$.

(b) Sketch $v_1(t)$ for $t \geq 0$.

(c) Find the impedance $Z_1(s)$ of the circuit N.

(d) Sketch the circuit N.

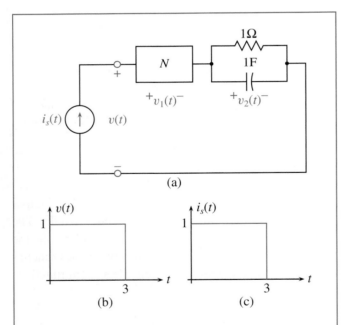

(a)

(b)

(c)

Fig. P7-40. Circuit for Problem P7-40 with the current excitation and voltage response shown. The voltage $v_1(t)$ and the circuit N are to be determined.

P7-41 Using an (ideal) operational amplifier, a single capacitor with a capacitance of $1\,\mu\text{F}$, and one or two (only) $1\,\text{M}\Omega$ resistors, design a circuit with each of the following system functions (i.e., perform 18 separate circuit designs). In each case, indicate the input and output terminals clearly .

(a) $H_a(s) = \frac{1}{s+1}$

(b) $H_b(s) = \frac{2}{s+2}$

(c) $H_c(s) = \frac{1}{s+2}$

(d) $H_d(s) = -\frac{1}{s+1}$

(e) $H_e(s) = -\frac{1}{s}$

(f) $H_f(s) = -\frac{2}{s}$

(g) $H_g(s) = s + 1$

(h) $H_h(s) = \frac{1}{2}(s + 2)$

(i) $H_i(s) = s + 2$

(j) $H_j(s) = -(s + 1)$

(k) $H_k(s) = -s$

(l) $H_l(s) = -\frac{1}{2}s$

(m) $H_m(s) = \frac{s}{s+1}$

(n) $H_n(s) = \frac{s}{s+2}$

(o) $H_o(s) = \frac{s+1}{s+2}$

(p) $H_p(s) = \frac{s+1}{s}$

(q) $H_q(s) = \frac{s+2}{s}$

(r) $H_r(s) = \frac{s+2}{s+1}$

CHAPTER 8

Sinusoidal Input Signals

Objectives

By the end of this chapter, you should be able to do the following:

1. *Find the response of a circuit to an input signal that is a complex exponential time function by working in the frequency domain.*

2. *Find the response of a circuit to an input that is a real sinusoid by using the real-part property.*

3. *Find the sinusoidal steady-state response by using Laplace transforms.*

4. *Find the frequency-domain Thévenin and Norton equivalent circuits for a two-terminal network containing a single sinusoidal source.*

5. *Calculate the average power absorbed in a circuit with a sinusoidal input.*

6. *Design a load impedance that will maximize the power that a sinusoidal source can deliver to a load.*

This chapter and the next look at how circuits behave when the source waveforms are sinusoids. This behavior is both useful and important for a number of reasons. First, these are signals that we see often. Power is distributed to our homes in the form of sinusoidal waveforms. Short segments of sinusoidal signals that vary in amplitude, frequency, or phase are also the means by which digital communications systems encode information. Sinusoids also form the basis for the waveforms used in radio, television, radar, and ultrasound systems. Even more important than the ubiquity of sinusoidal signals, however, is the fact that (almost) *all* signals can be expressed as *superpositions* of sinusoids. The Fourier transform and the Fourier series provide the mathematical basis for this representation. For a linear, time-invariant system, such as a circuit at initial rest, knowledge of how the system responds to each sinusoidal component of a more complex waveform tells us how it will respond to that waveform; it completely describes the input-output behavior of the circuit. For this reason, a circuit is often designed solely on the basis of how it behaves when the input signal is a sinusoid. Fortunately, we can usually discover the sinusoidal response of a circuit easily, using methods that are very similar to ones that we have already developed.

8-1   The Sinusoidal Steady State

8-1-1   Sinusoidal Signals

There are four closely related types of signals that we collectively call *sinusoids*. These were formally introduced in Chapter 5, but it is appropriate to review their definitions here. The first group consists of the *complex exponential time functions* (or complex sinusoidal signals) defined by

$$x(t) = Xe^{j\omega t}, \quad -\infty < t < \infty. \quad (8.1)$$

X is called the *complex amplitude* of the sinusoid and ω, which is real, is called its *frequency*. These signals are very convenient mathematical idealizations that we can use to analyze a circuit , but they are not signals that we can input to an actual circuit that we build, for two reasons: (1) they are complex; and (2) they begin in the infinite past. Notice that the complex sinusoids are also *periodic* signals, because

$$x(t) = x(t + \frac{2\pi}{\omega}) \quad \text{for all } t.$$

At the other extreme are the *switched real sinusoids*, which have the mathematical form

$$x(t) = A\cos(\omega t + \phi)u(t). \quad (8.2)$$

Here A is called the *amplitude*, ω is called the *frequency*, and ϕ is called the *phase*. All three of these parameters have real values. Since these signals are real and begin at $t = 0$, they can be input to any circuit that we can

build, and the response to them can be measured. For analysis purposes, however, they are less tractable than the complex exponential time functions. Fortunately, the behavior of a circuit in response to an input that is a switched real sinusoid is closely related to its behavior when the input is a complex exponential time function, as we shall see shortly.

Notice that we have used the word frequency and the symbol ω to describe similar parameters for both the complex exponential time functions and also the real switched sinusoids. In both cases, the frequency is related to the rate at which the signal repeats itself, but there is a slight difference. For a real sinusoid, the frequency is usually positive; for a complex exponential time function, it can be either positive or negative. In both cases, however, it is real.

Intermediate between these two extremes we will also have occasion to refer to *(unswitched) real sinusoids*

$$x(t) = A\cos(\omega t + \phi), \quad \text{for all } t. \quad (8.3)$$

which are real sinusoids that go infinitely back in time (and are periodic) and *switched complex exponential time functions*

$$x(t) = Xe^{\omega t}u(t). \quad (8.4)$$

As before, the term *switched* denotes a signal that is zero before $t = 0$, and *unswitched* denotes one that begins at $t = -\infty$.

    DEMO: *Sinusoids*

8-1-2   Physical Circuits and Mathematical Models

For most of the next two chapters, we will restrict the inputs to our circuits to being complex exponential time functions, although a more proper statement would be to say that we will be restricting the inputs to our circuit *models* to complex exponential time functions. The distinction is important, and we need to be careful that we understand what we are doing, why we are doing it, and what the results mean.

First, we need to recognize that, up to now, we have been operating in two worlds without making a careful distinction between them. There is the physical world of circuits that we design, build, and use, and there is the mathematical world of models. We have begun to see some of the power of the mathematical models; they allow us to predict certain behavior that we can observe in the physical world, but could not otherwise prove. To cite just one example, we have proven mathematically that when a circuit is linear and time invariant (i.e., when it is constructed from linear resistors, inductors, and capacitors at initial rest) and when $y(t)$ is known to be the response to the input $x(t)$, then $\frac{dy(t)}{dt}$ will always be the response to the input $\frac{dx(t)}{dt}$. Using our mathematical models, we can prove that this result must be true. In the physical world, we could observe that the result is valid for specific cases, but we could not prove that it would always be valid because we could not exhaustively test its validity for all possible input signals.

The components of our mathematical models have been chosen carefully to capture the behavior of circuits and devices in the physical world. Kirchhoff's current law, for example, is a mathematical embodiment of the conservation of current flow. The element relations of the mathematical models for devices capture their physical behavior (approximately). As a result, we have been able to move between these two worlds more or less at will. In the physical world of circuits, all of the voltages and currents must necessarily be real (i.e., not complex) time functions, and they cannot begin at $t = -\infty$; we cannot actually build a voltage source whose voltage waveform is a complex exponential time function. Our mathematical models, on the other hand, have no such restrictions. All that the models require is that there be a set of element and source voltages and currents that

satisfy the complete set of element relations and that satisfy KCL and KVL wherever appropriate. When the source variables in the model are complex, this merely means that some or all of the element variables will also be complex. In none of our derivations to date have we explicitly restricted the voltages and currents in our models to being real. This means that all of the properties of our mathematical models are equally valid when the sources and variables are complex, even though some of these results may not be measurable in physical circuits.

This having been said, the purpose of the mathematical models is still to enable us to understand real circuits. More specifically, in this case, we want to use the knowledge of how a circuit responds to an input that is a complex exponential time function to predict how it will respond to one that is a switched real sinusoid. Using the extended capability of the mathematical model will permit us to get some very useful characterizations of real circuits.

In the remainder of this section, develop a relationship between the response of an arbitrary circuit to a switched real sinusoid and its response to a complex exponential time function. We do this in two steps. First, we relate the response of a circuit to a real sinusoid and its response to a complex exponential time function. Then we relate the responses to the switched signals to the corresponding unswitched responses.

8-1-3 Responses to Real Sinusoidal Inputs

In this section, we derive an important property of our mathematical model for linear circuits at initial rest when the (only) input is complex and the component values are real. This last condition means that the resistance, inductance, and capacitance values are all (positive) real numbers and that any dependent-source gains are also real. Let $x(t)$ be the complex source waveform for such a circuit model and let $y(t)$ be its output (i.e., the element variable, node potential, or mesh current of interest). In mathematical shorthand, we write

$$x(t) \longrightarrow y(t). \qquad (8.5)$$

First, we would like to show that, under these conditions,

$$x^*(t) \longrightarrow y^*(t), \qquad (8.6)$$

where $x^*(t)$ is the complex conjugate of $x(t)$ and $y^*(t)$ is the complex conjugate of $y(t)$. To see this, we write our usual set of KCL equations at all but one of the nodes of the basic network, KVL equations at all of the meshes of the basic network, and element relations for all of the elements, placing all of the element variables on the left-hand sides and the source variables on the right. However, for this circuit, the only source term that will appear on the right-hand side is the single source $x(t)$ and possibly some of its derivatives. We solve these equations to get the variable of interest (output), $y(t)$. Let us call these equations, whose solution is $y(t)$, the *first set* of equations.

Next we replace the input $x(t)$ by its complex conjugate, $x^*(t)$, and solve the resulting *second* set of equations for the new output variable, $\hat{y}(t)$. The goal is now to show that $\hat{y}(t) = y^*(t)$.

One way to accomplish this is to take the complex conjugate of each of the original equations. Since all of the component values are real and all of the initial conditions are zero, the effect of this operation will be to conjugate all of the element variables and to replace $x(t)$ and any of its derivatives by their complex conjugates on the right-hand sides. This final set of equations is the *third set*. The second and third sets of equations are exactly the same, except the element variables have different names. Since $y^*(t)$ is the solution to the third set and $\hat{y}(t)$ is the solution to the second set, $\hat{y}(t) = y^*(t)$. This establishes (8.6).

The circuit is at initial rest, so it is linear. Using superposition and the input–output relations in (8.5) and (8.6) yields

$$\frac{1}{2}[x(t) + x^*(t)] \longrightarrow \frac{1}{2}[y(t) + y^*(t)],$$

or

$$\Re e[x(t)] \longrightarrow \Re e[y(t)]. \qquad (8.7)$$

Similarly,

$$\Im m[x(t)] \longrightarrow \Im m[y(t)]. \qquad (8.8)$$

These are called, respectively, the *real-part property* and the *imaginary-part property* of a circuit with real element values. They say that, if we know the response of a circuit model to a complex signal, we know the response to that signal's real and/or imaginary parts.

Inasmuch as $\cos \omega t = \Re e[e^{j\omega t}]$, we can use this result to compute the response of a circuit to a real sinusoid, if we know the response to the appropriate complex exponential time function. Let $e(t)$ be the circuit output when the input signal is a complex exponential

$$e^{j\omega t} \longrightarrow e(t);$$

then

$$\cos \omega t = \Re e[e^{j\omega t}] \longrightarrow \Re e[e(t)]. \qquad (8.9)$$

In a similar fashion, we can derive the response of a linear, time-invariant circuit to a switched real sinusoid from its response to a switched complex exponential time function.

Example 8-1   Finding the Response to a Switched Sinusoid

As an alternative to using the technique that we have just derived, let us illustrate the direct method of using the Laplace transform to find the output of a circuit when the input is a switched real sinusoid for the simple *RC* circuit

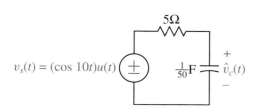

Fig. 8-1. An *RC* network with a single switched-sinusoidal voltage source.

shown in Figure 8-1. We can use a voltage divider to find the system function for the circuit, which is

$$H(s) = \frac{V_c(s)}{V_s(s)} = \frac{\frac{50}{s}}{5 + \frac{50}{s}} = \frac{10}{s + 10}.$$

The Laplace transform of the source waveform is

$$V_s(s) = \mathcal{L}\{\cos 10t\} = \frac{s}{s^2 + 100}.$$

Therefore,

$$V_c(s) = V_s(s)H(s) = \frac{s}{s^2 + 100} \cdot \frac{10}{s + 10}$$

$$= \frac{10s}{(s - j10)(s + j10)(s + 10)}$$

$$= \frac{A}{s - j10} + \frac{A^*}{s + j10} + \frac{B}{s + 10}.$$

We complete the partial-fraction expansion by using limits, to evaluate the unknown expansion coefficients:

$$A = \lim_{s \to j10} \frac{10s}{(s + j10)(s + 10)} = \frac{1}{2(1 + j)}$$

$$= \frac{1}{2\sqrt{2}} e^{-j\pi/4},$$

$$B = \lim_{s \to -10} \frac{10s}{s^2 + 100} = -\frac{1}{2}.$$

Therefore,

$$v_c(t) = \frac{1}{2\sqrt{2}} e^{-j\pi/4} e^{j10t}$$

$$+ \frac{1}{2\sqrt{2}} e^{+j\pi/4} e^{-j10t} - \frac{1}{2} e^{-10t}, \quad t > 0$$

$$= \frac{1}{\sqrt{2}} \cos(10t - \frac{\pi}{4}) - \frac{1}{2} e^{-10t}, \quad t > 0.$$

Since the circuit is at initial rest, however, we know that the response must be zero for $t < 0$. Therefore, the complete output waveform is

$$v_c(t) = \left[\frac{1}{\sqrt{2}} \cos(10t - \frac{\pi}{4}) - \frac{1}{2} e^{-10t} \right] u(t). \quad (8.10)$$

∎

To check your understanding, try Drill Problem P8-1.

DEMO: *Switched Sinusoids*

Example 8-2   Exploiting the Real-Part Property

As an alternative to the direct method for finding the solution to the circuit of Example 8-1, we can use the real-part property instead. This involves first solving the circuit in Figure 8-2, which is identical to the circuit in Example 8-1 except for the change in the source waveform and the different name given to the output. Since

$$\cos(10t)u(t) = \Re[e^{j10t}u(t)],$$

it follows that

$$v_c(t) = \Re[\hat{v}_c(t)],$$

where $v_c(t)$ is the response to the real (switched) sinusoidal input signal.

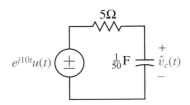

Fig. 8-2. The *RC* network of Figure 8.1 with the input signal replaced by a complex exponential time function.

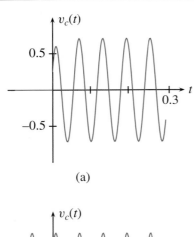

(a)

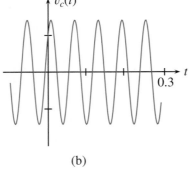

(b)

Fig. 8-3. (a) Complete response to the circuit in Figure 8-1 to a switched-sinusoidal input signal. (b) Response of the circuit to an unswitched sinusoid.

For the circuit in Figure 8-2, the system function is unchanged, because only the input signal is different. Therefore, we have

$$\hat{V}_c(s) = \frac{1}{s - j10} \cdot \frac{10}{s + 10} = \frac{\hat{A}}{s - j10} + \frac{\hat{B}}{s + 10}$$

$$\hat{A} = \lim_{s \to j10} \frac{10}{s + 10} = \frac{1}{1 + j} = \frac{1}{\sqrt{2}} e^{-j\pi/4}$$

$$\hat{B} = \lim_{s \to -10} \frac{10}{s - j10} = -\frac{1}{\sqrt{2}} e^{-j\pi/4}.$$

Substituting for $\hat{A}$ and $\hat{B}$ and computing the inverse Laplace transform gives us $\hat{v}_c(t)$:

$$\hat{v}_c(t) = \left[\frac{1}{\sqrt{2}} e^{j(10t - \pi/4)} - \frac{1}{\sqrt{2}} e^{-j\pi/4} e^{-10t} \right] u(t).$$

Then, taking the real part provides the solution to the original problem; that is,

$$v_c(t) = \Re[\hat{v}_c(t)] = \left[\frac{1}{\sqrt{2}} \cos(10t - \frac{\pi}{4}) - \frac{1}{2} e^{-10t} \right] u(t),$$

(8.11)

which is readily seen to be the same as before. This waveform is shown in Figure 8-3(a). ∎

 To check your understanding, try Drill Problem P8-2.

8-1-4   Responses to Unswitched Signals

We are half-way toward our goal of understanding how the responses of a circuit to the different types of sinusoidal input signals are related. We know that the response of a linear circuit to a real sinusoid is equal to the real part of its response to a complex exponential time function at the same frequency. We also know that the response to a switched real sinusoid is the real part of its response to a switched complex exponential time function. It remains to learn how the switched responses are related to the unswitched ones.

The key to understanding this relationship is a result that we derived in Chapter 7. There we showed that the response of a linear circuit to a switched exponential consisted of a sum of two components—a forced response and a transient response. Specifically, we showed that, if $x_{in}(t) = e^{s_p t} u(t)$, then the output is

$$y_{out}(t) = \underbrace{H(s_p)e^{s_p t}}_{\substack{\text{forced} \\ \text{response}}} + \underbrace{\sum_i B_i e^{p_i t}}_{\substack{\text{transient} \\ \text{response}}}, \qquad (8.12)$$

where the p_i are the poles of the system function. Since there was no assumption made in the derivation that s_p or $x_{in}(t)$ had to be real, the result applies equally to complex input signals. This means that we can set $s_p = j\omega$. By doing this and acknowledging that the output will be zero for $t < 0$, we see that

$$e^{j\omega t} u(t) \longrightarrow H(j\omega)e^{j\omega t} u(t) + \text{transients}. \qquad (8.13)$$

The forced response is a switched complex exponential time function just like the input signal. The exact shape of the transient component of the output waveform depends on the locations of the poles of the system function, but the important fact about these terms is that, *if the circuit is stable, these transient terms will die out.* Notice, however, that for this input signal, the forced response

does not decay. Therefore, after a sufficiently long time, the output will contain only the forced response:

$$y_{out}(t) \approx H(j\omega)e^{j\omega t}, \qquad t \gg 0.$$

This is known as the *sinusoidal steady-state response*.

If the circuit is at initial rest, so that it is time invariant, then we can advance the time at which the switched complex exponential is turned on. In particular, let $x_T(t) = e^{j\omega t} u(t + T)$ be a complex exponential time function that is switched at $t = -T$. When $T = 0$, we have the same signal as before; and when $T = \infty$, we have an (unswitched) complex exponential time function (with no transient terms, since these would have died out in the distant past). Thus,

$$y_\infty(t) = H(j\omega)e^{j\omega t},$$

and the response of a circuit to a periodic complex exponential time function consists only of the forced response

$$e^{j\omega t} \longrightarrow H(j\omega)e^{j\omega t}, \quad \text{for all } t. \qquad (8.14)$$

For a real sinusoid, we apply the real-part property to the forgoing:

$$\cos \omega t \longrightarrow \Re e[H(j\omega)e^{j\omega t}]$$
$$= |H(j\omega)| \cos(\omega t + \angle H(j\omega)).$$

For switched sinusoids, this is the *sinusoidal steady-state response*. The complete response is zero for $t < 0$, and it also includes the transient terms.

Figure 8-3 has already been used to illustrate the response of the circuit in Figure 8-1 to a switched real sinusoid. That response is illustrated in Figure 8-3(a). Below that, in Figure 8-3(b), we indicate the response to the real (unswitched) sinusoid $v_s(t) = \cos 10t$, for all t. Notice that the two responses are virtually identical after $t \approx 0.1s$, when the transient has virtually disappeared, and that both look like sinusoids after that time.

When the input signal is an unswitched sinusoid, the output is also an unswitched sinusoid at the same frequency as the input. When the input signal is a switched sinusoid, the forced response, or the sinusoidal steady-state response, is also a sinusoid at the same frequency. The function $H(j\omega)$, which corresponds to the system function evaluated at a purely imaginary value of its argument, plays a key role. It is called the *frequency response* of the circuit. We will look at it in some detail in the next chapter. In the remainder of this chapter, we want to look at how circuits behave when the input is a sinusoid at a *fixed* frequency.

http:// **DEMO:** *Transient and Steady-State Responses*

8-2   Analyzing Circuits by Using Phasors

In this section, we demonstrate an approach that can be used to solve a network that contains a single source when that source is a complex exponential time function. We have already seen one approach that can be used. If

$$x_{\text{in}}(t) = e^{j\omega_0 t}$$

then

$$y_{\text{out}}(t) = H(j\omega_0)e^{j\omega_0 t}, \qquad (8.15)$$

where $H(s)$ is the system function that relates the input and output signals. This relation is particularly easy to use when we are interested in a single output and the system function is known.

Finding the system function, however, is an algebraic procedure that becomes complicated when a circuit contains many elements, particularly when we are interested in sinusoidal inputs at only a single frequency and when we are interested in all or many of the element variables. In these cases, it is easier to work with the complex amplitudes (phasors) of the complex exponential time functions directly. When these conditions apply, the circuit can be analyzed by solving only linear equations with complex coefficients, a task that can be performed readily by a computer (using, for example, MATLAB.) Although the terminology is not standard, we shall refer to this procedure as circuit analysis in the *phasor domain*. The phasor domain is very closely related to the Laplace domain, and, when the frequency ω_0 is considered as a variable, it is also called the *frequency domain*. We illustrate the procedure with a simple example before moving on to the general case.

8-2-1   A Simple Circuit with a Complex Exponential Input

Initially, we limit our attention to the simple series *RC* circuit shown in Figure 8-4 that we have seen before. When the single independent source is a complex exponential time function and the network is stable, we know that all of the variables in the network are also complex exponential time functions (CETFs) at the same frequency as the source (from Eq. (8.15)). Thus, for this circuit, where $\omega_0 = 10$, each element variable can be written in the form

$$v_k(t) = V_k e^{j10t} \quad \text{or} \quad i_k(t) = I_k e^{j10t}. \qquad (8.16)$$

The phasors associated with these CETFs could be computed from the system functions that relate each element variable to the input signal, if we choose to calculate these. Here, we assume that we have not yet solved for the complete set of system functions, and we treat the phasors of each element variable as unknowns. Once we know these phasors, we can substitute into (8.16) and be done.

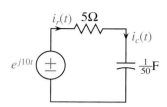

Fig. 8-4. A series RC network with a single complex exponential voltage source.

The circuit of Figure 8-4 contains two elements (a resistor and a capacitor) and a single voltage source. Since each of the four element variables is a complex exponential time function, we can write

$$v_r(t) = V_r e^{j10t}; \qquad v_c(t) = V_c e^{j10t};$$
$$i_r(t) = I_r e^{j10t}; \qquad i_c(t) = I_c e^{j10t}. \qquad (8.17)$$

Although we know the functional form for these variables, we do not know the values of the four phasors until we solve for them. Since the basic network contains two nodes and one mesh, we can write one KCL equation, one KVL equation, and two element relations as the four equations to be solved for the complete solution.

We begin by writing the KCL equation at the node where the resistor and the capacitor are connected:

$$i_r(t) - i_c(t) = 0.$$

After we substitute the known functional form for these time functions from (8.17), we get

$$I_r e^{j10t} - I_c e^{j10t} = (I_r - I_c)e^{j10t} = 0.$$

We can divide both sides by the exponential, since it is never zero. The result is a simpler equation that must be satisfied by the phasors:

$$I_r - I_c = 0$$
$$\implies I_r = I_c \qquad (8.18)$$

Notice that all of the time dependence drops out of the equation.

The KVL equation simplifies similarly. We begin with the time-domain statement

$$v_r(t) + v_c(t) = v_s(t).$$

Substituting as before and dividing through by the exponential term gives an equation involving only the phasors for the voltages:

$$V_r + V_c = 1. \qquad (8.19)$$

Next, we need to see what happens to the two element relations. For the resistor, Ohm's law states that

$$v_r(t) = 5i_r(t).$$

Substituting the known functional forms from (8.17) yields

$$V_r e^{j10t} = 5I_r e^{j10t}$$
$$\implies V_r = 5I_r. \qquad (8.20)$$

This equation, too, reduces to an equation involving only the phasors; the phasor associated with the voltage is five times the phasor of the current. For the capacitor, a similar thing happens. We know that the current is proportional to the derivative of the voltage:

$$i_c(t) = \frac{1}{50}\frac{d}{dt}v_c(t).$$

Substituting from (8.17) gives

$$I_c e^{j10t} = \frac{1}{50}\frac{d}{dt}\left(V_c e^{j10t}\right),$$

or, after we evaluate the derivative,

$$I_c e^{j10t} = j\frac{1}{5}V_c e^{j10t}.$$

If we again cancel the exponentials, we get a constraint between the voltage and current phasors that has no time dependence:

$$I_c = j\frac{1}{5}V_c.$$

Equivalently,

$$V_c = -j5I_c. \tag{8.21}$$

The phasor associated with the capacitor voltage is proportional to the complex amplitude of its current, just as for the resistor. There are two differences, however. First the proportionality constant depends on the frequency of the source; second, it is purely imaginary. In the phasor domain, as in the Laplace domain, differential equations become algebraic equations.

All four of the equations that define the equilibrium solution are algebraic equations that involve only the phasors of the element variables and the phasor of the source, all of which are complex constants. We summarize those equations by writing them in matrix–vector form:

$$\begin{bmatrix} 0 & 0 & 1 & -1 \\ 1 & 1 & 0 & 0 \\ 1 & 0 & -5 & 0 \\ 0 & 1 & 0 & j5 \end{bmatrix} \begin{bmatrix} V_r \\ V_c \\ I_r \\ I_c \end{bmatrix} = \begin{bmatrix} 0 \\ 1 \\ 0 \\ 0 \end{bmatrix}.$$

Solving for the phasors gives

$$V_r = \frac{1+j}{2}$$
$$V_c = \frac{1-j}{2}$$
$$I_r = I_c = \frac{1+j}{10}.$$

Now that we know the phasors, we know the element variables, because of (8.17). All of these are complex exponential time functions. They are as follows:

$$v_r(t) = \left[\frac{1+j}{2}\right]e^{j10t} \tag{8.22}$$

$$v_c(t) = \left[\frac{1-j}{2}\right]e^{j10t} \tag{8.23}$$

$$i_r(t) = i_c(t) = \left[\frac{1+j}{10}\right]e^{j10t}. \tag{8.24}$$

WORKED SOLUTION: *Complex Exponentials*

Before we move on to treat the more general case, let us see how we can use the result that we derived in the previous section to find the response of the circuit to a *real* sinusoidal source waveform, as in Figure 8-5.

We know that, if

$$x(t) \longrightarrow y(t),$$

then

$$\Re[x(t)] \longrightarrow \Re[y(t)].$$

We derived this result by using the input–output viewpoint of the circuit. Here we are looking for all of the element variables in the circuit and not simply for one output signal, but the same result applies. Let $\hat{v}_c(t)$,

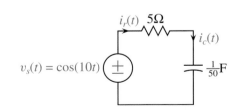

Fig. 8-5. An *RC* network containing a single sinusoidal voltage source.

$\hat{v}_r(t)$, $\hat{i}_c(t)$, and $\hat{i}_r(t)$ denote the element variables when the voltage source is the CETF $\hat{v}_s(t)$. Then, if the source voltage is replaced by $v_s(t) = \Re[\hat{v}_s(t)]$, the element variables become

$$v_r(t) = \Re[\hat{v}_r(t)]; \qquad i_r(t) = \Re[\hat{i}_r(t)]$$
$$v_c(t) = \Re[\hat{v}_c(t)]; \qquad i_c(t) = \Re[\hat{i}_c(t)].$$

We can derive this result by letting each of the element variables be the output variable in turn. Since we already know the waveforms $\hat{v}_c(t)$, $\hat{v}_r(t)$, $\hat{i}_c(t)$, and $\hat{i}_r(t)$ from our previous analysis of the circuit, we can write

$$v_r(t) = \Re[\hat{v}_c(t)] = \Re\left[\left(\frac{1+j}{2}\right)e^{j10t}\right]$$

$$= \frac{1}{\sqrt{2}}\cos(10t + \frac{\pi}{4}), \tag{8.25}$$

$$v_c(t) = \Re[\hat{v}_r(t)] = \Re\left[\left(\frac{1-j}{2}\right)e^{j10t}\right]$$

$$= \frac{1}{\sqrt{2}}\cos(10t - \frac{\pi}{4}), \tag{8.26}$$

$$i_r(t) = i_c(t) = \Re[\hat{i}_c(t)] = \Re\left[\left(\frac{1+j}{10}\right)e^{j10t}\right]$$

$$= \frac{1}{5\sqrt{2}}\cos(10t + \frac{\pi}{4}). \tag{8.27}$$

When the source is a real sinusoid, the element variables are also real sinusoids at the same radian frequency, but with different amplitudes and different phases.

The amplitudes and phases of the element variables vary with the frequency of the source. To make this fact evident, let the frequency of the source be a variable ω_0. As before, we will initially let the source be a complex exponential time function at this frequency. The

equations that define the phasors which determine the solution become

$$\begin{bmatrix} 0 & 0 & 1 & -1 \\ 1 & 1 & 0 & 0 \\ 1 & 0 & -5 & 0 \\ 0 & 1 & 0 & j50/\omega_0 \end{bmatrix} \begin{bmatrix} V_r \\ V_c \\ I_r \\ I_c \end{bmatrix} = \begin{bmatrix} 0 \\ 1 \\ 0 \\ 0 \end{bmatrix}.$$

As functions of ω_0, the new values of the phasors are

$$V_r = \frac{j\omega_0}{10 + j\omega_0}$$

$$V_c = \frac{10}{10 + j\omega_0}$$

$$I_r = I_c = \frac{j\omega_0/5}{10 + j\omega_0}.$$

These values for the phasors mean that the actual element variable waveforms are as follows:

$$v_r(t) = \frac{\omega_0}{\sqrt{100 + \omega_0^2}}\cos(\omega_0 t + \frac{\pi}{2} - \tan^{-1}\frac{\omega_0}{10})$$

$$v_c(t) = \frac{10}{\sqrt{100 + \omega_0^2}}\cos(\omega_0 t - \tan^{-1}\frac{\omega_0}{10})$$

$$i_r(t) = i_c(t) = \frac{1}{5}v_r(t).$$

As an aside, it is reassuring to notice that, when we compare $v_c(t)$ with $i_c(t)$, we confirm that

$$i_c(t) = \frac{1}{50}\frac{dv_c(t)}{dt}.$$

We could also verify, albeit tediously, that these time waveforms satisfy KVL around the mesh (i.e., that $v_s(t) = v_c(t) + v_r(t)$).

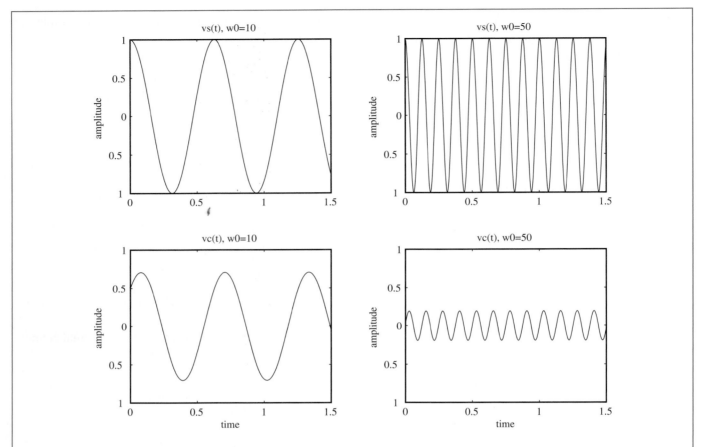

Fig. 8-6. Behavior of the circuit in Figure 8-5. Top row: $v_s(t)$. Bottom row: $v_c(t)$. Left column: $\omega_0 = 10$ rad/s. Right column: $\omega_0 = 50$ rad/s.

Figure 8-6 shows the source and corresponding capacitor voltages for two different frequencies. Notice that the amplitude of the higher-frequency output is reduced relative to the amplitude of the lower-frequency one.

8-2-2   The General Case

The previous section showed that we could find the complete solution of a circuit containing a single

sinusoidal source by solving a series of algebraic equations for the phasors of the element voltages and currents—at least for one circuit. In this section, we demonstrate that this approach will work for any RLC circuit. The steps involved are the same as those that we saw in the example.

We consider only circuits that contain a *single independent source* that is a complex exponential time function. This is not a serious limitation, because we can use source superposition when multiple sources are present or when the source consists of a superposition of complex exponential time functions.

Complex exponential source waveforms with radian frequency ω_0 and complex amplitude (phasor) X have the form

$$x(t) = Xe^{j\omega_0 t} \quad \text{for all } t.$$

If the circuit is stable, all of the other voltages and currents in the circuit will also be complex exponential time functions at the same radian frequency (ω_0), but with different phasors. As in the example in the previous section, those voltages and currents will be completely known once we know those phasors. This allows us to write the voltage and current of the kth element variable in terms of the unknown phasors as

$$v_k(t) = V_k e^{j\omega_0 t}; \quad i_k(t) = I_k e^{j\omega_0 t}. \quad (8.28)$$

As in the example, we can simply substitute this expected form into the various time-domain equations that define the circuit laws and turn these into equations that constrain the phasors. As in the example, all of the time variation will drop out.

Kirchhoff's Current Law: Consider a node in the circuit, like the one shown in Figure 8-7. By KCL, we

know that the sum of the currents entering the node is zero. Thus,

$$i_1(t) + i_2(t) + i_3(t) + i_4(t) = 0.$$

The phasors for these four currents must satisfy a similar equation:

$$I_1 + I_2 + I_3 + I_4 = 0.$$

This means that KCL is satisfied not only for the time-domain currents that enter a node (or by extension to all currents that enter a closed surface) but *also for the phasors of the currents that enter any node (or closed surface)!*

Kirchhoff's Voltage Law: Now consider any closed path in the network, such as the one in Figure 8-8. A similar observation can be made for KVL when there is a single complex exponential source. If we write a KVL equation for the voltages around the loop, then

$$v_1(t) + v_2(t) - v_3(t) - v_4(t) = 0,$$

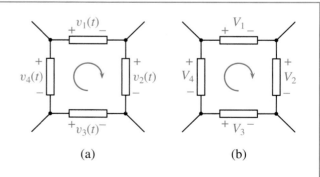

(a)   (b)

Fig. 8-8. Kirchhoff's voltage law. (a) The sum of the voltages around any closed path is zero. (b) The sum of the complex amplitudes of the voltages around any closed path is zero.

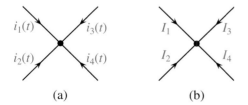

(a)   (b)

Fig. 8-7. Kirchhoff's current law. (a) The net current entering a node is zero. (b) The sum of the complex amplitudes of the currents entering a node is zero.

and, for the phasors,

$$V_1 + V_2 - V_3 - V_4 = 0.$$

So KVL is satisfied, not only for the time-domain voltages encountered on any closed path, but also *for the complex amplitudes of the voltages that lie along any closed path in the circuit!*

Element Relations: We saw in the earlier example that when there is a single complex exponential source in the network, the phasors of the element variables satisfy a relation that resembles Ohm's law: The phasor associated with the voltage across an element is proportional to the phasor of the current passing through it. This is true, not only for resistors, but also for capacitors and, by extension, for inductors. It is also true in circuits that are more general than that example.

   Thus, for an ideal resistor with a resistance R, a voltage $v_r(t)$, and a current $i_r(t)$, defined in a way that is consistent with the default sign convention, the voltage is directly proportional to the current:

$$v_r(t) = R i_r(t).$$

If we substitute for these variables, using (8.28), then

$$V_r e^{j\omega_0 t} = R I_r e^{j\omega_0 t}.$$

Dividing through by the complex exponential reveals the phasor relation for a resistor:

$$V_r = R I_r,$$

which is a phasor-domain statement of Ohm's law.

   Next, consider an inductor with an inductance L, voltage $v_\ell(t)$, and current $i_\ell(t)$. From Chapter 1, we know that these waveforms are related in the time domain by

$$v_\ell(t) = L \frac{di_\ell(t)}{dt}.$$

Once again we substitute for the voltage and current:

$$V_\ell e^{j\omega_0 t} = L \frac{d}{dt}\left(I_\ell e^{j\omega_0 t}\right)$$
$$= j\omega_0 L I_\ell e^{j\omega_0 t}.$$

Cancelling the complex exponentials gives

$$V_\ell = (j\omega_0 L) I_\ell.$$

Again, the phasor associated with the voltage is proportional to the phasor of the current. The constant of proportionality is $j\omega_0 L$, which depends upon the frequency of the complex exponential; this constant is also complex. The fact that the derivative of an exponential time function is also an exponential time function is key here. It is this fact that allows us to remove the time dependence from the equation. Clearly, this step is not possible for other kinds of signals, which is why we are limiting our attention to complex exponential signals.

   Next, we consider a capacitor having capacitance C, voltage $v_c(t)$, and current $i_c(t)$. In the time domain, these quantities are related by

$$i_c(t) = C \frac{dv_c(t)}{dt}.$$

Using a derivation that is nearly identical to the one that we used for the inductor, we see that

$$I_c = j\omega_0 C V_c$$

or

$$V_c = \frac{1}{j\omega_0 C} I_c.$$

For both the inductor and the capacitor, the "resistance" is a purely imaginary function of the frequency of the source.

Finally, we consider a dependent source. For a dependent source, the source variable is proportional to either a voltage or a current somewhere else in the circuit. That controlling signal either is an element variable or can be expressed as a linear combination of element variables. Without sacrificing any generality, let us consider the case of a voltage-controlled voltage source for which

$$v_o(t) = \beta v_i(t).$$

Since the controlling variable has the functional form $v_i(t) = V_i e^{j\omega_0 t}$, we substitute and then observe that

$$v_o(t) = \beta V_i e^{j\omega_0 t}.$$

Thus, the source variable is also a complex exponential, and its phasor is given by

$$V_o = \beta V_i.$$

This means that the time-domain specification of a dependent source is equally valid in the complex-amplitude domain.

 DEMO: *Elements in the Comp. Ampl. Domain*

Solving the Equations: Before going any further, we should summarize the major points that we have developed for networks containing a single source that is a complex exponential time function.

- A complex exponential time function, $x(t) = X e^{j\omega_0 t}$, is *completely specified* by its associated phasor, X, and its radian frequency, ω_0.
- *All* of the element variables in the network will be complex exponential time functions at the same radian frequency as the source.
- The KCL and KVL equations that relate the voltages and currents in the network reduce to algebraic equations for which the variables are the phasors associated with the voltage and current variables.

- The phasors associated with the voltages and currents of any element are proportional to one another; that is, they obey a relation that is similar to Ohm's law.
- The phasors for the element variables can be found by solving a series of algebraic equations.

The equations that define the complete solution for the phasors are no different from the ones that we encountered in Chapters 1–3 for resistive circuits or in Chapter 6 for RLC circuits in the Laplace domain. We have KCL equations, KVL equations, and algebraic element relations that constrain the phasors of the element voltages to be proportional to phasors of the element currents. The number of variables is exactly the same as for the all-resistor case; the number of independent KCL equations is exactly as it was for the all-resistor case; and the number of independent KVL equations is exactly as it was for the all-resistor case. The "element variables" are now complex numbers instead of time waveforms, but the methodologies that we use to solve for them have not changed. All of the shortcuts that we developed, which depended only on the existence of KCL, KVL, and Ohm's law, are still valid—voltage and current dividers, source substitutions, network equivalent two-terminal networks, and so on—provided that we apply them properly and only in the domain for which the Ohm's law–like relation is valid.

Impedances in the Frequency Domain: Our development of techniques for analyzing circuits in the frequency (phasor) domain should look very familiar. It differs very little from the Laplace-domain derivations in Chapter 6. There are some important similarities and differences between the two domains, however, and we should comment on these before we go any further. Let's begin with some of the differences.

The primary distinction between the two domains is when they are used. The frequency domain is limited to situations in which the input signal is a sinusoid, or—by exploiting linearity—when it is a superposition of sinusoids. The Laplace domain, on the other hand, is applicable to any source signal, although it is particularly useful for understanding the transient behavior of a circuit. There is an implied assumption with the Laplace transform that the inductor currents and capacitor voltages are known at a particular time, usually $t = 0$.

If we care about the behavior of a circuit only at a *single* frequency, the equations to be solved for the phasors associated with the element variables have (complex) coefficients with numerical values. In the Laplace domain, on the other hand, some of the coefficients are functions of s. (This distinction goes away, however, if we wish to understand the sinusoidal behavior of a circuit for *all* frequencies.) Furthermore, once the Laplace transforms of the output variables have been constructed, a final inverse Laplace transform needs to be performed. The implication of these statements is that when we care about the circuit behavior only at a single frequency, the phasor domain will require less analysis effort than the Laplace domain, although computer-aided analysis tools sometimes blur this distinction. In short, getting the complete solution in the Laplace domain can require more effort, but it will also provide different types of information. We will come back to this issue shortly.

The similarities between the frequency domain and the Laplace domain are more obvious. Kirchhoff's voltage and current laws are valid in both domains, and in both domains the element relations for the linear circuit elements resemble Ohm's law. For example, for a capacitor with no initial voltage in the Laplace domain, we saw that

$$I_c(s) = sC \cdot V_c(s);$$

in the complex-amplitude domain,

$$I_c = j\omega_0 C \cdot V_c.$$

We have defined the impedance of an element as the ratio of the Laplace transform of its voltage to the Laplace transform of its current:

$$Z_c(s) \triangleq \frac{V_c(s)}{I_c(s)} = \frac{1}{Cs}.$$

In the frequency domain, we can define the impedance of an element to be the ratio of the phasor of its voltage to the phasor of its current:

$$\frac{V_c}{I_c} = \frac{1}{j\omega_0 C} \triangleq Z_c(j\omega_0)$$

This definition of the impedance is consistent. Similarly, for an inductor,

$$\frac{V_\ell}{I_\ell} = j\omega_0 L \triangleq Z_\ell(j\omega_0).$$

The concept of an impedance extends across both domains. In the Laplace domain, it relates the Laplace transforms of the element variables; in the frequency domain, it relates their phasors. Furthermore, the impedance functions are also the same, provided that we set $s = j\omega_0$ when going from the Laplace domain to the frequency domain. The time-domain and phasor-domain representations of resistors, inductors, and capacitors are shown in Figure 8-9.

Example 8-3   Solving a Circuit in the Phasor Domain

As an example, we can use the phasor domain to solve a circuit with a single sinusoidal source. Consider the

Signal Operations	*Phasor Operations*
R ⌇ $\quad v_r(t) = R i_r(t)$	R ⌇ $\quad V_r = R I_r$
L ⌇ $\quad v_\ell(t) = L \dfrac{d}{dt} i_\ell(t)$	$j\omega_0 L$ ⌇ $\quad V_\ell = j\omega_0 L I_\ell$
C ⊣⊢ $\quad i_c(t) = C \dfrac{d}{dt} v_c(t)$	$\dfrac{1}{j\omega_0 C}$ ⊣⊢ $\quad V_c = \dfrac{1}{j\omega_0 C} I_c$

Fig. 8-9. Element relations for resistors, inductors, and capacitors. The column on the left indicates the familiar operations on the time signals themselves. The column on the right indicates the equivalent operations on the phasors when the input signals are complex exponential time functions. The reference directions for the voltages and currents follow the default sign convention.

simple parallel RL circuit shown in Figure 8-10. We know that every element variable in the circuit will be of the form

$$v_k(t) = V_k e^{j3t}, \quad i_k(t) = I_k e^{j3t};$$

hence, we can draw a phasor-domain version of the circuit, in which each signal is represented by its phasor and each element is replaced by the appropriate impedance. Each source is still a source, but it is now labelled with its phasor. Using this complex-amplitude-domain circuit, we can write the familiar KCL, KVL, and

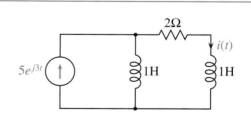

Fig. 8-10. Circuit with an input that is a complex exponential time function.

Ohm's law–style element relations, but in terms of the phasors instead of the time waveforms. Since they only depend upon KCL, KVL, and Ohm's law, we can also use

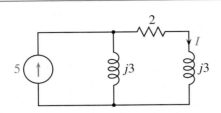

Fig. 8-11. Mapping of the network of Figure 8-10 to the phasor domain.

our familiar circuit simplifications, such as voltage and current dividers, when appropriate. Figure 8-11 shows the phasor-domain representation for this circuit.

Having mapped the circuit, we can now use the familiar current divider to solve for I:

$$I = \frac{j3}{2 + j3 + j3}5 = \frac{j15}{2 + j6} = \frac{9 + j3}{4}. \qquad (8.29)$$

Knowing I, we know $i(t)$:

$$i(t) = \frac{9 + j3}{4}e^{j3t}. \qquad (8.30)$$

∎

 To check your understanding, try Drill Problem P8-3.

       WORKED SOLUTION: *Phasor Domain*

Example 8-4   **Using the Frequency Domain with Real-Sinusoidal Input Signals**

It is more common to encounter real sinusoids as input signals to circuits than complex ones. In these cases, we combine the approach of the previous problem with the real- or imaginary-part property. This example demonstrates the approach.

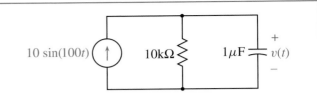

Fig. 8-12. An *RC* circuit with a real sinusoidal source.

We shall solve for the capacitor voltage in the circuit in Figure 8-12. Initially, we replace the sinusoidal source waveform by a complex exponential source at the same frequency,

$$i_s(t) = \Im\mathrm{m}(10e^{j100t}) = \Im\mathrm{m}(\hat{i}_s(t)),$$

and solve that problem via the same approach as in the previous example. At the end, we shall simply take the imaginary part of the result. The first step is to redraw the circuit in the phasor domain. This is shown in Figure 8-13. Notice that the impedance of the capacitor is

$$\frac{1}{jC\omega_0} = \frac{1}{j10^{-6}\cdot 100} = -j10^4.$$

Since V is the phasor of both the capacitor voltage and the resistor voltage, applying KCL at the upper node of the phasor-domain circuit yields

$$\frac{V}{10^4} + \frac{V}{-j10^4} = 10,$$

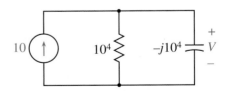

Fig. 8-13. Circuit of Figure 8-12 mapped to the complex-amplitude domain.

from which it follows that

$$V = \frac{10^5}{1+j} = \frac{10^5}{\sqrt{2}} e^{-j\frac{\pi}{4}}.$$

This means that

$$v(t) = \Im\left(\frac{10^5}{\sqrt{2}} e^{-j\frac{\pi}{4}} e^{j100t}\right)$$

$$= \frac{10^5}{\sqrt{2}} \sin(100t - \frac{\pi}{4}).$$

The output is a real sinusoid at the same frequency as the input, but with an amplitude and phase that are determined by the circuit, as we have seen before for similar examples. ∎

 To check your understanding, try Drill Problem P8-4.

At this point, one might question why we need to work in the phasor domain at all. Why not simply solve the circuit of the previous example (Figure 8-12) by familiar Laplace transforms? The simple answer is that we *can* use Laplace transforms, but that they are likely to require more effort than would working with phasors. We also need to be particularly careful when we use Laplace transforms, because of the way that they deal with initial conditions.

In order to map a circuit containing capacitors or inductors to the Laplace domain, we need to know the voltages on the capacitors (or the currents flowing through the inductors) at $t = 0$. These are then modelled as auxiliary voltage sources (or current sources). When the source waveform begins at $t = -\infty$, which is the case for an unswitched sinusoid, we do not know the values of the element variables at $t = 0$ until after we have solved the circuit—an apparent dilemma. We have already seen

some clues that suggest a resolution to this difficulty. We know that the solution of the circuit will consist of two parts: a forced response and a transient response. The transient arises from the solution of the homogeneous differential equation, and its role is to guarantee that the initial conditions are satisfied; change the initial conditions and the transient response changes. The forced response, on the other hand, does not depend on the initial conditions. The response to an *unswitched* sinusoidal input, however, contains *only* the forced response. This suggests the following strategy for solving a circuit with a sinusoidal input via Laplace transforms. (1) Map the circuit to the Laplace domain without auxiliary sources. (This is equivalent to setting the initial conditions to zero.) (2) Find the variable(s) of interest. (3) Discard the transient component of the solution. (The transient component consists of everything except the sinusoidal component of the output that has the same frequency as the source waveform.)

Example 8-5 **Finding the Sinusoidal Steady State by Using Laplace Transforms**

To illustrate how Laplace transforms can be used to find the sinusoidal steady-state response, we shall resolve the circuit of the previous example, which is shown in Figure 8-12. Under the assumption that the initial conditions are zero, the Laplace-domain circuit is shown in Figure 8-14.

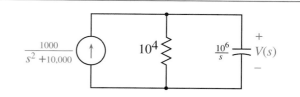

Fig. 8-14. The circuit of Figure 8-12 mapped to the Laplace domain under the assumption of zero initial conditions.

We have

$$V(s) = \frac{(10^4)(\frac{10^6}{s})}{(10^4 + \frac{10^6}{s})} \cdot \frac{10^3}{(s^2 + 10^4)}$$

$$= \frac{10^9}{s + 100} \cdot \frac{1}{s^2 + 10^4}$$

$$= \frac{A}{s - j100} + \frac{A^*}{s + j100} + \frac{B}{s + 100}.$$

The value of B is not needed, since it is part of the transient solution, which will be discarded. On the other hand,

$$A = \lim_{s \to j100} \frac{10^9}{(s + 100)(s + j100)}$$

$$= \frac{10^5}{2(-1 + j)} = \frac{10^5}{2\sqrt{2}} e^{-j\frac{3\pi}{4}}.$$

Therefore,

$$v(t) = \frac{10^5}{\sqrt{2}} \cos(100t - \frac{3\pi}{4}) = \frac{10^5}{\sqrt{2}} \sin(100t - \frac{\pi}{4}),$$

as before. Both methods give the same result, but working in the phasor domain requires less effort and is less likely to result in errors.   ∎

8-2-3   Networks with Multiple Sources

In the preceding discussions, we very carefully limited our treatment to circuits in which there was one independent source that was a complex exponential. What happens when there is more than one such source? We need to consider two cases: the case where the multiple sources all have the same frequency ω_0 and the case where they have different frequencies (e.g., one source has frequency ω_0 and another has frequency ω_1). Neither of these cases is particularly difficult.

In the first case, where there are multiple complex exponential sources but all have the same frequency, we do not need to modify our procedure at all, although we do need to be careful with the relative phases of the sources. (See Example 8-6.) All of the element variables in the circuit will again be complex exponentials with the common frequency. We can solve for the unknown phasors by setting up and solving an appropriate set of linear equations. This situation is identical to that purely resistive case in which there are multiple sources.

As an example of the second case, in which the sources have different frequencies, consider a (stable) circuit in which there are two sources: a voltage source with the waveform

$$v_s(t) = V_s e^{j\omega_0 t}$$

and a current source with the waveform

$$i_s(t) = I_s e^{j\omega_1 t}.$$

Notice that the frequencies are different. Then each element variable in the network will have the form

$$x_i(t) = X_{0i} V_s e^{j\omega_0 t} + X_{1i} I_s e^{j\omega_1 t}.$$

The phasor X_{0i} depends on the value of ω_0; X_{i1} depends on the value of ω_1. We can solve for those two phasors by using source superposition; we turn off the current source and solve the resulting network to get X_{0i}, then we turn the current source back on and turn off the voltage source to solve for X_{1i}. This approach generalizes in the obvious way when there are more than two independent sources. (See Example 8-7.)

Example 8-6   Two Sinusoidal Sources at the Same Frequency

The circuit in Figure 8-15(a) has two voltage sources at the same frequency, although the phases of the two sources are different. This means that the entire circuit can be mapped to the phasor domain, but we need to be careful because of the phases. The redrawing of the circuit in the phasor domain is shown in Figure 8-15(b). Notice that

$$2 \sin 5t = 2 \cos(5t - \frac{\pi}{2})$$

$$= \Re\{2e^{j(5t-\frac{\pi}{2})}\} = \Re(-j2e^{j5t}).$$

Because we intend to use the real-part property, the phasor attached to the voltage source on the right is $-j2$.

There are many methods available to solve for V_c. One approach is to write a single KCL equation at the node in the top middle of the circuit (in the phasor domain). This resembles the node method. The result is

$$\frac{V_c - 1}{2} + \frac{V_c}{-j2} + \frac{V_c + j2}{2} = 0,$$

which simplifies to

$$V_c(2 + j) = 1 - j2.$$

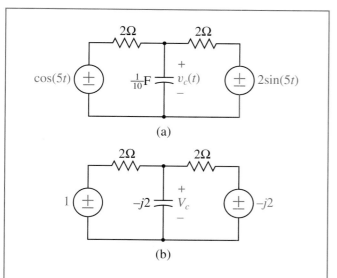

(a)

(b)

Fig. 8-15. (a) A circuit with two independent sources at the same frequency. (b) Phasor-domain representation of the circuit in (a).

Thus,

$$V_c = \frac{1 - j2}{2 + j} = \frac{(1 - j2)(2 - j)}{(2 + j)(2 - j)} = \frac{-j5}{5} = -j$$

and

$$v_c(t) = \Re\{-je^{j5t}\}$$

$$= \Re\{-j(\cos 5t + j \sin 5t)\} = \sin 5t.$$

∎

 To check your understanding, try Drill Problem P8-5.

Example 8-7   **Two Sinusoidal Sources at Different Frequencies**

The approach that we used in Example 8-6 will not work when there are multiple independent sinusoidal sources in the network, not all of the same frequency. This is because the inductors and capacitors in the circuit will have different impedance values for the different frequencies. The recommended approach in this case is to use superposition of sources. This will enable us to analyze the circuit by solving a number of simpler circuits, each of which contains a single sinusoidal source, a situation that we can readily handle. To illustrate the approach, consider the circuit in Figure 8-16.

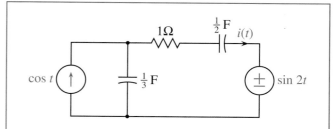

Fig. 8-16. A circuit with two independent sinusoidal sources at different frequencies.

To use superposition of sources, we first replace the voltage source by a short circuit and find the component of the current $i(t)$ that is caused by the current source; then we reinsert the voltage source and replace the current source by an open circuit to evaluate the component of $i(t)$ that is caused by the voltage source. The phasor-domain representations of the two cases are shown in Figure 8-17. Notice that the impedances of the capacitors are different in the two circuits, because the frequencies of the sources are different and the impedance of a capacitor is frequency dependent. For the circuit on the top, we will let $i_1(t) = \Re\{I_1 e^{jt}\}$; for the one on the bottom, we will take $i_2(t) = \Im\{I_2 e^{j2t}\}$. Because we are solving the problem by using source superposition, we can use the real-part property for one part of the solution and the imaginary-part property for the other. Thus, we let phasor of $\sin 2t$ be 1. This is in contrast to what we did in the previous example, where we solved the circuit with both sources turned on at the same time.

For the circuit in Figure 8-17(a), we can find I_1 by using a current divider:

$$I_1 = \frac{-j3}{1 - j5} = \frac{15 - j3}{26}.$$

Therefore,

$$i_1(t) = \Re\{\frac{(15 - j3)}{26} e^{jt}\}$$

$$= \frac{\sqrt{234}}{26} \cos(t - \tan^{-1}\frac{1}{5}). \qquad (8.31)$$

This is not the complete solution, of course, merely one component of it. To get the other component, we must find I_2 in Figure 8-17(b). From a KVL equation written around the single complete path, we have

$$(1 - j - j\frac{3}{2})I_2 = -1,$$

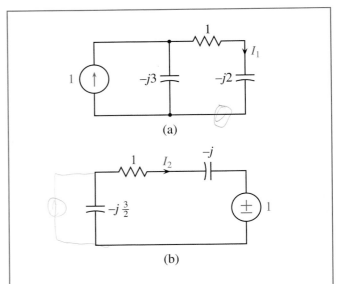

Fig. 8-17. (a) Phasor-domain representation of the circuit in Figure 8-16 with the voltage source turned off. (b) Same with the current source turned off.

so that

$$I_2 = -\frac{1}{1 - j\frac{5}{2}} = -\frac{4 + j10}{29}.$$

From the phasor, we know that

$$i_2(t) = \Im\{-\frac{4 + j10}{29} e^{j2t}\}$$

$$= -\frac{\sqrt{116}}{29} \sin(2t + \tan^{-1}\frac{5}{2}). \qquad (8.32)$$

Finally,

$$i(t) = i_1(t) + i_2(t)$$

$$= \frac{\sqrt{234}}{26} \cos(t - \tan^{-1}\frac{1}{5})$$

$$- \frac{\sqrt{116}}{29} \sin(2t + \tan^{-1}\frac{5}{2}). \qquad (8.33)$$

There were sinusoidal inputs at two different frequencies, and the output contains a component at each of these frequencies.   ■

 To check your understanding, try Drill Problem P8-6.

              WORKED SOLUTION: *Superposition*

8-3   Frequency-Domain Circuit Simplifications

8-3-1   Series and Parallel Connections

The impedance functions that describe element relations in the frequency domain and the Laplace domain are the same, provided that we set $s = j\omega$. In the Laplace domain, the impedance can be interpreted as the ratio of the Laplace transform of the voltage to the Laplace transform of the current; in the frequency domain, it represents the ratio of the phasors of the voltage and the current when there is a single complex exponential time function source. It should not be surprising that many of the shortcuts that we have learned to appreciate in the Laplace and time domains, such as equivalent impedances for series and parallel connections of elements and voltage and current dividers, carry over to the new domain directly. We have already been making use of some of these in the examples that were discussed earlier in the chapter.

Figure 8-18 shows a two-terminal network consisting of two impedances connected in series. These are equivalent to a single impedance. We can see this by expressing the phasor associated with the terminal voltage in terms of the phasor of the current:

$$V = Z_1(j\omega_0)I + Z_2(j\omega_0)I = [Z_1(j\omega_0) + Z_2(j\omega_0)]I.$$

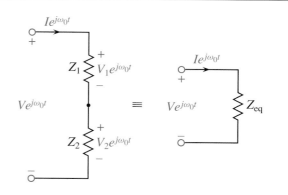

Fig. 8-18. Two impedances connected in series are equivalent to a single impedance in the frequency domain: $Z_{eq} = Z_1 + Z_2$.

Thus,

$$Z_{eq}(j\omega) = Z_1(j\omega) + Z_2(j\omega). \tag{8.34}$$

The voltage-divider relationship also applies in the frequency domain. For the impedances in Figure 8-18, it states that

$$V_1 = Z_1(j\omega_0)I = Z_1(j\omega_0)\left(\frac{V}{Z_{eq}(j\omega_0)}\right)$$

$$= \frac{Z_1(j\omega_0)}{Z_1(j\omega_0) + Z_2(j\omega_0)}V \tag{8.35}$$

and that

$$V_2 = \frac{Z_2(j\omega_0)}{Z_1(j\omega_0) + Z_2(j\omega_0)}V.$$

Two impedances connected in parallel , as in Figure 8-19, also behave like a single impedance. In the Laplace domain, the equivalent admittance is given by

$$Y_{eq}(s) = Y_1(s) + Y_2(s).$$

Therefore, in the frequency domain,

$$Y_{eq}(j\omega_0) = Y_1(j\omega_0) + Y_2(j\omega_0)$$

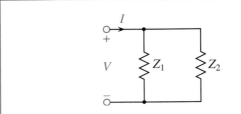

Fig. 8-19. A parallel connection of two impedances.

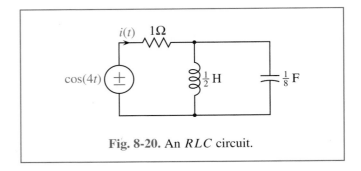

Fig. 8-20. An RLC circuit.

or

$$Z_{eq}(j\omega_0) = \cfrac{1}{\cfrac{1}{Z_1(j\omega_0)} + \cfrac{1}{Z_2(j\omega_0)}}.$$

Thus,

$$I = V[Y_1(j\omega_0) + Y_2(j\omega_0)].$$

The current divider also extends straightforwardly to the frequency domain. Let I_1 be the phasor associated with the current passing through impedance Z_1; then

$$I_1 = VY_1(j\omega_0) = \frac{I}{Y_{eq}(j\omega_0)} \cdot Y_1(j\omega_0)$$

$$= \frac{Y_1(j\omega_0)}{Y_1(j\omega_0) + Y_2(j\omega_0)} I$$

$$= \frac{Z_2(j\omega_0)}{Z_1(j\omega_0) + Z_2(j\omega_0)} I.$$

Example 8-8   **Using Circuit Simplifications to Find the Sinusoidal Response**

As an example, consider the circuit (consisting of a resistor, inductor, and capacitor) in Figure 8-20,which we wish to solve for the current $i(t)$. Since the source is a real sinusoid with a frequency of 4 rad/s, we first solve the circuit for a complex exponential time function input signal, then take the real parts of the complex element variable waveforms that result. In the frequency domain,

this circuit looks like the one in Figure 8-21. Notice that, at this particular frequency, the impedance of the inductor is $2j$, that of the capacitor $-2j$. The equivalent admittance of the parallel connection is therefore zero, and its equivalent impedance is infinite. This means that $I = 0$ and that no current flows out of the voltage source. This is not to say that the circuit is at rest. Since $I = 0$, $V_\ell = V_c = 1$ and $I_c = -I_\ell = j/2$. Therefore,

$$i_c(t) = \Re(\frac{j}{2}e^{j4t}) = -\frac{1}{2}\sin 4t.$$

There is a sinusoidal current that flows through the mesh involving the inductor and the capacitor. This behavior can occur only at this particular frequency; if we were to perturb the frequency of the source slightly, the circuit would behave quite differently. At the frequency for which the impedance of the inductor is equal to the

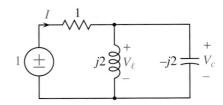

Fig. 8-21. The RLC circuit of Figure 8-20 mapped to the frequency domain. At this frequency the circuit is at resonance.

negative of the impedance of the capacitor, we say that the circuit is in *resonance*, and that frequency is called the *resonant frequency*. We will have more to say about resonance in Chapter 9.   ∎

 To check your understanding, try Drill Problem P8-7.

8-3-2   Thévenin and Norton Equivalent Circuits

Thévenin and Norton equivalent networks are circuit simplifications that allow us to replace two-terminal subnetworks by simpler structures that are network equivalent. Recall that when a two-terminal network contains only resistors and sources, its Thévenin equivalent network is a voltage source connected in series with a resistor. When it also contains inductors and capacitors, it can be modelled *in the Laplace domain* as a voltage source connected in series with an impedance. (The Norton equivalent network is a current source connected in parallel with either a resistance or an impedance.) In the Laplace domain, however, these networks are much more difficult to work out, and the simplifications that result from a reduced number of elements are largely offset by the complexity of the expressions for the impedance and voltage source. In the phasor domain, however, when we care only about how the circuit behaves when excited by a sinusoidal excitation at a single frequency, the Thévenin and Norton equivalent circuits become simpler and more useful again.

Consider a circuit containing linear resistors, inductors, and capacitors. It can also contain any number of dependent sources, but it should have either a single independent source with a CETF source waveform or multiple independent sources, all of which have complex exponential time function source waveforms that oscillate *at the same frequency,* ω_0. Next, we identify a subnetwork that is connected to the remainder of the

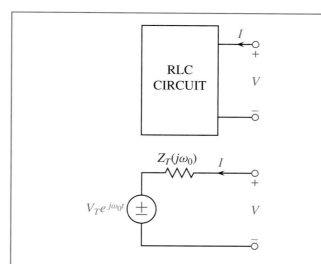

Fig. 8-22. Any circuit containing resistors, inductors, capacitors, dependent sources, and a single independent source whose waveform is a complex exponential time function is network equivalent to an impedance connected in series with a voltage source. The voltage-source waveform is also a complex exponential time function.

circuit at two terminals, as shown in Figure 8-22. The phasor associated with the voltage between the terminals is V; that of the current entering its positive terminal is I. The subnetwork could contain (or not contain) the single independent source. Then, the phasor-domain V–I relation will always have the form

$$V = Z_T I + V_T. \qquad (8.36)$$

In general, both the Thévenin equivalent impedance, Z_T, and the phasor of the Thévenin equivalent voltage, V_T, depend upon the frequency of the complex exponential source, ω_0. We will not prove (8.36) here. In Chapter 3, we argued that it followed for the resistive case from KCL, KVL, and Ohm's law. Since the voltage and current complex amplitudes in the phasor domain satisfy these same relations, the derivation would be identical. Equation (8.36) also defines the terminal behavior of the network shown on the right-hand side of Figure 8-22. Therefore, these two two-terminal networks are network equivalent (at frequency ω_0.)

The components of the Thévenin equivalent circuit can be found from the phasors of the open-circuit voltage, V_{oc}, and of the short-circuit current, I_{sc}. As was done for resistive circuits, the equations follow by equating the measurements made on the original two-terminal network with the corresponding measurements made on the Thévenin (or Norton) equivalent:

$$V_T = V_{\text{oc}} \qquad (8.37)$$

$$Z_T = -\frac{V_{\text{oc}}}{I_{\text{sc}}}. \qquad (8.38)$$

 WORKED SOLUTION: *Thevenin Equivalents*

Example 8-9   **Finding a Thévenin Equivalent Network in the Phasor Domain**

To illustrate the procedure for finding a Thévenin equivalent network for a circuit with reactive elements, consider the circuit in Figure 8-23. The mapping of the two-terminal network to the phasor domain is shown in Figure 8-24. All of the variables have been replaced by their phasors, and the elements have been replaced by impedances. The impedance values are valid for the

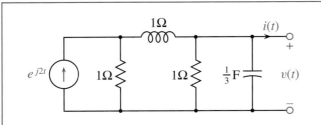

Fig. 8-23. A circuit for which we would like to determine the Thévenin equivalent circuit.

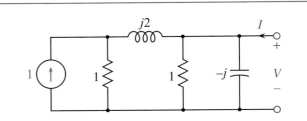

Fig. 8-24. The circuit of Figure 8-23, but mapped to the phasor domain.

specific value of the frequency of the sinusoidal source, which is 2 rad/s.

We have seen several methods for figuring out the Thévenin equivalent for a resistive circuit, and it might be appropriate to explore the possibilities before jumping in. One possibility is to attach a current source with a current I at the terminal pair and solve the resulting circuit for the voltage V; this gives the required frequency domain V–I relation, from which we can construct the Thévenin equivalent circuit. Alternatively, we could find any two of the following three quantities: the (phasor associated with the) open-circuit voltage, the short-circuit current, and the equivalent impedance of the two-terminal network when the current source is turned off. As a third possibility, we could first try to simplify the circuit and then continue with one of the other methods. This latter approach might be promising here, since this circuit has a couple of obvious simplifications. For example, we can replace the two parallel impedances with an equivalent impedance, and we can replace the current source that is connected in parallel with the 1-Ω resistor by a voltage source connected in series with a 1-Ω resistor (Thévenin-to-Norton (source) substitution), which will create a series connection that can be reduced. The simplified circuit that results is shown in Figure 8-25. It is straightforward to compute Z_T and V_{oc} directly for this simplified circuit. To calculate the Thévenin impedance,

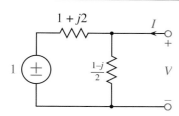

Fig. 8-25. The circuit of Figure 8-24 after simplifications to reduce the number of elements.

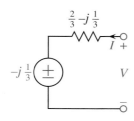

Fig. 8-26. The Thévenin equivalent to the circuit of 8-23 at a frequency of 2 rad/s.

we can turn off the voltage source (replace it by a short circuit) and measure the equivalent impedance of the parallel connected elements. This approach gives

$$Z_T = \frac{\left(\frac{1-j}{2}\right)(1 + j2)}{\frac{1}{2} - j\frac{1}{2} + 1 + j2} = \frac{2}{3} - j\frac{1}{3}.$$

We can get V_{oc} by using a voltage divider:

$$V_{oc} = \frac{\frac{1-j}{2}}{\frac{1}{2} - j\frac{1}{2} + 1 + j2} \cdot 1 = -j\frac{1}{3}.$$

The Thévenin equivalent circuit that results is shown in Figure 8-26.

∎

 To check your understanding, try Drill Problem P8-8.

8-3-3   The Node and Mesh Methods in the Phasor Domain

When a network contains a single independent source whose independent variable is a complex exponential time function, both the node method and the mesh method can be used to compute the phasors associated with all of the element variables in the circuit. This is because the node potentials and the mesh currents will themselves be complex exponential time functions that are completely specified by their phasors. The process of setting up equations and solving them for the phasors of the node potentials (or mesh currents) mimics the procedure that we used earlier for finding the node potentials and mesh currents in resistive circuits in the time domain. The differences from the earlier procedures are those that we are coming to expect: The phasors of the node potentials and mesh currents do not vary with time; they are functions of the frequency of the exponential source; and the equations that need to be solved have complex coefficients. We illustrate the approach by an example.

Example 8-10   The Node Method in the Phasor Domain

To illustrate the complete procedure, consider the circuit drawn in Figure 8-27. In this circuit, the frequency ω_0 is not specified, so we treat it as a variable. Since this circuit contains a single complex exponential source, we know

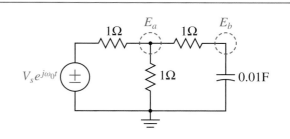

Fig. 8-27. A circuit with a single complex exponential source, used to illustrate the node method.

that all of the element variables in the circuit are complex exponential time functions. The node potentials are also complex exponential time functions, because each node potential is equal to the sum of the element voltages of the elements that lie on a path connecting that node to the ground node, and the sum of complex exponential time functions at a common frequency is also a complex exponential time function at that frequency.

Defining a supernode encircling the voltage source, we observe that the network in this example contains this supernode and two isolated nodes. Therefore, the node method requires that we set up and solve two KCL equations; the variables in these equations must be the potentials at the two nonground nodes. Denote these node potentials $e_a(t)$ and $e_b(t)$. Since each is a complex exponential time function, it follows that

$$e_a(t) = E_a e^{j\omega_0 t}; \quad e_b(t) = E_b e^{j\omega_0 t}, \qquad (8.39)$$

where E_a and E_b are the two unknown phasors.

We choose nodes a and b for writing our KCL equations. At node a,

$$[e_a(t) - v_s(t)] + e_a(t) + [e_a(t) - e_b(t)] = 0. \quad (8.40)$$

Similarly, at node b,

$$[e_b(t) - e_a(t)] + 0.01\frac{de_b(t)}{dt} = 0. \qquad (8.41)$$

Recall that the current through a capacitor is proportional to the derivative of its voltage.

As an alternative to solving (8.40) and (8.41) directly, which involves a differential equation, we substitute the known functional form for the variables from (8.39). This converts the two node equations into

$$[E_a e^{j\omega_0 t} - V_s e^{j\omega_0 t}] + E_a e^{j\omega_0 t}$$
$$+ [E_a e^{j\omega_0 t} - E_b e^{j\omega_0 t}] = 0$$

$$[E_b e^{j\omega_0 t} - E_a e^{j\omega_0 t}] + j\omega_0(0.01)E_b e^{j\omega_0 t} = 0. \quad (8.42)$$

Notice that we can divide both of these equations by the common exponential term. This will produce two *algebraic* equations in the unknown phasors E_a and E_b. After rearrangement, these equations can be rewritten as

$$3E_a - E_b = V_s$$
$$-E_a + [1 + j\omega_0(0.01)]E_b = 0. \qquad (8.43)$$

Solving for E_a and E_b gives

$$E_a = \frac{1}{2 + j(0.03)\omega_0}V_s$$

$$E_b = \frac{1 + j(0.01)\omega_0}{2 + j(0.03)\omega_0}V_s.$$

Now that we know the phasors of the node potentials, we can express each node potential as a function of time by using (8.39). This gives

$$e_a(t) = \frac{1}{2 + j(0.03)\omega_0}V_s e^{j\omega_0 t}$$

$$e_b(t) = \frac{1 + j(0.01)\omega_0}{2 + j(0.03)\omega_0}V_s e^{j\omega_0 t},$$

which we can use to compute any of the element variables in the circuit.   ■

 To check your understanding, try Drill Problem P8-9.

We can use an observation from this example to simplify the procedure for later problems. Specifically, when a circuit contains one or more complex exponential sources, all at the same frequency ω_0, the node method can be formulated directly in terms of the impedances of the elements and the phasors of the node potentials. We can begin the procedure at equation (8.42). After we divide through by the common complex exponential terms, the KCL equations simplify to

$$[E_a - V_s] + E_a + [E_a - E_b] = 0$$
$$[E_b - E_a] + j\omega_0(0.01)E_b = 0. \qquad (8.44)$$

8-3 FREQUENCY-DOMAIN CIRCUIT SIMPLIFICATIONS325

These KCL equations are identical in form to the equations produced by the node method for a resistive circuit, provided that we (1) replace the node potential waveforms by their phasors, and (2) use the impedance values for a phasor-domain statement of Ohm's law. As we did for the time-domain node method, again we can state the method as a procedure.

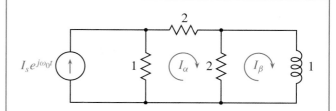

Fig. 8-28. A circuit with a single complex exponential source, used to illustrate the mesh method. The resistances are measured in ohms and the inductances are measured in henrys.

The Node Method with Exponential Sources:

1. *Define a supernode encircling each voltage source in the network.*
2. *Select one of the nodes of the network as the ground node.*
3. *Define $n - 1$ node-potential phasors at the remaining nodes/supernodes of the network.*
4. *Set up and solve KCL equations for the phasors of the currents crossing the surfaces associated with $n - 1$ nodes/supernodes in the network. The phasors of the currents in these equations must be expressed in terms of the phasors of the node potentials and the element impedances.*
5. *Solve those $n-1$ equations for the $n-1$ complex phasors of the node potentials.*
6. *Compute the element voltages and currents of interest from the phasors of the node potentials.*

The mesh method can be similarly reformulated in terms of phasors. Consider the circuit in Figure 8-28. Since this circuit contains a single complex exponential source, we could proceed as before by developing the mesh equations in the time domain, observing that the mesh currents are complex exponential time functions,

substituting the known functional form for those currents into the mesh equations, and then simplifying the resulting terms by cancelling out the complex exponential terms. The equations that would result would be exactly the same as if we wrote mesh equations for the phasors of the mesh currents in the frequency domain directly, and that is the approach that we shall use.

The circuit contains the two meshes that do not incorporate exterior current sources, as indicated by colored arrows. Let I_α and I_β denote the phasors of the two mesh currents. Writing the KVL equations for these two meshes directly in the phasor domain gives

$$(I_\alpha - I_s) + 2I_\alpha + 2(I_\alpha - I_\beta) = 0$$
$$2(I_\beta - I_\alpha) + j\omega_0 I_\beta = 0.$$

Regrouping the terms to simplify the solution gives the equivalent equations

$$5I_\alpha - 2I_\beta = I_s$$
$$-2I_\alpha + (2 + j\omega_0)I_\beta = 0,$$

which yield the solution

$$I_\alpha = \frac{1}{3 + j\frac{5}{2}\omega_0} I_s$$

$$I_\beta = \frac{1 + j\frac{1}{2}\omega_0}{3 + j\frac{5}{2}\omega_0} I_s.$$

As with the node method, we can formally state the phasor version of the mesh method as a procedure.

The Mesh Method with Complex Exponential Sources:

1. *Define ℓ mesh current phasors at the appropriate meshes or supermeshes of the network.*
2. *Set up and solve KVL equations over the paths associated with these meshes/supermeshes. The phasors of the voltages in these equations must be expressed in terms of the phasors of the mesh currents and the element impedances.*
3. *Solve those ℓ equations for the ℓ phasors of the mesh currents.*
4. *Compute the element currents and voltages of interest from the phasors of the mesh currents.*

This section concludes with several additional examples that further explore the use of the node and mesh methods for circuits containing inductors and capacitors, along with source waveforms that are sinusoids.

Example 8-11   The Node Method with a Real-Sinusoidal Source

The source in the circuit in Figure 8-29 is a real sinusoid. This condition does not create a problem, because we can use the real-part property; we simply solve the problem with a complex exponential time function for the source and then take the real part of the result. The fact that we use the node method to solve the intermediate circuit does not invalidate this approach. Since

$$20 \cos(100t) = \Re\mathrm{e}\left(20\, e^{j100t}\right),$$

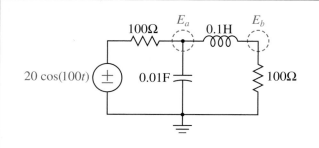

Fig. 8-29. An RLC circuit with a single sinusoidal source.

we replace the voltage source waveform by

$$\hat{v}_s(t) = 20\, e^{j100t}.$$

After defining a supernode to encircle the voltage source, we see that the circuit contains the supernode and two additional isolated nodes; it also contains two meshes, which means that the node and mesh methods require the same level of effort. Here we use the node method and leave solving the circuit via the mesh method as an exercise. (See Problem P8-11.) We choose to write the equations at the two circled nodes for which the phasors of the node potentials are denoted E_a and E_b. We will not explicitly redraw the circuit in the phasor domain, but we note that the impedance of the inductor at this frequency is $j(0.1)(100) = j10$ and that the impedance of the capacitor is $1/[j(100)(0.01)] = -j$, which means that its admittance is j. We thus have the following equations:

$$\textit{node a:} \quad \frac{E_a - 20}{100} + jE_a + \frac{E_a - E_b}{j10} = 0$$

$$\textit{node b:} \quad \frac{E_b - E_a}{j10} + \frac{E_b}{100} = 0.$$

These equations can be rearranged and expressed in matrix–vector form:

$$\begin{bmatrix} 0.01 + j - j(0.1) & j(0.1) \\ j(0.1) & 0.01 - j(0.1) \end{bmatrix} \begin{bmatrix} E_a \\ E_b \end{bmatrix} = \begin{bmatrix} 0.2 \\ 0 \end{bmatrix}.$$

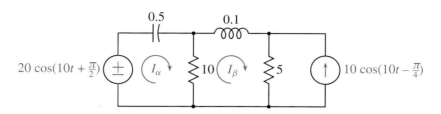

Fig. 8-30. A circuit illustrating the use of the mesh method. The resistances are measured in ohms, the capacitance in farads, and the inductance in henrys.

Using MATLAB to solve the equations gives the solution

$$\begin{bmatrix} E_a \\ E_b \end{bmatrix} = \begin{bmatrix} 0.2002e^{-j1.5509} \\ 0.1992e^{-j1.6505} \end{bmatrix}.$$

Thus, for the exponential source, we get

$$\hat{e}_a(t) = 0.2002e^{-j1.5509}e^{j100t}$$
$$\hat{e}_b(t) = 0.1992e^{-j1.6505}e^{j100t},$$

and for the original cosine wave source, we simply take the real parts:

$$e_a(t) = 0.2002\cos(100t - 1.5509)$$
$$e_b(t) = 0.1992\cos(100t - 1.6505).$$

∎

 To check your understanding, try Drill Problem P8-10.

Example 8-12   Mesh Method with a Real-Sinusoidal Source

As an example of using the mesh method, consider the circuit in Figure 8-30. This circuit contains two sinusoidal sources, but both are at the same frequency (10 radians/s), so the analysis is still straightforward. Since

$$v_s(t) = 20\cos(10t + \frac{\pi}{2}) = \Re(20e^{j\pi/2}e^{j10t})$$
$$i_s(t) = 10\cos(10t - \frac{\pi}{4}) = \Re(10e^{-j\pi/4}e^{j10t}),$$

we initially solve the circuit with the sources replaced by the complex exponential sources

$$\hat{v}_s(t) = V_s e^{j10t}$$
$$\hat{i}_s(t) = I_s e^{j10t},$$

where

$$V_s = 20e^{j\pi/2}; \quad I_s = 10e^{-j\pi/4}.$$

Writing KVL equations around the two indicated meshes, we get

$$\alpha: \quad \frac{2}{j10}I_\alpha + 10(I_\alpha - I_\beta) = V_s$$
$$\beta: \quad 10(I_\alpha - I_\beta) + j(0.1)10I_\beta + 5(I_\beta + I_s) = 0.$$

To aid the solution, we put these in the matrix–vector form

$$\begin{bmatrix} 10 - j(0.2) & -10 \\ -10 & 15 + j \end{bmatrix}\begin{bmatrix} I_\alpha \\ I_\beta \end{bmatrix} = \begin{bmatrix} 20e^{j\pi/2} \\ 50e^{-j\pi/4} \end{bmatrix},$$

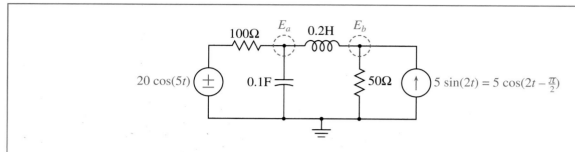

Fig. 8-31. An example containing two sources at different frequencies.

from which we can find the solution with complex exponential sources:

$$I_\alpha = 14.8519e^{j1.9515}$$

$$I_\beta = 13.0027e^{j1.9858}.$$

For the original problem with sinusoidal sources, we must again take the real parts:

$$i_\alpha(t) = 14.8519\cos(10t + 1.9515)$$

$$i_\beta(t) = 13.0027\cos(10t + 1.9858).$$

■

To check your understanding, try Drill Problem P8-11.

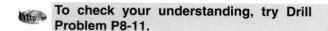

WORKED SOLUTION: *Mesh Method*

Example 8-13   Node Method with Multiple Sinusoidal Inputs

As a final example, consider the circuit in Figure 8-31, for which the variables of interest are the node potentials $e_a(t)$ and $e_b(t)$. This represents the most general case, since it contains multiple sources that operate at different frequencies. The network can be solved via source superposition, as before, but this approach must be used carefully, since source superposition, though valid for time waveforms, is not valid for phasors unless all of the

sources have the same frequency, because impedances vary with the source frequency.

First, we turn off the current source and let the node potentials be denoted as $e_{a1}(t)$ and $e_{b1}(t)$. In the phasor domain, the circuit reduces to the one pictured in Figure 8-32(a). The impedance values reflect the fact that, for this source, $\omega = 5$ rad/s.

We can now write the node equations:

$$node\ a: \quad \frac{E_{a1} - 20}{100} + \frac{E_{a1}}{-j2} + \frac{E_{a1} - E_{b1}}{j} = 0$$

$$node\ b: \quad \frac{E_{b1} - E_{a1}}{j} + \frac{E_{b1}}{50} = 0.$$

We then put them in matrix–vector form, namely,

$$\begin{bmatrix} 0.01 - j(0.5) & +j \\ j & 0.02 - j \end{bmatrix} \begin{bmatrix} E_{a1} \\ E_{b1} \end{bmatrix} = \begin{bmatrix} 0.2 \\ 0 \end{bmatrix},$$

and solve them:

$$E_{a1} = 0.0239 - j0.3989 = 0.3996e^{-j(1.5108)}$$

$$E_{b1} = 0.0160 - j0.3992 = 0.3995e^{-j(1.5308)}.$$

This gives, for the time-domain waveforms corresponding to the node potentials,

$$e_{a1}(t) = 0.3996\cos(5t - 1.5108)$$

$$e_{b1}(t) = 0.3995\cos(5t - 1.5308).$$

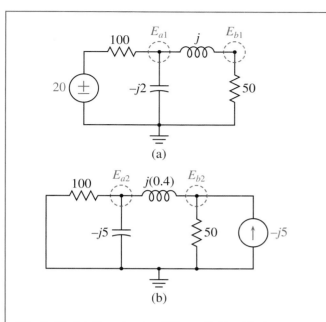

Fig. 8-32. (a) The circuit of Figure 8-31 with the current source turned off and hence $\omega = 5$ rad/s. (b) The circuit of Figure 8-31 with the voltage source turned off and hence $\omega = 2$ rad/s.

This is not the complete solution to the problem; it is merely the contribution to the solution created by the voltage source. We need to add the contribution from the current source. To compute that contribution, we turn off the voltage source and turn on the current source. For the current source, the frequency is $\omega = 2$, which gives the equivalent circuit shown in Figure 8-32(b) in the phasor domain. Notice that the two circuits look quite different.

As before, we can write the node equations, cast them in matrix–vector form, and then solve them. These steps follow. We begin with

$$\textit{node a:} \quad \frac{E_{a2}}{100} + \frac{E_{a2}}{-j5} + \frac{E_{a2} - E_{b2}}{j(0.4)} = 0$$

$$\textit{node b:} \quad \frac{E_{b2} - E_{a2}}{j(0.4)} + \frac{E_{b2}}{50} = -j5.$$

Putting these equations in matrix–vector form gives

$$\begin{bmatrix} 0.01 - j(2.3) & +j2.5 \\ j2.5 & 0.02 - j2.5 \end{bmatrix} \begin{bmatrix} E_{a2} \\ E_{b2} \end{bmatrix} = \begin{bmatrix} 0 \\ -j5 \end{bmatrix}.$$

We can now use MATLAB to solve for the phasors:

$$E_{a2} = -24.4965 - j3.4771 = 24.7420 e^{-j(3.0006)}$$

$$E_{b2} = -22.5228 - j3.2969 = 22.7629 e^{-j(2.9962)},$$

These give, for the components of the node potentials due to the current source,

$$e_{a2}(t) = 24.7420 \cos(2t - 3.0006)$$

$$e_{b2}(t) = 22.7629 \cos(2t - 2.9962).$$

Lastly, we can apply superposition to get our final answer:

$$e_a(t) = e_{a1}(t) + e_{a2}(t)$$

$$= 0.3996 \cos(5t - 1.5108)$$

$$+ 24.7420 \cos(2t - 3.0006)$$

$$e_b(t) = e_{b1}(t) + e_{b2}(t)$$

$$= 0.3995 \cos(5t - 1.5308)$$

$$+ 22.7629 \cos(2t - 2.9962).$$

∎

8-3-4   Operational-Amplifier Circuits

Our approach to analyzing circuits that contain operational amplifiers is to use a variation of the node method. When such a circuit contains a single source that is a complex exponential time function, then all of the voltages and currents in the circuit will also be complex exponential time functions that are completely specified by their phasors, and we can use this approach again. The special conditions that must be imposed on operational amplifiers translate directly to the phasor domain. We know, for example, that the voltages at the two input terminals of an ideal operational amplifier must be equal; this implies that their phasors must also be equal. We also know that the current entering the two terminals must be zero, a condition that can also be imposed on the phasors. As in the time domain, when we write KCL equations in the phasor domain, we are forbidden to write them at the output terminal of the operational amplifier or at the ground.

Example 8-14   An Opamp in the Phasor Domain

Consider the circuit shown in Figure 8-33, which contains an operational amplifier, a capacitor, and a single source that is a complex exponential time function. The first step in finding the circuit output is to map the circuit to the phasor domain. Because that mapping is topologically equivalent to the original circuit, we omit redrawing it here. The voltage source has amplitude V_s, the resistor becomes an impedance of R, the capacitor becomes an impedance of $1/(jC\omega)$, and the opamp remains an opamp.

We next write a KCL equation at the inverting input to the opamp. Notice that the phasor associated with the node potential at that node is zero, because of the virtual ground:

$$\frac{1}{R} V_{\text{in}} + jC\omega V_{\text{out}} = 0.$$

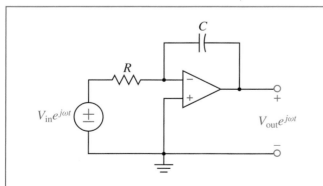

Fig. 8-33. A circuit containing a single operational amplifier and a single complex exponential source.

We can solve this equation for V_{out}:

$$V_{\text{out}} = \frac{1}{jRC\omega} V_{\text{in}}.$$

Therefore, the output signal is

$$v_{\text{out}}(t) = \frac{1}{jRC\omega} V_{\text{in}} e^{j\omega t}.$$

∎

 To check your understanding, try Drill Problem P8-12.

8-4   Power in the Phasor Domain

8-4-1   Average Power

In Chapter 1, we defined the *instantaneous power* absorbed by an element as

$$P_{\text{inst}}(t) = v(t)i(t),$$

where the reference directions for $v(t)$ and $i(t)$ are governed by the default sign convention. Any circuit element in a network that has a single sinusoidal source will have an instantaneous power that fluctuates rapidly as a function of time. For example, if the frequency of the real-sinusoidal source is ω_0, we know that the voltage across a resistor in the circuit must have the form

$$v_r(t) = |V_r| \cos(\omega_0 t + \phi). \tag{8.45}$$

The instantaneous power absorbed by that resistor is

$$P_{r,\text{inst}}(t) = \frac{v_r^2(t)}{2R} = \frac{|V_r|^2}{R} \cos^2(\omega_0 t + \phi)$$

$$= \frac{|V_r|^2}{2R} [1 + \cos(2\omega_0 t + 2\phi)]. \tag{8.46}$$

The instantaneous value of the absorbed power oscillates between the values zero and $|V_r|^2/R$. Notice that the frequency of oscillation of the instantaneous power is twice that of the voltage or current.

The *average power* absorbed is a more meaningful quantity for many problems in which the waveforms are sinusoidal. It is directly related to the battery life in a circuit or to the amount of fuel that must be consumed in a power plant. The average power absorbed is defined as

$$P_{\text{ave}} = \lim_{T \to \infty} \frac{1}{T} \int_0^T P_{\text{inst}}(t)\, dt.$$

Power ratings for electrical tools and appliances generally refer to average power. For the resistor that we considered, the average power absorbed is $|V_r|^2/(2R)$. Notice that the average power does not vary with time.

The average power absorbed by an inductor or a capacitor excited by a sinusoidal excitation is zero. To see this, we evaluate the instantaneous power:

$$P_{\ell,\text{inst}}(t) = i_\ell(t) v_\ell(t) = L i_\ell(t) \frac{d i_\ell}{dt}$$

If the current $i_\ell(t)$ is a sinusoid, $i_\ell(t) = A\cos(\omega_0 t + \phi)$; then

$$P_{\ell,\text{inst}}(t) = -L\omega_0 \cos(\omega_0 t + \phi) \sin(\omega_0 t + \phi)$$

$$= -\frac{L}{2}\omega_0 \sin(2\omega_0 + 2\phi);$$

and

$$P_{\ell,\text{ave}} = 0.$$

Similarly, the average power absorbed in a capacitor is also zero. Even though the average power absorbed or supplied by the reactive elements is zero, their presence in a circuit can alter the total amount of power that is absorbed by the resistors. This is illustrated in a later example.

 DEMO: *Average Power*

8-4-2   RMS Values

What is the resistance of a 60-W light bulb? The "60-W" designation refers to the average power dissipated by the bulb in the form of heat and light. Turning the lamp on connects the light bulb, which can be modelled as a resistor, across a voltage source whose waveform is a real 60-Hz sinusoid. It turns out that the phase of this waveform is not important, so, for the sake of answering the question at the beginning of this section, we will set it to zero. On the other hand, the amplitude of that sinusoid is very important—but what is it?

It is well known that the electrical service available in North America is typically 120 V, but it turns out that this is not exactly the amplitude of the sinusoidal voltage. We will come back to this question shortly; for the moment, we will simply let the amplitude of the sinusoidal voltage be A. The voltage applied across the terminals of the bulb is then

$$v(t) = A\cos(120\pi t).$$

We can calculate the average power dissipated by the bulb in terms of the amplitude of the voltage, A, and the resistance of the bulb, R, as

$$60(W) = \frac{1}{R}\left\{ \lim_{T \to \infty} \frac{1}{T} \int_0^T v^2(t)\, dt \right\}$$

$$= \frac{A^2}{R}\left\{ \lim_{T \to \infty} \frac{1}{T} \int_0^T \cos^2(120\pi t)\, dt \right\}$$

$$= \frac{A^2}{2R}\left\{ \lim_{T \to \infty} \frac{1}{T} \int_0^T [1 + \cos(240\pi)]\, dt \right\}.$$

The integral of a sum is the sum of the integrals, as long as both are well defined. The limit operation similarly distributes with respect to addition as long as both limits exist. Therefore,

$$60 = \frac{A^2}{2R} \left\{ \lim_{T \to \infty} \frac{1}{T} \int_0^T dt + \lim_{T \to \infty} \frac{1}{T} \int_0^T \cos(240\pi t)\, dt \right\}$$

$$= \frac{A^2}{2R}.$$

The integral of $\cos(240\pi t)$ is a sinusoid. Since this function is bounded, when it is divided by T and T is allowed to become large, that term will go to zero.

Now we need to address the amplitude, A, of the sinusoid. If you were to use an oscilloscope to measure the amplitude of the voltage waveform at a wall socket, you would see that it is not 120 V; in fact, it is quite a bit larger. 120 V is actually the *rms (root mean square) value of the voltage*. This means that the amplitude of the sinusoid is adjusted so that the power delivered to a device is the same as would be delivered by a 120-V DC (i.e., constant) voltage. Therefore,

$$\frac{A^2}{2R} = \frac{120^2}{R} \implies A = 120\sqrt{2} \approx 170 \text{ V}.$$

and

$$60W = \frac{120^2}{R}$$

from which it follows that

$$R = \frac{120^2}{60} = 240\Omega.$$

Rms values are very convenient for power calculations. For a resistor,

$$P_{\text{inst}}(t) = v(t)i(t) = Ri^2(t) = \frac{1}{R}v^2(t)$$

and

$$P_{\text{ave}} = V_{\text{rms}}I_{\text{rms}} = RI_{\text{rms}}^2 = \frac{1}{R}V_{\text{rms}}^2.$$

Notice that the functional forms of these equations are identical.

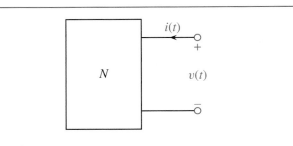

Fig. 8-34. An arbitrary two-terminal RLC network containing no independent sources.

8-4-3   Power in Circuits with Sinusoidal Excitations

When a circuit is excited by a single sinusoidal source, we can derive expressions for the average power absorbed in the circuit directly in terms of the phasors of the sinusoidal element variables.

Consider an arbitrary two-terminal network containing resistors, inductors, capacitors, and dependent sources, but lacking independent sources, as shown in Figure 8-34. This network is connected to an external circuit that contains a single real-sinusoidal source at a frequency ω_0. Let $v(t)$ and $i(t)$ be the voltage and current measured at the terminals of the two-terminal network.

Recall that, in a circuit with a single real-sinusoidal source, all of the variables in the circuit will be sinusoids at the same frequency, but having different amplitudes and phases. Therefore, we know that

$$v(t) = |V| \cos(\omega t + \phi) = \frac{1}{2}\left[V e^{j\omega t} + V^* e^{-j\omega t} \right]$$

$$i(t) = |I| \cos(\omega t + \eta) = \frac{1}{2}\left[I e^{j\omega t} + I^* e^{-j\omega t} \right],$$

where

$$V = |V|e^{j\phi}; \quad I = |I|e^{j\eta}.$$

By substituting, we can compute the instantaneous power flowing into the terminals as the product

$$P_{\text{inst}}(t) = v(t)i(t)$$

$$= \frac{1}{4}[VI^* + V^*I + VIe^{j2\omega t} + V^*I^*e^{-j2\omega t}]$$

$$= \frac{1}{2}\Re e[VI^*] + \frac{1}{2}\Re e[VIe^{j2\omega t}]$$

$$= \frac{1}{2}\Re e[VI^*] + \frac{1}{2}|V||I|\cos(2\omega t + \phi + \eta).$$

The second term is a sinusoid with a frequency 2ω, which is twice the frequency of the voltage or current. Since the average value of this second term is zero, the average power absorbed by the two-terminal element is given by the first term:

$$P_{\text{ave}} = \frac{1}{2}\Re e[VI^*]. \qquad (8.47)$$

A sketch of the waveform of $P_{\text{inst}}(t)$ is shown in Figure 8-35. There is a component of the instantaneous power that is sinusoidal with frequency 2ω and amplitude $\frac{1}{2}|V||I|$, and there is a constant component whose height is equal to the average value, $\Re e(VI^*)$. Since $\Re(VI^*) \leq |V||I|$, the instantaneous power either becomes negative during part of each period or becomes tangent to the axis. Energy flows out of the terminals of the network during the time intervals indicated by the shaded regions; it flows into the terminals at the other times.

Since the two-terminal subnetwork in Figure 8-34 does not contain any independent sources, it is equivalent to an impedance. Let the value of that impedance be

$$Z = |Z|e^{j\theta}.$$

Then, if V and I are the phasors associated with the voltage and current at the terminals, we have

$$VI^* = ZII^* = |Z||I|^2e^{j\theta}$$
$$= |V||I|e^{j\theta}$$

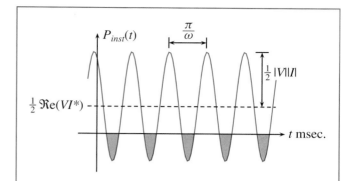

Fig. 8-35. The instantaneous power at a pair of terminals with a sinusoidal voltage and current.

and

$$P_{\text{ave}} = \frac{1}{2}\Re e[VI^*] = \frac{1}{2}|V||I|\cos\theta.$$

The power flowing into the network can then be written as

$$P_{\text{inst}}(t) = \frac{1}{2}|V||I|\cos\theta + \frac{1}{2}|V||I|\cos(2\omega t + \phi + \eta).$$

The constant $\cos\theta$ determines the relative amount of energy that is returned to the source in each cycle (i.e., it determines the relative size of the shaded regions in Figure 8-35). If $\cos\theta = 1$, which corresponds to an impedance that is a pure resistance, then $P_{\text{ave}} = \frac{1}{2}|V||I|$, and the amplitude of the sinusoidal component of the power is equal to the average power, On the other hand, when $\theta = \pm\pi/2$, then $P_{\text{ave}} = 0$, and power flows into the terminals during one-half of each cycle and out of them during the other half. In this case, the power flow is like that of an inductor or a capacitor. The constant $\cos\theta$ is called the *power factor* of the network. For a given voltage and current amplitude, it determines the average power that is absorbed.

If we define the *complex power* as

$$P = \frac{1}{2}VI^*,$$

then

$$P_{\text{ave}} = \Re\mathrm{e}(P).$$

We denote the imaginary part of P as Q_{ave}, which we call the *reactive power*. Thus, the complex power can be written as

$$P = P_{\text{ave}} + jQ_{\text{ave}}. \qquad (8.48)$$

Q_{ave} is closely related to the stored energy in a circuit, just as P_{ave} is closely related to the dissipated energy.

To understand the relationship between Q_{ave} and stored energy, recall our derivation of conservation-of-power from Section 2-6. In that derivation, we showed that

$$\sum_k v_k(t)i_k(t) = 0,$$

where the summation is over all the elements and sources of the network and where the voltages and currents associated with an element are defined consistently with respect to the default sign convention. The derivation of this conservation-of-power result depended only on the facts that the voltages satisfied KVL over all closed paths and that the currents satisfied KCL at all nodes. In a circuit with a single sinusoidal source, the phasors associated with the voltages satisfy KVL and the phasors associated with the currents satisfy KCL. Therefore, if we were to rederive the conservation-of-power relation in the phasor domain, we could show that

$$\sum_k \frac{1}{2} V_k I_k^* = 0. \qquad (8.49)$$

In this derivation, the only requirement that we impose on an element is that it have two terminals through which a current is defined and across which a voltage is defined. In a subnetwork with terminals, each terminal pair satisfies this requirement, and thus we can apply the conservation-of-power result to a two-terminal subnetwork by treating its terminal pair as an element. The sign convention for

a terminal pair, however, is the opposite from the default sign convention for an element. Incorporating this fact makes (8.49) become

$$-\sum_{\text{terminals}} \frac{1}{2} V I^* + \sum_{\text{elements}} \frac{1}{2} V_k I_k^* = 0.$$

Thus,

$$P = \sum_{\text{terminals}} \frac{1}{2} V I^* = \sum_{\text{elements}} \frac{1}{2} V_k I_k^*.$$

The complex power absorbed by the network is equal to the sum of the complex powers absorbed by each of the elements. The summation over the elements can be decomposed into three summations—one over the resistors, one over the inductors, and one over the capacitors. Thus,

$$P = \sum_R \frac{1}{2} V_k I_k^* + \sum_L \frac{1}{2} V_\ell I_\ell^* + \sum_C \frac{1}{2} V_m I_m^*.$$

Substituting the element relations gives

$$P = \sum_R \frac{1}{2} R_k |I_k|^2 +$$
$$j2\omega \left[\sum_L \frac{1}{4} L_\ell |I_\ell|^2 - \sum_C \frac{1}{4} C_m |V_m|^2 \right]. (8.50)$$

However, the average energy stored in an inductor is $\frac{1}{4} L |I|^2$, and the average energy stored in a capacitor is $\frac{1}{4} C |V|^2$. (See Problem P8-42.) Therefore, the last two summations in (8.50) represent the stored energy in the inductors and capacitors, respectively. The first summation is P_{ave}. This permits the following interpretation for (8.50):

$$P = P_{\text{ave}} + jQ_{\text{ave}} = P_{\text{ave}} + j2\omega[T_{\text{ave}} - U_{\text{ave}}].$$

Here, T_{ave} is the average stored energy in the inductors and U_{ave} is the average stored energy in the capacitors. The reactive power is 2ω times the difference between the average stored energy in the inductors (magnetic stored energy) and the average stored energy in the capacitors (electric stored energy). When the reactive power is zero, the stored magnetic and electric energy are balanced; when the reactive power is positive, there is an excess of magnetic stored energy; when the reactive power is negative, there is an excess of electric stored energy.

Example 8-15   Effect of a Reactive Element on Power Dissipation

Consider a resistor connected across a sinusoidal voltage source, as in Figure 8-36. The average power delivered by the source is

$$P_r = \frac{1}{2}\Re\mathrm{e}[1 \cdot \frac{1}{R}] = \frac{1}{2R},$$

which is independent of the frequency of the sinusoidal excitation.

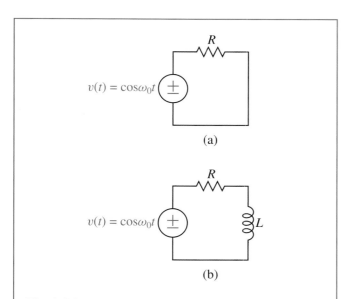

(a)

(b)

Fig. 8-36. (a) A source driving a resistive load. (b) The same source driving a load, consisting of a resistor and an inductor.

Next, consider the case where an inductive component is added to the load, as shown in part (b). The complex power is given by

$$P_{r\ell} = \frac{1}{2}VI^* = \frac{1}{2} \cdot 1 \cdot \frac{1}{Z_{r\ell}^*(j\omega_0)}$$

$$= \frac{1}{2}\left[\frac{1}{R - jL\omega_0}\right]$$

$$= \frac{1}{2}\left[\frac{R + jL\omega_0}{R^2 + L^2\omega_0^2}\right].$$

The average power dissipated in the resistor is the real part of the complex power, or

$$P_{r\ell,\mathrm{ave}} = \frac{R}{2(R^2 + L^2\omega_0^2)}.$$

This is always less than the average power dissipated by the load when the inductor is absent, except for the degenerate case where $\omega_0 = 0$. This means that the source delivers less power to the load. This is unfortunate because, for many applications, we want to be able to deliver as much power as possible to the load (e.g., when driving an electric motor or an antenna). In the next section, we look at methods for maximizing the amount of power delivered to the load.

The *reactive power* delivered to the load can be found from the imaginary part of the complex power:

$$Q_{r\ell,\mathrm{ave}} = \Im\mathrm{m}(P_{r,\ell}) = \frac{L\omega_0}{2(R^2 + L^2\omega_0^2)}$$

$$= 2\omega_0 \cdot \frac{1}{4}\frac{L}{R^2 + L^2\omega_0^2}.$$

To verify that this is $2\omega_0$ times the average stored energy in the inductor, we note that

$$T_{\mathrm{ave}} = \frac{1}{4}L|I|^2 = \frac{1}{4}L\left|\frac{1}{R + jL\omega_0}\right|^2$$

$$= \frac{1}{4}\frac{L}{R^2 + L^2\omega_0^2}.$$

Why is there less power delivered to the load even though no power is dissipated by the inductor? It is because the inductor shifts the phase of the voltage with respect to the current, and the peaks of the two waveforms occur at different times instead of simultaneously. To see how this reduces the average power, let the voltage to the two-terminal network be

$$v(t) = A \cos(\omega_0 t + \phi)$$

and let the current be

$$i(t) = B \cos(\omega_0 t + \eta);$$

then the average power is

$$P_{\text{ave}} = \frac{1}{2}\Re e[V I^*] = \frac{1}{2}\Re e[A e^{j\phi} \cdot B e^{-j\eta}]$$

$$= \frac{1}{2}|A| \cdot |B| \cos(\phi - \eta). \tag{8.51}$$

This is maximized when the voltage and current are in phase ($\phi = \eta$). Notice further that, when they *are* in phase, the reactive power is zero.   ∎

8-5   Maximum Power Transfer

Power is expensive. If we want to design a circuit to deliver power to a device, it is important that the circuit be as efficient as possible; we want the power to get to the load, not to be wasted in the driving circuitry. Examples of applications where it is important to get power to a device, which we will call a load, would be in the design of a power amplifier to drive a loudspeaker, an amplifier to drive the antenna in a wireless telephone, or compensating circuitry for driving an electric motor. Maximum efficiency means that the driving circuit needs to be matched carefully to the load. Exactly what this means should become clearer shortly. We begin with the resistive case, then later treat the more general case of a reactive load.

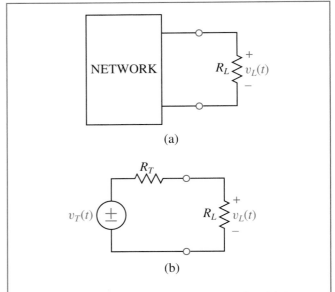

Fig. 8-37.   (a) An arbitrary network driving a load. (b) Same with the network replaced by its Thévenin equivalent.

8-5-1   Resistive Circuits

Consider a network attached to a purely resistive load, R_L, as shown in Figure 8-37(a). We want to find the value of the load resistance R_L that will enable the maximum amount of power to be delivered. The problem becomes a little less abstract and simpler if we replace the network by its Thévenin equivalent network, as in Figure 8-37(b). The Thévenin resistance R_T represents the resistance that is inherent to the source. We assume that the waveform of the independent source, which must supply the power, is fixed. Clearly, if R_L is very large, then the power delivered to the load will be small, because it is equal to $v_L^2(t)/R_L$. If R_L is very small, the power will also be small, because the voltage divider will limit the value of $v_L(t)$. We can derive an expression for the actual power delivered to the load by first using a voltage divider to evaluate the load voltage, $v_L(t)$:

$$v_L(t) = \frac{R_L}{R_L + R_T} v_T(t);$$

then the power delivered to the load (i.e., absorbed and possibly dissipated by R_L) is

$$P_L = \frac{R_L^2}{(R_L + R_T)^2} \frac{v_T^2(t)}{R_L}$$

$$= \frac{R_L}{(R_L + R_T)^2} v_T^2(t).$$

Figure 8-38 shows the power absorbed in the load as a function of R_L. Notice that this is zero when R_L is zero and also in the limit as it becomes infinitely large, but that there is a well-defined value of R_L by which the power delivered to the load is maximized.

To find the value of R_L by which the power is maximized, we set the derivative with respect to R_L to zero:

$$\frac{\partial P_L}{\partial R_L} = 0 = v_T^2(t) \cdot \frac{(R_L + R_T)^2 - 2R_L(R_L + R_T)}{(R_L + R_T)^4}.$$

The numerator is zero when

$$(R_T + R_L)^2 - 2(R_T + R_L)R_L = 0,$$

or

$$R_L = R_T.$$

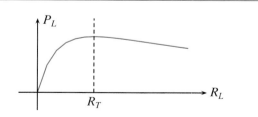

Fig. 8-38. A plot of the relative power delivered to a load resistor as a function of the load resistance.

Thus, the power delivered to the load is maximized when the load resistance is equal to the source resistance. In this case, the voltage across R_T is equal to the voltage across R_L. Since these voltages are equal, this means that one-half of the power supplied by the source is absorbed internally and does not get delivered to the load (one reason why power amplifiers get hot).

   DEMO: *Power Transfer in Resistive Circuits*

8-5-2   General Circuits

Now consider the more general case where the Thévenin equivalent model for the source is an impedance with a reactive component. The phasor-domain representation of the equivalent circuit is shown in Figure 8-39. The excitation is assumed to be sinusoidal at a fixed frequency, ω_0.

We begin by writing Z_T and Z_L in terms of their real and imaginary parts:

$$Z_T = R_T + jX_T$$

$$Z_L = R_L + jX_L.$$

The task is to select the values of R_L and X_L that maximize the power transmitted to the load. The value of R_L is constrained to be positive, but the value of X_L

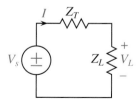

Fig. 8-39. The phasor-domain Thévenin equivalent representation of a source with a reactive component driving a reactive load, Z_L.

can be either positive or negative. The phasor associated with the current flowing through the load is

$$I = \frac{V_s}{Z_T + Z_L},$$

and that of the voltage across it is given by

$$V_L = Z_L I = \frac{Z_L V_s}{Z_T + Z_L}.$$

Therefore, the average power delivered to the load is

$$
\begin{aligned}
P_L &= \frac{1}{2}\Re e[V_L I^*] \\
&= \frac{1}{2}\Re e\left[\frac{Z_L V_s}{(Z_T + Z_L)} \cdot \frac{V_s^*}{(Z_T^* + Z_L^*)}\right] \\
&= \frac{1}{2}|V_s|^2 \frac{\Re e[Z_L]}{|Z_T + Z_L|^2} \\
&= \frac{1}{2}|V_s|^2 \frac{R_L}{|Z_T + Z_L|^2} \\
&= \frac{1}{2}|V_s|^2 \cdot \frac{R_L}{(R_T + R_L)^2 + (X_T + X_L)^2}.
\end{aligned}
$$

Again, we need to choose R_L and X_L to maximize P_L. X_L is fairly easy, since it appears only in the denominator. For any value of R_L, if we select $X_L = -X_T$, we minimize the denominator. Once we make this choice, however, the resulting power expression is the same as the one that we saw in the real case. From that earlier derivation, we know that the optimal value of R_L is $R_L = R_T$. This means that we choose

$$R_L = R_T; \quad X_L = -X_T,$$

or equivalently,

$$Z_L = Z_T^*.$$

It is informative to look at the complex power that is supplied by the voltage source under the conditions corresponding to maximum power delivery to the load. This is given by

$$
\begin{aligned}
P_{\text{complex}} &= \frac{1}{2}V_s I_L^* \\
&= \frac{1}{2}V_s \cdot \frac{V_s^*}{Z_T + Z_L} \\
&= \frac{|V_s|^2}{4R_T},
\end{aligned}
$$

which is purely real. Thus, when the power delivered to the load is maximized, the reactive power supplied by the source is zero, which corresponds to electric and magnetic stored energy that is balanced. Matching the load to the source implies adding additional electrical stored energy to the load if the source has excess magnetic stored energy, or adding additional magnetic stored energy to the load if the source has excess electric stored energy.

The importance of matching both the real and imaginary parts of the load impedance is illustrated by the following example.

Example 8-16   Matching the Load

Consider the model for a 60-Hz generator and connecting cabling driving the 100-Ω load depicted in Figure 8-40. In order to deliver 1000 watts to the load, the generator must produce a current, I_L, such that

$$\frac{1}{2} \cdot 100 \cdot |I_L|^2 = 1000.$$

This requires that $|I_L| = \sqrt{20}$. Of course, to do this the generator itself must produce 2000 watts, since 1000 W will be absorbed in R_s.

The picture is even bleaker, however, if there is a reactive component in the load, as shown in Figure 8-41. Notice, in this case, that the load and the source are not

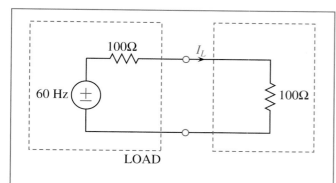

Fig. 8-40. A generator attached to a resistive load.

optimally matched to maximize the power delivery. Now we have

$$Z_L = \frac{1}{\frac{1}{100} + j\frac{1.2\pi}{2\pi(60)}} = \frac{1}{\frac{1}{100} + j\frac{1}{100}} = 50(1-j).$$

In order to provide 1000 W to the load, $|I_L|$ must be larger. In order to see how much larger it will have to be, we set

$$1000 = \frac{1}{2} \cdot \Re \left\{ 50(1-j)I_L \cdot I_L^* \right\},$$

from which it follows that

$$|I_L|^2 = 40.$$

Because of the reactive component in the load, we must supply additional current.

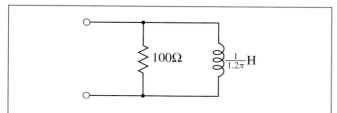

Fig. 8-41. An alternative load with a reactive component.

This example illustrates a major challenge for electric-power utilities. In order to minimize the total amount of power that they need to generate to meet customer needs, they need to have an accurate model for the effective impedance of the total user load. With such a model they can implement compensation by varying the reactive component of the source impedance Z_s. ∎

DEMO: *Load Matching*

8-6 Chapter Summary

8-6-1 Important Points Introduced

- When a circuit has only one independent source, which is a complex exponential time function, all of the element variables in the circuit will be complex exponential time functions.

- If the response of a circuit to a complex input signal is known, the response to the real (imaginary) part of that input will be the real (imaginary) part of the known output.

- The sinusoidal steady-state response (the forced response after the transients have died out) to a switched sinusoid is the same as the response to an unswitched sinusoid.

- The phasors associated with the element variables in a circuit whose input is a complex exponential time function satisfy KCL at all nodes, KVL on all closed paths, and Ohm's law for all elements.

- In the phasor domain, the impedances of the inductors and capacitors in a circuit vary with the frequency of the sinusoidal excitation.

- The average power absorbed in a circuit excited with a sinusoidal source can be computed from the phasors of the element variables as $P_{\text{ave}} = \frac{1}{2}\Re[V I^*]$.

- The reactive power $Q_{ave} = \frac{1}{2}\Im m[VI^*]$ is proportional to the difference between the amount of energy stored in the inductors and the amount of energy stored in the capacitors of a circuit.

- The maximum power delivery to a load occurs when the impedance of the load is matched to the impedance of the source.

8-6-2   New Abilities Acquired

You should now be able to do the following:

(1) Express real-sinusoidal signals as the real (imaginary) parts of complex exponential time functions.

(2) Find the response of a circuit to an input that is a real sinusoid by using the real-part property.

(3) Find the response of a circuit to an input signal that is a complex exponential time function by working in the phasor domain.

(4) Find the sinusoidal steady-state response, using Laplace transforms.

(5) Compute the element variable waveforms in a circuit that has multiple sinusoidal inputs, by superposition.

(6) Apply the node and mesh methods in the phasor domain.

(7) Find the phasor-domain Thévenin and Norton equivalent circuits for a two-terminal network containing a single sinusoidal source.

(8) Calculate the average power absorbed in a circuit.

(9) Calculate the average reactive power delivered to a circuit.

(10) Design a load impedance that will maximize the power that a source can deliver to the load.

8-7   Problems

8-7-1   Drill Problems

P8-1 Find $i_\ell(t)$ in the circuit of Figure P8-1, using Laplace transforms in a manner that is similar to Example 8-1. Let $i_s(t) = (\sin 5t)u(t)$.

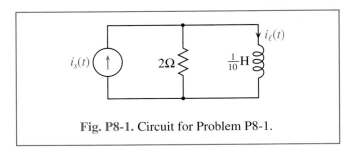

Fig. P8-1. Circuit for Problem P8-1.

P8-2 Rework Problem P-8.1, but this time use the imaginary-part property, in a manner that is similar to Example 8-2.

P8-3 The circuit in Figure P8-3 has a single input, which is the complex exponential time function $v_s(t) = 10e^{j3t}$. Compute the current flowing through the inductor $i_\ell(t)$.

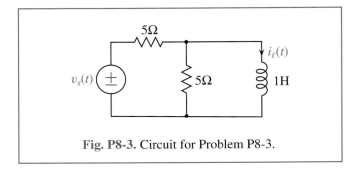

Fig. P8-3. Circuit for Problem P8-3.

P8-4

(a) Map the circuit in Figure P8-4 to the phasor domain. Sketch your result.

(b) Compute the phasor associated with the output voltage, $v_\ell(t)$.

(c) Determine $v_\ell(t)$ for all t. Express your answer in the form $v_\ell(t) = A \cos(\omega_0 t + \phi)$.

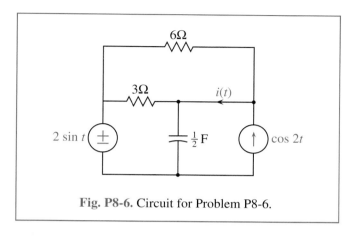

Fig. P8-6. Circuit for Problem P8-6.

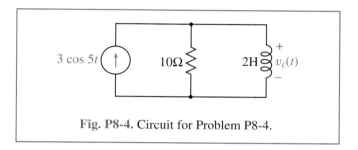

Fig. P8-4. Circuit for Problem P8-4.

P8-5 Find the current $i(t)$ in the circuit of Figure P8-5.

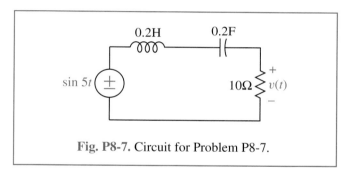

Fig. P8-7. Circuit for Problem P8-7.

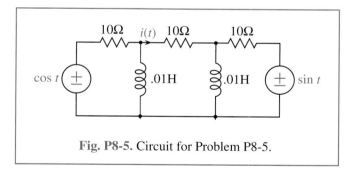

Fig. P8-5. Circuit for Problem P8-5.

P8-8 Find the phasor-domain Thévenin equivalent of the two-terminal network shown in Figure P8-8.

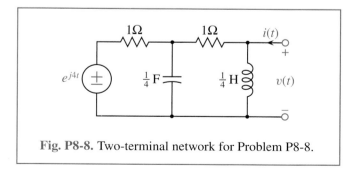

Fig. P8-8. Two-terminal network for Problem P8-8.

P8-6 Compute the current $i(t)$ for the circuit in Figure P8-6.

P8-7 Find the output of the circuit in Figure P8-7, $v(t)$, when the input is the source voltage $\sin 5t$.

P8-9 Find the two node voltages $e_a(t)$ and $e_b(t)$ in the circuit in Figure P8-9.

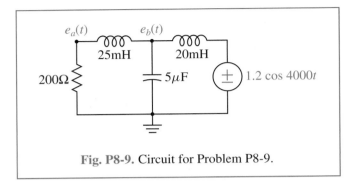

Fig. P8-9. Circuit for Problem P8-9.

P8-10 Solve the circuit of Example 8-12, using the node method.

P8-11 Solve the circuit of Example 8-11, using the mesh method.

P8-12 In the circuit of Figure P8-12, find $v_{out}(t)$ for all t, if $v_s(t) = \cos(100t)$ for $-\infty < t < \infty$.

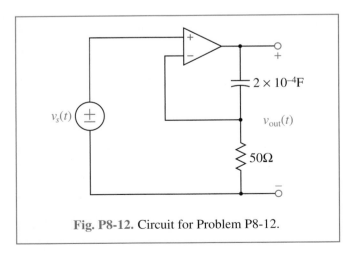

Fig. P8-12. Circuit for Problem P8-12.

8-7-2   Basic Problems

P8-13 Being able to manipulate sums of sinusoids is a useful skill when dealing with the sinusoidal responses of circuits. Because the sum of two sinusoids that have the same frequency is also a sinusoid at that frequency, there are many expressions that correspond to equivalent sinusoids, but which appear to be quite different. It can be very helpful to recognize some of these alternative forms.

(a) Show that the relation

$$K \cos(\omega_0 t + \phi) = A \cos \omega_0 t + B \sin \omega_0 t$$

is true by expressing A and B in terms of K and ϕ.

(b) This expression can be used to go both ways. Express K and ϕ in terms of A and B.

(c) If

$$x(t) = A_1 \cos(\omega_0 t + \phi_1) + A_2 \cos(\omega_0 t + \phi_2),$$

show that $x(t)$ can be written in the form

$$x(t) = B \cos(\omega_0 t + \theta).$$

Express B and θ in terms of A_1, A_2, ϕ_1, and ϕ_2.

P8-14 The purpose of this problem is to verify the real- and imaginary-part properties. Consider the circuit in Figure P8-14, which is at initial rest.

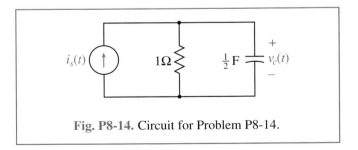

Fig. P8-14. Circuit for Problem P8-14.

(a) Find the response of the circuit, $v_c(t)$ when the current-source waveform is the step $i_s(t) = u(t)$.

(b) Find the response of the circuit when the current source is a switched exponential $i_s(t) = e^{-t}u(t)$.

(c) Now calculate the response of the circuit to the source waveform $i_s(t) = (1 + je^{-t})u(t)$. Verify that your solutions to (a) and (b) are the real and imaginary parts of your result to (c), respectively.

P8-15

(a) Find the value of the voltage $v(t)$ in the circuit of Figure P8-15, when the current $i_s(t)$ is the complex exponential time function $i_{s1}(t) = e^{j\omega t}$.

(b) Repeat for the current-source waveform $i_{s2}(t) = e^{-j\omega t}$. Is this result equal to the complex conjugate of your result from part (a)?

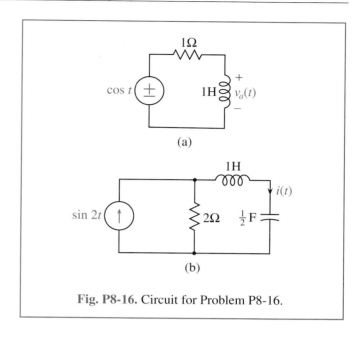

(a)

(b)

Fig. P8-16. Circuit for Problem P8-16.

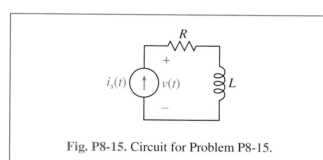

Fig. P8-15. Circuit for Problem P8-15.

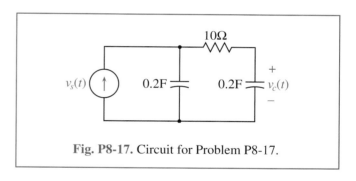

Fig. P8-17. Circuit for Problem P8-17.

P8-16 The input signals to the circuits in Figure P8-16 are sinusoidal. In each case, find the waveform corresponding to the indicated output variable.

P8-17 Find $v_c(t)$, if $v_s(t) = \cos 5t$ for all t for the circuit in Figure P8-17.

P8-18 For the circuit in Figure P8-18, find $v(t)$ when $v_s(t) = \cos(\omega t)$.

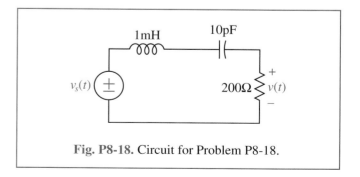

Fig. P8-18. Circuit for Problem P8-18.

P8-19 For the circuit in Figure P8-19, find $i(t)$ when $v_s(t) = \sin(\omega t)$.

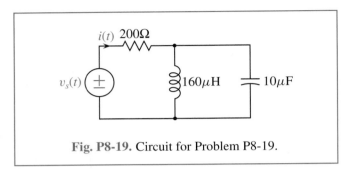

Fig. P8-19. Circuit for Problem P8-19.

P8-20 Determine $v_{out}(t)$ for all t for the circuit in Figure P8-20.

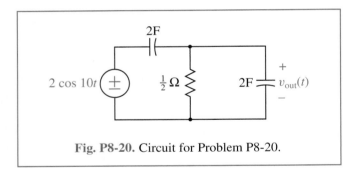

Fig. P8-20. Circuit for Problem P8-20.

P8-21 The input voltage to the circuit in Figure P8-21 is $i_s(t) = \sin 5t$. Compute the voltage $v_{out}(t)$ for all t.

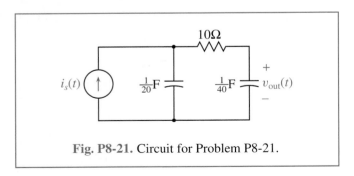

Fig. P8-21. Circuit for Problem P8-21.

P8-22 Calculate $i(t)$ for all t for the circuit in Figure P8-22.

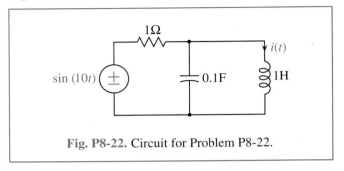

Fig. P8-22. Circuit for Problem P8-22.

P8-23 Compute $i_\ell(t)$ for the circuit in Figure P8-23, if $i_s(t) = 120 \cos(40t)$.

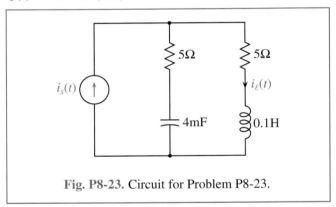

Fig. P8-23. Circuit for Problem P8-23.

P8-24 Calculate $v_c(t)$ for all t in the circuit in Figure P8-24.

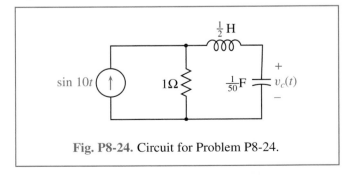

Fig. P8-24. Circuit for Problem P8-24.

P8-25 Compute $i(t)$ in the circuit in Figure P8-25.

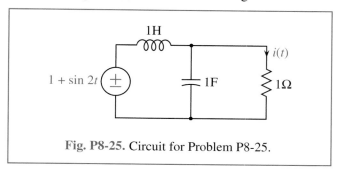

Fig. P8-25. Circuit for Problem P8-25.

P8-26 Assuming $v_s(t) = \cos t + 2\sin(2t + \pi/4)$ for the circuit in Figure P8-26, find $v_{out}(t)$.

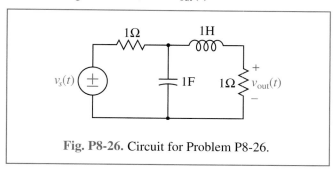

Fig. P8-26. Circuit for Problem P8-26.

P8-27 Find $v_{out}(t)$ for the circuit in Figure P8-27, if $v_{in}(t) = 3 + 4\sin(1000t)$. The terminals are *open-circuited*.

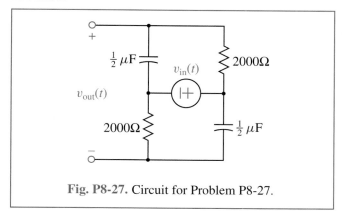

Fig. P8-27. Circuit for Problem P8-27.

P8-28 Find $v_{out}(t)$ in the circuit in Figure P8-28 if the input is $i_{in} = \sin(4t) + 2\cos(2t)$.

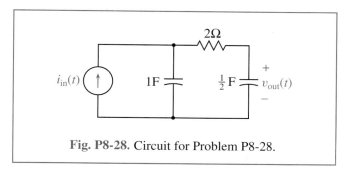

Fig. P8-28. Circuit for Problem P8-28.

P8-29 Compute the voltage $v(t)$ indicated in Figure P8-29.

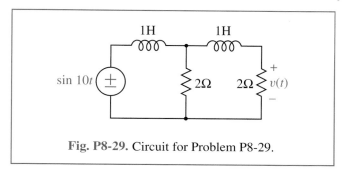

Fig. P8-29. Circuit for Problem P8-29.

P8-30 Find the current $i(t)$ in the circuit of Figure P8-30.

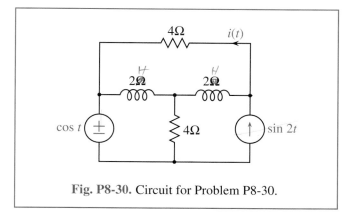

Fig. P8-30. Circuit for Problem P8-30.

P8-31 A circuit with input source waveform $x_{in}(t)$ and output $y_{out}(t)$ has the system function

$$H(s) = \frac{s}{(s+2)^2 + 100}.$$

Find $y_{out}(t)$ if

$$x_{in}(t) = 3\cos(10t).$$

P8-32

(a) Find the system function of the circuit in Figure P8-32.
(b) Find $v_c(t)$ for all t.

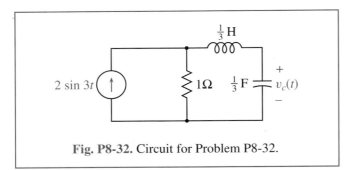

Fig. P8-32. Circuit for Problem P8-32.

P8-33 For the circuit in Figure P8-33, find $v(t)$.

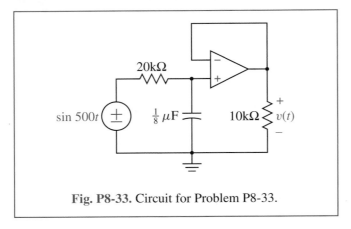

Fig. P8-33. Circuit for Problem P8-33.

P8-34 Calculate the average power absorbed in the resistor R in the circuit in Figure P8-34 for each of the following sets of source waveforms.

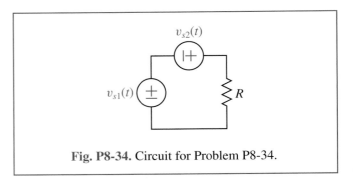

Fig. P8-34. Circuit for Problem P8-34.

(a) $v_{s1}(t) = \cos 3\omega t$ and $v_{s2}(t) = 2\cos \omega t$
(b) $v_{s1}(t) = 2\cos \omega t$ and $v_{s1}(t) = \sin \omega t$
(c) $v_{s1}(t) = \sin 2\omega t$ and $v_{s1}(t) = \sin 2\omega t$

P8-35

(a) Draw the circuit in Figure P8-35, but in the phasor domain.
(b) Find the phasor I associated with the current $i(t)$.
(c) Compute the current $i(t)$ generated by the voltage source.
(d) Calculate the average power supplied by the voltage source.

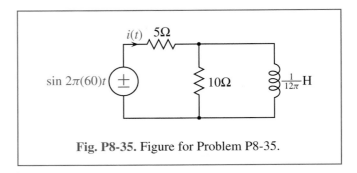

Fig. P8-35. Figure for Problem P8-35.

P8-36 For the circuit in Figure P8-36, what is the average power supplied by the current source?

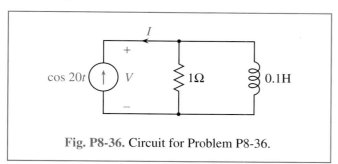

Fig. P8-36. Circuit for Problem P8-36.

P8-37

(a) Compute the impedance, $Z(j\omega)$, of the two-terminal network shown in Figure P8-37 at $\omega = 2$ radians/s.

(b) For a voltage excitation $v(t) = \cos 2t$, compute the average power absorbed.

(c) For the excitation in (b), compute the average magnetic and electric stored energy.

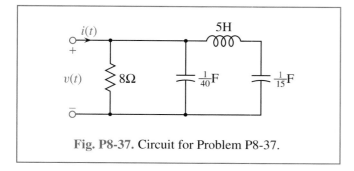

Fig. P8-37. Circuit for Problem P8-37.

8-7-3 Advanced Problems

P8-38 The real-part property is applicable only to circuits that are linear and time-invariant, a condition that excludes circuits that are not at initial rest. This problem looks at extensions of that property. Consider the circuit

in Figure P8-38, which is the same circuit that was used in Problem P-8.7.

(a) Find $v_{ca}(t)$, if the initial value of the capacitor voltage is $v_{ca}(0) = x_0$ and $i_{sa}(t) = u(t)$.

(b) Find $v_{cb}(t)$, if $v_{cb}(0) = y_0$ and $i_{sb}(t) = e^{-t}u(t)$.

(c) Find $v_{cc}(t)$, if $v_{cc}(0) = z_0$ and $i_{sc}(t) = i_{sa}(t)+i_{sb}(t)$.

(d) What relationship must hold between x_0, y_0, and z_0 in order that $v_{cc}(t) = v_{ca}(t) + v_{cb}(t)$?

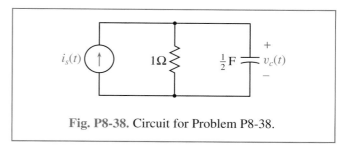

Fig. P8-38. Circuit for Problem P8-38.

P8-39 Example 8-5 illustrates a procedure by which the Laplace transform can be used to find the sinusoidal steady-state response (i.e., a procedure for using the response to an *unswitched* sinusoid to find the response to a *switched* one). There always exists a set of (usually nonzero) initial conditions for which these two responses will be the same for $t \geq 0$. To explore this point, consider the circuit in Figure P8-39.

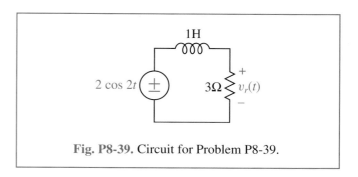

Fig. P8-39. Circuit for Problem P8-39.

(a) Let the current flowing through the inductor at $t = 0$ be i_0. Compute $v_r(t)$ for $t > 0$.

(b) For what value of i_0 will the transient component of the response that you calculated in (a) be zero?

(c) Compute the voltage drop across the resistor when the source voltage is the *unswitched* voltage $v_s(t) = \sin 2t$. Compare your result to the result in part (a), when i_0 is set to the value that you computed in part (b).

(d) Using phasors, compute the value of inductor current at $t = 0$ when the input signal is the same as in part (c). How does this result compare with the result from part (b)?

P8-40 The system function of a certain RLC circuit is of the form

$$H(s) = K \frac{s + a}{s^2 + bs + c}.$$

When the input to the circuit is $x(t) = [\cos 2t]\, u(t)$, the output variable for $t > 0$ is

$$y(t) = Ae^{-t}\cos(2t + \phi) + 2\cos 2t,$$

where A and ϕ are real constants. Find values for K, a, b, and c.

P8-41 If $x(t)$ is a periodic signal with period T—that is, it $x(t + T) = x(t)$ for all t—then the root-mean-square (rms) value of $x(t)$ is defined as

$$X_{\text{rms}} = \sqrt{\frac{1}{T}\int_0^T x^2(t)\, dt}.$$

(a) Assume that the potential difference $v(t)$ across a resistor with resistance R is periodic with period T, but not necessarily sinusoidal. Let the current through that resistor be $i(t)$. Derive a relation that relates the quantities V_{rms} and I_{rms}.

(b) Derive an expression for the average power absorbed in the resistor, namely,

$$P_{\text{ave}} = \frac{1}{T}\int_0^T v(t) i(t)\, dt,$$

(i) in terms of I_{rms} and R;
(ii) in terms of V_{rms} and R;
(iii) in terms of V_{rms} and I_{rms}.

(c) A constant voltage is applied across the same resistor. What should the value of this voltage be so that the power absorbed is the same as P_{ave}, which you derived in (b).

P8-42

(a) If the current flowing through an inductor having inductance L is $i(t) = |I| \cos(\omega_0 t + \phi)$, show that the average stored energy is $\frac{1}{4}L|I|^2$.

(b) If the voltage across the terminals of a capacitor having capacitance C is $v(t) = |V| \cos(\omega_0 t + \phi)$, show that the average stored energy is $\frac{1}{4}C|V|^2$.

P8-43 The input to the circuit in Figure P8-43 is a sinusoid $i_s(t) = \cos(\omega t)$.

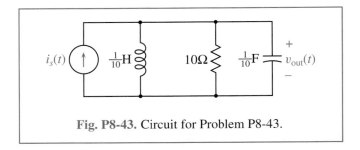

Fig. P8-43. Circuit for Problem P8-43.

(a) Express the output $v_{\text{out}}(t)$ as a function of ω.

(b) At what frequency ω is the average power supplied by the current source a maximum?

P8-44 Consider a two-terminal network constructed from resistors, inductors, and capacitors (with positive values), but lacking sources, as in Figure P8-44(a).

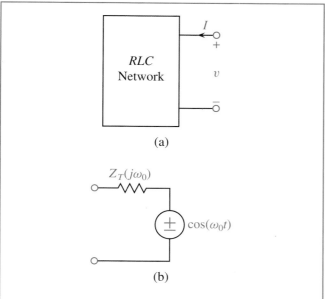

(a)

(b)

Fig. P8-44. (a) Two-terminal network for Problem P8-44. (b) Two-terminal network to be attached to the network in Figure P8-44(a).

(a) Let the impedance of the two-terminal network at a particular frequency be $Z = |Z|e^{j\theta}$. Show that the average power absorbed can be written as

$$P_{\text{ave}} = \frac{1}{2}|V| \cdot |I| \cos\theta.$$

The quantity $\cos\theta$ is called the *power factor*.

(b) The fact that the average power absorbed in a network that contains no sources can never be negative constrains the value of θ. State this constraint.

(c) The constraint that you derived in (b), in turn, places a condition on the real part of the impedance $Z(j\omega)$ (at any frequency). State this condition.

(d) Now connect the network to the two-terminal network shown in Figure P8-44(b). Show that, if $Z_T(j\omega_0) = Z^*(j\omega_0)$, then $\theta = 0$.

8-7-4 Design Problems

P8-45 Find those values of R and L for the circuit in Figure P8-45 that cause the maximum amount of power to be delivered to the load.

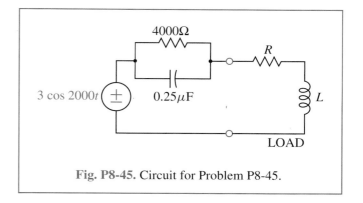

Fig. P8-45. Circuit for Problem P8-45.

P8-46 Find those values of L and R that will maximize the power delivered to the load (enclosed by the dashed box) in the circuit in Figure P8-46.

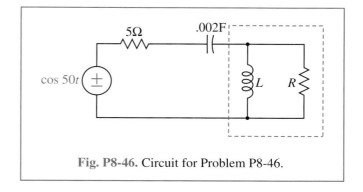

Fig. P8-46. Circuit for Problem P8-46.

P8-47 The imaginary part of an impedance is called its *reactance*; the imaginary part of an admittance is called its *susceptance*. In power-transmission applications it is often useful to modify a network so that its reactance or susceptance at a particular frequency, ω_0, is zero.

(a) If the impedance of a circuit at a particular frequency is

$$Z(j\omega_0) = A + jB,$$

derive an expression for the susceptance at that frequency in terms of A and B.

(b) What type of element (R, L, or C) should be connected in parallel with a network whose susceptance at a particular frequency is negative in order that the resulting admittance will be real at that frequency? How does the real part of the admittance change before and after the parallel element is added?

(c) What type of element (R, L, or C) should be connected in parallel with a network whose susceptance at a particular frequency is positive in order that the resulting admittance will be real at that frequency?

(d) What type of element (R, L, or C) should be connected in series with a network whose reactance at a particular frequency is negative in order that the resulting impedance will be real at that frequency? How does the real part of the impedance change before and after the series element is added?

(e) What type of element (R, L, or C) should be connected in series with a network whose reactance at a particular frequency is positive in order that the resulting impedance will be real at that frequency?

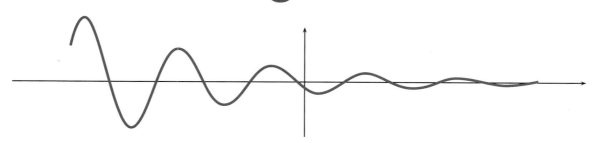

Frequency Responses of Circuits

Objectives

By the end of this chapter, you should be able to do the following:

1. *Evaluate the frequency response of a linear time-invariant circuit.*

2. *Find the spectrum of the output of a linear circuit from the spectrum of the input and the frequency response of the circuit.*

3. *Compute and plot the magnitude and phase response of a circuit.*

4. *Sketch the magnitude and phase response of a circuit from its pole–zero plot.*

5. *Draw the Bode magnitude and phase plots for a circuit.*

Chapter 8 showed that it is straightforward to understand how a linear circuit behaves when the input signal is a sinusoid or a complex exponential time function at a single frequency, but such signals are somewhat limited. What happens when the input signal is more interesting? In this chapter, we generalize our earlier analysis to understand how circuits react to more complex circuit excitations containing sinusoidal components at many different frequencies. This allows us to deal with much richer signal sets.

Before looking at the circuits, we begin with the signals themselves. Most signals can be constructed out of sinusoids. For example, the Fourier-series construction demonstrates that almost any *periodic* signal can be synthesized as a weighted sum of complex exponential time functions. Using the Fourier series, the fact that a circuit at initial rest is linear and time invariant, and the

351

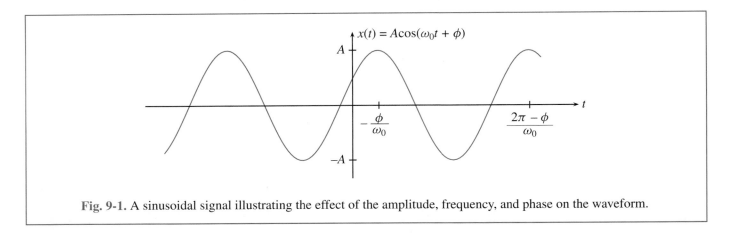

Fig. 9-1. A sinusoidal signal illustrating the effect of the amplitude, frequency, and phase on the waveform.

known response of a circuit to sinusoidal inputs that we explored in Chapter 8, we can predict how a circuit will respond to (almost) any periodic input signal. The Fourier transform allows us to take things one step further. Using it, we can represent (almost) any *nonperiodic* signal as a (linear) superposition of sinusoids. When we use the Fourier transform as a tool, we can predict how a circuit will respond to *any* input signal. To make use of either of these tools, however, we need to learn how a circuit behaves in response to sinusoidal input components at a number of different frequencies. This is most easily done by treating the frequency of the sinusoidal input (complex exponential time function) as a variable.

The behavior of the output of a circuit (one or more of the element variables) as a function of the frequency of a sinusoidal input is called the *frequency response* of the circuit. This is an extremely powerful tool for describing circuit behavior. In fact, the frequency response and the transient response, embodied in the impulse response and step response of the circuit, which were discussed in Chapter 7, are the two most important means for describing the behavior of a linear circuit. Control of these two complementary aspects of a circuit's behavior forms the basis for most circuit-design procedures. Before we discuss the frequency response itself, however,

we shall first look at how useful input signals can be constructed as superpositions of sinusoids.

9-1   Spectrum Representations of Periodic Signals

9-1-1   Spectrum of a Sum of Sinusoids

We encountered sinusoidal signals in Chapter 8. A sinusoidal signal

$$x(t) = A \cos(\omega_0 t + \phi) = \Re e(A e^{j\phi} e^{j\omega_0 t})$$

is completely defined by its *amplitude*, A, its *frequency*, ω_0, and its *phase*, ϕ. If we know these three numbers, we can draw the waveform, as in Figure 9-1. In Chapter 8 we assumed that the input signal was a single sinusoid at a fixed frequency. In this section, we want to look at signals that are constructed as sums of sinusoids.

If we know that a signal consists of a sum of sinusoidal components, we can create that signal if we know the amplitude, frequency, and phase of each component. Such a representation is called the *spectrum* of the signal.

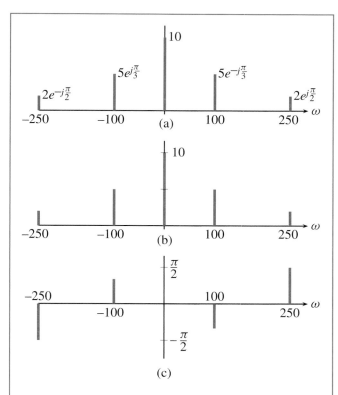

Fig. 9-2. The spectrum of a signal with five complex exponential components. (a) Complex spectrum. (b) Magnitude Spectrum. (c) Phase Spectrum.

Figure 9-2 shows two graphical representations of the spectrum of the waveform

$$x(t) = 10 + 10\cos(100t - \frac{\pi}{3}) - 4\sin(250t)$$

$$= 10 + 5e^{-j\frac{\pi}{3}}e^{j100t} + 5e^{j\frac{\pi}{3}}e^{-j100t}$$

$$+ 2e^{j\frac{\pi}{2}}e^{j250t} + 2e^{-j\frac{\pi}{2}}e^{-j250t}.$$

We have plotted the spectrum two ways, but each is based on the representation of the signal as a sum of sinusoids. The graph at the top shows the *complex spectrum*. In it, each complex exponential component is represented by a line located at the frequency of that component; the phasor associated with the component appears as a label

next to the spectral line. The lengths of the spectral lines are arbitrary, but we have drawn them with lengths that are proportional to the magnitudes of the phasors. Notice that a spectral line is drawn for each *complex exponential* component. A real sinusoid, such as $x(t) = \cos t$, has two spectral lines, one at the frequency $\omega = 1$ and one at the frequency $\omega = -1$. This is because a real sinusoid can be constructed from two complex exponentials as

$$\cos t = \frac{1}{2}e^{jt} + \frac{1}{2}e^{-jt},$$

which we recall from the inverse Euler relation.

The second graphical representation of the spectrum consists of two parts: a magnitude spectrum and a phase spectrum. The middle plot in Figure 9-2 displays the *magnitude spectrum*. Here the length of each line is proportional to the magnitude of the sinusoidal component, and its position on the frequency axis is again determined by its frequency. The bottom plot is the *phase spectrum*. Here the lengths of the lines indicate the phases of the sinusoids. The magnitude and phase spectra together contain the same information as the complex spectrum. The time-domain representation of $x(t)$ is shown in Figure 9-3.

 DEMO: *Complex Spectrum*

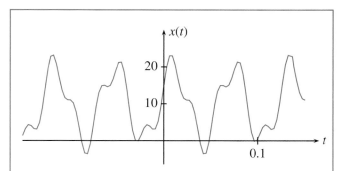

Fig. 9-3. The time-domain waveform corresponding to the spectrum in Figure 9-2.

9-1-2 Periodic Waveforms

A signal $x(t)$ is *periodic* with period $T > 0$ if

$$x(t + T) = x(t) \quad \text{for all } t.$$

For example, the waveform $x(t) = \cos t$ is periodic with period 2π. We require that the period be positive, since all waveforms would be trivially periodic with period 0, if that value were allowed. A periodic signal repeats itself endlessly, both forward and backward in time.

The period, T, is not unique as we have defined it. With a little thought, we notice that $x(t) = \cos t$ is periodic not only with period 2π but also with periods 4π, 6π, etc. The minimum value of the period, which is unique, is called the *fundamental period*. The fundamental period of $x(t) = \cos t$ is 2π. A constant signal $x(t) = K$ is somewhat special in that it is periodic for any value of T, but it does not have a fundamental period.

A set of sinusoidal signals is said to be *harmonically related* if they share a common period. For example, the sinusoidal signals

$$x_k(t) = \cos(k\omega_0 t), \quad k = 0, 1, 2, \ldots$$

for integer values of k are harmonically related, since each has a period of $2\pi/\omega_0$. (Notice that they have different *fundamental* periods, since the fundamental period of the kth member of the set is $2\pi/(k\omega_0)$.) The frequencies of a set of harmonically related sinusoids are all integer multiples of a single frequency, ω_0, which is called the *fundamental frequency*. There may or may not be a sinusoid in the set at the fundamental frequency. Similarly, a set of complex exponential time functions is harmonically related if they have a common period. Thus, the complex exponentials

$$e_k(t) = e^{jk\omega_0 t}, \quad k = 0, \pm 1, \pm 2, \ldots$$

are harmonically related. Notice with complex exponential time functions that the frequencies of the individual signals are allowed to be negative; with real sinusoids, the frequencies are limited to positive nonzero values.

Any linear combination of harmonically related sinusoids with a common period T is also periodic with period T. This means that any signal $x(t)$ which can be expressed as

$$x(t) = \sum_k X_k e^{j\frac{2\pi}{T}kt} \tag{9.1}$$

is periodic with period T for any values of the complex coefficients X_k. Figure 9-4 illustrates the construction of a periodic signal by summing harmonically related sinusoids. The signal at the top,

$$x_a(t) = 1 + 2\cos 2\pi t = e^{-j2\pi t} + 1 + e^{j2\pi t},$$

contains three complex exponential terms, each of which has a period of one second. The two other waveforms are formed from this one by adding additional harmonics. Specifically,

$$x_b(t) = x_a(t) + e^{-j4\pi t} + e^{j4\pi t}$$

$$x_c(t) = x_b(t) + e^{-j8\pi t} + e^{-j6\pi t} + e^{j6\pi t} + e^{j8\pi t}.$$

With each additional pair of complex harmonics that is added, the waveform becomes more spiked—the peak values grow, and the frequency intervals where the magnitude is large become narrower. The converse to (9.1) is also true. Almost all periodic signals can be decomposed into sums of harmonically related sinusoids. This is the essence of the *Fourier-series representation* for a periodic signal.

Our treatment of the Fourier series and the Fourier transform will necessarily be brief and incomplete. Our primary goal is to motivate looking at the frequency response of a linear time-invariant circuit and to demonstrate the effect that it has on the spectrum of an input signal. These are topics that will be treated more fully in later courses.

 DEMO: *Periodic Signals*

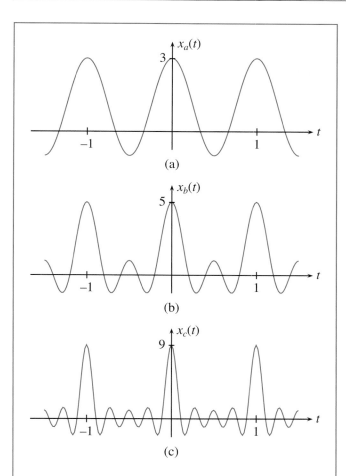

Fig. 9-4. An illustration of the construction of a periodic signal as a sum of harmonically related sinusoids.
(a) $x_a(t) = 1 + 2\cos 2\pi t$.
(b) $x_b(t) = 1 + 2\cos 2\pi t + 2\cos 4\pi t$.
(c) $x_c(t) = 1 + 2\cos 2\pi t + 2\cos 4\pi t + 2\cos 6\pi t + 2\cos 8\pi t$.

9-1-3 Fourier-Series Representations

In 1807, Jean-Baptiste Fourier demonstrated that nearly all periodic waveforms with period T could be expressed as an infinite sum of sinusoids of the form

$$x(t) = \sum_{k=-\infty}^{\infty} X_k e^{j\frac{2\pi}{T}kt}. \tag{9.2}$$

There are some requirements that $x(t)$ must satisfy in order to have a Fourier-series representation, but almost all signals that we encounter satisfy these conditions:

1. $x(t)$ must be a single-valued function except at possibly a countable number of points.

2. The integral of the magnitude of $x(t)$ over a period must be finite:

$$\int_{\tau}^{\tau+T} |x(t)| \, dt < \infty.$$

3. $x(t)$ must have at most a finite number of discontinuities within each period.

4. $x(t)$ must have at most a finite number of maxima and minima within each period at points where it is not constant.

The converse of this statement is also true: any waveform $x(t)$ that can be expressed as in (9.2) is periodic with period T. We see this by actually evaluating $x(t + T)$:

$$x(t + T) = \sum_{k=-\infty}^{\infty} X_k e^{j\frac{2\pi}{T}k(t+T)}$$

$$= \sum_{k=-\infty}^{\infty} X_k e^{j\frac{2\pi}{T}kt} e^{j2\pi}$$

$$= \sum_{k=-\infty}^{\infty} X_k e^{j\frac{2\pi}{T}kt} = x(t).$$

Given the signal $x(t)$, how can we find the complex amplitudes (harmonic coefficients), X_k? This is fairly straightforward. We begin by multiplying both sides of (9.2) by a complex exponential time function $e^{-j\frac{2\pi}{T}mt}$ and integrating both sides over one period:

$$\int_0^T x(t) e^{-j\frac{2\pi}{T}mt} \, dt = \int_0^T \left\{ \sum_{k=-\infty}^{\infty} X_k e^{j\frac{2\pi}{T}kt} \right\} e^{-j\frac{2\pi}{T}mt} \, dt.$$

The next step involves interchanging the order of the integration and summation. Without going into mathematical details, this is allowable if $x(t)$ satisfies the conditions enumerated earlier. The result is

$$\int_0^T x(t)e^{-j\frac{2\pi}{T}mt}\,dt = \sum_{k=-\infty}^{\infty} X_k \int_0^T e^{j\frac{2\pi}{T}(k-m)t}\,dt. \quad (9.3)$$

Now consider the inner integral on the right-hand side of (9.3). When $k = m$, the integrand is 1 and the integral is equal to T. When $k \neq m$, we are integrating a complex exponential time function over exactly an integer number of periods. By actually performing the integration, we see that this integral is zero. Therefore,

$$\int_0^T e^{j\frac{2\pi}{T}(k-m)t}\,dt = T\delta_{km}, = \begin{cases} T, & k = m \\ 0, & k \neq m \end{cases}.$$

(δ_{km} is the Kronecker delta.) With this observation, (9.3) becomes

$$\int_0^T x(t)e^{-j\frac{2\pi}{T}mt}\,dt = \sum_{k=-\infty}^{\infty} X_k T\delta_{km}.$$

The term inside the summation is zero unless $k = m$. This means that the sum is equal to its single nonzero term and

$$\int_0^T x(t)e^{-j\frac{2\pi}{T}mt}\,dt = TX_m.$$

We can use this result as a formula for determining the harmonic coefficients X_k, more commonly called the *Fourier-series coefficients*:

$$X_m = \frac{1}{T}\int_0^T x(t)e^{-j\frac{2\pi}{T}mt}\,dt. \quad (9.4)$$

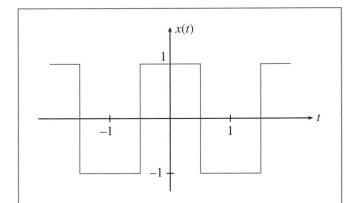

Fig. 9-5. A periodic square wave whose period is two seconds.

Example 9-1   Finding Fourier-Series Coefficients

As an example, let us express the periodic square wave drawn in Figure 9-5 as a Fourier series (i.e., as a sum of harmonically related complex exponential time functions). To evaluate the kth Fourier-series coefficient, we use (9.4). Since the period, T, for this waveform is two seconds, the integral becomes

$$X_k = \frac{1}{2}\int_0^2 x(t)e^{-j\pi kt}\,dt.$$

When we substitute for $x(t)$, the integral falls naturally into three parts:

$$X_k = \frac{1}{2}\int_0^{\frac{1}{2}} e^{-j\pi kt}\,dt - \frac{1}{2}\int_{\frac{1}{2}}^{\frac{3}{2}} e^{-j\pi kt}\,dt + \frac{1}{2}\int_{\frac{3}{2}}^2 e^{-j\pi kt}\,dt$$

$$= -\frac{1}{j2\pi k}e^{-j\pi kt}\Big|_0^{\frac{1}{2}} + \frac{1}{j2\pi k}e^{-j\pi kt}\Big|_{\frac{1}{2}}^{\frac{3}{2}} - \frac{1}{j2\pi k}e^{-j\pi kt}\Big|_{\frac{3}{2}}^2$$

$$= \frac{1}{j2\pi k}\left[1 - e^{-j\frac{\pi}{2}k} + e^{-j\frac{3\pi}{2}k} - e^{-j\frac{\pi}{2}k} - e^{-j2\pi k} + e^{j\frac{3\pi}{2}k}\right].$$

Recognizing that $e^{-j\pi k/2} = (-j)^k$, $e^{-j3\pi k/2} = (j)^k$, and $e^{j2\pi k} = 1$ allows us to rewrite this expression in the simpler form

$$X_k = \frac{1}{j\pi k}[(j)^k - (-j)^k].$$

When k is even, but not zero, this evaluates to zero. (When $k = 0$ it is indeterminate.) For odd values of k, the coefficients are real, alternate in sign, and are inversely proportional to $|k|$. We go back to the definition to evaluate X_0 (the indeterminate case):

$$X_0 = \frac{1}{2}\int_0^2 x(t)\,dt.$$

This integral is seen to be zero, since $x(t)$ has a value of one for exactly one-half of the period and a value of -1 for the other half. Putting everything together gives

$$X_k = \begin{cases} 0, k \text{ even} \\ \frac{2}{\pi k}, k = \ldots, -7, -3, 1, 5, 9, \ldots \\ -\frac{2}{\pi k}, k = \ldots, -5, -1, 3, 7, 11, \ldots \end{cases} \quad (9.5)$$

We can now use the Fourier-series coefficients to express $x(t)$ as a superposition of sinusoids; that is,

$$x(t) = \sum_{m=0}^{\infty} \frac{4(-1)^m}{\pi(2m+1)} \cos([2m+1]\pi t), \quad (9.6)$$

where $k = 2m + 1$. ∎

To check your understanding, try Drill Problem P9-1.

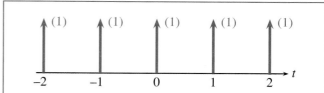

Fig. 9-6. A periodic impulse train whose period is 1 second.

Example 9-2   Fourier Series of an Impulse Train

As a second example, consider the periodic impulse train illustrated in Figure 9-6. The formula for determining the Fourier-series coefficients that we derived requires that we evaluate the integral

$$X_k = \int_0^1 x(t)e^{-j2\pi kt}\,dt;$$

however, that is difficult to do, since there are impulse functions at both the lower and upper limits of integration. Should these impulses be included in the integral or not? To resolve this question we go back to our derivation of that formula. The key step is the integration over one period in (9.3) to isolate one of the harmonic terms. We chose that period to extend from 0 to T for convenience, but it was an arbitrary choice. Had we chosen the range of integration to extend from $-T/2$ to $T/2$, we would have gotten an equally valid formula for the Fourier coefficients, and the value of those coefficients would have been exactly the same. For this example, we exploit this flexibility to replace the period of the waveform used for integration by a more convenient choice. If we use the interval $[0.5, 0.5)$ for our basic period, the kth Fourier-series coefficient can be evaluated by using

$$X_k = \int_{-0.5}^{0.5} \delta(t)e^{-j2\pi kt}\,dt = 1 \quad \text{for all } k. \quad (9.7)$$

(Recall the special techniques for manipulating singularity functions that we saw in Chapter 5.) Therefore,

$$x(t) = \sum_{k=-\infty}^{\infty} e^{j2\pi t}.$$

Figure 9-4 shows some of the partial sums from this expression—the ones containing three, five, and nine of the central harmonics. As more harmonics are added to the partial sums in that figure, we see the waveforms looking more and more like a periodic sequence of impulses—the peaks become taller and narrower and the region between the peaks becomes more nearly zero. ■

 To check your understanding, try Drill Problem P9-2.

9-1-4   Response of a Circuit to a Periodic Input

When the input, $x(t)$, to a single-input, single-output linear circuit is periodic, the output signal, $y(t)$, must also be periodic. This follows from the time invariance of the system. We have already shown that a circuit at initial rest is time invariant.[1] Consider the input–output system representation of the circuit shown in Figure 9-7, in which $x(t)$ denotes the input variable (the waveform associated with an independent source) and $y(t)$ denotes the output variable (one of the element variables in the

---

[1] Initial rest is not an issue when the input signal is periodic, since the input signal begins at $t = -\infty$.

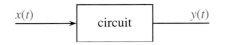

Fig. 9-7. A generic linear circuit with input $x(t)$ and output $y(t)$.

circuit). The fact that $y(t)$ is the output when $x(t)$ is the input can be acknowledged via the notation

$$x(t) \longrightarrow y(t). \tag{9.8}$$

The fact that $x(t)$ is periodic with period T means that

$$x(t) = x(t - T). \tag{9.9}$$

Using (9.8) and time invariance, we know that

$$x(t - T) \longrightarrow y(t - T).$$

The fact that $x(t)$ is periodic, however, implies that

$$x(t) \longrightarrow y(t - T), \tag{9.10}$$

but the same input cannot produce two different outputs ((9.8) and (9.10)). Therefore, these two outputs must be the same, and

$$y(t) = y(t - T),$$

which establishes that $y(t)$ must also be periodic with period T.

Since $x(t)$ and $y(t)$ are both periodic signals with the same period, each can be written as a sum of harmonically related complex exponentials by using a Fourier-series:

$$x(t) = \sum_{k=-\infty}^{\infty} X_k e^{-j\frac{2\pi}{T}kt}$$

$$y(t) = \sum_{k=-\infty}^{\infty} Y_k e^{-j\frac{2\pi}{T}kt}$$

This leads us to the key question for this section: How are the Fourier-series coefficients for the output signal Y_k related to the coefficients for the input signal X_k?

We are already very close to having the answer to this question. First, let the input to the circuit be a single complex exponential with frequency $\omega_k = \frac{2\pi k}{T}$ and phasor X_k for some integer k. We map the circuit to the phasor domain and compute the phasor of the output signal; call this value Y_k. This establishes the input–output pair

$$x(t) = X_k e^{j\frac{2\pi kt}{T}} \longrightarrow y(t) = Y_k e^{j\frac{2\pi kt}{T}}, \qquad (9.11)$$

where

$$\frac{Y_k}{X_k} = H(j\frac{2\pi k}{T}). \qquad (9.12)$$

$H(j\omega)$ is the *frequency response* of the circuit. It is also equal to the system function evaluated at $s = j\omega$. Equation (9.12) says that the phasors associated with the kth components of the input and output are related by the value of the frequency response evaluated at the frequency of that component.

Now let the input signal be a sum of complex exponential time functions. Because the circuit is linear, the output of the circuit when the input is a superposition of components must be the sum of the individual component outputs. The components of our input signal all have the form (9.11) for different values of the integer k:

$$x(t) = \sum_{k=-\infty}^{\infty} X_k e^{j\frac{2\pi kt}{T}}.$$

By linearity, the output is

$$y(t) = \sum_{k=-\infty}^{\infty} Y_k e^{j\frac{2\pi kt}{T}} = \sum_{k=-\infty}^{\infty} H(j\frac{2\pi k}{T}) X_k e^{j\frac{2\pi kt}{T}}. \qquad (9.13)$$

Notice that the output signal is a sum of harmonically related sinusoids; this fact confirms our earlier statement that $y(t)$ is periodic with the same period as the input. Furthermore, the Fourier-series coefficients of the input and output signals are related by the frequency response of the system:

$$Y_k = X_k H(j\frac{2\pi}{T}k), \quad \text{for all } k. \qquad (9.14)$$

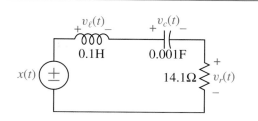

Fig. 9-8. A second-order RLC circuit that acts like a bandpass filter.

Each Fourier-series coefficient is a product of two terms, one of which tells how much of that component is present in the input and the other of which tells how the circuit modifies that component. If a particular component is missing in the input signal, it cannot appear in the output.[2]

To illustrate how a circuit can alter the blend of sinusoidal components in its input, consider the circuit in Figure 9-8. We let the input $x(t)$ be the signal

$$x(t) = 10 + 10\cos(100t - \frac{\pi}{3}) - 4\sin(250t)$$

$$= 10e^{0t} + 5e^{-j\frac{\pi}{3}}e^{j100t} + 5e^{j\frac{\pi}{3}}e^{-j100t}$$

$$+ 2e^{-j\frac{\pi}{2}}e^{j250t} + 2e^{j\frac{\pi}{2}}e^{-j250t}, \qquad (9.15)$$

which contains five complex exponential components (at frequencies $\omega = 0$, $\omega = \pm 100$ rad/s, and $\omega = \pm 250$ rad/s). The signal was drawn in Figure 9-3. The output signal is the voltage across the resistor, $v_r(t)$.

 WORKED SOLUTION: *Frequency Response*

We can write down the output signal $y(t)$ once we know its Fourier-series coefficients, Y_k, which we can compute from (9.14). That formula, in turn, requires that we know the Fourier-series coefficients of the input signal

---

[2]This is a feature of *linear* circuits only. If a circuit contains nonlinear elements, the output may contain harmonics that are not present in the input.

X_k and the frequency response of the circuit $H(j\omega)$. We can read off the Fourier-series coefficients of the input from (9.15). There are only five of these that are nonzero:

$$X_{-5} = 2e^{j\frac{\pi}{2}}; \quad X_{-2} = 5e^{j\frac{\pi}{3}}; \quad X_0 = 10;$$
$$X_2 = 5e^{-j\frac{\pi}{3}}; \quad X_5 = 2e^{-j\frac{\pi}{2}}. \tag{9.16}$$

Notice that the fundamental frequency for this input is $\omega_0 = 50$ rad/s, which corresponds to $T = \pi/25$. The frequencies of all of the sinusoids are integer multiples of this frequency, and this is the largest positive value for which that statement can be made. There is no input component at the fundamental frequency, however.

The next step is to compute the frequency response of the circuit, $H(j\omega)$, at the frequencies of the harmonics. One straightforward approach to use here is to exploit the relationship between the system function and the frequency response, namely,

$$H(j\omega) = H(s)|_{s=j\omega},$$

to compute the frequency response for all frequencies. We can then get the gains for the various harmonics by selecting the appropriate frequency values. We get the system function by using a voltage divider:

$$H(s) = \frac{R}{Ls + R + \frac{1}{Cs}}$$

$$= \frac{141s}{s^2 + 141s + 10,000}. \tag{9.17}$$

Combining (9.14), (9.16), and (9.17) gives the Fourier-series coefficients, Y_k:

$$Y_{-5} = X_{-5}H(-j250) = \left(2e^{j\frac{\pi}{2}}\right)\left(0.5574e^{j(0.9745)}\right)$$

$$= 1.114e^{j2.5503}$$

$$Y_{-2} = X_{-2}H(-j100) = \left(5e^{j\frac{\pi}{3}}\right)(1) = 5e^{j(1.0472)}$$

$$Y_0 = X_0 H(0) = (10)(0) = 0$$

$$Y_2 = X_2 H(j100) = \left(5e^{-j\frac{\pi}{3}}\right)(1) = 5e^{-j(1.0472)}$$

$$Y_5 = X_5 H(j250) = \left(2e^{-j\frac{\pi}{2}}\right)\left(0.5574e^{-j(0.9745)}\right)$$

$$= 1.114e^{-j2.5503}.$$

Constructing the output signal from its Fourier-series coefficients gives

$$y(t) = v_r(t)$$
$$= 10\cos(100t - 1.0472)$$
$$+ 2.228\cos(250t - 2.5503).$$

This signal is plotted in Figure 9-9(c). A similar computation for the voltages across the capacitor and the inductor gives the frequency responses

$$H_c(j\omega) = \frac{10^4}{-\omega^2 + j141\omega + 10^4}$$

$$H_\ell(j\omega) = \frac{-\omega^2}{-\omega^2 + j141\omega + 10^4},$$

from which we can compute the voltage waveforms:

$$v_c(t) = 10 - 7.09\cos(100t - 0.5231)$$
$$- 0.63\sin(250t - 2.5503) \tag{9.18}$$
$$v_\ell(t) = 7.09\cos(100t + 0.5231)$$
$$- 3.95\sin(250t + 0.5913). \tag{9.19}$$

These waveforms and their spectra are also shown in Figure 9-9.

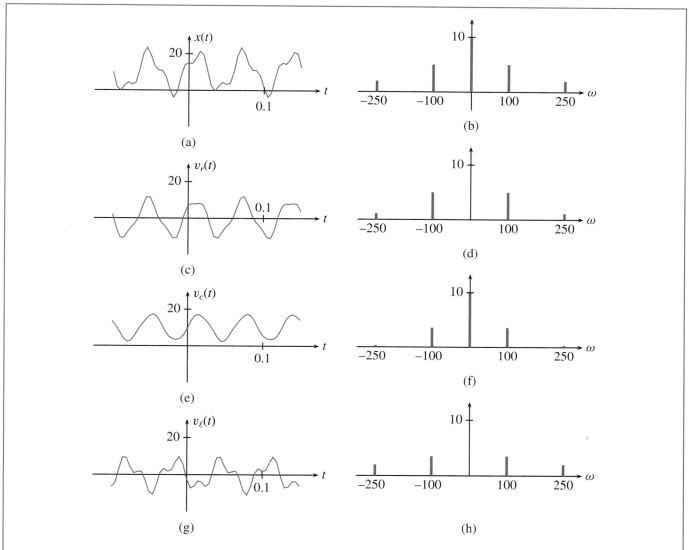

Fig. 9-9. Voltages in the circuit in Figure 9-8. (a) Input signal, $x(t)$. (b) Magnitude spectrum of $x(t)$. (c) $v_r(t)$. (d) Magnitude spectrum of $v_r(t)$. (e) $v_c(t)$. (f) Magnitude spectrum of $v_c(t)$. (g) $v_\ell(t)$. (h) Magnitude spectrum of $v_\ell(t)$.

The average value of a sinusoid with a nonzero frequency is zero; for one-half of the period it is negative, and for the other half it is positive. This is not the case for a zero-frequency sinusoid, which has a constant value. When several sinusoids at different frequencies are added, the average value of the sum will thus be the same as the coefficient of the zero-frequency sinusoid. This is called the *DC (direct current) value* for historical reasons. The DC value can readily be seen in the spectrum. Notice that $x(t)$ and $v_c(t)$ have average (DC) values of ten, while $v_r(t)$ and $v_\ell(t)$ have DC values of zero. This is readily seen in the time-domain plot of those waveforms in Figure 9-9. The components of the waveforms at ± 250 rad/s are the highest-frequency components. This component is largest for $v_\ell(t)$, which is the least smooth of the three element voltages, and it is smallest for $v_c(t)$, which is the smoothest of the three.

9-2   Spectrum Representations for Aperiodic Signals

9-2-1   The Fourier Transform

Most aperiodic signals can also be represented as superpositions of complex exponential time functions. This is the basis for the *Fourier-transform representation of a signal.* In this section, we are going to touch only lightly on the topic of Fourier transforms. Our intent is solely to provide additional motivation for looking at the frequency responses of circuits. The Fourier transform, like the Fourier-series, will be studied more extensively in later courses.

Let $x(t)$ be an absolutely integrable signal—that is, one for which the integral

$$S = \int_{-\infty}^{\infty} |x(t)|\, dt$$

is finite. Then $x(t)$ can be written as a superposition of sinusoids:

$$x(t) = \frac{1}{2\pi} \int_{-\infty}^{\infty} X(\omega) e^{j\omega t}\, d\omega. \qquad (9.20)$$

$X(\omega)$ is called the *Fourier-transform of $x(t)$.* Intuitively, it tells how the sinusoids are weighted and delayed to construct $x(t)$. $X(\omega)$ can be computed from $x(t)$ by evaluating the integral

$$X(\omega) = \int_{-\infty}^{\infty} x(t) e^{-j\omega t}\, dt. \qquad (9.21)$$

Often, (9.21) is called the *forward Fourier-transform integral* and (9.20) is called the *inverse Fourier-transform integral*. Notationally, we use a two-headed arrow to denote a Fourier transform pair. Thus,

$$x(t) \longleftrightarrow X(\omega)$$

means that $x(t)$ and $X(\omega)$ are related by (9.20) and (9.21).

If we replace the quantity $j\omega$ by the variable s, the Fourier transform integral looks very much like the Laplace transform. One visible difference, however, concerns the lower limit of integration. This gives the two transforms somewhat different properties. A second difference concerns the fact that the inverse Fourier transform can be computed by evaluating an integral, whereas computing an inverse Laplace transform was done by performing a partial-fraction expansion and consulting a table of known transform pairs. An even bigger difference, however, concerns the way in which the two transforms are used. The Fourier transform is used primarily to describe signals, whereas the Laplace transform is used primarily to understand properties of systems (circuits). It is important to notice that when $x(t) = 0$ for $t < 0$, the two transforms become identical after the change of variables. This is frequently useful. For example, it allows us to compute inverse Fourier transforms by using partial-fraction expansions, when appropriate.

There are two properties of the Fourier transform that we need in order to turn it into a tool for analyzing linear circuits. First the transform is linear. If $x_1(t)$ and $x_2(t)$ are two signals with Fourier transforms $X_1(\omega)$ and $X_2(\omega)$, respectively, then

$$ax_1(t) + bx_2(t) \longleftrightarrow aX_1(\omega) + bX_2(\omega)$$

for any (possibly complex) values of a and b. In addition, if the Fourier transform of $x(t)$ is $X(\omega)$, then the Fourier transform of its derivative is

$$\frac{dx(t)}{dt} \longleftrightarrow j\omega X(\omega).$$

This can be proven by evaluating the derivative of the inverse-Fourier-transform definition (9.20) with respect to the time variable.

9-2-2   Circuit Analysis by Using Fourier Transforms

The two properties of the Fourier transform enumerated above are all that we need to establish the fact that, in the frequency domain, the Fourier transforms of the element variables in a circuit satisfy KCL, KVL, and Ohm's Law. Establishing these facts is very similar to the equivalent derivations that were done for find the Laplace transform in Section 6-2 or for the sinusoid at a fixed frequency in Section 8-2-2. We will not go through the details for establishing KCL and KVL of the behavior of resistors, because they are so similar. Instead, we consider only the V–I relations for inductors and capacitors, where there are slight differences.

For an inductor with the element relation

$$v_L(t) = L\frac{di_L(t)}{dt},$$

we compute the Fourier transform of both sides, using the derivative property, to see that

$$V_L(\omega) = j\omega L I_L(\omega). \tag{9.22}$$

Similarly, for a capacitor, the frequency-domain element relation has the form

$$I_C(\omega) = j\omega C V_C(\omega). \tag{9.23}$$

Thus, in the frequency domain, all of our familiar elements have a voltage variable that is proportional to the current variable:

$$V(\omega) = Z(j\omega)I(\omega). \tag{9.24}$$

The impedances for the three element types are

$$Z_R(j\omega) = R; \quad Z_L(j\omega) = j\omega L; \quad Z_C(j\omega) = \frac{1}{j\omega C}.$$

These are similar to the impedances in the Laplace domain, except for the change of variables $s = j\omega$. Notice that, in (9.22) and (9.23), the frequency and Laplace domains are slightly different because there is no auxiliary term in the frequency-domain expressions to account for initial conditions. This is because the component signals that are used to synthesize waveforms in the frequency (Fourier) domain are complex exponential time functions, which begin at $t = -\infty$.

Now consider mapping an arbitrary linear circuit from the time domain to the frequency domain. Each time waveform is replaced by its Fourier transform, and each element is replaced by an impedance with the appropriate value that varies as a function of ω. To find all of the element variables (in the frequency domain), we write out the usual set of KCL equations, KVL equations, and element relations and solve. If instead we mapped the circuit to the *Laplace* domain with all initial conditions set to zero, we would get a topologically equivalent circuit, except that the s would appear in the Laplace domain everywhere that $j\omega$ occurred in the frequency domain. The KCL equations would be identical, the KVL equations would be identical, and the element relations would be identical. As a result, the solutions to the two systems must be identical! In the Laplace domain, we know that the Laplace transform of the output variable is equal to the Laplace transform of the input variable multiplied by the system function

$$Y(s) = X(s)H(s).$$

Therefore, in the frequency domain, the (Fourier transform of the) output variable is equal to the (Fourier transform of the) input variable multiplied by the system function with s replaced by $j\omega$:

$$Y(\omega) = X(\omega)H(j\omega). \qquad (9.25)$$

Notice that while $H(j\omega)$ is equal to the system function with the change of variables $s = j\omega$, the Fourier transform of the input variable $X(\omega)$ is not necessarily equal to $X(s)$ with the same change of variables. The Laplace transform and the Fourier transform were defined with different lower limits of integration, which means that these two transforms will be quite different if $x(t)$ is nonzero for negative values of t.

The inverse Fourier transform in (9.20) told us that most signals could be expressed as superpositions of complex exponential time functions. The Fourier transform $X(\omega)$ tells us how these sinusoids in the input signal are weighted before they are combined. $Y(\omega)$ provides similar information for the output signal. Equation (9.25) says that these two weighting functions are related by the frequency response of the circuit. Equation (9.14) provides similar information for periodic input and output signals.

Example 9-3   A First-Order Lowpass Filter

Consider the simple RC circuit shown in Figure 9-10. To find the frequency response $H(j\omega)$ that relates the

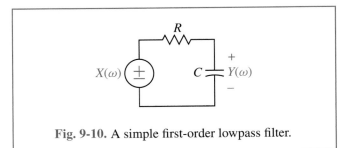

Fig. 9-10. A simple first-order lowpass filter.

Fourier-transform of the output $Y(\omega)$ to the Fourier transform of the input $X(\omega)$, we use a voltage divider:

$$Y(\omega) = X(\omega) \cdot \frac{Z_C(j\omega)}{Z_C(j\omega) + Z_R(j\omega)} = X(\omega) \cdot H(j\omega)$$

$$H(j\omega) = \frac{Y(\omega)}{X(\omega)} = \frac{\frac{1}{jC\omega}}{\frac{1}{jC\omega} + R} = \frac{1}{1 + jRC\omega}. \qquad (9.26)$$

H is a complex function of ω. Its magnitude and phase as a function of frequency are known as the *magnitude response* and *phase response* of the system, respectively. The magnitude and phase response of this network for this output signal are given by

$$|H(j\omega)| = \frac{1}{\sqrt{1 + (RC\omega)^2}}$$

$$\angle H(j\omega) = \tan^{-1}\frac{\Im m(H(j\omega))}{\Re e(H(j\omega))} = -\tan^{-1}(RC\omega),$$

which are plotted in Figure 9-11.

This is an example of a *lowpass filter*, because low-frequency components (those for which $\omega \ll \frac{1}{RC}$) appear in the output with a magnitude gain that is close to 1, while high-frequency components (those for which $\omega \gg \frac{1}{RC}$) have their magnitudes attenuated. The cutoff frequency, $1/RC$, is controlled by the time constant of the filter. This is a *first-order* filter because it contains only a single reactive element. ∎

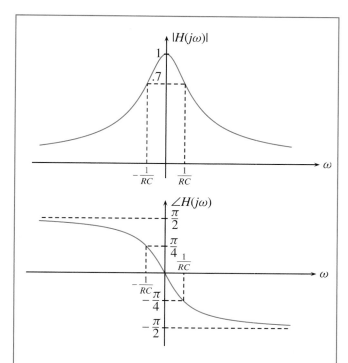

Fig. 9-11. The magnitude and phase response of the first-order lowpass filter shown in Figure 9-10.

    DEMO: *First-Order Lowpass filter*

    To check your understanding, try Drill Problem P9-3.

Example 9-4   A First-Order Highpass Filter

If we let the output variable be the voltage across the resistor instead of that across the capacitor in the previous example, we get a first-order highpass filter. This is complementary to the lowpass of the first example, in the sense that the sum of the frequency responses of the lowpass and the highpass is exactly equal to one. The circuit is redrawn in Figure 9-12.

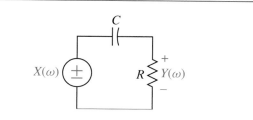

Fig. 9-12. A simple first-order highpass filter.

Again, we can compute the frequency response by using a voltage divider:

$$H(j\omega) = \frac{Y(\omega)}{X(\omega)} = \frac{Z_R(j\omega)}{Z_C(j\omega) + Z_R(j\omega)}$$

$$= \frac{R}{\frac{1}{jC\omega} + R} = \frac{jRC\omega}{1 + jRC\omega}.$$

This has the magnitude response

$$|H(j\omega)| = \frac{RC|\omega|}{\sqrt{1 + (RC\omega)^2}}$$

and the phase response

$$\angle H(j\omega) = \frac{\pi}{2}\text{sgn}(\omega) - \tan^{-1}(RC\omega),$$

which are plotted in Figure 9-13. The signum function, sgn(x), is defined by

$$\text{sgn}(x) = \begin{cases} 1, & x > 0 \\ -1, & x < 0. \end{cases}$$

This is called a *highpass filter*, because the high-frequency components of the input signal appear in the output with a gain that is approximately one, while the low-frequency components are attenuated.   ∎

    To check your understanding, try Drill Problem P9-4.

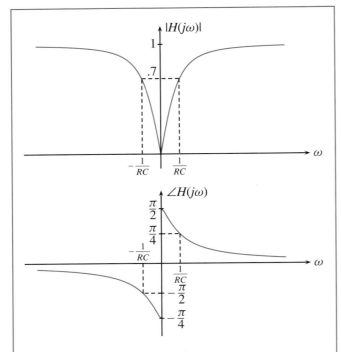

Fig. 9-13. The magnitude and phase response of the first-order highpass filter shown in Figure 9-12.

9-3 The System Function and the Frequency Response

It should be clear by now that the system function is key to understanding circuit behavior. By itself, it is a useful analysis tool. Through its specialization to the frequency response, it also tells how a circuit will respond to any superposition of complex exponential time functions. The inverse Laplace transform of the system function, which is the impulse response, provides important clues about the time-domain behavior of the circuit. Its linear combination of exponentials characterizes the transient behavior of the circuit. The exponential terms that make up the transient behavior, in turn, come from the poles of the system, which (no surprise) can also be derived from

the system function.

Because the transient behavior of a circuit and its frequency response are both tied to the system function, these two attributes of circuit behavior are not independent. This is often a source of frustration in those filter-design problems in which we want to control both the time-domain and the frequency-domain aspects of a circuit's behavior. In Chapter 7, we looked at the impulse and step responses of several first- and second-order circuits and saw how those depended on the locations of the system poles and zeros. Now we would like to relate the poles and zeros to the frequency response.

9-3-1 Graphical Interpretation of Frequency Response

Consider the system function that relates the input and output of a linear circuit. This is a rational function of the variable s:

$$H(s) = \frac{B(s)}{A(s)}$$

$$= \frac{b_m s^m + b_{m-1} s^{m-1} + \cdots + b_1 s + b_0}{a_n s^n + a_{n-1} s^{n-1} + \cdots + a_1 s + a_0}$$

$$= K \frac{(s - z_1)(s - z_2) \cdots (s - z_m)}{(s - p_1)(s - p_2) \cdots (s - p_n)}. \qquad (9.27)$$

To be completely general, we have assumed that the system contains n poles and m zeros; thus, it is an nth-order system. Equation (9.27) expresses $H(s)$ directly in terms of its poles and zeros. The gain constant K is equal to b_m/a_n. The pole–zero plot is shown in Figure 9-14. As in Chapter 7, the zeros are represented by circles and the poles by crosses. The coordinate axes are $\sigma = \Re\mathrm{e}(s)$ and $\omega = \Im\mathrm{m}(s)$. Since the coefficients of $B(s)$ and $A(s)$ are real (because they are products of element values, which are real), for every pole or zero that is complex, its complex conjugate will also be a pole or zero. (See Problem P5-18.) This means that the pole–zero plot will always display symmetry across the σ-axis.

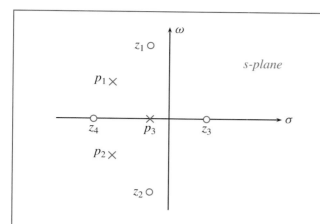

Fig. 9-14. An example of a pole–zero plot of a stable circuit.

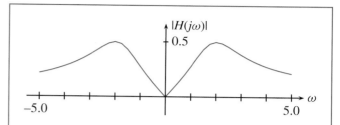

Fig. 9-15. The magnitude response of a second-order bandpass filter.

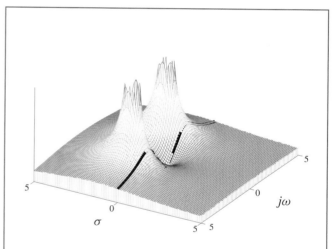

Fig. 9-16. s-plane plot of $|H(s)|$ for a second-order bandpass filter.

Furthermore, if the circuit is stable, the real parts of all of the pole values must be negative. This places all of the poles in the left half of the s-plane. The pole–zero plot in Figure 9-14 displays both of these features.

The frequency response is equal to the system function when $s = j\omega$. This corresponds to evaluating $H(s)$ on the ω-axis of the s-plane. Consider the simple system function

$$H(s) = \frac{s}{s^2 + 2s + 4}.$$

This is an example of a second-order bandpass filter , because it attenuates both the low and high frequencies; the gain is significant only for an intermediate range of frequencies. This is seen in the plot of $|H(j\omega)|$ in Figure 9-15.

A three-dimensional plot of $|H(s)|$ shows the expected behavior (Figure 9-16); the magnitude of $H(s)$ becomes infinite at the poles and is pinned to zero at the zero. The rest of the s-plane looks like an elastic membrane that has been deformed by the poles and zeros. The frequency response is equal to $|H(s)|$ evaluated on the ω-axis. That particular contour is shown highlighted. It is clear that as

the poles are moved closer to the $j\omega$-axis, the maximum value of the frequency response becomes higher; as they are moved farther away, it decreases.

The magnitude plot in Figure 9-15 was produced by using the function `freqs` in MATLAB to evaluate an expression like (9.27). This works well when we need an accurate plot, but we can also produce a simple sketch of the frequency response by exploiting the relationship between the frequency response and the pole–zero plot. Not only does this enable us to develop some intuition

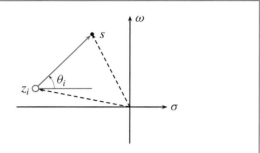

Fig. 9-17. The vector corresponding to $(s - z_i)$ in the complex plane is the difference between the vector corresponding to the point s and the vector corresponding to z_i.

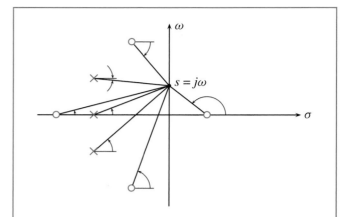

Fig. 9-18. Vector interpretation of the frequency response.

about how a particular circuit behaves, it often allows us to design circuits by carefully specifying the locations where they should have poles and zeros.

Each of the factors of the form $(s - z_i)$ or $(s - p_i)$ that appears in (9.27) can be interpreted as a complex number with a magnitude and angle that vary as s varies. Figure 9-17 shows a portion of the s-plane containing a single zero at location z_i and an arbitrary point in the plane denoted by the point at s. Consider the points z_i and s to be vectors in the complex s-plane that extend from the origin to the points z_i and s, respectively. These are shown as dashed lines on that figure. Then $(s - z_i)$ is the vector that extends from the point z_i to the point s shown in the figure. Its length is $|s - z_i|$, and $\angle(s - z_i)$ is its angle θ_i, measured with respect to the σ-axis. As s is allowed to move around the plane, the magnitude and angle vary, but we can track these by tracking the magnitude (length) and angle of the difference vector.

Now consider a general circuit with the system function in (9.27). To get the frequency response, we set $s = j\omega$. If we factor the numerator and denominator polynomials and compute the magnitude, we arrive at the

following expression for the magnitude response:

$$|H(j\omega)| = |K|\frac{|j\omega - z_1| \cdot |j\omega - z_2| \cdots |j\omega - z_m|}{|j\omega - p_1| \cdot |j\omega - p_2| \cdots |j\omega - p_n|}$$

$$= \frac{\text{product of magnitudes of vectors from zeros}}{\text{product of magnitudes of vectors from poles}}.$$

These magnitudes are the lengths of the lines in Figure 9-18. To find the frequency response at any frequency, we find the point on the $j\omega$-axis corresponding to that frequency, compute the product of the distances from that point to each of the zeros, and divide the result by the product of the distances to each of the poles. The graphical computation of the phase response is similar. From (9.27),

$$\angle H(j\omega) = \angle K + \sum \text{angles of vectors from zeros to } j\omega$$

$$- \sum \text{angles of vectors from poles to } j\omega.$$

Graphical computation of the frequency response from the pole–zero plot is particularly useful when the poles or zeros lie close to the $j\omega$-axis. In this case, the local behavior is most closely controlled by the closest pole.

Consider the following two examples, the first of which is a first-order lowpass filter.

 DEMO: *Graphical Interpretation*

Example 9-5   **A First-Order Lowpass Filter**

A first-order lowpass filter has a single real pole and a system function of the form

$$H(s) = \frac{a}{s+a}.$$

This is a generalization of the circuit we examined in Example 9-3. Its pole–zero plot is shown in Figure 9-19. Since this system has only a single pole, its frequency response is inversely proportional to the length of the vector drawn from the pole to the point on the imaginary axis that corresponds to the particular frequency ω. The phase is equal to $-\theta$, which also varies with ω.

The minimum distance from the pole to the imaginary axis occurs when $\omega = 0$. Therefore, the frequency response has a maximum value at this frequency, and the maximum value is equal to $a/a = 1$. Furthermore, we see that when $\omega = 0$, $\theta = 0$. As ω moves up the ω-axis, the distance to the pole increases monotonically and the angle, θ, approaches $\pi/2$. At $\omega = \infty$, the length of the vector is infinite, the magnitude of the frequency response is zero, and the phase is equal to $-\pi/2$. At $\omega = a$, the triangle with its vertices at the pole, at the point $\omega = a$, and at the origin is a $45°$ right triangle. From the Pythagorean theorem, its hypotenuse has length $a\sqrt{2}$. Therefore, at $\omega = a$, the magnitude of the frequency response is

$$|H(ja)| = \frac{a}{\sqrt{2}a} = \frac{1}{\sqrt{2}}$$

and the phase is $-\pi/4$. This is called the *half-power frequency* of the low-pass filter, because the square of the magnitude of the frequency response is one-half of its

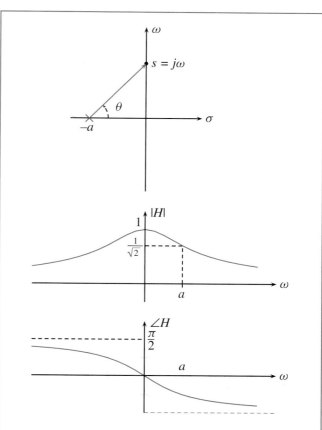

Fig. 9-19. The pole–zero plot of a first-order lowpass filter is shown at the top, and its magnitude and phase responses are shown below. The magnitude response is inversely proportional to the length of the vector from the pole to the point $j\omega$ as that point moves up and down the imaginary axis. The phase response tracks the negative of its angle (because it comes from a pole).

maximum value at this frequency. As a is made smaller, the half-power frequency becomes smaller and the filter rejects more of the high frequencies; as a is made larger, it passes more of them. ∎

 To check your understanding, try Drill Problem P9-5.

Example 9-6   A Second-Order Bandpass Filter

As a second example, consider a second-order bandpass filter with the system function

$$H(s) = \frac{2as}{(s+a)^2 + b^2}$$

($b \gg a$) and having the pole–zero plot shown in Figure 9-20. To avoid cluttering up the pole–zero diagram, none of the vectors connecting the poles and zeros to an arbitrary point on the $j\omega$-axis have been plotted, but the frequency response at any value of ω is still proportional to the distance from that point to the zero divided by the product of the distances to the two poles. To get a feeling for how the frequency response behaves, we need consider only three parts of the frequency axis: frequencies near zero, frequencies near $\omega = b$, and very large frequencies. We also need consider only the positive frequencies, because the magnitude response is a symmetric (even) function of ω and the phase response is an antisymmetric (odd) function. Let's deal with the magnitude response first.

At $\omega = 0$, the zero guarantees that the magnitude response is zero, and this zero dominates the magnitude response in this region. As ω increases, the magnitude response increases linearly, because the distance to the zero is $|\omega|$ and the relative change in the distances to the poles is small, so these distances are nearly constant.

Now consider the magnitude response at a very large frequency $\omega = \omega_\infty$, where $\omega_\infty \gg b$. The distance to the zero is ω_∞, and the distance to each pole is approximately ω_∞ as well. Therefore,

$$H(\omega_\infty) \approx \frac{2a\omega_\infty}{\omega_\infty^2} = \frac{2a}{\omega_\infty},$$

and we would expect the magnitude response to decrease as $1/\omega$ for large frequencies.

More interesting behavior occurs at frequencies near $\omega = b$. In particular, consider how the frequency

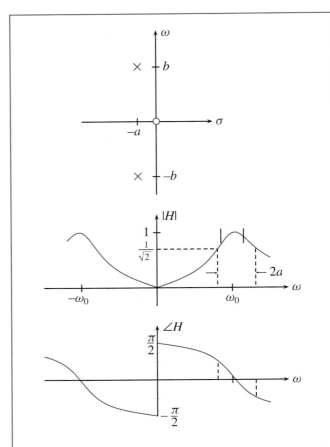

Fig. 9-20. The pole–zero plot, magnitude response, and phase response of a second-order bandpass filter, an illustration of the real and imaginary parts of the pole locations affect the frequency response.

response changes as ω varies from $b - a$ to $b + a$. Since $b \gg a$, the distance to the zero is approximately b and the distance to the farther pole is approximately $2b$. The effect of these two poles is to impose a nearly constant gain of about 0.5 over the entire interval. The distance to the closer pole, however, is far from constant and it has a major effect on the frequency response. The distance to that pole reaches a minimum value at $\omega = b$, and thus we expect the frequency response to have its maximum value at approximately that frequency. When the effect of the more distant pole and the zero are taken into effect, the exact frequency at which the maximum value of the frequency response occurs is seen to be

$$\omega_0 = \sqrt{a^2 + b^2},$$

but $\omega_0 \approx b$, if $a \ll b$. At $\omega = b \pm a$, the distance to the upper pole is $a\sqrt{2}$; thus, these points are approximately equal to the two positive half-power frequencies. We control the frequency at which the filter has its maximum gain by varying b, the imaginary part of the pole location, and we control the range of frequencies over which the magnitude response is significant (called the *passband of the filter*) by controlling a, the real part of the pole location.

Now consider the phase response. At large frequencies, the angles from the zero and from both of the poles are approximately the same and equal to $\pi/2$. Therefore, the phase approaches the value $-\pi/2\,(=\pi/2 - 2(\pi/2))$. At $\omega = b$, the angle from the zero is $\pi/2$, the angle from the farther pole is approximately $\pi/2$, and the angle from the nearer pole is zero. Therefore, the phase value should be approximately zero. (It is exactly zero at $\omega = \omega_0$.) Furthermore, the phase associated with the nearer pole will vary from $-\pi/4$ radians to $\pi/4$ radians as ω moves from $b - a$ to $b + a$, the two half-power frequencies. The closer the pole lies to the $j\omega$-axis, the more rapid will be the phase transition (i.e, its two positive half-power frequencies will be closer together). At $\omega = 0$, the phase response is discontinuous. This is because the contribution to the phase made by the zero is $\pi/2$ for all positive values of ω and is $-\pi/2$ for all negative values of ω. This discontinuity in the phase does not cause a problem, since the magnitude response is zero at this frequency.

The important points to notice from this example are that (1) the behavior of the frequency response is dominated by the poles and zeros that lie close to or on the $j\omega$-axis and (2) that the effect of those poles and zeros is most important at the frequencies that lie close to them.

■

 To check your understanding, try Drill Problem P9-6.

Some additional examples of pole–zero plots and their associated magnitude responses are shown in Figure 9-21. The gain factor applied to the magnitude plots is arbitrary. The corresponding phase plots are left as an exercise. (See Problem P9-15.)

                    DEMO: *Pole-Zero Plots*

9-3-2   Resonance

First-order circuits contain only a single independent inductor or capacitor. Although they may have an unlimited number of resistors and dependent sources, Thévenin's theorem tells us that all first-order circuits are network equivalent to a circuit with a single reactive element, a single resistor, and a single independent source. We have seen examples of first-order circuits that have lowpass and highpass frequency responses, but, beyond these, the types of frequency responses that we can get with a first-order circuit are quite limited.

To get a bandpass frequency response requires at least a second-order circuit. We looked at a pole–zero pattern associated with this behavior in Example 9-6, but we haven't looked at the circuits themselves. One example of a second-order bandpass filter is the parallel RLC circuit shown in Figure 9-22. To learn its system function, we apply a current divider:

$$H(s) = \frac{I_{\text{out}}(s)}{I_{\text{in}}(s)} = \frac{\frac{1}{R}}{\frac{1}{R} + \frac{1}{Ls} + Cs}$$

$$= \frac{\frac{1}{RC}s}{s^2 + \frac{1}{RC}s + \frac{1}{LC}}. \qquad (9.28)$$

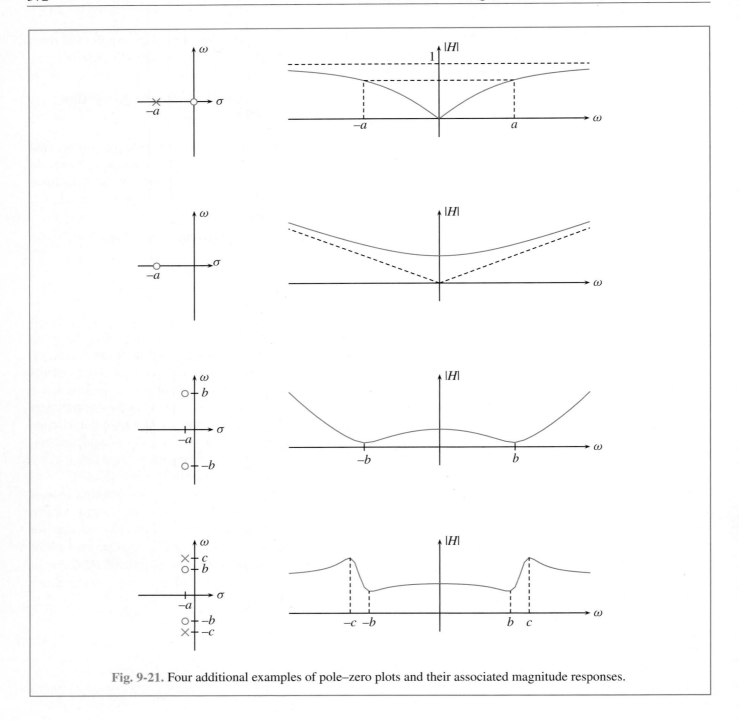

Fig. 9-21. Four additional examples of pole–zero plots and their associated magnitude responses.

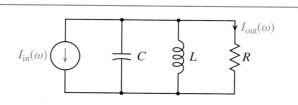

Fig. 9-22. A second-order RLC bandpass filter.

Depending on the values of R, L, and C, this system can have two distinct real poles, a double real pole, or two complex poles that are complex conjugates of each other. The conditions for these three types of behavior are summarized in the table below.

distinct real poles	$R < \frac{1}{2}\sqrt{\frac{L}{C}}$
double real pole	$R = \frac{1}{2}\sqrt{\frac{L}{C}}$
complex poles	$R > \frac{1}{2}\sqrt{\frac{L}{C}}$

As we stated in Chapter 7, when there are two real poles the system is said to be *overdamped*; when there are two complex poles, it is said to be *underdamped*; and when there is a double pole, it is said to be *critically damped*.

The most interesting case is the underdamped one, when there are two complex poles. In this situation, the circuit is also said to be *in resonance*. If we take the inverse Laplace transform, we see that the impulse response has the form of a switched damped sinusoid, as shown in Figure 9-23. If we define

$$\omega_d^2 = \frac{1}{LC} - \left(\frac{1}{2RC}\right)^2,$$

then the precise expression for the impulse response is

$$h(t) = \frac{1}{RC}e^{-t/2RC}\cos\left(\omega_d t + \tan^{-1}\frac{1}{2RC\omega_d}\right)u(t).$$

$$(9.29)$$

  DEMO: *Resonance*

We can understand a little bit about the behavior of the circuit by looking at the underlying physics. The circuit stores energy alternately in the magnetic field of the inductor and in the electric field of the capacitor; that energy is passed back and forth between the two devices, and on each cycle a fraction is absorbed through the resistor and dissipated as heat. In this respect, the circuit is analogous to a roller coaster, in which the total energy in the car alternates between potential and kinetic energy, but where each peak has lower total energy than the previous one because of frictional losses. The rate at which the signal decays is determined by the time constant of the exponential envelope in (9.29), $\tau = 2RC$. The smaller the value of RC, the faster the decay of the impulse response. Intuitively, the more of the inductor current that flows to the resistor, the faster the dissipation of the total energy. Since the capacitor and the resistor are connected in parallel, more current will flow to the resistor if its impedance is reduced (i.e., if R is made small) or if the impedance of the capacitor is increased (i.e., if C is made small, since the impedance of a capacitor is inversely proportional to its capacitance).

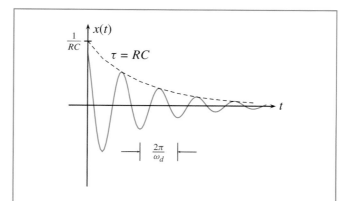

Fig. 9-23. The impulse response of the second-order filter shown in Figure 9-22.

In the underdamped case, this circuit has a pole–zero plot that is similar to the example in Figure 9-20. There is a single zero at $s = 0$ and a pair of complex poles at

$$s_{1,2} = -\frac{1}{2RC} \pm j\sqrt{\frac{1}{LC} - \left(\frac{1}{2RC}\right)^2}.$$

To relate this circuit to that figure, we notice that

$$a = \frac{1}{2RC}$$

$$b = \omega_d = \sqrt{\frac{1}{LC} - \left(\frac{1}{2RC}\right)^2}$$

$$\omega_0 = \sqrt{a^2 + b^2} = \frac{1}{\sqrt{LC}}.$$

We can obtain the frequency response of the circuit if we let $s = j\omega$ in (9.28). It can be written in the useful form

$$H(j\omega) = \frac{1}{1 + jR(\omega C - \frac{1}{\omega L})}.$$

The maximum value of the frequency response occurs when the imaginary part of the denominator is zero—that is, when

$$C\omega = \frac{1}{L\omega}.$$

Solving for ω gives

$$\omega = \frac{1}{\sqrt{LC}} = \omega_0.$$

This is called the **undamped natural frequency** or the **resonant frequency** of the circuit. It is the frequency at which the impulse response would oscillate if R were infinite. At this frequency, the frequency response is one (i.e., the magnitude is 1 and the phase is 0). The quantity ω_d is sometimes called the **damped natural frequency of the circuit.**

The width of the passband for this filter is defined as the difference between the two half-power points.

$$\mathrm{BW} = (\omega_0 + a) - (\omega_0 - a) = 2a = \frac{1}{RC}$$

We see that this is directly related to the exponential envelope of the impulse response. The narrower the passband of the filter, the slower is the decay of the impulse response and the greater is the degree of resonance. In practical terms, this means that circuits that are highly selective, such as narrow bandpass filters, have transient responses that are relatively long.

A good measure of the selectivity or "sharpness" of the peak in a resonant system is the ratio of the resonant frequency to the bandwidth. This ratio is called the **quality factor**, Q, of the network. For this circuit,

$$Q = \frac{\omega_0}{1/RC} = RC\omega_0 = R\sqrt{\frac{C}{L}}. \qquad (9.30)$$

The bandpass filter in Figure 9-22 was constructed from only resistors, inductors, and capacitors. As such, it is an example of a **passive** circuit. A passive circuit has the following attributes:

- If the input and output variables are both voltages or both currents, $|H(j\omega)|$ is almost always less than one for all ω.
- The filter is guaranteed to be stable, even if the values of the circuit elements deviate from their nominal values.
- The circuit does not require any sources of power. The only power delivered to the circuit comes from the input source.
- To have a bandpass frequency response, a passive circuit must contain at least one inductor *and* one capacitor.

The first three of these characteristics are normally considered to be advantages, but the latter can be a disadvantage, because inductors are bulky components that often have nonideal characteristics. Large inductors are difficult to build on integrated circuits.

To avoid using inductors, we can build an active second-order bandpass filter, using only resistors, capacitors, and an operational amplifier. Figure 9-24 shows an example of an active second-order bandpass filter. It is left as an exercise for the reader to verify that the system function of this circuit is

$$H(s) = \frac{V_{\text{out}}(s)}{V_{\text{in}}(s)} = \frac{\frac{k}{RC}s}{s^2 + \frac{3-k}{RC}s + \frac{1}{R^2C^2}}. \qquad (9.31)$$

Its natural frequency is

$$\omega_0 = \frac{1}{RC},$$

and its quality factor is

$$Q = \frac{1}{3-k}.$$

This is not a passive filter, because the presence of the operational amplifier creates a need for it to be

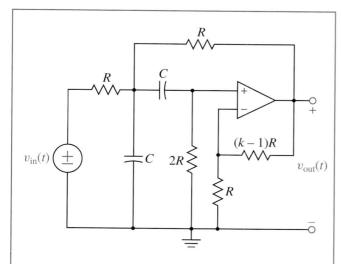

Fig. 9-24. A more complex second-order bandpass circuit containing an operational amplifier.

connected to a source of power, and that need might be a disadvantage. The maximum value of the frequency response is

$$H_{\text{max}} = \frac{k}{3-k} > 1, \text{ if } k > 1.5,$$

and the filter will be unstable if $k \geq 3$. Notice that we must have $k > 1$, since the feedback resistor associated with the opamp must have a positive resistance.

9-4   Bode Plots

Plots of the magnitude and phase responses of a circuit, as we have been making them, while useful for some purposes, are inadequate for others. They have two particular shortcomings: they often fail to capture critical information when the range of frequencies of interest varies over several orders of magnitude, and they often fail to quantify the stopband (the range of frequencies where the magnitude of the frequency response is close to zero) behavior of the filter; for example it might not be enough for the stopband response to simply be *near* zero. A magnitude of 0.001 might be acceptable, whereas one of 0.01 might not, but this information is not visible on a plot of the magnitude response. A human being with normal hearing can hear sinusoidal signals with a frequency as low as 20Hz and as high as 20kHz, a variation of three orders of magnitude. Manufacturers of audio equipment, such as amplifiers and loudspeakers, often use plots of the magnitude response of their equipment to describe its behavior and to allow a consumer to compare their product with that of a competitor. These plots are generally log–log plots: A quantity proportional to the logarithm of the magnitude response is plotted as a function of the logarithm of the frequency. The logarithmic frequency axis allows the frequency response to vary over several orders of magnitude on one plot. The fact that human perception is also logarithmic in its sensitivity adds to the relevance of these plots.

A Bode plot is a plot of the log-magnitude and phase of the frequency response of a circuit versus $\log_{10} \omega$ (or $\log_{10} f$). Only positive frequency values are plotted. Base-10 logarithms are used, for historical reasons, and the log-magnitude value is plotted in decibels (dB); the value of the log-magnitude in dB is given by $20 \log_{10}(|H(j\omega)|)$. Thus, the Bode magnitude plot is a plot of $20 \log_{10}(|H(j\omega)|)$ vs. $\log_{10} \omega$, and the Bode phase plot is a plot of $\angle H(j\omega)$ vs. $\log_{10} \omega$. The phase can be plotted either in degrees or in radians; we shall use radians.

     DEMO: *Bode Plots*

9-4-1   First-Order Systems

As a first example, consider the familiar first-order RC lowpass filter shown in Figure 9-25. We have already seen that the frequency response of this circuit is

$$H(j\omega) = \frac{1}{1 + jRC\omega}.$$

Therefore, its magnitude and phase response are

$$|H(j\omega)| = \sqrt{H(j\omega)H^*(j\omega)} = \frac{1}{\sqrt{1 + \omega^2 R^2 C^2}}$$

$$\angle H(j\omega) = \tan^{-1}\left(\frac{\Im m(H(j\omega))}{\Re e(H(j\omega))}\right) = -\tan^{-1}(\omega RC).$$

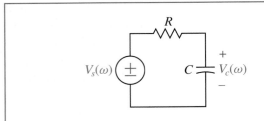

Fig. 9-25. A first-order lowpass filter.

On a dB scale, we can express the magnitude response as

$$20 \log_{10}(|H(j\omega)|) = 20 \log_{10}\left(\frac{1}{\sqrt{1 + \omega^2 R^2 C^2}}\right)$$
$$= -10 \log_{10}(1 + \omega^2 R^2 C^2).$$

$$(9.32)$$

For very small values of ω, where $\omega \ll \frac{1}{RC}$ (i.e., $\omega RC \ll 1$),

$$20 \log_{10}(|H(j\omega)|) \approx -10 \log_{10} 1 = 0 \text{ dB}, \qquad (9.33)$$

which is (nearly) constant. This describes the low-frequency asymptotic behavior of the plot. For relatively large values of ω, where $\omega \gg \frac{1}{RC}$ (i.e., $\omega RC \gg 1$),

$$20 \log_{10}(|H(j\omega)|) \approx -10 \log_{10}(\omega^2 R^2 C^2)$$
$$= -20 \log_{10} \omega - 10 \log_{10}(R^2 C^2). \qquad (9.34)$$

For these frequencies, a graph of $20 \log_{10}(|H(j\omega)|)$ vs. $\log_{10} \omega$ will be a straight line with a slope of -20 dB/decade—that is, for every factor-of-ten increase in ω, the magnitude will decrease by 20 dB. This linear relation defines the high-frequency asymptotic behavior of the plot.

The low- and high-frequency asymptotes are plotted in Figure 9-26, along with the actual magnitude response. Notice that the asymptotes pretty well define the overall shape of the log-magnitude response. If the low-frequency and high-frequency asymptotes are extended, they meet at the frequency $\frac{1}{RC}$, which is the half-power frequency. At this point, the actual value of the magnitude response is $1/\sqrt{2}$, which corresponds to a gain of $-10 \log_{10} 2 \approx -3$ dB. This is sometimes called the ***breakpoint*** or ***break frequency*** of the Bode plot. It is the point on the plot where the actual magnitude response is farthest away from the asymptotes, and this maximum excursion is 3 dB. For hand-drawn Bode plots, the procedure is to first identify the breakpoints, then plot in the asymptotes, and finally draw a smooth curve that is consistent with these asymptotes.

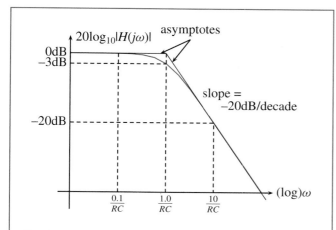

Fig. 9-26. Bode magnitude plot for the first-order lowpass filter shown in Figure 9-25.

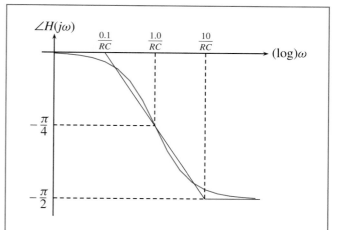

Fig. 9-27. Bode phase plot for the first-order lowpass filter shown in Figure 9-25.

The phase response of this filter is $\angle H(j\omega) = -\tan^{-1}(RC\omega)$, which is plotted in Figure 9-27. Observe that for low frequencies ($\omega \ll \frac{1}{RC}$), $-\tan^{-1}(RC\omega) \approx -\tan^{-1}(0) = 0$. For high frequencies ($\omega \gg \frac{1}{RC}$), $-\tan^{-1}(RC\omega) \approx -\tan^{-1}(\infty) = -\pi/2$. Also, when $\omega = \frac{1}{RC}$, the phase is equal to $-\pi/4$. As we did with the magnitude plot, the phase plot can be approximated by a piecewise linear curve that captures the asymptotic behavior. For the phase plot, we use two breakpoints, one at $\omega = 0.1/RC$ (one decade below the magnitude breakpoint) and one at $\omega = 10/RC$ (one decade above the magnitude breakpoint). Using these breakpoints allows the phase response to be approximated by three straight-line segments, the upper and lower of which are horizontal and attached to the two breakpoints and the center of which connects the two breakpoints. These asymptotes are included in Figure 9-27.

As a second example, consider the first-order highpass filter shown in Figure 9-28. Using a current divider shows the frequency response to be

$$H(j\omega) = \frac{I_{\text{out}}}{I_{\text{in}}} = \frac{j\omega \frac{L}{R}}{1 + j\omega \frac{L}{R}}.$$

The break frequency occurs at $\omega = R/L$. If $\omega \ll \frac{R}{L}$, then

$$H(j\omega) \approx \frac{j\omega L}{R}$$

$$20 \log_{10} |H(j\omega)| \approx 20 \log_{10}(\omega) + 20 \log_{10}(\frac{L}{R}).$$

The graph of this portion of the log-magnitude response is a straight line with a slope of $+20$ dB/decade. The low-frequency asymptote for the phase response has a constant value of $\pi/2$ radians.

For high frequencies, where $\omega \gg \frac{R}{L}$,

$$H(j\omega) \approx \frac{j\omega \frac{L}{R}}{j\omega \frac{L}{R}} = 1$$

$$20 \log_{10} |H(j\omega)| \approx 0.$$

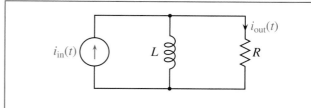

Fig. 9-28. A first-order highpass filter.

The high-frequency phase asymptote is approximately zero.

At the break frequency $\omega = \frac{R}{L}$, the frequency response is

$$H\left(j\frac{R}{L}\right) = \frac{j}{1+j} = \frac{1}{\sqrt{2}}e^{j\frac{\pi}{4}},$$

which corresponds to a log-magnitude response of -3 dB and a phase of $\pi/4$. The Bode magnitude and phase plots are shown in Figure 9-29.

WORKED SOLUTION: *First-Order Bode Plots*

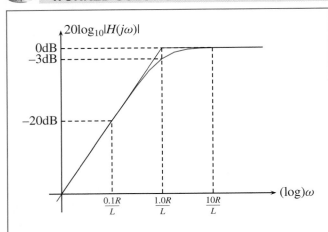

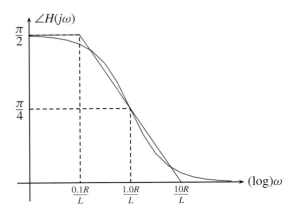

Fig. 9-29. Complete Bode plot for the first-order highpass filter shown in Figure 9-28.

9-4-2   Second-Order Systems

The procedure is similar for a second-order circuit with complex poles. We can get a quick approximation to the Bode plot by looking at its asymptotes. In the case of a second-order circuit, however, how the actual plot behaves near the break frequency will depend upon the quality factor, Q, of the pole pair. Consider the parallel RLC circuit that we saw earlier in Figure 9-22, except that now we will consider the output to be the current flowing through the inductor rather than the current flowing through the resistor. The effect of changing the output variable is to change from a second-order bandpass frequency response to a second-order lowpass frequency response. (If the output variable were changed to the current flowing through the capacitor, it would become a second-order highpass.)

With this choice of the output variable, the system function is

$$H(s) = \frac{\frac{1}{Ls}}{Cs + \frac{1}{R} + \frac{1}{Ls}} = \frac{\frac{1}{LC}}{s^2 + \frac{1}{RC}s + \frac{1}{LC}}$$

$$= \frac{\omega_0^2}{s^2 + 2\zeta\omega_0 s + \omega_0^2}. \qquad (9.35)$$

In the last expression, ω_0 is the undamped natural frequency that we saw before, namely,

$$\omega_0 = \frac{1}{\sqrt{LC}},$$

and ζ is the *damping ratio* that we saw in Chapter 7. For this circuit,

$$\zeta = \frac{1}{2R}\sqrt{\frac{L}{C}}.$$

If the circuit has complex poles, ζ is inversely proportional to the quality factor, Q:

$$\zeta = \frac{1}{2Q}.$$

For $0 < \zeta < 1$, the system is underdamped (two complex poles); for $1 < \zeta < \infty$, it is overdamped (two real poles);

and for $\zeta = 1$, it is critically damped (two identical real poles). The asymptotic behavior of the Bode plot depends only on ω_0, but how the actual magnitude response hugs the asymptotes also depends upon ζ.

Let's begin with the asymptotic behavior of the log-magnitude response. We can get the frequency response by setting $s = j\omega$ in the system function (9.35):

$$H(j\omega) = \frac{1}{-\left(\frac{\omega}{\omega_0}\right)^2 + j2\zeta\left(\frac{\omega}{\omega_0}\right) + 1} \quad (9.36)$$

For very small values of ω, we have

$$H(j\omega) \approx \frac{\omega_0^2}{\omega_0^2} = 1,$$

so the log magnitude will have a value of 0 dB and the phase will be approximately zero radians. For very large values of ω, the ω^2 term dominates the denominator and we have

$$H(j\omega) \approx \frac{\omega_0^2}{-\omega^2}$$

$$20\log_{10}|H(j\omega)| \approx 40\log_{10}\omega_0 - 40\log_{10}\omega$$

$$\angle H(j\omega) \approx -\pi.$$

This graph is linear in $\log\omega$, but the slope is -40 dB/decade rather than the -20 dB/decade that we saw for a first-order lowpass. The two asymptotes intersect at $\omega = \omega_0$ as before. Figure 9-30 shows the complete Bode plot for several different values of the damping ratio, ζ. These share the same asymptotes, but, as ζ is decreased, the magnitude response becomes increasingly peaked and the phase transition becomes sharper.

9-4-3 Multifactor Bode Plots

When the frequency response of a system (circuit) can be decomposed into a product of factors, the Bode plot for the complete system can be derived by summing the Bode plots of the individual factors. This is because the logarithm of a product is equal to the sum of the logarithms of the factors. Consider a system with several

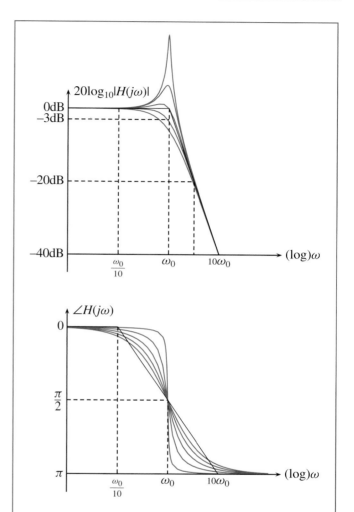

Fig. 9-30. The Bode magnitude and phase plots for a second-order lowpass filter. Separate curves are plotted (top to bottom) for $\zeta = 0.05, 0.25, 0.50, 0.75, 1.00$.

poles and zeros. Its frequency response can be written in the unfamiliar form

$$H(j\omega) = \frac{K(j\omega)^L \prod\limits_{i=1}^{M}(1 - j\frac{\omega}{z_i})}{\prod\limits_{k=1}^{N}(1 - j\frac{\omega}{p_k})}.$$

(The symbol that resembles a capital Greek *pi* (Π) denotes a product of identical terms.) To be completely general, we have allowed for L zeros (if $L > 0$) or poles (if $L < 0$) at the origin, for M additional zeros at possibly complex points z_i, and for N possibly complex poles at locations p_k. Computing the log-magnitude of the frequency response gives

$$20 \log_{10} |H(j\omega)| = 20 \log_{10} |K| + 20L \log_{10} \omega$$

$$+ \sum_{i=1}^{M} 20 \log_{10} \left| 1 - \frac{j\omega}{z_i} \right| - \sum_{k=1}^{N} 20 \log_{10} \left| 1 - \frac{j\omega}{p_k} \right|,$$

which we can identify as the sum of the log-magnitude functions of the individual terms. The phase similarly decomposes into a sum:

$$\angle H(j\omega) = \angle K + L\frac{\pi}{2} + \sum_{i=1}^{M} \angle(1 - \frac{j\omega}{z_i}) - \sum_{k=1}^{N} \angle(1 - \frac{j\omega}{p_k}).$$

We can construct the Bode plot for an arbitrary system if we understand the Bode plots of four types of components: the constant term K, zeros or poles at the origin, real-valued zeros and poles, and complex conjugate pairs of zeros and poles. Let's look at each of these contributions separately.

Constant Gain, K: A constant-gain term has the form

$$H_K(j\omega) = K.$$

By definition, neither the magnitude nor the phase of a constant-gain varies with frequency:

$$20 \log_{10} |H_K(j\omega)| = 20 \log_{10} |K|, \qquad \forall \omega$$

$$\angle H_K(j\omega) = \begin{cases} 0 & \forall \omega, \text{ if } K > 0 \\ \pi & \forall \omega, \text{ if } K < 0. \end{cases}$$

Thus, the effect of changing the gain on a system function is to add a constant value to its Bode plot. This, in turn, simply shifts the Bode plot up (if $|K| > 1$) or down (if

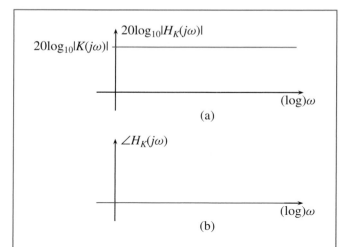

Fig. 9-31. The Bode plot associated with a constant gain. (a) Magnitude plot. (b) Phase plot. If the gain is negative, the phase plot will have the constant value π.

$|K| < 1$). The Bode plot of a constant term is shown in Figure 9-31.

Zeros or Poles at the Origin: L zeros at the origin contribute a multiplicative term of the form

$$H_{z0}(j\omega) = (j\omega)^L$$

to the frequency response. In turn, this introduces an additive term of the form

$$20 \log_{10} |H_{z0}(j\omega)| = 20L \log_{10} \omega$$

to the log-magnitude plot and a constant term

$$\angle H_{z0}(j\omega) = L\frac{\pi}{2}$$

to the phase plot. One or more poles at the origin have a similar effect, except that, for a pole, L is negative, which changes the slope of the term added to the magnitude plot and subtracts, rather than adds, a constant term to the phase. The Bode magnitude plot for a single pole or zero at the origin is shown in Figure 9-32. Notice that the gain is 0 dB at $\omega = 1$ rad/s.

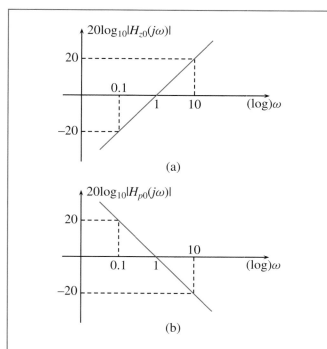

Fig. 9-32. Bode magnitude plot for (a) a single zero and (b) a single pole located at the origin.

Real Zero or Pole: A real zero corresponds to a term like

$$H_{zr}(j\omega) = 1 - j\frac{\omega}{z_k} \qquad (9.37)$$

in the frequency response. Such a factor corresponds to a zero located at $z = z_k$, a real number. A real pole will contribute a similar term to the denominator of the factored frequency response, but if the system is stable, the pole location will always be negative. For a zero, z_k can be either positive or negative. The Bode plot of a term like the one in (9.37) can be found by using the same approach that we used for the first-order lowpass filter.

For very small values of ω, such that $\omega \ll |z_k|$, we get the low-frequency asymptotic values

$$H_{zr}(j\omega) \approx 1$$

$$20\log_{10}|H_{zr}(j\omega)| \approx 0$$

$$\angle H_{zr}(j\omega) \approx 0.$$

For $\omega \gg |z_k|$, we see, in the high-frequency limit, that

$$H_{zr}(j\omega) \approx -j\frac{\omega}{z_k}$$

$$20\log_{10}|H_{zr}(j\omega)| \approx 20\log_{10}\omega - 20\log_{10}|z_k|$$

$$\angle H_{zr}(j\omega) \approx \begin{cases} -\frac{\pi}{2} \text{ if } z_k > 0, \\ \frac{\pi}{2} \text{ if } z_k < 0. \end{cases}$$

The log-magnitude response is, therefore, constant with a value of 0 dB for low frequencies and grows linearly with $\log|\omega|$ with a slope of 20 dB/decade for high frequencies. These two asymptotes meet at the break frequency $\omega = |z_k|$. The actual gain at this frequency is +3 dB.

The Bode magnitude and phase plots for a real pole look like the negatives of those for a real zero. Figure 9-33 shows the Bode magnitude and phase plots associated with a real zero and a real pole.

Complex-Conjugate Zeros or Poles: We have already seen the contribution of a complex-conjugate pole pair when we examined a second-order lowpass filter in the previous section. The Bode plot of such a section was given in Figure 9-30. The Bode plot of a pair of complex-conjugate zeros will be identical, except that both the log-magnitude and phase plots will be multiplied by -1. Notice that both the actual plots and the asymptotes are multiplied by -1. As a result, the log-magnitude plot from a pair of complex zeros will contain a dip at the break frequency and the amount of that dip will depend upon the damping ratio ζ. The smaller the value of ζ, the deeper the dip.

The Bode-plot asymptotes resemble those of the real pole and zero, except that all of the slopes are doubled. These are shown in Figure 9-34. They are labelled in terms of the break frequency ω_i, which is approximately equal to the imaginary part of the zero (pole) locations.

A Confusing Issue: Normalized Forms for System Functions

In computing an inverse Laplace transform or a Bode plot, the first step usually involves factoring the denominator polynomial, and possibly the numerator polynomial, to identify the poles and zeros of the system. When doing this, one is wise to express the factors in forms that facilitate the subsequent computations, as we have consistently done. Writing the system functions involving factors in a particular form is a process that we generally call **normalization**. Normalizing the system function reduces the likelihood of careless errors. Unfortunately, the normalization that is best for computing inverse Laplace transforms is different from the normalization that is best for computing Bode plots. This is not a serious problem once we are aware that the normalizations are different, but it is appropriate to reemphasize the differences and comment on why each normalization is appropriate.

In computing an inverse Laplace transform it is helpful to express the system function in the form

$$H(s) = \frac{B(s)}{(s - p_1)(s - p_2) \cdots (s - p_n)}, \qquad (9.38)$$

where p_k is the kth pole, in order to facilitate the partial-fraction expansion, which is the key step of the operation. The PFE consists of converting this representation to one of the form (for Case I)

$$H(s) = \sum_{k=1}^{n} \frac{A_k}{s - p_k},$$

where

$$A_k = \lim_{s \to p_k} (s - p_k) H(s).$$

Then,

$$h(t) = \sum_{k=1}^{n} A_k e^{p_k t} u(t).$$

Always writing the denominator factors the same way means that limits can always be evaluated in the same way and the impulse response can always be written (mechanically) in the same way. Notice that the normalization applied to (9.38) writes each denominator factor as a monic first-order polynomial.

When preparing a Bode plot, we again factor the denominator, but we express the terms in the different form

$$H(j\omega) = \frac{K(j\omega)^L \prod_{i=1}^{M} (1 - j\frac{\omega}{z_i})}{\prod_{k=1}^{N} (1 - j\frac{\omega}{p_k})}.$$

Here each denominator factor is not monic, but instead has a normalized constant term. In this fashion each denominator factor contributes a gain of 0 dB below its break frequency, which makes it considerably easier to add the components of the multifactor Bode plot together. Had we used the normalization of (9.38), each term in the composite Bode plot would have a different gain for low frequencies, and adding the graphs together would be somewhat more involved (although still manageable).

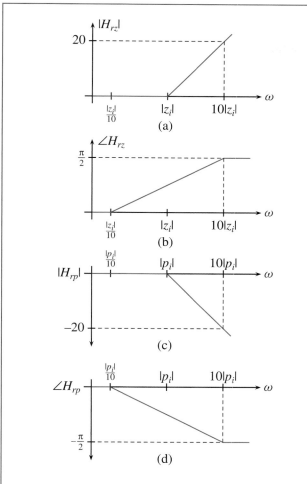

Fig. 9-33. (a) Bode magnitude plot for a single real zero at $s = z_i$ (negative). (b) Bode phase plot for single real zero. (c) Bode magnitude plot for a single real pole at $s = p_i$ (negative). (d) Bode phase plot for single real pole.

Example 9-7   A Multifactor Bode Plot

As an example to illustrate the construction of a Bode plot for a system function that has many factors, consider the example

$$H(s) = \frac{100(s + \frac{1}{10})}{(s + 10)(s^2 + 25s + 10000)}.$$

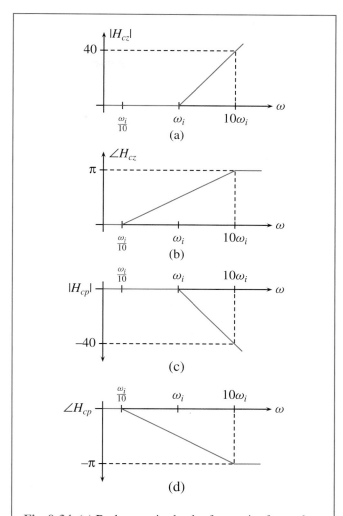

Fig. 9-34. (a) Bode magnitude plot for a pair of complex-conjugate zeros (ω_i = break frequency). (b) Bode phase plot for complex-conjugate-zero pair. (c) Bode magnitude plot for a pair of complex-conjugate poles. (d) Bode phase plot for a pair of complex-conjugate poles.

We begin by replacing $s = j\omega$ and writing the factors in their normalized form. This latter step is not strictly necessary, but it can be helpful if the terms corresponding to first- and second-order zeros are written so that their

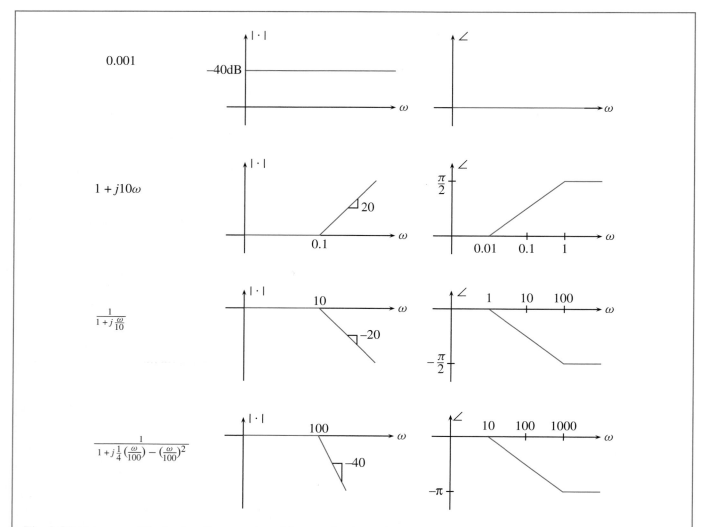

Fig. 9-35. Component Bode plots for a system with several poles and zeros. The log-magnitude plots are shown in the center column, the phase plots on the right.

gain at low frequencies is 0 dB. The result after we perform these operations is

$$H(j\omega) = \frac{0.01(1 + j10\omega)}{(1 + j\frac{\omega}{10})(1 + j\frac{1}{4}(\frac{\omega}{100}) - (\frac{\omega}{100})^2)}.$$

We identify four contributions to the Bode plot: the constant, the real zero at $s = -1/10$, the real pole at $s = -10$, and the complex pole pair, for which $\omega_0 = 100$ and $\zeta = 1/8$. The Bode plots of these four components are shown in thumbnail form in Figure 9-35. The complete

Bode plot is given in Figure 9-36. The asymptotes are also shown. When there are many poles and zeros, the asymptotes track the actual signal accurately only for the very low and very high frequencies; in the middle range they give only a very crude picture of what is going on. This is particularly true in this example. For a more accurate plot, MATLAB is helpful. ∎

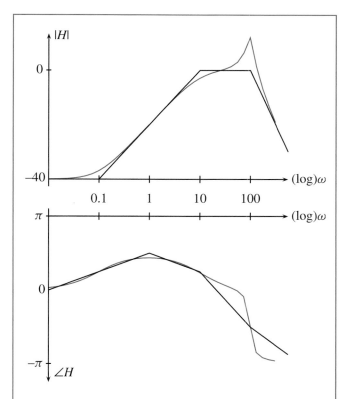

Fig. 9-36. The Bode magnitude and phase plots for the multifactor system whose components are shown in Figure 9-35.

9-5 Chapter Summary

9-5-1 Important Points Introduced

- Any periodic signal can be represented as a superposition of harmonically related complex exponential time functions.

- The output of a linear circuit in response to a periodic input is periodic. The Fourier-series coefficients of the output signal are equal to the Fourier-series coefficients of the input signal multiplied by the frequency response at the appropriate frequency.

- The frequency response of a circuit can be found by mapping the circuit to the frequency domain and then applying KCL, KVL, and the frequency-domain element relations.

- The frequency response and the impulse response of a circuit both depend upon the locations of its poles and zeros.

- Lowpass and highpass frequency responses can be obtained from simple first-order circuits.

- Second-order circuits can have lowpass, bandpass, and highpass frequency responses.

- A Bode plot is a plot of the logarithm (base 10) of the magnitude response and the phase response vs. the logarithm of the frequency.

- Bode magnitude and phase plots can be approximated by asymptotic plots made up from straight-line segments.

- The Bode plot of a system function constructed from several factors is the sum of the Bode plots of the factors.

9-5-2 New Abilities Acquired

You should now be able to do the following:

(1) Evaluate the spectrum of a periodic signal; that is, decompose an arbitrary periodic signal into a superposition of complex exponential time functions.

(2) Find the Fourier-series coefficients of a periodic signal.

(3) Evaluate the frequency response of a linear time-invariant circuit.

(4) Find the spectrum of the output of a linear circuit from the spectrum of the input and the frequency response of the circuit.

(5) Compute and plot the magnitude and phase response of a circuit.

(6) Sketch the magnitude and phase response of a circuit from its pole–zero plot.

(7) Draw the Bode magnitude and phase plots for a circuit.

9-6 Problems

9-6-1 Drill Problems

P9-1 Express the signal in Figure P9-1 as a Fourier series.

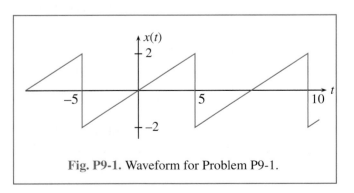

Fig. P9-1. Waveform for Problem P9-1.

P9-2 Express the signal in Figure P9-2 as a Fourier series. Note that the impulses in that figure have areas that alternate in sign.

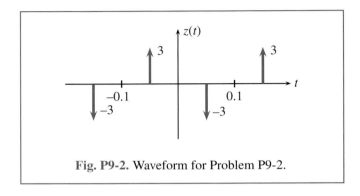

Fig. P9-2. Waveform for Problem P9-2.

P9-3 Compute and sketch the magnitude and phase of the frequency response of the circuit in Figure P9-3. The input to the circuit is the source voltage $v_s(t)$, and the output is the current $i(t)$.

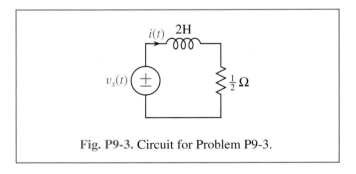

Fig. P9-3. Circuit for Problem P9-3.

P9-4 Compute and sketch the magnitude and phase of the frequency response of the circuit in Figure P9-4. The input to the circuit is the source current $i_s(t)$, and the output is the voltage $v(t)$.

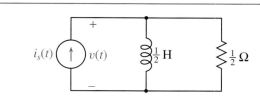

Fig. P9-4. Circuit for Problem P9-4.

P9-5 A circuit has the system function

$$H(s) = s + 10.$$

(a) Draw the pole–zero plot of the circuit.
(b) Produce a labelled sketch of the magnitude response of the circuit, using the pole–zero plot as a guide, as in Example 9-5.
(c) Produce a labelled sketch of the phase response of the circuit.

P9-6 A circuit has the system function

$$H(s) = s^2 + 2as + (a^2 + b^2).$$

(a) Draw the pole–zero plot of the circuit.
(b) Produce a labelled sketch of the magnitude response of the circuit, using the pole–zero plot as a guide, as in Example 9-6.
(c) Produce a labelled sketch of the phase response of the circuit.

9-6-2 Basic Problems

P9-7 A signal waveform $x(t)$ is a sum of sinusoids:

$$x(t) = -2 + 2\cos(10t) - 3\cos(15t - \frac{\pi}{3})$$

$$+ \sin(20t + \frac{\pi}{4}).$$

(a) What is the fundamental frequency of $x(t)$?
(b) Sketch the magnitude and phase spectra of $x(t)$.

P9-8 Find the signal $x(t)$ that has the spectrum drawn in Figure P9-8.

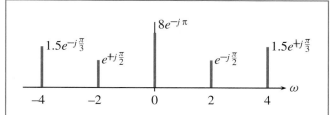

Fig. P9-8. The complex spectrum of a signal to be found for Problem P9-8.

P9-9 A signal $x(t)$ has the magnitude and phase spectra illustrated in Figure P9-9.

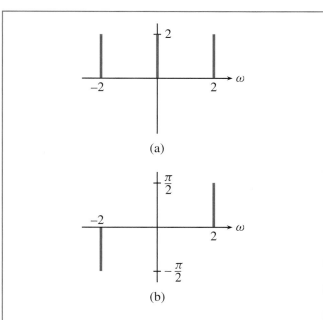

(a)

(b)

Fig. P9-9. Spectra for the signal in Problem P9-9. (a) Magnitude spectrum. (b) Phase spectrum.

Sketch the magnitude and phase spectra of the following signals that are derived from $x(t)$.

(a) $y_a(t) = 3x(t)$

(b) $y_b(t) = -x(t)$

(c) $y_c(t) = x(t - \frac{\pi}{2})$

P9-10 A 1-H inductor, a 1-F capacitor, and a voltage source are connected in parallel. The source voltage is

$$x(t) = \cos(50t + \frac{\pi}{3}) + \sin(150t - \frac{\pi}{6}) - \sin(50t + \frac{\pi}{6}).$$

(a) Draw the complex spectrum of the current flowing through the inductor.

(b) Draw the complex spectrum of the current flowing through the capacitor.

(c) Draw the complex spectrum of the current flowing through the voltage source.

P9-11 A periodic square wave $x(t)$ with a period of one second is input to the circuit shown in Figure P9-11. Express the capacitor ~~current, $i_c(t)$,~~ as a Fourier series.

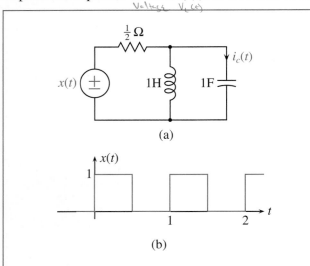

(a)

(b)

Fig. P9-11. (a) Circuit for Problem P9-11. (b) Input waveform, $x(t)$.

P9-12

(a) Calculate the frequency response of the circuit shown in Figure P9-12.

(b) Plot the magnitude response.

(c) Plot the phase response.

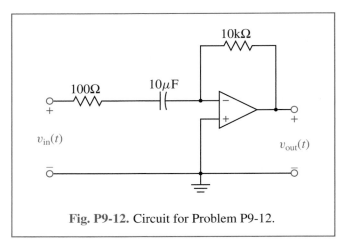

Fig. P9-12. Circuit for Problem P9-12.

P9-13 The elements in each of the networks in Figure P9-13 can have any positive real values except zero and infinity. .

(a) For which, if any, of the three two-terminal networks could the pole–zero plot of the impedance resemble the one in Figure P9-13(a)?

(b) For which, if any, of the three two-terminal networks could the pole–zero plot of the impedance resemble the one in Figure P9-13(b)?

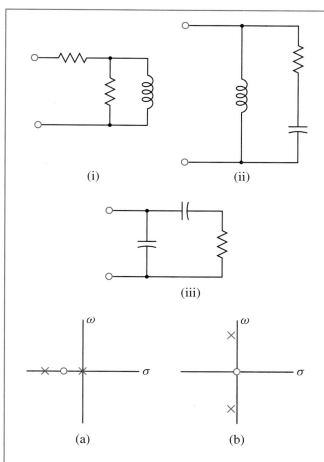

Fig. P9-13. (i)–(iii) Networks for Problem P9-13. All of the element values are finite and positive. (a)–(b) Pole-zero plots for Problem P9-13.

P9-14 A circuit has the system function

$$H(s) = \frac{s^2 + b^2}{(s + a)^2 + b^2},$$

where $b \gg a$.

(a) Draw the pole–zero plot. Clearly label the locations of the poles and zeros on your plots.

(b) Sketch the magnitude of the frequency response of the filter.

P9-15 Produce the phase plots that correspond to the pole–zero plots and magnitude responses in Figure 9-21 in the text.

P9-16

(a) Which, if any, of the three two-terminal networks shown in Figure P9-16 could have an input *admittance* whose magnitude response, as a function of frequency, would have the form shown in Figure P9-16(a)? The network elements can have any positive real values except zero and infinity.

(b) Which, if any, of the three two-terminal networks shown in Figure P9-16 could have an input admittance whose *phase* behavior, as a function of frequency, would have the form shown in Figure P9-16(b)?

P9-17 A circuit has the pole–zero plot shown in Figure P9-17.

(a) Graph the magnitude response of the filter. Indicate key frequencies on your plot.

(b) Compute the phase at the following frequencies:
   (i) $\omega = 0$
   (ii) $\omega = \infty$
   (iii) $\omega = a + b$

P9-18 Sketch and label the magnitude response, $|H(j\omega)|$, of the circuit in Figure P9-18 as a function of ω (not a Bode plot).

P9-19 Poles and zeros affect the time-domain behavior as well as the frequency response of a circuit. Which of the three networks shown in Figure P9-13 could produce a voltage waveform (at the terminals) of the form $v(t) = Ae^{-3t} + Be^{-t}$ for $t > 0$ if a current source is applied to the terminals with the current $i(t) = u(t)$?

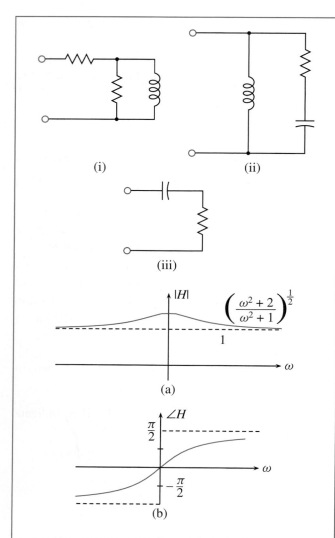

(i)

(ii)

(iii)

(a)

$$\left(\frac{\omega^2 + 2}{\omega^2 + 1}\right)^{\frac{1}{2}}$$

(b)

Fig. P9-16. (i)–(iii) Three two-terminal networks for Problem P9-16. The elements can have any positive values except zero and infinity. (a) Possible magnitude behavior for the admittance of one or more of the two-terminal networks in Figure P9-16. (b) Possible phase behavior for the admittance of one or more of the two-terminal networks in Figure P9-16.

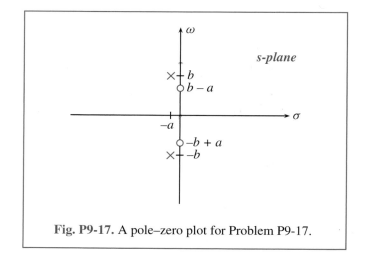

Fig. P9-17. A pole–zero plot for Problem P9-17.

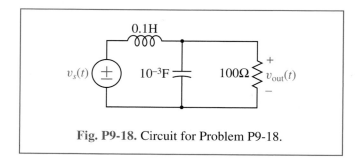

Fig. P9-18. Circuit for Problem P9-18.

P9-20 State the following attributes of the resonant circuit depicted in Figure P9-20.

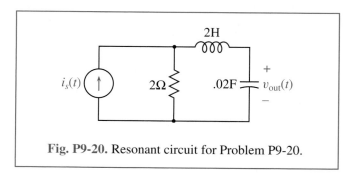

Fig. P9-20. Resonant circuit for Problem P9-20.

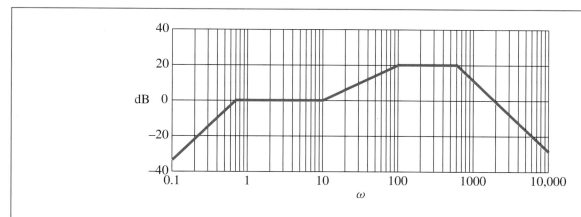

Fig. P9-21. Bode magnitude plot for Problem P9-21.

(a) Construct the system function of this circuit. The input is $i_s(t)$ and the output is $v_{out}(t)$.

(b) Draw the pole–zero plot corresponding to this circuit. Carefully label all of the pole and zero locations.

(c) What is the (undamped) natural frequency?

(d) What is the damped natural frequency?

(e) What is the time constant of the exponential envelope of the impulse response?

(f) What is the quality factor, Q, for this circuit?

P9-21 Construct $H(j\omega)$ from the asymptotic Bode plot shown in Figure P9-21. Assume that all poles and zeros lie in the left half of the s-plane (i.e., assume that the real parts of the pole and zero locations are all negative).

P9-22 Sketch Bode magnitude and phase plots for the following system functions.

(a) $H(s) = \frac{1}{s+10}$

(b) $H(s) = 1 - 10s$

(c) $H(s) = \frac{s-20}{s+200}$

P9-23 Sketch Bode magnitude and phase plots for the following system functions.

(a) $H_a(s) = \frac{s+5}{s^2+100s}$

(b) $H_b(s) = \frac{s-10}{s+10}$

(c) $H_c(s) = \frac{s(s+10)}{(s+1)(s+100)^2}$

P9-24 Plot the asymptotic Bode plot for the magnitude of the frequency response of the circuit in Figure P9-24.

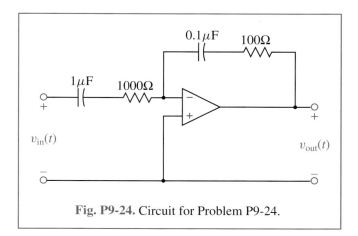

Fig. P9-24. Circuit for Problem P9-24.

P9-25

(a) Compute the system function of the circuit in Figure P9-25.

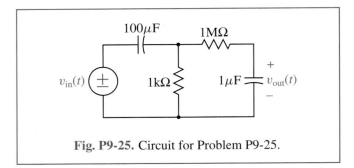

Fig. P9-25. Circuit for Problem P9-25.

(b) Sketch the magnitude and phase of this circuit on a Bode plot.

P9-26 If the system function of a circuit is

$$H(s) = \frac{1000s}{(s+1)(s+100)},$$

(a) draw and label the asymptotes of the Bode plot of the magnitude of the frequency response;

(b) draw and label the Bode phase asymptotes.

P9-27 A circuit with the system function

$$H(s) = \frac{ks}{(s+a)(s+b)}$$

has the Bode magnitude plot shown in Figure P9-27. Compute the values of k, a, and b.

P9-28

(a) Calculate the system function of the circuit in Figure P9-28.

(b) Draw the Bode magnitude plot for the circuit.

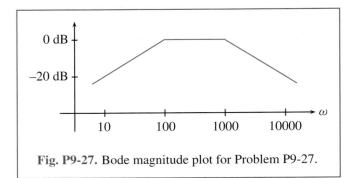

Fig. P9-27. Bode magnitude plot for Problem P9-27.

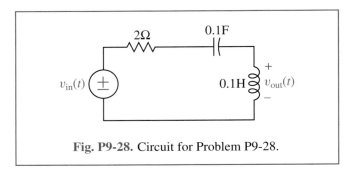

Fig. P9-28. Circuit for Problem P9-28.

P9-29 The Bode magnitude plot for a circuit is shown in Figure P9-29. We are told that all of the poles of the system and all of its zeros have negative real parts (i.e., they lie in the left half of the s-plane). Find the system function of the circuit.

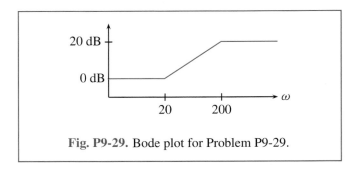

Fig. P9-29. Bode plot for Problem P9-29.

P9-30 A practical opamp implementation of an integrator includes a resistor shunted across the capacitor to limit the low-frequency gain, as shown in Figure P9-30. (Otherwise the small, but inevitable unbalanced offset voltages and bias currents would be integrated, leading eventually to saturation of the opamp.)

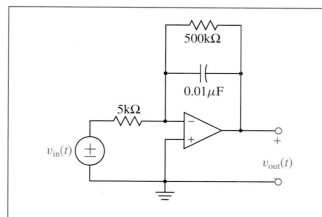

Fig. P9-30. A practical integrator implementation for Problem P9-30.

(a) Find the system function of this circuit.

(b) Draw a Bode magnitude plot of the frequency response of the integrator.

9-6-3 Advanced Problems

P9-31 Figure P9-31 shows two waveforms, each of which is periodic and has a period of 2 seconds.

(a) Compute the Fourier-series coefficients of $y(t)$ in Figure P9-31(a). How are these related to the Fourier-series coefficients of $x(t)$ of Figure 9-5?

(b) Repeat part (a) for the periodic signal $z(t)$ drawn in Figure P9-31(b).

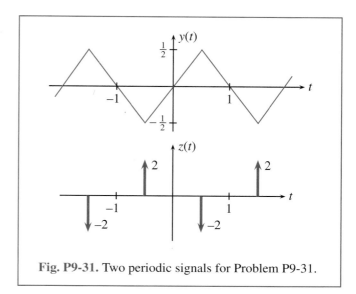

Fig. P9-31. Two periodic signals for Problem P9-31.

(c) How is $y(t)$ related to $x(t)$? How is $z(t)$ related to $x(t)$?

(d) In general, if $f(t)$ is a periodic signal and $g(t) = \frac{df(t)}{dt}$, how are the Fourier-series coefficients of $g(t)$ related to the Fourier-series coefficients of $f(t)$?

P9-32 Figure P9-32(a)–(c) shows the pole–zero plots corresponding to three linear circuits. For each of these, select the appropriate magnitude response from Figure P9-32(i)–(iv).

P9-33 For each of the pole–zero plots in Figure P9-32, select the appropriate phase response from Figure P9-33.

P9-34 An oscillator is a circuit whose system function has a pair of complex-conjugate poles located on the $j\omega$-axis in the s-plane. It is capable of supporting a sustained oscillation: Once started, it will continue to produce a sinusoidal output $v_{out}(t)$, measured across its output terminals. In the network in Figure P9-34, find the value of K for which the system will oscillate, and compute the frequency of oscillation.

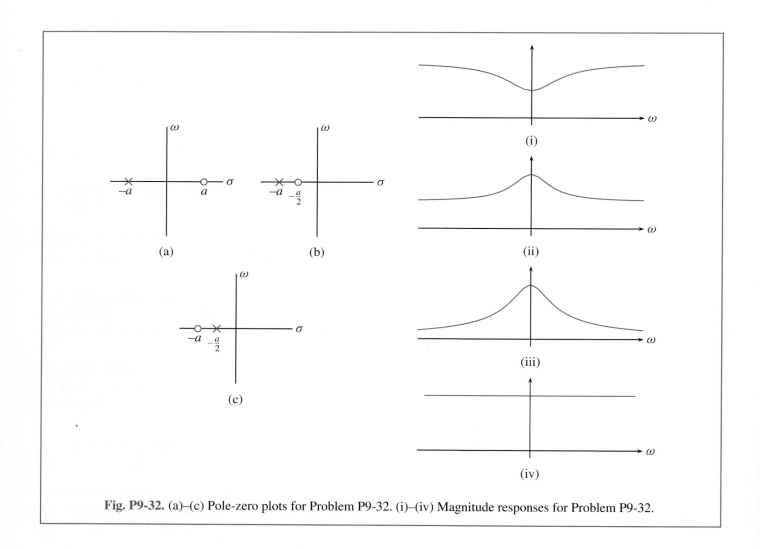

Fig. P9-32. (a)–(c) Pole-zero plots for Problem P9-32. (i)–(iv) Magnitude responses for Problem P9-32.

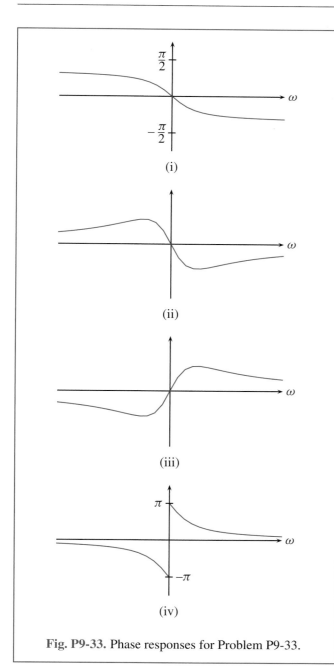

Fig. P9-33. Phase responses for Problem P9-33.

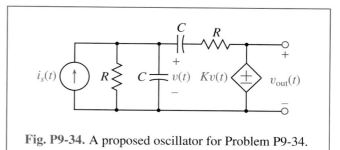

Fig. P9-34. A proposed oscillator for Problem P9-34.

P9-35 A stable circuit (i.e., one with all of its poles in the left half of the s-plane) has the Bode magnitude plot shown in Figure P9-35.

(a) State the locations of all of the poles and zeros of the circuit.

(b) Construct the system function of the circuit, $H(s)$.

(c) Draw the Bode phase plot (asymptotes only). Label your plot clearly.

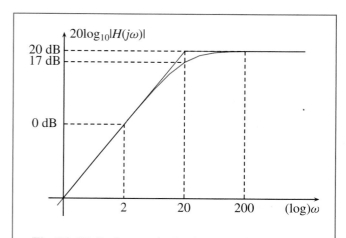

Fig. P9-35. Bode magnitude plot for Problem P9-35.

P9-36 The second stage of an instrumentation amplifier is shown in Figure P9-36.

(a) Draw the asymptotic Bode plot for the magnitude of the frequency response of the circuit if the difference $v_2(t) - v_1(t)$ is considered to be the input signal.

(b) Calculate the differential gain, $V_o/(V_2 - V_1)$, at 50 Hz.

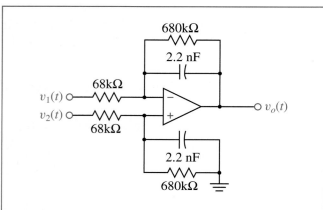

Fig. P9-36. Instrumentation amplifier for Problem P9-36.

9-6-4   Design Problems

P9-37

(a) Compute the system function of the circuit in Figure P9-37.

(b) Select convenient values of R, R_1, and C to obtain a lowpass filter with a cutoff frequency (half-power frequency) of $\omega_0 = 2000$ rad/s and $Q = 8$.

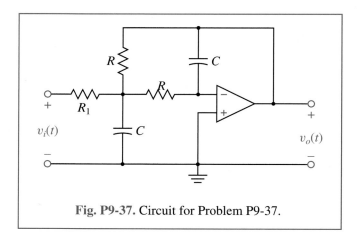

Fig. P9-37. Circuit for Problem P9-37.

P9-38   Many loudspeaker systems consist of two loudspeakers: the woofer, which reproduces the low-frequency part of the signal, and the tweeter, which reproduces the high-frequency part of the signal. A crossover network is used to select the high-frequency part of the signal and feed it into the tweeter. Such a network functions as a highpass filter. The entire audio signal is applied at the terminals a–a' in Figure P9-38.

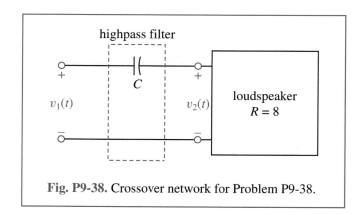

Fig. P9-38. Crossover network for Problem P9-38.

(a) Assuming that the equivalent circuit for the tweeter consists of just a resistor with a resistance R, plot the pole–zero pattern of the system function that relates $v_2(t)$ to $v_1(t)$ and sketch the frequency response curves (magnitude and angle).

(b) If $R = 8\,\Omega$, find the value of the capacitance C that will make the half-power frequency of the highpass filter 5 kHz ($= 2\pi(5000)$ rad/s).

P9-39

(a) Find the frequency response, $H(j\omega)$, of the circuit in Figure P9-39(a).

(b) Find values of R_1, R_2, C_1, and C_2 such that the asymptotes of the Bode magnitude plot for this circuit will resemble the ones in Figure P9-39(b).

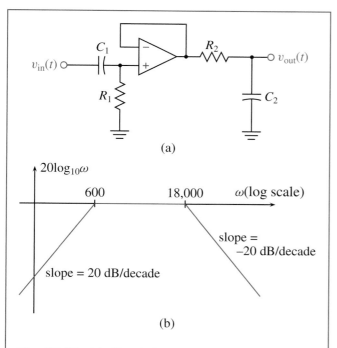

(a)

(b)

Fig. P9-39. (a) Circuit for Problem P9-39. (b) Bode magnitude plot for Problem P9-39.

P9-40 In the circuit in Figure P9-40, the value of R_1 is $10\,\mathrm{k\Omega}$.

(a) Find values of R_2 and R_3 such that the gain (magnitude of the frequency response) at low frequencies is 5 and the gain at high frequencies is 2.

(b) Find the frequency at which the gain is midway between these two values (i.e., the frequency at which the gain is 3.5).

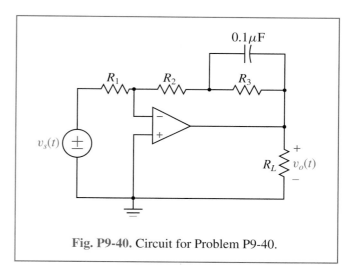

Fig. P9-40. Circuit for Problem P9-40.

Filter Circuits

Objectives

By the end of this chapter, you should be able to do the following:

1. *Find the system function corresponding to a Butterworth or Chebyshev (type 1) lowpass filter.*

2. *Find the order of a Butterworth filter that will meet a set of frequency-domain specifications.*

3. *Implement a high-order lowpass filter by cascading sections of second order.*

4. *Select the element values in a Sallen–Key filter to locate the poles at a desired location.*

5. *Modify a system function to frequency-scale its frequency response by an arbitrary scale factor.*

6. *Impedance-scale a circuit to reduce the currents in it and the amount of power absorbed.*

Until now, our attention has focused on tools for analyzing circuits; relatively little attention has been directed to their design. This chapter looks at some of the issues associated with designing filters to produce certain frequency-response characteristics and at some practical methods for modifying existing designs to make them easier to implement. We have seen examples of first- and second-order circuits with lowpass, highpass, and bandpass frequency responses, but sometimes these simple filters are not good enough. In those cases, can we find filters that have frequency responses that are closer to ideal lowpass or bandpass filters? How can we modify the element values in a circuit to change the cutoff frequency or the gain of a filter? In most of the examples in this book, the element values were chosen to be simple, in order to minimize the amount of calculation. This resulted in element values that were often not practical—capacitors and inductors that were generally too large, and resistors that were too small. How can we modify a basic design to allow more practical component values to be used? Currents measured in amperes are reasonable in household wiring, but are too large for many electronic circuits. For circuits, such as cellular telephones, that operate on batteries, current values measured in microamperes are more reasonable. Can we modify the element values in a circuit to reduce the amount of power that it consumes without altering its frequency response? These are among the questions that we will answer in this chapter.

This chapter is definitely not the final word on filter design, but it does illustrate the essential steps, which are the following: (1) finding a system function $H(s)$ that satisfies a set of basic design constraints (e.g., a satisfactory approximation to a desired frequency response, impulse response, or step response); (2) finding an implementation (circuit) that has that system function; and (3) adjusting the component values to optimize other important design criteria, such as using readily available element values, reducing power dissipation, and fine-tuning the frequency response.

10-1 Lowpass Filters

Applications usually dictate what behavior a circuit should have and what constraints it must meet. For example, if we are building an amplifier, the specifications might tell us what the gain should be, what range of frequencies in the input signal should be amplified, what the maximum allowable level of noise can be, and what are the expected amplitudes of the input signals. To illustrate the steps involved in one class of design problems, we shall assume in the first part of this chapter that we want a circuit that acts like a *lowpass filter* (i.e., one that will approximate a lowpass frequency response, such as the one shown in Figure 10-1).

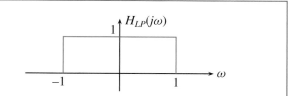

Fig. 10-1. An ideal lowpass filter with a gain of 1 and a cutoff frequency of 1 rad/s. Sinusoidal components of the input with frequencies below 1 rad/s in magnitude are unaltered by the filter; components with frequencies above this value are rejected.

One important use of lowpass filters is in conjunction with analog-to-digital and digital-to-analog converters. These are the devices that convert a continuous-time signal into a discrete-time signal and vice versa. They are typically used in a configuration like the one illustrated in Figure 10-2. The lowpass filters are the boxes labelled LPF. The one on the left, called an *antialiasing filter*, removes from the input signal $i(t)$ high-frequency components that would otherwise be aliased by the sampling operation. It should reject all signal components above one-half the sampling frequency. The lowpass filter on the right is the *reconstruction filter*. Its purpose is to remove high-frequency replications of the signal spectrum introduced by the D/A converter. Its frequency response should be similar to that of the antialiasing filter, although it might have a different gain.

 DEMO: *Lowpass Filters*

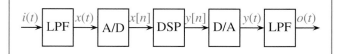

Fig. 10-2. A system for processing an analog (continuous-time) signal by using digital signal processing and then producing a continuous-time result.

10-1-1   Ideal vs. Buildable Filters

The frequency response illustrated in Figure 10-1 is an example of an *ideal lowpass filter*. This particular filter has a *cutoff frequency* of 1 rad/s. This value is convenient for illustrating the design problem, but unrealistic; most applications would require a cutoff frequency that is much higher. We shall see later in the chapter that changing the cutoff frequency after the filter has been designed can be done easily by making a straightforward adjustment of its element values.

Unfortunately, we cannot built an ideal lowpass filter like the one in Figure 10-1. Among other limitations, it does not correspond to an $H(s)$ that is a rational function of s. Filters that do have rational system functions—that is, filters that are buildable—differ from this ideal in four significant ways. For a buildable circuit whose frequency response passes some frequencies and rejects others, (1) the magnitude response cannot be constant over $[-1, 1]$, (2) the frequency response cannot be identically zero over a band of frequencies, (3) the magnitude response cannot be discontinuous, and (4) the phase response cannot be zero. Nevertheless, it is possible to build a filter that has a frequency response that *approximates* this ideal lowpass filter. Constraints on the accuracy of that approximation are usually stated as part of the filter-design problem.

For example, we might be willing to accept any phase behavior for the filter, provided that the magnitude response satisfies the following inequalities:

$$1 - \epsilon_p \leq |H(j\omega)| \leq 1, \qquad |\omega| \leq \omega_p < 1$$
$$0 \leq |H(j\omega)| \leq 1 + \epsilon_p, \qquad \omega_p \leq |\omega| \leq \omega_s$$
$$|H(j\omega)| \leq \epsilon_s, \qquad |\omega| \geq \omega_s > 1.$$

These inequalities define a filter as an acceptable lowpass as long as its magnitude response stays within the region delimited by the shaded barriers in Figure 10-3. To counter the limitation that the magnitude response cannot be constant in the *passband* (those frequencies

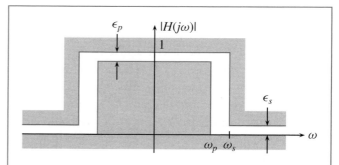

Fig. 10-3. User-controllable bounds that define an acceptable frequency response for a lowpass filter. ϵ_p and ϵ_s are the maximum errors in the passband and stopband of the filter, respectively. ω_p and ω_s are the passband and stopband cutoff frequencies.

whose magnitude is not significantly attenuated by the filter), these specifications allow the response in the passband to vary from the ideal value—namely, 1—but not by more than ϵ_p. By making ϵ_p very small, we can limit the amount of passband deviation. To counter the limitation that the magnitude response at high frequencies cannot be identically zero, there is a similar deviation ϵ_s allowed in the *stopband* (that band of frequencies that are significantly attenuated by the filter). To combat the limitation that the frequency response must be continuous, we replace the single cutoff frequency of the ideal by a pair of cutoff frequencies: ω_p, which defines the upper extent of the passband, and ω_s, which defines the lower extent of the stopband.

The parameters that define an acceptable magnitude response for the lowpass filter can be tied directly to the requirements of a particular application. For example, for an antialiasing filter, ϵ_s controls the amount of aliasing that will be introduced by sampling and ϵ_p affects the amount of magnitude distortion that will be introduced by the filter. If the application is more concerned about aliasing than distortion, we would choose $\epsilon_p > \epsilon_s$. An antialiasing filter should remove all frequency components (measured in Hz) above one-half the sampling rate, f_s. If the important components of the signal are those with a frequency below f_0 (Hz), then we should choose $\omega_p = 2\pi f_0$ and $\omega_s = \pi f_s$. Increasing the sampling rate (oversampling the signal) increases the width $\omega_s - \omega_p$ of the transition region and makes the filter easier and cheaper to build. The tighter the design constraints (i.e., the smaller the values of ϵ_p, ϵ_s, and $\omega_s - \omega_p$), the more difficult it is to find an acceptable $H(s)$. Designing one usually can be done only by making the order of the filter larger, and that approach adds greatly, and in many ways, to the difficulty of implementing the filter. Generally, we are forced to trade off among these competing requirements, relaxing the values of less critical parameters to enable tight values for the more important ones.

10-1-2 Butterworth and Chebyshev Lowpass Filters

During the 1930's, filter designers developed analytical procedures for finding the system functions that would provide a good approximation to an ideal lowpass-filter magnitude response. Many of these designs were provably optimal according to some criterion. One class of such filters is the class of **Butterworth filters**. Butterworth filters are optimal in the sense that, for a given passband and stopband cutoff frequency, they have the narrowest transition band among all filters with a *monotonic* magnitude response. The magnitude response of an nth order Butterworth lowpass filter with a cutoff frequency of 1 rad/s is

$$|H(j\omega)| = \frac{1}{\sqrt{1 + \omega^{2n}}}. \qquad (10.1)$$

A quick look at the formula shows that, for values of $\omega \ll 1$, the magnitude response is approximately 1; for values

of $\omega > 1$, the magnitude response decreases rapidly with increasing ω. Furthermore, for all n,

$$|H(j1)| = \frac{1}{\sqrt{2}}.$$

As the order of the filter n is increased, the magnitude response of the filter more nearly approximates the ideal lowpass filter. This is seen in the frequency response plot in Figure 10-4.

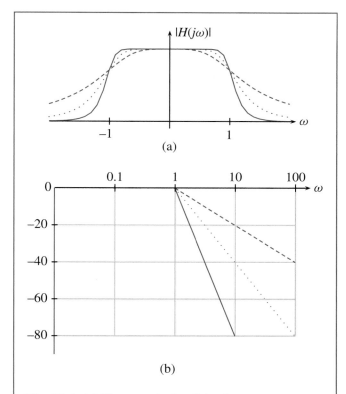

Fig. 10-4. (a) The magnitude of the frequency response of a Butterworth lowpass filter for $n = 2$ (dashed), $n = 4$ (dotted), and $n = 8$ (solid). (b) The Bode magnitude plots for the same three filters. It should be noted that as the order is increased, the magnitude response of the filter more closely approximates that of the ideal lowpass.

Where are the poles and zeros of an nth order Butterworth filter? To find these, we need to know $H(s)$. The standard way to learn it is to manipulate the frequency response until we can use the familiar substitution $s = j\omega$. If we square both sides of (10.1), we find that

$$|H(j\omega)|^2 = \frac{1}{1 + \omega^{2n}} = H(j\omega)H^*(j\omega)$$
$$= H(j\omega)H(-j\omega). \qquad (10.2)$$

The latter equality is true if $h(t)$ is real. Replacing $j\omega$ by s gives

$$H(s)H(-s) = \frac{1}{1 + \left(\dfrac{s}{j}\right)^{2n}} = \frac{(-1)^n}{s^{2n} + (-1)^n}. \qquad (10.3)$$

This function has $2n$ poles, located on a circle of radius 1 in the s-plane at equispaced angles. For the filter to be stable, we know that all of its poles must have negative real parts. Thus, we form $H(s)$ from the n poles with negative real parts (the other poles are associated with $H(-s)$). When n is odd, there will be a single real pole at $s = -1$ and all of the other poles will be complex. When n is even, all of the poles will be complex. These two situations are illustrated in Figure 10-5.

The calculations involved in finding the system function for a Butterworth filter are straightforward, but tedious. There are a number of software packages available to relieve the tedium and the likelihood of computational errors. For example, in the Signal Processing Toolbox of MATLAB, the coefficients of the Butterworth system function can be computed by using the function `butter`. Executing the command `help butter` will provide details on how it should be called. In the next two examples, we design two Butterworth filters, using only a calculator.

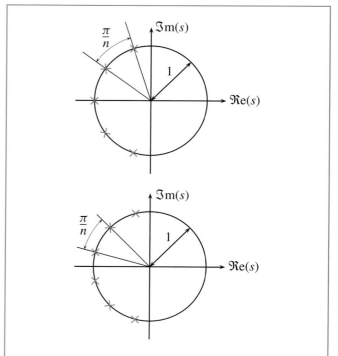

Fig. 10-5. Pole locations for an nth-order Butterworth lowpass filter with a cutoff frequency $\omega_c = 1$ rad/s. (top) n odd. (bottom) n even. The poles are located on a semi-circle of radius 1 at equally spaced angles.

DEMO: *Butterworth Filter*

Example 10-1   Finding $H(s)$ for a Butterworth Filter

What is the system function of a sixth-order Butterworth filter with a cutoff frequency of 1 rad/s? We can work this out from the known locations of the poles of this filter. This filter will have six poles, all of which lie on a circle in the s-plane of radius one. Because the angular spacing between adjacent poles must be $\pi/6$ radians and the poles must form complex conjugate pairs, we see from Figure 10-5 that they will be located at the complex locations

$$s_{1,2} = e^{\pm j \frac{11\pi}{12}}$$

$$s_{3,4} = e^{\pm j \frac{9\pi}{12}}$$

$$s_{5,6} = e^{\pm j \frac{7\pi}{12}}.$$

Having the poles, we can write down the system function, which is

$$H(s) = \frac{1}{(s^2 + 2\cos(\frac{11\pi}{12})s + 1)}$$

$$\cdot \frac{1}{(s^2 + 2\cos(\frac{9\pi}{12})s + 1)} \cdot \frac{1}{(s^2 + 2\cos(\frac{7\pi}{12})s + 1)}$$

$$= \frac{1}{(s^2 + 1.932s + 1)}$$

$$\cdot \frac{1}{(s^2 + 1.414s + 1)} \cdot \frac{1}{(s^2 + 0.518s + 1)}. \quad (10.4)$$

Each second-order factor realizes one of the pole pairs. For implementing the filter, it is often helpful to leave the expression for $H(s)$ in factored form rather than multiplying out the denominator polynomial, as we shall see shortly. ∎

 To check your understanding, try Drill Problem P10-1.

Example 10-2   **Finding the order of a Butterworth Filter**

The previous example illustrated the procedure for finding the system function of a Butterworth filter if the order of the filter, n, is known. More commonly, however, it is necessary to work out the necessary order from the filter specifications. To illustrate the procedure, consider the following problem:

Find the minimum-order Butterworth filter with a nominal cutoff frequency of 1 rad/s such that the maximum passband deviation is 0.1 for frequencies in the range $0 \leq \omega \leq 0.8$ and the maximum stopband deviation is 0.1 for frequencies greater than 2.0 rad/s.

We know that if the cutoff frequencies are fixed, then the passband and stopband deviations decrease monotonically with increasing filter order, and that the maximum passband deviation occurs at the edge of the passband (at $\omega = \omega_p = 0.8$). We also know that the order must be an integer. One approach is to find the (non-integer) order so that the gain is exactly 0.9 at $\omega = 0.8$ and then round up this value to the next larger integer. We then do a similar computation at $\omega = 2.0$, where the maximum stopband deviation occurs. Since both conditions must be satisfied, the necessary order will be the larger of these two values.

The passband constraint thus requires that the magnitude response at $\omega = 0.8$ be larger than 0.9, or

$$\frac{1}{1 + (0.8)^{2n}} \geq (0.9)^2.$$

This inequality requires that $n \geq 3.249$. The minimum integer that satisfies this constraint is $n = 4$.

The stopband constraint requires that

$$\frac{1}{1 + (2.0)^{2n}} \leq (0.1)^2,$$

which is satisfied for $n \geq 3.3146$. This condition will also be met by a fourth-order filter. Therefore, the design constraints can be met with a fourth-order Butterworth filter.

The poles of a fourth-order Butterworth lowpass filter whose cutoff frequency is 1 rad/s are located at

$$s_{1,2} = e^{\pm j \frac{7\pi}{8}}$$

$$s_{3,4} = e^{\pm j \frac{5\pi}{8}},$$

and its system function is given by

$$H(s) = \frac{1}{(s^2 + 2\cos(\frac{7\pi}{8})s + 1)} \cdot \frac{1}{(s^2 + 2\cos(\frac{5\pi}{8})s + 1)}$$

$$= \frac{1}{(s^2 + 1.848s + 1)} \cdot \frac{1}{(s^2 + 0.765s + 1)}.$$

∎

 To check your understanding, try Drill Problem P10-2.

We have already mentioned that Butterworth filters are optimal in the sense that, for a given passband and stopband cutoff frequency, they have the narrowest transition band among all filters with a *monotonic* magnitude response. We can get sharper filters, however, if we relax the monotonicity constraint. Chebyshev (type 1) filters achieve comparable behavior from a lower-order filter by allowing oscillations, or ripples, in the passband magnitude. What we mean by this is seen in Figure 10-6. Notice that, for this fourth-order filter, there are four peaks (and three valleys) in the passband magnitude function, but that the magnitude response in the passband, nevertheless, is confined to the range $0.9 \le |H(e^{j\omega})| \le 1$. The magnitude response is exactly 1 at each of the peaks and, for this example, it is exactly 0.9 at each of the valleys. The stopband behavior of a type-1

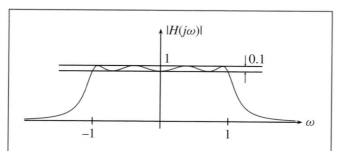

Fig. 10-6. A fourth-order (type I) Chebyshev filter with a passband cutoff frequency of 1 rad/s.

Chebyshev filter is monotonic. In general, the maximum passband gain is 1 and the minimum gain in the passband, which is related to the maximum passband deviation, is $1/\sqrt{1 + \epsilon^2}$. The parameter ϵ is user definable.

Chebyshev lowpass filters, like Butterworth lowpass filters, have no zeros. The n poles lie on an ellipse in the s-plane. For a filter with a passband cutoff frequency of 1 radian/second, their locations can be found by means of the following calculations. First, the maximum deviation in the passband is used to compute the parameter ϵ. It, in turn, is used to calculate an auxiliary parameter

$$\alpha = \frac{1}{\epsilon} + \sqrt{1 + \frac{1}{\epsilon^2}}. \tag{10.5}$$

The parameter α, along with the filter order n, is then used to find the minor (a) and major (b) axes of the ellipse on which the poles lie:

$$a = \frac{1}{2}\left(\alpha^{1/n} - \alpha^{-1/n}\right) \tag{10.6}$$

$$b = \frac{1}{2}\left(\alpha^{1/n} + \alpha^{-1/n}\right). \tag{10.7}$$

The ith pole of the filter is located at

$$s_i = a\cos\theta_i + jb\sin\theta_i, \tag{10.8}$$

where θ_i is the same as the angle of the ith pole of a Butterworth filter of the same order. Finally, the numerator constant of the filter needs to be set so that the filter will have the proper gain. The filter gain at $\omega = 0$ will be 1 if the filter order is odd and $1/\sqrt{1 + \epsilon^2}$ if the filter order is even, since $\omega = 0$ is located on a peak for an odd-length filter, but in a valley for an even-length one.

Software implementations of these computations are readily available. In the Signal Processing Toolbox of MATLAB, the appropriate function is `cheby1`. The following example illustrates the details of the calculations.

 DEMO: *Chebyshev Filter*

Example 10-3   **Finding a Chebyshev Lowpass Filter**

As an example, let us find the system function of the filter shown in Figure 10-6, which is a fourth-order Chebyshev lowpass filter with a maximum passband deviation of 0.1 and a passband cutoff frequency of 1 radian/second. We begin by finding the four poles of the filter. We know that the parameters θ_i needed to find the pole locations in (10.8) are the same as those required for a fourth-order Butterworth filter. Therefore, they are

$$\theta_{1,2} = \pm \frac{7\pi}{8}$$

$$\theta_{3,4} = \pm \frac{5\pi}{8}.$$

Finding the axes of the ellipse, a and b, however, requires a little more work.

First, we need to compute the parameter ϵ. Since the desired minimum frequency-response magnitude in the passband of the filter is 0.9, we can write

$$\frac{1}{\sqrt{1 + \epsilon^2}} = 0.9,$$

which we can solve for ϵ. The result is $\epsilon = 0.4843$. We next use the value of ϵ to find the parameter α, using (10.5):

$$\alpha = \frac{1}{0.4843} + \sqrt{1 + \frac{1}{(0.4843)^2}} = 4.3589.$$

From this value, we can find a and b, using (10.7):

$$a = 0.5 \left(\alpha^{1/4} - \alpha^{-1/4} \right) = 0.3764$$

$$b = 0.5 \left(\alpha^{1/4} + \alpha^{-1/4} \right) = 1.0685$$

Therefore, from (10.8), the four poles are located at

$$s_{1,2} = 0.3764 \cos (7\pi/8) \pm j1.0685 \sin (7\pi/8)$$

$$= -0.3478 \pm j0.4089$$

$$s_{3,4} = 0.3764 \cos (5\pi/8) \pm j1.0685 \sin (5\pi/8)$$

$$= -0.1441 \pm j0.9872.$$

The pole locations tell us the system function to within a constant multiplier:

$$H(s) = \frac{K}{(s - s_1)(s - s_2)(s - s_3)(s - s_4)}$$

$$= \frac{K}{(s^2 + 0.6956s + 0.2881)(s^2 + 0.2882s + 0.9952)}.$$

We can figure out the constant K from the fact that $H(j0) = 0.9$. (The filter is of even order.) Thus,

$$\frac{K}{(0.2881)(0.9952)} = 0.9 \implies K = 0.2580,$$

so

$$H(s) = \frac{0.2580}{s^2 + 0.6956s + 0.288}$$

$$\cdot \frac{1}{s^2 + 0.2882s + 0.9952}.$$

The plot of the magnitude response is shown in Figure 10-6. ∎

 To check your understanding, try Drill Problem P10-3.

One can gain insight by examining the frequency responses of the Butterworth and Chebyshev filters in light of the pole locations, using the geometric reasoning that we developed in Chapter 9. There, we saw that a pair of complex poles introduced a bandpass type of frequency response. The distance of the poles from the $j\omega$-axis controlled the bandwidth of the response caused by the pair. For both the Butterworth and the Chebyshev filters, the pole pairs work together to maintain the passband of the filter. The poles of the Chebyshev filter are closer to the $j\omega$-axis than the poles of the Butterworth. The resulting narrower bandwidths of the responses from these poles are responsible for the ripples in the passband response and also for the narrower transition bands that these filters enjoy as a consequence. For the Butterworth, the pole locations are such that the frequency response contributions from the various poles compensate for each other in a way that ripples are avoided.

10-1-3   Realizations of Lowpass Second-Order Sections That Use Operational Amplifiers

The previous section has shown how we can find a system function whose magnitude response approximates that of an ideal lowpass filter, but this solves only part of the problem. An important question remains: how can we find an actual circuit that will have that system function? There are three approaches that have been used and that can be made to work. One is to simply lay out a candidate circuit containing a number of resistors, inductors, and capacitors and then to adjust the element values until the poles and zeros sit at their proper locations. This trial-and-error method is the approach that we have been using for simple first- and second-order circuits. For a sixth-order Butterworth lowpass filter, however, the circuit will need to have three inductors, three capacitors, and at least one resistor. The possibilities for false starts and the labor involved with solving the non-linear equations for the appropriate element values, assuming that we guess the right circuit topology, suggest that this should be the approach of last resort.

Another approach would be to design three second-order circuits, each of which implements one of the pairs of complex conjugate poles. These circuits can be connected together via buffer amplifiers, as shown in Figure 10-7. Breaking the circuit up into sections greatly simplifies the design. One problem with the resulting design, however, is the fact that it still requires inductors. It is often (but not always) desirable to avoid inductors. They do not integrate easily, particularly for large inductance values, and real inductors often deviate considerably from their ideal models.

A third approach is similar, in that it involves cascading second-order filters together, but it uses active

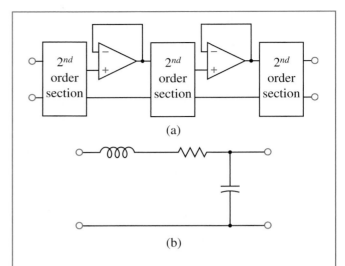

(a)

(b)

Fig. 10-7. (a) An implementation of a sixth-order Butterworth or Chebyshev filter as a cascade of three second-order sections. (b) One of the second-order lowpass sections.

sections that can be implemented with operational amplifiers, resistors, and capacitors only. Because of the active components, loading between the sections can be kept low. One such section is the Sallen–Key section illustrated in Figure 10-8, which we have seen before.

This particular circuit has a system function with two complex poles and no zeros. To see how the pole locations depend upon the component values, we need to compute its system function, as we shall now do. As with any operational amplifier circuit, we write KCL equations at all of the nodes except for the output of the operational amplifier and the ground. In this case, there are three such nodes, which are encircled with dashed circles. In Figure 10-8, the two nodes connected to the two terminals of the operational amplifier have the same node potential $f(t)$ (or $F(s)$ in the Laplace domain). Let the potential at the other encircled node be $e(t)$ ($E(s)$) and the potential at the output of the operational amplifier be $V_{\text{out}}(t)$ ($V_{\text{out}}(s)$).

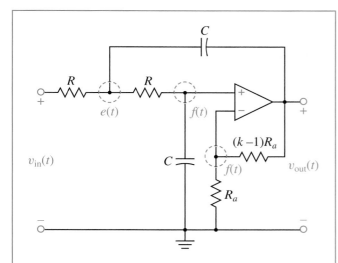

Fig. 10-8. A two-pole Sallen–Key second-order section. The input signal is applied as a potential difference between the two terminals on the left, and the output is the voltage measured between the terminals on the right.

Since no current flows into the inverting input of the opamp, the Laplace-domain KCL equation at its node reduces to a voltage-divider relation that can be solved for $F(s)$:

$$F(s) = \frac{R_a}{R_a + (k-1)R_a} V_{\text{out}}(s) = \frac{1}{k} V_{\text{out}}(s). \quad (10.9)$$

Notice that the result is independent of the value of R_a, so this value can be chosen for convenience.

The two other KCL equations are

$$\frac{E(s) - V_{\text{in}}(s)}{R} + Cs[E(s) - V_{\text{out}}(s)]$$
$$+ \frac{E(s) - F(s)}{R} = 0 \quad (10.10)$$

and

$$\frac{F(s) - E(s)}{R} + CsF(s) = 0. \quad (10.11)$$

From (10.11), we have

$$F(s)\left[\frac{1}{R} + Cs\right] = \frac{1}{R} E(s)$$

$$\implies E(s) = F(s)[1 + RCs] = \frac{1}{k}[1 + RCs]V_{\text{out}}(s).$$

Substituting this result and (10.9) into (10.10) gives

$$V_{\text{out}}(s)\frac{1}{k}[1 + RCs][2 + RCs]$$

$$- V_{\text{out}}(s)RCs - \frac{1}{k}V_{\text{out}}(s) = V_{\text{in}}(s).$$

If we group the appropriate terms, this can be written as

$$V_{\text{out}}(s)[2 + 3RCs + (RCs)^2 - kRCs - 1] =$$
$$kV_{\text{in}}(s),$$

from which it follows that

$$\frac{V_{\text{out}}(s)}{V_{\text{in}}(s)} = \frac{k}{(RCs)^2 + (3-k)RCs + 1}. \quad (10.12)$$

Therefore,

$$H(s) = \frac{\frac{k}{R^2C^2}}{s^2 + \frac{3-k}{RC}s + \frac{1}{R^2C^2}}. \quad (10.13)$$

As promised, this system function has one complex-conjugate pole pair. The distance of the poles from the origin of the s-plane is $1/RC$, and the real part of the pole locations can be adjusted by varying the value of k. Thus, we have the ability to place the poles wherever we wish, subject to some constraints. The gain of the section at $\omega = 0$ is k. We cannot control this gain, since the value of k is fixed once we place the poles. Furthermore,

$$1 < k < 3.$$

The lower limit is imposed by the fact that the feedback resistor for the opamp must have a positive resistance. The upper limit is necessary because, for $k \geq 3$, the real parts of the pole locations will be nonnegative, so the filter will be unstable. This means that the DC gain of the section (the magnitude of the frequency response at $\omega = 0$) must lie between one and three. If this limitation creates a problem, we can always cascade an additional amplifier (or a voltage divider) to reduce the gain.

To implement the sixth-order Butterworth filter of Example 10-1, which has three complex pole pairs, requires three of these sections. If the parameters for these sections are (R_i, C_i, k_i) for $i = 1, 2, 3$, then, by comparing (10.4) with (10.13), we immediately see that we must choose

$$R_1C_1 = R_2C_2 = R_3C_3 = 1. \qquad (10.14)$$

Notice that only the products of the resistance and capacitance values matter; we can choose convenient values for the capacitance values and then select the appropriate resistance values. The gain constants must be chosen so that

$$3 - k_1 = 0.518 \Longrightarrow k_1 = 2.482$$
$$3 - k_2 = 1.414 \Longrightarrow k_2 = 1.586$$
$$3 - k_3 = 1.932 \Longrightarrow k_3 = 1.068.$$

Notice that all three of these values fall within the required range. The overall gain of the filter is

$$k_1k_2k_3 = (2.482)(1.586)(1.068) = 4.2041.$$

The ordering of the three sections is arbitrary.

Even though we have taken some real filter specifications, approximated them by filters with implementable system functions, and identified at least a couple of circuits that are capable of realizing those system functions, we have still ignored several important real-world considerations. A thorough discussion of these is not appropriate here, but we should at least mention what they are. First, there are some limitations on component values. Resistances, inductances, and capacitances cannot be selected to have arbitrary values.

In general, they come only in standard values. Resistors, for example, can be purchased with resistance values that increase in increments of approximately 10%. Furthermore, two resistors that have the same nominal resistance, in fact, will not be identical and the resistance could change as a function of temperature. As a consequence, after we complete a design and measure the performance of the resulting circuit, there will be some deviation from the ideal. An implementation based on the Sallen–Key structure could be less accurate than one based on a different structure, when implemented with real component values. It may also require more power, since it requires operational amplifiers. The fact that this structure was used for this illustration should not be interpreted as a statement that this is a best choice. Indeed, the best choice would depend upon the constraints of a particular application.

10-2   Transforming Basic Filters

The methods in the last section showed us how to find circuits that have good lowpass frequency responses with a cutoff frequency of 1 rad/s. These are quite limiting, but they hint at other possibilities. How can we change the cutoff frequency of the filter? How can we get a design that uses more realistic element values? How should we choose the element values if we want to reduce the power absorbed by the circuit? How can we design a highpass filter? There are answers to all of these questions. Some require that we find different system functions; others that we find different implementations. For those that require new system functions, we could go back to first principles and work out a best design for each case as it is encountered. For those questions whose solution requires a better implementation, we could use trial and error until we identify a realization that meets our needs. In both cases, however, these are brute-force approaches. With a little insight, we can address all of these questions simply by modifying the filter designs that we have already found. In Section 10-2-1, we shall explore ways to do so. We begin by looking at ways to change the cutoff frequency of the filter after it has been designed.

10-2-1   Frequency Scaling

The simplest way to change a filter after it has been designed is to scale its frequency axis. For a lowpass filter, this has the effect of changing its cutoff frequency. Let $H(s)$ be the system function of a lowpass filter with a cutoff frequency of 1 rad/s, and define a new system function

$$G(s) = H(\frac{s}{\omega_0}). \qquad (10.15)$$

To find the frequency response of the new filter, we set $s = j\omega$. Then

$$G(j\omega) = H(j\frac{\omega}{\omega_0}).$$

Since $|H(j\omega)| \approx 1$ for $|\omega| < 1$, it follows that

$$|G(j\omega)| \approx 1 \qquad \text{for } \left|\frac{\omega}{\omega_0}\right| < 1$$

or

$$|G(j\omega)| \approx 1 \qquad \text{for } |\omega| < \omega_0.$$

By similar reasoning, $|G(j\omega)| \approx 0$ for $|\omega| > \omega_0$. Thus, $G(s)$ is the system function of a lowpass filter with a cutoff frequency of ω_0 rad/s.

 DEMO: *Frequency Scaling*

Example 10-4   Scaling the Cutoff Frequency

A second-order Butterworth filter with a cutoff frequency of 1 rad/s has the system function

$$H(s) = \frac{1}{s^2 + 1.414s + 1}.$$

Find an implementation of a second-order Butterworth lowpass filter with a cutoff frequency of 1000 rad/s.

In the approach outlined in Section 10-2-1, the first step is to find the appropriate system function, $G(s)$. Using (10.15), we have

$$G(s) = H(\frac{s}{1000}) = \frac{1}{\left(\frac{s}{1000}\right)^2 + 1.414\left(\frac{s}{1000}\right) + 1}$$

$$= \frac{1000^2}{s^2 + 1.414(1000)s + 1000^2}.$$

Comparing this result with (10.13), we find that we can implement the filter with a single Sallen–Key section if we select its parameters so that they have the values

$$\frac{1}{RC} = 1000 \implies RC = \frac{1}{1000}$$

$$3 - k = 1.414 \implies k = 1.586.$$

$\blacksquare$

 WORKED SOLUTION: *Frequency Scaling*

 To check your understanding, try Drill Problem P10-4.

The approach implied by the preceding example is a two-step procedure. First we applied frequency scaling to find the system function of the transformed system, then we found an implementation for the new system function. There is also a single-step approach. To understand it, recall that the system function for a circuit is derived by computing the ratio

$$H(s) = \frac{Y(s)}{X(s)}.$$

To find $Y(s)$, either we write KCL equations, KVL equations, and element relations and solve these explicitly, or we use one or more of the shortcuts that we have derived, such as voltage and current dividers, that do this same procedure implicitly. However we solve the circuit, the variable s is introduced into the analysis only through the element relations for the inductors and capacitors. This suggests that an alternative to applying frequency scaling to the system function is to apply it to the reactive elements in the circuit directly. But applying frequency scaling to the impedance of an inductor or capacitor is equivalent to modifying the inductance or capacitance values in a systematic way, as the following reasoning shows:

$$Ls \longrightarrow L\left(\frac{s}{\omega_0}\right) = \left(\frac{L}{\omega_0}\right)s = L's$$

$$Cs \longrightarrow C\left(\frac{s}{\omega_0}\right) = \left(\frac{C}{\omega_0}\right)s = C's.$$

This demonstrates that we can frequency-scale a circuit by replacing all of the inductors by new inductors whose inductances are smaller by the factor ω_0 and replacing all of the capacitors by new capacitors whose capacitances are similarly scaled.

Figure 10-9 shows two second-order RLC circuits and their magnitude responses. The top filter has element values chosen so that the filter will have a Chebyshev (type 1) response with a passband cutoff frequency of 1 rad/s:

$$H_a(s) = \frac{1.0324}{s^2 + 1.1376s + 1.1471}$$

For the second filter, the inductance and capacitance values have each been reduced by the factor 100 and the resistance has been left unchanged. The frequency response of this filter also has a Chebyshev characteristic, but now with a cutoff frequency of 100 rad/s as seen in the plot of its magnitude response. The system function of the second filter is

$$H_b(s) = \frac{10,324}{s^2 + 113.76s + 11,471}.$$

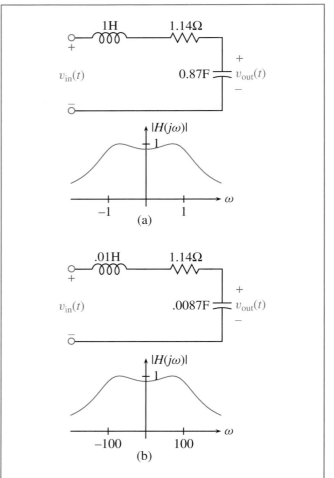

Fig. 10-9. An illustration of frequency scaling. (a) A second-order RLC circuit and its frequency response. (b) The circuit and its frequency response after the inductance and capacitance are each reduced by the factor 100.

10-2-2   Lowpass-to-Highpass Transformations

The idea of changing a system function by transforming the frequency variable can also be used to change the shape of the frequency response. For example, we can convert a lowpass filter with the system function $H(s)$

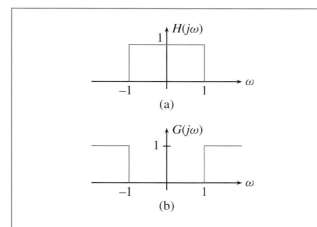

Fig. 10-10. (a) Lowpass filter. (b) Highpass filter that results from the transformation $s \to 1/s$.

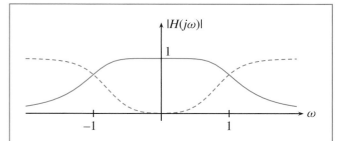

Fig. 10-11. The magnitude of the frequency response of a sixth-order Butterworth lowpass filter (solid) and the corresponding Butterworth highpass filter (dashed).

into a highpass filter if we apply the transformation

$$G(s) = H(\frac{1}{s}).$$

Letting $s = j\omega$ reveals that

$$G(j\omega) = H(-j\frac{1}{\omega}).$$

When ω is large, $1/\omega$ is small. The high-frequency behavior of $G(j\omega)$, therefore, resembles the low-frequency behavior of $H(j\omega)$, and vice versa. Thus, if $H(j\omega)$ passes low frequencies and rejects high ones, then $G(j\omega)$ will pass high frequencies and reject low ones. Furthermore, $G(j1) = H(-j1)$. So, if the lowpass filter has a cutoff frequency of ± 1 rad/s, the highpass filter will also have a cutoff frequency of ± 1 rad/s. Figure 10-10 illustrates this case.

Example 10-5   Butterworth Highpass Filter

As an example, let us design a sixth-order Butterworth highpass filter with a cutoff frequency of 1 rad/s. We begin with the sixth-order Butterworth lowpass filter that we

designed in Example 10-1. In that example, we showed its system function to be

$$H(s) = \frac{1}{(s^2 + 1.932 + 1)} \cdot \frac{1}{(s^2 + 1.414s + 1)}$$
$$\cdot \frac{1}{(s^2 + 0.518s + 1)}.$$

To get the system function of the highpass, we replace s by $1/s$, which gives

$$G(s) = \left[\frac{1}{(\frac{1}{s})^2 + 1.932(\frac{1}{s}) + 1} \right] \left[\frac{1}{(\frac{1}{s})^2 + 1.414(\frac{1}{s}) + 1} \right]$$
$$\cdot \left[\frac{1}{(\frac{1}{s})^2 + 0.518(\frac{1}{s}) + 1} \right]$$
$$= \left[\frac{s^2}{s^2 + 1.932s + 1} \right] \left[\frac{s^2}{s^2 + 1.414s + 1} \right]$$
$$\cdot \left[\frac{s^2}{s^2 + 0.518s + 1} \right].$$

The frequency responses of the two filters are superimposed in Figure 10-11. Notice that the lowpass and the highpass have the same poles, but that the highpass adds zeros at $s = 0$. This is the same behavior that we observed for first- and second-order examples in Chapter 9. ∎

To check your understanding, try Drill Problem P10-5.

As was the case for a lowpass filter, a highpass filter of high order can be implemented by cascading a number of second-order filters. Figure 10-12 shows a Sallen–Key second-order highpass section. (Notice that it is different from the lowpass section in Figure 10-8.) It has the system function

$$H(s) = \frac{\frac{ks^2}{R^2C^2}}{s^2 + \frac{3-k}{RC}s + \frac{1}{R^2C^2}}. \tag{10.16}$$

The derivation of this system function is left as a homework problem. (See Problem P10-10.)

The lowpass-to-highpass transformation can be combined with frequency scaling to permit complete flexibility between the cutoff frequencies of the two filters. This can be accomplished in a single step by using the transformation function

$$G(s) = H(\frac{\omega_0}{s}). \tag{10.17}$$

With this transformation, a lowpass filter with a cutoff frequency of one radian/second maps to a highpass filter of the same order with a cutoff frequency of ω_0 radians/second.

10-2-3 Lowpass-to-Bandpass Transformations

A bandpass filter can also be designed by using a lowpass prototype filter and then transforming the design. Let $H_{lp}(j\omega)$ be the system function of the lowpass filter with a cutoff frequency of W radians/second, such as the one whose frequency response is idealized in Figure 10-13(a). Then, if $G_{bp}(j\omega)$ is defined as

$$G_{bp}(s) = H_{lp}\left(\frac{s^2 + \omega_c^2}{s}\right), \tag{10.18}$$

it will be a bandpass filter with a passband that extends in frequency from approximately $\omega_c - W/2$ to $\omega_c + W/2$.

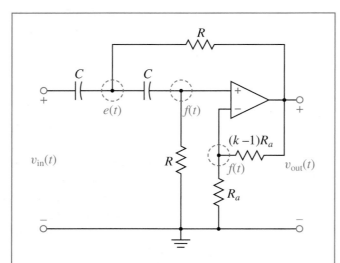

Fig. 10-12. A two-pole, two-zero Sallen–Key second-order highpass section.

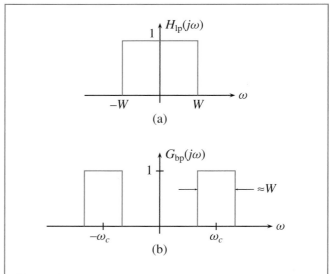

Fig. 10-13. Illustration of a lowpass-to-bandpass transformation. (a) Prototype lowpass filter. (b) Resulting bandpass filter when the transformation of (10.18) is applied.

To see this, it is helpful to sample $G(j\omega)$ at a few key frequencies:

$$G_{bp}(0) = H_{lp}(-j\infty) = 0$$

$$G_{bp}(\pm j\infty) = H_{lp}(\mp j\infty) = 0$$

$$G_{bp}(\pm j\omega_c) = H_{lp}(0) = 1.$$

The cutoff frequencies for $G_{bp}(s)$ will occur when

$$\frac{-\omega^2 + \omega_c^2}{j\omega} = \pm jW.$$

This quadratic equation has the solutions

$$\omega = \frac{\pm W \pm \sqrt{W^2 + 4\omega_c^2}}{2}.$$

For the case $\omega_c \gg W$, the cutoff frequencies are at $\pm\omega_c \pm W/2$, as shown on Figure 10-13(b).

If the order of the lowpass prototype filter is n, the order of the bandpass filter will be $2n$. This is clearly seen in the following example.

Example 10-6   **A Fourth-Order Butterworth Bandpass Filter**

A second-order Butterworth lowpass filter with a cutoff frequency of W rad/s has the system function

$$H_{lp}(s) = \frac{W^2}{s^2 + \sqrt{2}Ws + W^2}. \qquad (10.19)$$

Applying the lowpass-to-bandpass transformation (10.18) gives

$$G_{bp}(s) = \frac{W^2}{\left(\frac{s^2+\omega_c^2}{s}\right)^2 + 2\sqrt{2}W\left(\frac{s^2+\omega_c^2}{s}\right) + W^2}$$

$$= \frac{W^2s^2}{s^4 + \sqrt{2}Ws^3 + (2\omega_c^2 + W^2)s^2 + \sqrt{2}W\omega_c^2 s + \omega_c^4}.$$

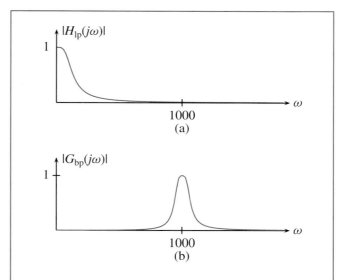

Fig. 10-14. (a) Second-order Butterworth prototype lowpass filter (positive frequencies only). (b) Bandpass filter designed from it using a lowpass-to-bandpass transformation with $\omega_c = 1000$.

This is a fourth-order filter. Substituting $W = 100$, $\omega_c = 1000$ gives a bandpass filter that passes frequencies between 950 rad/s and 1050 rad/s. Its system function is

$$G_{bp}(s) =$$

$$\frac{10^4 s^2}{s^4 + 100\sqrt{2}s^3 + (201 \times 10^4)s^2 + \sqrt{2} \times 10^8 s + 10^{12}},$$

and the frequency response is plotted in Figure 10-14 for both the lowpass prototype and the bandpass filter. Notice that the passband of the bandpass filter is only one-half the width of the passband of the lowpass prototype.

To implement this filter, we can factor the system function into the product of two second-order sections:

$$G_{bp}(s) = \left(\frac{100}{s^2 + 73.2107s + 1.07330 \times 10^6}\right)$$

$$\cdot \left(\frac{100s^2}{s^2 + 68.2107s + 0.9317 \times 10^6}\right).$$

In this form, it can be implemented by cascading a Sallen–Key lowpass section (Figure 10-8) and a Sallen–Key highpass section (Figure 10-12), if the element values are appropriately chosen. It can also be implemented by cascading two Sallen–Key bandpass sections. (See Problem P10-12 and Figure 9-24.)   ■

10-2-4   Impedance Scaling

Whenever the system function of a circuit is a ratio of voltages (i.e., when the input to the circuit is a voltage source and the output of the circuit is a voltage), we can scale all of the impedances in the circuit by the factor K without affecting the system function. This is also true when the system function is a ratio of currents.

To increase all of the impedances in a circuit by the factor K, we

- increase all the resistances by the factor K,
- increase all the inductances by the factor K, and
- decrease all the capacitances by the factor K.

For a circuit whose input is a voltage source, the net effect of this impedance scaling will be to reduce all of the currents in the circuit by the factor K and to reduce the power absorbed by the circuit by the factor K, while keeping all of the voltages in the circuit the same. This can also be achieved in a circuit whose input is a current source, if the current source is also scaled downward by the factor K.

Impedance scaling is an important technique to enable a circuit to be realized with realistic element values and to limit the power absorbed by a circuit. This is illustrated in the following example.

Example 10-7   Working with Limited Component Values

In Figure 10-13, we applied frequency scaling to a lowpass filter with a cutoff frequency of 1 rad/s to get a

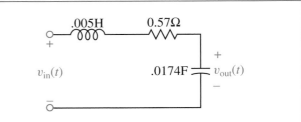

Fig. 10-15. Circuit with the same frequency response as that in Figure 10-13, but to which impedance scaling has been applied.

filter with a cutoff frequency of 100 rad/s. Unfortunately, we discover that the only inductor that is readily available to implement the latter filter has an inductance of 5mH (=0.005H). In order to use it, we will need to adjust the values of the other components. Let R, L, and C denote the values of the components before scaling and R', L', and C' denote their values after the scaling. Then $R = 1.14\Omega$, $L = 0.01H$, and $C = 0.0087F$. Furthermore, $R' = KR$, $L' = KL$, and $C' = C/K$. We require that

$$L' = 0.005 = KL = K(0.01),$$

which means we must have

$$K = \frac{0.005}{0.01} = 0.5.$$

Therefore,

$$R' = 0.5(1.14) = .72\Omega \qquad (10.20)$$

$$C' = 0.0087/0.5 = 0.0174F. \qquad (10.21)$$

The final circuit is shown in Figure 10-15.   ■

To check your understanding, try Drill Problem P10-6.

10-3 Chapter Summary

10-3-1 Important Points Introduced

- Butterworth lowpass filters are a family of system functions with monotonic magnitude responses that approximate an ideal lowpass filter.
- The poles of a Butterworth filter lie on a semicircle in the left half of the s-plane. The poles of a Chebyshev (type I) filter lie on half of an ellipse.
- A high-order filter can be implemented by cascading sections of second order.
- Active (opamp) realizations of circuits allow systems with complex poles to be implemented without inductors.
- A circuit can be frequency scaled by dividing all of its inductances and capacitances by the scaling factor.
- The system function of a lowpass filter can be mapped into the system function of a highpass or bandpass filter by using an appropriate transformation of variables.
- Impedance scaling can be used to reduce the power absorbed by a circuit or to change its element values.

10-3-2 New Abilities Acquired

You should now be able to do the following:

(1) Find the system function corresponding to a Butterworth or Chebyshev (type 1) lowpass filter.

(2) Find the order of a Butterworth filter that will meet a set of frequency-domain specifications.

(3) Implement a high-order lowpass filter by cascading sections of second order.

(4) Select the element values in a Sallen–Key filter so as to locate the poles at a desired location.

(5) Modify a system function to frequency scale its frequency response by an arbitrary scale factor.

(6) Modify the inductor and capacitor values in a circuit to frequency scale its frequency response by an arbitrary factor.

(7) Convert the system function of a lowpass filter into the system function of a highpass or bandpass filter with arbitrary cutoff frequencies.

(8) Impedance scale a circuit to reduce the currents in it and the amount of power dissipated.

10-4 Problems

10-4-1 Drill Problems

P10-1 Design an eighth-order Butterworth filter with a cutoff frequency of 1 rad/s.

P10-2 Find the system function of the Butterworth filter of minimum order that satisfies the following specifications: The cutoff frequency is 1 rad/s; the maximum passband deviation is 0.01 for frequencies in the range $0 \leq \omega \leq 0.7$; and the maximum stopband deviation is 0.01 for frequencies greater than 3.0 rad/s.

P10-3 Design a sixth-order Chebyshev (type 1) lowpass filter with a cutoff frequency of 1 rad/s and a maximum passband deviation of 0.05.

P10-4 The sixth-order Butterworth lowpass filter that we derived in Example 10-1 had a nominal cutoff frequency of 1 rad/s and the system function

$$H(s) = \frac{1}{(s^2 + 1.932 + 1)} \cdot \frac{1}{(s^2 + 1.414s + 1)} \cdot \frac{1}{(s^2 + 0.518s + 1)}.$$

(a) Use this as a prototype filter to construct the system function of a sixth-order Butterworth highpass filter with a nominal cutoff frequency of 100 rad/s.

(b) For what range of frequencies is

$$|H(j\omega)| \leq 0.05?$$

(c) For what range of frequencies is

$$1 \geq |H(j\omega)| \geq 0.95?$$

P10-5 A third-order Chebyshev (type 1) lowpass filter with a passband cutoff frequency of 1 rad/s ($\epsilon = 0.1526$) has the system function

$$H_{\mathrm{lp}}(s) = \frac{1.6381}{s^3 + 1.9388s^2 + 2.6295s + 1.6381}.$$

Find the system function of a Chebyshev (type 1) highpass filter with a passband cutoff frequency of 5 rad/s for the same value of ϵ.

P10-6 Through a process of trial and error, it has been learned that, when $R = 1000\,\Omega$, $C = 10^{-5}\,\mathrm{F}$, and $L = 10^{-5}\,\mathrm{H}$, the circuit in Figure P10-6 will have just the right system function to meet a specific need. Unfortunately, the capacitance is not a value that is readily available. Use impedance scaling to find values for the three elements, subject to the constraints

$$10^{-9}\,\mathrm{F} \leq C \leq 10^{-6}\,\mathrm{F}$$
$$10^{-6}\,\mathrm{H} \leq L \leq 10^{-3}\,\mathrm{H},$$

that will minimize the power absorbed by the circuit while leaving the system function unchanged.

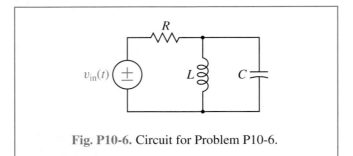

Fig. P10-6. Circuit for Problem P10-6.

10-4-2 Basic Problems

P10-7 Let $H(s)$ be the system function of a fourth-order Butterworth filter with a cutoff frequency (3-dB frequency) of 1 rad/s.

(a) If the frequencies in the range $0 \leq \omega \leq 0.9$ form the passband of the filter, find the maximum passband deviation from unity, ϵ_p.

(b) If the stopband of the filter consists of all frequencies above 1.5 rad/s, find the maximum stopband deviation from unity, ϵ_s.

(c) Repeat parts (a) and (b) if the filter order is increased to six.

P10-8] Let $H(s)$ be the system function of a fourth-order Butterworth filter with a cutoff frequency (3-dB frequency) of 1 rad/s.

(a) Define the passband of the filter as the range of frequencies $0 \leq \omega \leq \omega_p$. Find the largest value of ω_p such that the maximum passband deviation, ϵ_p, is 0.1.

(b) Define the stopband of the filter as the range of frequencies above ω_s ($\omega \geq \omega_s$). If the maximum stopband deviation $\epsilon_p = 0.01$, find the minimum value of ω_s.

(c) Repeat parts (a) and (b) if the filter order is increased to six.

P10-9 A second-order Butterworth lowpass filter with a cutoff frequency of 8000 rad/s will have an $H(s)$ described by the pole–zero plot shown in Figure P10-9(b), where the output signal is the voltage across the terminals of the capacitor.

(a) Calculate analytically the quantity $|H(j2\pi f)|/H(0)$ from the pole–zero plot, and plot it as a function of f.

(b) Find values of L and C to realize $H(s)$, using the form of the circuit drawn in Fig P10-9(a).

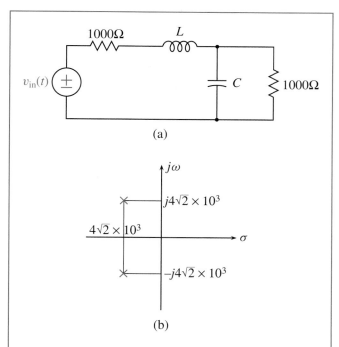

(a)

(b)

Fig. P10-9. (a) Circuit for Problem P10-9. (b) Pole–zero plot.

P10-10 Show that the system function of the Sallen–Key second-order *highpass* section shown in Figure 10.12 is

$$H(s) = \frac{\frac{ks^2}{R^2C^2}}{s^2 + \frac{3-k}{RC}s + \frac{1}{R^2C^2}}.$$

P10-11 A second-order Chebyshev (type 1) lowpass filter with a cutoff frequency of 2 rad/s has the system function

$$H(s) = \frac{3.9305}{s^2 + 2.1955s + 4.4100}.$$

(a) Construct the system function of second-order Chebyshev (type 1) highpass filter with a cutoff frequency of 1000 rad/s.

(b) Exhibit the system function of a fourth-order Chebyshev (type 1) bandpass filter that will pass frequencies in the range $1000 < |\omega| < 2000$ and attenuate other frequencies as much as possible.

P10-12 The circuit in Figure P10-12 is a second-order Sallen–Key *bandpass* section. Compute its system function

$$H_{bp}(s) = \frac{V_{out}(s)}{V_{in}(s)}.$$

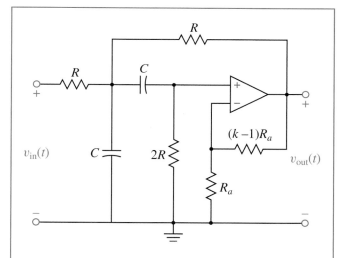

Fig. P10-12. A Sallen–Key second-order bandpass section for Problems P10-12 and P10-13.

P10-13 In Example 10-5, we derived the system function of a fourth-order bandpass filter. In factored form, the result was

$$G_{bp}(s) = \frac{10^2 s}{s^2 + 68.2s + 931,700}$$

$$\cdot \frac{10^2 s}{s^2 + 73.2s + 1,073,300}.$$

(a) Draw an implementation of the filter as a cascade of a Sallen–Key lowpass section and a Sallen–Key highpass section. Include all of the element values in your drawing.

(b) Draw an implementation of the filter as a cascade of two Sallen–Key bandpass sections. (See Problem P10-12.)

P10-14 A prototype filter with the system function $H(s)$ is used to create a new filter with the system function

$$G(s) = H\left(\frac{100}{s}\right).$$

The magnitude of the frequency response of $H(s)$ is shown in Figure P10-14. Sketch the magnitude of the frequency response of $G(s)$.

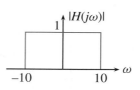

Fig. P10-14. Magnitude of the frequency response of the filter $H(s)$ in Problem P10-14.

P10-15 A first-order lowpass filter has the system function

$$H_{lp}(s) = \frac{1}{s + a}.$$

(a) This filter is used to design the highpass filter $G_{hp}(s) = H_{lp}(\omega_c/s)$. Plot the pole–zero diagram for $G_{hp}(s)$.

(b) Now let the same filter be used to design the bandpass filter

$$G_{bp}(s) = H_{lp}\left(\frac{s^2 + \omega_c^2}{s}\right).$$

Plot the pole–zero diagram for $G_{bp}(s)$, if $\omega_c \gg |a|$.

P10-16 The starting point of this problem is the ideal lowpass filter shown in Figure P10-16.

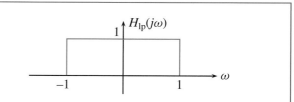

Fig. P10-16. An ideal lowpass filter with a gain of 1 and a cutoff frequency of 1 rad/s for Problem P10-16.

(a) The filter $H_{hp}(s)$ is generated from this lowpass filter by using the transformation

$$H_{hp}(s) = H_{lp}\left(\frac{3}{s}\right).$$

Plot the resulting frequency response, $H_{hp}(j\omega)$.

(b) The filter $H_{bp}(s)$ is generated from the lowpass filter $H_{lp}(s)$ by using the transformation

$$H_{bp}(s) = H_{lp}\left(\frac{s^2 + 16}{s}\right).$$

Plot the resulting frequency response $H_{bp}(j\omega)$.

(c) A new filter $H_{bs}(s)$ is derived by applying the transformation of part (b) to the filter designed in part (a):

$$H_{bs}(s) = H_{hp}\left(\frac{s^2 + 16}{s}\right).$$

Plot the resulting frequency response $H_{bs}(j\omega)$.

(d) Find a single transformation function $T(s)$ so that the filter in part (c) can be obtained from the original lowpass filter in a single step; that is, T should satisfy the relationship

$$H_{bs}(s) = H_{lp}(T(s)).$$

10-4-3 Advanced Problems

P10-17 If $H(s)$ is a rational function (i.e., a ratio of polynomials in s with real coefficients), then it is easy to show that $|H(j\omega)|^2$ will be a rational function in ω^2. We can go backwards from $|H(j\omega)|^2$ to $H(s)$ via the following procedure:

(a) Observe that

$$|H(j\omega)|^2 = H(j\omega)H^*(j\omega) = H(j\omega)H(-j\omega)$$
$$= [H(s)H(-s)]_{s=j\omega}.$$

Hence, if we substitute $\omega^2 = -s^2$ in $|H(j\omega)|$, we can identify the result as $H(s)H(-s)$.

(b) Produce a pole–zero plot of $H(s)H(-s)$. In general, the complex poles and zeros will occur in quadruplets, and the real poles and zeros will occur in pairs, as shown in Figure P10-17.

(c) Identify the poles in the left half-plane with $H(s)$. Their images in the right half-plane then belong to $H(-s)$. The zeros of $H(s)$ may be taken (in complex-conjugate pairs) from either half-plane; then their mirror images go to $H(-s)$.

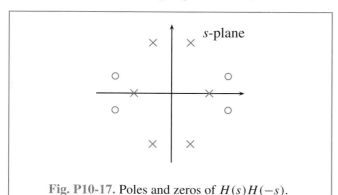

Fig. P10-17. Poles and zeros of $H(s)H(-s)$.

Find the system functions, $H(s)$, corresponding to the following squared-magnitude functions.

(a) $|H(j\omega)|^2 = \frac{1+\omega^2}{1+\omega^4}$

(b) $|H(j\omega)|^2 = \frac{\omega^4+5\omega^2+4}{\omega^4-6\omega^2+25}$

P10-18 The circuit in Figure P10-18 can enable an amplifier (modelled by its Thévenin equivalent network to the left of the terminals) to drive a high-frequency tweeter (modelled as the lower resistor on the right) and a low-frequency woofer (modelled by the upper resistor on the right).

(a) Show that the impedance $Z(s)$ seen by the amplifier (measured to the right of the terminals) is, in fact, constant and has a value that is independent of frequency.

(b) Find the system functions relating each of the outputs to the input if $L = \sqrt{2}R/\omega_0$ and $C = 1/(\sqrt{2}R\omega_0)$.

(c) Show that each of these is a second-order Butterworth filter with a cutoff frequency of ω_0.

(d) Why is it important that all three of the resistance values should be the same?

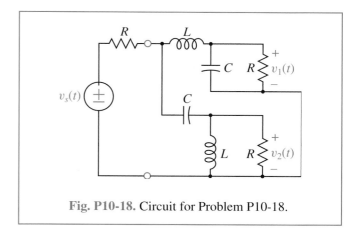

Fig. P10-18. Circuit for Problem P10-18.

P10-19 In the implementations of the filters in this chapter, we have relied on the fact that second-order sections can be cascaded together when there is no loading between them and that the overall system function will be the product of their individual system functions. In this problem, we wish to establish this fact.

Figure P10-19 shows two four-terminal networks connected together in cascade, with an operational amplifier as a buffer to isolate them. If the system function of the first network is

$$H_1(s) = \frac{V_2(s)}{V_1(s)}$$

and that of the second is

$$H_2(s) = \frac{V_4(s)}{V_3(s)},$$

show that the overall system function $H(s) = V_4(s)/V_1(s)$ is equal to $H_1(s)H_s(s)$.

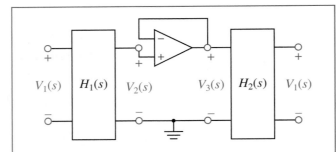

Fig. P10-19. Two four-terminal networks connected in cascade for Problem P10-19.

Show that if a circuit is connected by isolating networks, the overall system function is the cascade.

P10-20 This problem asks you to find the minimum-order Butterworth lowpass filter that satisfies the following constraints on its magnitude response:

$$1 \geq |H(j\omega)| \geq 0.9 \ \text{for} \ |\omega| \leq 800$$

$$0.1 \geq |H(j\omega)| \quad \text{for} \ |\omega| \geq 1200.$$

You should also compute the half-power (cutoff) frequency, ω_c, of the filter. Because the filter order n must be an integer, ω_c is not unique.

P10-21 Throughout this problem, assume that $H(j\omega) \approx 0$ for all $|\omega| > W$ and that $W \ll \omega_0$.

(a) One simple form of *lowpass–bandpass transformation* replaces ω by $(\omega^2 - \omega_0^2)/\omega$ wherever it appears. Argue that, under the assumed conditions,

$$H\left(j\frac{\omega^2 - \omega_0^2}{\omega}\right) \approx H(j[\omega - \omega_0]) + H(j[\omega + \omega_0]).$$

(b) Evaluate approximately the impulse response of the bandpass filter in terms of $h(t)$, the impulse response of the lowpass filter.

(c) Suppose that the system function of the lowpass filter, which is constructed from resistors, inductors, and capacitors, is $H(s)$. Argue that an implementation of the bandpass filter can be obtained by making the following substitutions:

  (i) Replace each capacitor having capacitance C in the lowpass by a *parallel* connection of an inductor having inductance L_1 and a capacitor having capacitance C_1.

  (ii) Replace each inductor having inductance L in the lowpass by a *series* connection of an inductor having inductance L_2 and a capacitor having capacitance C_2.

Find the values of L_1, C_1, L_2, and C_2 in terms of L, C, and ω_0.

P10-22 Consider the circuit in Figure P10-22, for which the single input is a voltage source and there are no dependent sources. The output of the circuit is a voltage, as shown on the right.

If all of the impedances in the circuit are increased by the factor K, what effect will this have on the total power consumed by the circuit?

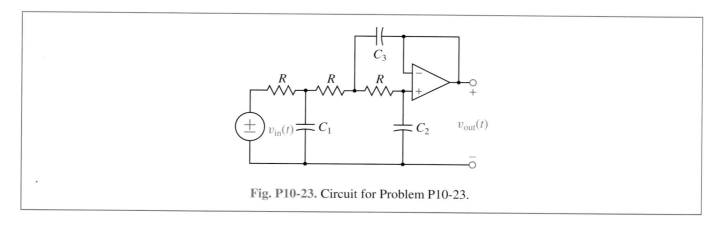

Fig. P10-23. Circuit for Problem P10-23.

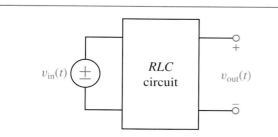

Fig. P10-22. Circuit for Problem P10-22, which contains only resistors, inductors, and capacitors.

P10-23

(a) Show that the system function of the circuit in Figure P10-23 is given by

$$H(s) =$$
$$\frac{1}{R^3 C_1 C_2 C_3 s^3 + 2R^2 C_2 (C_1 + C_3) s^2 + R(C_1 + 3C_2)s + 1}.$$

(b) Locate the poles of $H(s)$ on a plot of the s-plane for the special case $C_1 = 0.0022\,\mu F$, $C_2 = 330\,pF$, $C_3 = 0.0056\,\mu F$, $R = 10k\Omega$. (You might want to use a computer to help in factoring the polynomial.) Notice that the poles are close to those of a third-order Butterworth lowpass filter.

(c) Describe how the pole locations and the cutoff frequency of the filter change if R is doubled with no change in the values of the capacitors. Repeat if R is halved. (You should be able to make these alterations without refactoring or recomputing the denominator polynomial.)

10-4-4 Design Problems

P10-24 A filter of odd order cannot be built by using only second-order sections. One possible implementation uses a single first-order section, which can then be cascaded with an appropriate number of second-order sections.

(a) Design an active implementation of a circuit that will have the system function

$$H(s) = \frac{1}{s + a}.$$

(b) Design a third-order Butterworth filter with a cutoff frequency of 1 rad/sec.

(c) Draw the implementation of your filter as a cascade of first- and second-order sections. Be sure to provide values for all of the circuit elements.

(d) Draw an implementation of a third-order Butterworth *highpass* filter.

P10-25 A straightforward method for designing a bandpass filter that has a wide passband is to cascade a lowpass filter with a highpass filter.

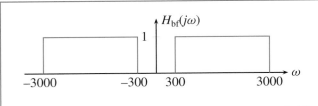

Fig. P10-25. An ideal bandpass filter for Problem P10-25.

(a) The ideal bandpass filter shown in Figure P10-25 can be realized as the cascade of a unity-gain ideal lowpass filter and an ideal highpass filter. Draw the frequency responses of these two filters.

(b) Design a second-order Butterworth lowpass and a second-order Butterworth highpass that will approximate the ideal bandpass filter in Figure P10-25. Assume that the two half-power frequencies for the filter are at 300 rad/s and 3000 rad/s.

(c) Draw the implementation of the filter, using two appropriate Sallen–Key sections. Be sure to specify values for all the resistors and capacitors. Include an additional operational amplifier (appropriately configured) so that the frequency response has a maximum magnitude of one.

(d) Find the range of frequencies for which the magnitude of the frequency response exceeds 0.9.

P10-26 An RLC bandpass filter has been constructed to pass signals in a narrow frequency band centered about 1.0 kHz. After construction of this filter, it was decided to modify it for use on a slightly different application in which the signal frequency is centered at 1.1 KHz instead of 1.0 kHz. Since the inductors in the filter are very expensive relative to the resistors and capacitors, it is desired to use the same inductors in the modified filter and to change the values of the resistors and capacitors. The filter is to be modified by using a combination of magnitude (impedance) scaling and frequency scaling. Without referring to any specific circuit, describe how the resistor and capacitor values should be changed.

P10-27 The circuit in Figure P10-27 can realize a third-order Butterworth lowpass filter if the values of L and C are appropriately chosen.

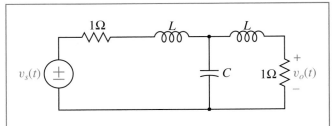

Fig. P10-27. A structure for realizing a third-order Butterworth lowpass filter.

(a) Using the procedure described in Problem P10-17, find the system function $H(s)$ of a third-order Butterworth filter with the squared magnitude function

$$|H(j\omega)|^2 = \frac{1}{1 + \omega^6}.$$

(b) Find values for L and C in the circuit of Figure P10-27 such that the circuit will have the system function that you derived in (a), except possibly for its gain.

(c) Now state values for a more practical filter that will use 1 kΩ resistors instead of 1 Ω ones and that will have a cutoff frequency of 1000 Hz (i.e., 2000π rad/s) instead of 1 rad/s.

APPENDIX A:
REVIEW OF COMPLEX NUMBERS

Complex numbers play an important role in the analysis of circuits. They occur naturally in the evaluation of inverse Laplace transforms (by using partial-fraction expansions) in Chapter 5. They occur even more fundamentally in understanding the response of circuits to sinusoidal source waveforms in Chapters 8 and 9. The ability to perform arithmetic and algebraic manipulations on complex quantities and to convert between the different representations of complex numbers needs to become automatic. This appendix reviews the important properties of complex numbers that we will need, and the problems at the end of the appendix serve as a self-test.

A-1 Definitions

Complex numbers represent an extension of the real numbers. A complex number s is an ordered pair of real numbers $s = [\sigma, \omega]$ where $\sigma = \Re e(s)$ is called the *real part of s* and $\omega = \Im m(s)$ is called the *imaginary part of s*. Real numbers are thus a special case of complex numbers whose imaginary parts are zero. Thus, since σ and ω are both real, we can write

$$\sigma = [\sigma, 0]; \quad \omega = [\omega, 0].$$

We add complex numbers by adding their respective real and imaginary parts:

$$s_1 + s_2 = [\sigma_1, \omega_1] + [\sigma_2, \omega_2]$$
$$\stackrel{\triangle}{=} [\sigma_1 + \sigma_2, \omega_1 + \omega_2]. \qquad (A.1)$$

Multiplication of two complex numbers is more complicated. It is defined by

$$s_1 s_2 = [\sigma_1, \omega_1] \cdot [\sigma_2, \omega_2]$$
$$\stackrel{\triangle}{=} [\sigma_1 \sigma_2 - \omega_1 \omega_2, \sigma_1 \omega_2 + \sigma_2 \omega_1]. \qquad (A.2)$$

Multiplication becomes simpler if we introduce the symbol j, where $j = [0, 1]$. This enables us to write

$$s = \sigma + j\omega.$$

Observe that $j^2 = -1$. With this notation complex numbers can be manipulated by using the familiar rules of algebra. For example, we can compute the product of s_1 and s_2 by

$$s_1 s_2 = (\sigma_1 + j\omega_1) \cdot (\sigma_2 + j\omega_2)$$
$$= \sigma_1 \sigma_2 + j\sigma_1 \omega_2 + j\sigma_2 \omega_1 + j^2 \omega_1 \omega_2$$
$$= (\sigma_1 \sigma_2 - \omega_1 \omega_2) + j(\sigma_1 \omega_2 + \sigma_2 \omega_1),$$

which is the same result as the definition in (A.2).

If two complex numbers are equal, their respective real parts and their respective imaginary parts must both be equal. Thus, $s_1 = s_2$ implies that

$$\sigma_1 = \sigma_2 \ \textit{and} \ \omega_1 = \omega_2.$$

We can use this fact to define the negative of the complex number s, denoted by $-s$, as that number such that

$$s + (-s) = 0, \qquad (A.3)$$

where $0 = [0, 0]$ is the complex number whose real and imaginary parts are both equal to zero. Equating the real and imaginary parts of (A.3), we have

$$\mathfrak{Re}(s) + \mathfrak{Re}(-s) = 0$$

$$\mathfrak{Im}(s) + \mathfrak{Im}(-s) = 0.$$

This means that

$$\mathfrak{Re}(-s) = -\mathfrak{Re}(s) = -\sigma$$

$$\mathfrak{Im}(-s) = -\mathfrak{Im}(s) = -\omega,$$

or $-s = [-\sigma, -\omega]$. We define the *difference of two complex numbers* as the first number plus the negative of the second:

$$s_1 - s_2 \stackrel{\triangle}{=} s_1 + (-s_2) = (\sigma_1 - \sigma_2) + j(\omega_1 - \omega_2).$$

We see that the real part of the difference of two complex numbers is the difference of their real parts and that the imaginary part of the difference is the difference of their imaginary parts. Therefore, we can add or subtract two complex numbers by adding or subtracting their real and imaginary parts.

In a similar fashion, we define the *reciprocal of a complex number* $(1/s)$ as the complex number such that

$$s \cdot (1/s) = 1 = 1 + j0.$$

We can compute the reciprocal by expressing both s and $(1/s)$ in terms of their real and imaginary parts, performing the indicated multiplication, and then equating the real and imaginary parts of the complex equation, as we did before. This gives

$$\mathfrak{Re}(s)\mathfrak{Re}(1/s) - \mathfrak{Im}(s)\mathfrak{Im}(1/s) = 1$$

$$\mathfrak{Re}(s)\mathfrak{Im}(1/s) + \mathfrak{Im}(s)\mathfrak{Re}(1/s) = 0.$$

Solving these two equations for $\mathfrak{Re}(1/s)$ and $\mathfrak{Im}(1/s)$ yields

$$\mathfrak{Re}(1/s) = \frac{\mathfrak{Re}(s)}{[\mathfrak{Re}(s)]^2 + [\mathfrak{Im}(s)]^2} = \frac{\sigma}{\sigma^2 + \omega^2}$$

$$\mathfrak{Im}(1/s) = \frac{-\mathfrak{Im}(s)}{[\mathfrak{Re}(s)]^2 + [\mathfrak{Im}(s)]^2} = \frac{-\omega}{\sigma^2 + \omega^2}.$$

Thus,

$$\frac{1}{s} = \frac{\sigma - j\omega}{\sigma^2 + \omega^2}. \tag{A.4}$$

Division of two complex numbers is defined as the product of the dividend with the reciprocal of the divisor. If we use (A.4) and formally perform the multiplication, this reveals the result

$$\frac{s_1}{s_2} \stackrel{\triangle}{=} s_1 \cdot \left(\frac{1}{s_2}\right)$$

$$= \frac{\sigma_1\sigma_2 + \omega_1\omega_2}{\sigma_2^2 + \omega_2^2} + j\frac{\sigma_2\omega_1 - \sigma_1\omega_2}{\sigma_2^2 + \omega_2^2}.$$

It is clear that when a complex number is expressed in terms of its real and imaginary parts, known as the *Cartesian representation* of the complex number, addition and subtraction are straightforward, but multiplication and division are complicated. These latter operations are easier to perform when the complex numbers are expressed in *polar form*. We define the polar form and the closely related complex exponential in the next two sections.

A-2   Graphical Interpretation of Complex Numbers

Complex numbers are often represented as points in a two-dimensional plane, called the *complex plane*, as shown in Figure A-1. The horizontal and vertical coordinates are the real and imaginary parts, respectively. We can also identify the complex number $s_0 = [\sigma_0, \omega_0]$

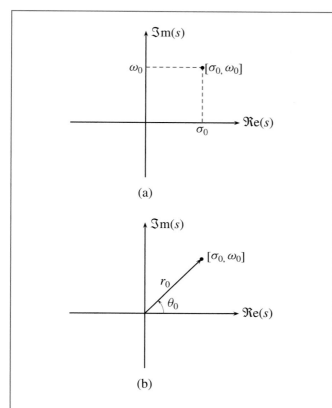

Fig. A-1. (a) Graphical representation of a complex number as a point using Cartesian coordinates. (b) As a vector with its polar variables indicated.

Figure A-2, we see that the real number $s = 1$ lies on the real axis in the complex plane. Its polar representation is $r = 1$, $\theta = 0$ and its Cartesian representation is [1,0]. The number $s = -1$ also lies on the real axis, but since it lies on the negative side of the axis, it has the different polar representation $r = |-1| = 1$, $\theta = \arg(-1) = \pi$. Similarly, j corresponds to the point $r = |j| = 1$, $\theta = \arg(j) = \pi/2$, and $-j$ corresponds to $r = |-j| = 1$, $\theta = \arg(-j) = -\pi/2$. In general, if a complex number s has the polar representation $s = r\angle\theta$, its negative has the representation $-s = r\angle(\theta \pm \pi)$.

Adding complex numbers is equivalent to vector addition. Graphically this is illustrated in Figure A-3. To add s_2 to s_1, we position the vector corresponding to s_1 with its tail at the origin, then position the vector corresponding to s_2 with its tail at the head of s_1. The sum $s_1 + s_2$ is then the vector that extends from the tail of s_1 to the head of s_2.

The polar and Cartesian representations for a complex number are equivalent and we can switch back and forth between them whenever it is convenient. To convert a number s from its Cartesian representation to its polar one, we calculate as follows:

$$r = |s| = \sqrt{\sigma^2 + \omega^2}$$

$$\theta = \arg(s) = \arctan\frac{\omega}{\sigma} + \begin{cases} \pi, & \text{if } \sigma < 0 \text{ and } \omega \geq 0 \\ 0, & \text{if } \sigma \geq 0 \\ -\pi, & \text{if } \sigma < 0 \text{ and } \omega < 0. \end{cases}$$

The principal value of the inverse tangent function is defined for angles between $-\frac{\pi}{2}$ and $\frac{\pi}{2}$. The vector corresponding to a complex number, however, can lie in any of the four quadrants of the complex plane, so its angle can range from $-\pi$ to π. As a result, we need to "correct" the value of the inverse tangent when the vector lies in quadrants two or three (i.e., when the real part of the complex number is negative). This correction is incorporated into the previous equation. We further observe that the argument can be stated only

with a two-dimensional vector. This vector is uniquely defined by its coordinates σ_0 and ω_0 or by its magnitude $|s_0| = r_0$ and angle, or argument, $\arg(s_0) = \theta_0$. The argument is measured as the angle between the real axis and the vector measured in a *counterclockwise* direction, as shown in Figure A-1(b). We will usually measure that angle in radians. The representation of a complex number s in terms of $|s| = r$ and $\arg(s) = \theta$ is called its *polar representation*. The representation in terms of the real and imaginary parts is called its *Cartesian*, or rectangular, representation. Referring to the examples in

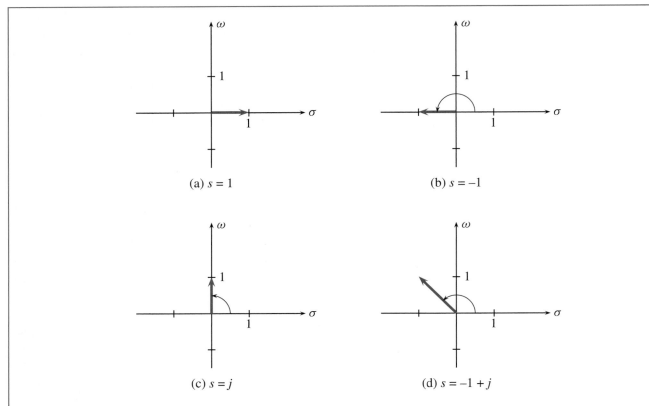

Fig. A-2. Graphical representations of four complex numbers. (a) $s = 1$; $|1| = 1$; $\arg(1) = 0$. (b) $s = -1$; $|-1| = 1$; $\arg(-1) = \pi$. (c) $s = j$; $|j| = 1$; $\arg(j) = \pi/2$. (d) $s = -1 + j$; $|-1 + j| = \sqrt{2}$; $\arg(-1 + j) = 3\pi/4$.

to within a multiple of 2π. For this reason, the value of $\arg(s)$ is normally restricted to lie within the range $-\pi < \arg(s) \leq \pi$. From the polar representation, we can convert back to the Cartesian representation by the following calculation:

$$\mathfrak{Re}(s) = \sigma = r \cos\theta = |s| \cos(\arg(s)) \quad (A.5)$$

$$\mathfrak{Im}(s) = \omega = r \sin\theta = |s| \sin(\arg(s)). \quad (A.6)$$

 DEMO: *Complex Numbers*

A-3   Complex Exponentials

The definition of the familiar exponential function, e^x, can be extended to the case where the exponent is complex by using a Taylor-series expansion. If we perform such an expansion for the function $e^{j\theta}$, group the purely real terms together, group the purely imaginary terms together, and recognize the two groupings as Taylor-series expansions for $\cos\theta$ and $\sin\theta$, respectively, we can show that

$$e^{j\theta} = \cos\theta + j \sin\theta. \quad (A.7)$$

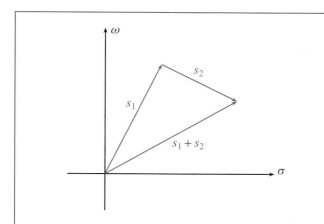

Fig. A-3. Graphical addition of two complex numbers s_1 and s_2.

This result is known as *Euler's relation*. It amounts to a Cartesian representation of the complex number $e^{j\theta}$. Its polar representation is seen to be

$$|e^{j\theta}| = \sqrt{\cos^2\theta + \sin^2\theta} = 1$$

$$\arg(e^{j\theta}) = \arctan\left(\frac{\sin\theta}{\cos\theta}\right) = \arctan(\tan\theta) = \theta.$$

By (A.7),

$$e^{j\,\arg(s)} = \cos(\arg(s)) + j\sin(\arg(s)).$$

If we multiply both sides of this expression by $|s|$, then

$$|s|e^{j\,\arg(s)} = |s|\cos(\arg(s)) + j|s|\sin(\arg(s))$$
$$= \sigma + j\omega = s.$$

This means that a complex number s can be written as the complex exponential

$$s = |s|e^{j\,\arg(s)}. \tag{A.8}$$

This is more than just a change in notation, as with the polar representation, because it means that the law of exponents holds for complex exponentials just as it does for real ones. This makes the evaluation of the product of two complex numbers that are expressed in polar form particularly straightforward:

$$(|s_1|e^{j\,\arg(s_1)}) \cdot (|s_2|e^{j\,\arg(s_2)}) = |s_1||s_2|e^{j[\arg(s_1)+\arg(s_2)]}.$$

The magnitude of the product of two complex numbers is the product of their magnitudes, and the argument of the product is the sum of their arguments. The reciprocation and division operations are also much simpler when the complex numbers are expressed in polar form, whereas the Cartesian form is simpler for addition and subtraction. In polar form,

$$\frac{1}{s} = \frac{1}{|s|}e^{-j\,\arg(s)} \tag{A.9}$$

$$\frac{s_1}{s_2} = \frac{|s_1|}{|s_2|}e^{j[\arg(s_1)-\arg(s_2)]}. \tag{A.10}$$

If we replace θ by $-\theta$ in (A.7), we see that

$$e^{-j\theta} = \cos(-\theta) + j\sin(-\theta) = \cos\theta - j\sin\theta. \tag{A.11}$$

Solving (A.7) and (A.11) together for $\cos\theta$ and $\sin\theta$ provides two very useful relations, known as the *inverse Euler Relations*.

$$\cos\theta = \frac{1}{2}\left(e^{j\theta} + e^{-j\theta}\right) \tag{A.12}$$

$$\sin\theta = \frac{1}{j2}\left(e^{j\theta} - e^{-j\theta}\right). \tag{A.13}$$

We use these relations often in this text.

   WORKED SOLUTION: *Complex Numbers*

Example A-1   Operations on Complex Numbers

Express the following complex numbers in Cartesian form:

(a) $s_a = 4 e^{-j(\pi/6)}$

(b) $s_b = (1 + j)^6$

(c) $s_c = \dfrac{1+j\sqrt{3}}{\sqrt{3}+j}$

For the first of these expressions, we can use Euler's relation:

$$s_a = 4 e^{-j(\pi/6)} = 4\cos(-\pi/6) + j4\sin(-\pi/6)$$

$$= 4 \cdot \frac{\sqrt{3}}{2} - j4 \cdot \frac{1}{2} = 2\sqrt{3} - j2.$$

For the second expression, one possibility is to actually perform the indicated multiplication, using (A.2). An easier method is to convert the quantity inside parentheses into polar form, exponentiate the result, and then convert the answer back into Cartesian form. This gives

$$s_b = (1 + j)^6 = \left(\sqrt{2}\, e^{j(\pi/4)}\right)^6 = 8 e^{j(3\pi/2)} = -j8.$$

For the third expression, there are also several possibilities. One is to express the numerator and denominator separately in polar form, perform the indicated division, and then convert the result back into Cartesian form:

$$s_c = \frac{1 + j\sqrt{3}}{\sqrt{3} + j} = \frac{2 e^{j(\pi/3)}}{2 e^{j(\pi/6)}} = e^{j(\pi/6)} = \frac{\sqrt{3}}{2} + j\frac{1}{2}.$$

■

To check your understanding, try Drill Problem PA-1.

Example A-2   More Operations on Complex Numbers

Express the following complex numbers in polar form:

(a) $s_a = (\sqrt{3} + j)(1 - j)$

(b) $s_b = \sqrt{2}\, e^{j(-3\pi/4)} + \sqrt{2}\, e^{j(3\pi/4)}$

(c) $s_c = (1 - j\sqrt{2})^3$

To calculate the first of these quantities, we can multiply the two factors and then convert the result to polar form, or we can perform the conversion first and then multiply the results. Using the latter approach gives

$$s_a = (\sqrt{3} + j)(1 - j) = (2 e^{j(\pi/6)})(\sqrt{2}\, e^{-j(\pi/4)})$$

$$= 2\sqrt{2}\, e^{-j(\pi/12)}.$$

As we have already noted, additions and subtractions are performed most easily when the complex numbers are expressed in Cartesian form; multiplications and divisions are easier in polar form. To compute the summation in (b), we first convert the quantities to Cartesian form, then perform the addition, and then reconvert the result:

$$s_b = \sqrt{2}\, e^{j(-3\pi/4)} + \sqrt{2}\, e^{j(3\pi/4)}$$

$$= (-1 - j) + (-1 + j) = -2 = 2 e^{j\pi}.$$

Remember that the magnitude is always positive and that negative real numbers have an angle of π radians.

Raising a number to a power is a multiplication operation, so this is most easily done when the numbers are in polar form:

$$(1 - j\sqrt{2})^3 = (\sqrt{3}\, e^{-j\tan^{-1}(\sqrt{2})})^3 = 3\sqrt{3}\, e^{-j3\tan^{-1}(\sqrt{2})}$$

$$= 5.152 e^{-j2.8659}.$$

■

To check your understanding, try Drill Problem PA-2.

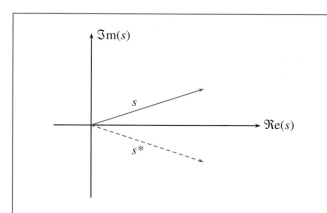

Fig. A-4. The vectors corresponding to a complex number s (solid) and its complex conjugate s^* (dashed).

A-4   The Complex Conjugate

450 The number $s^* = \Re\mathrm{e}(s) - j\Im\mathrm{m}(s) = \sigma - j\omega$ is called the *complex conjugate* of s. Notice that the real part of s^* is the same as the real part of s and that the imaginary part of s^* is $-\Im\mathrm{m}(s)$. In polar form,

$$s^* = |s|e^{-j\arg(s)}.$$

The graph of s^* is the reflection of the graph of s across the real axis, as depicted in Figure A-4.

If a complex number is added to its complex conjugate, the result is always real. In fact,

$$s + s^* = (\sigma + j\omega) + (\sigma - j\omega) = 2\sigma = 2\Re\mathrm{e}(s).$$

Similarly, since

$$s - s^* = (\sigma + j\omega) - (\sigma - j\omega) = j2\omega = j2\Im\mathrm{m}(s),$$

it follows that

$$\Re\mathrm{e}(s) = \frac{1}{2}(s + s^*)$$

$$\Im\mathrm{m}(s) = \frac{1}{2j}(s - s^*).$$

The product of a complex number with its complex conjugate is also real:

$$s \cdot s^* = (\sigma + j\omega)(\sigma - j\omega)$$
$$= \sigma^2 + \omega^2 = \Re\mathrm{e}^2(s) + \Im\mathrm{m}^2(s) = |s|^2.$$

This relation is particularly useful as a means for computing the magnitude of a complex function.

A-5   Drill Problems

PA-1   Express the following complex numbers in Cartesian form:

(a) $s_a = je^{j(9\pi/4)}$

(b) $s_b = (1 + j)(1 - j2)$

(c) $s_c = e^{j(\pi/3)} + e^{j(2\pi/3)}$

(d) $s_d = (\sqrt{3} - j)2\sqrt{2}e^{-3\pi/4}$

(e) $s_e = \frac{e^{j(\pi/3)} - 1}{1 + j\sqrt{3}}$

(f) $s_f = \frac{2 + j(6/\sqrt{3})}{2 - j(6/\sqrt{3})}$

PA-2 Express the following complex numbers in polar form:

(a) $s_a = 3 - j4$

(b) $s_b = (1 + j)(1 - j2)$

(c) $s_c = e^{j(\pi/3)} + e^{j(2\pi/3)}$

(d) $s_d = \frac{j(2 + j)}{(1 + j)(2 - j)}$

Give numerical values for the magnitude and for the angle (or phase) in radians.

PA-3 Express each of the following in Cartesian form:

(a) $s_a = 4e^{-j(\pi/6)}$

(b) $s_b = \sqrt{2}e^{j(27\pi/4)}$

(c) $s_c = (1+j)^7$

(d) $s_d = \frac{1-j\sqrt{2}}{\sqrt{2}-j}$

(e) $s_e = je^{j(3\pi/4)}$

(f) The six roots of the equation $s^6 + 1 = 0$

Give numerical values for the real and imaginary parts.

PA-4 Express each of the following in polar form:

(a) $s_a = e^{j(3\pi/4)} + e^{j(\pi/4)}$

(b) $s_b = (1+j)^7$

(c) $s_c = \dfrac{1 - j\sqrt{2}}{\sqrt{2} - j}$

(d) $s_d = \dfrac{1 - j}{j(1 + j)}$

Give numerical values for the magnitude and for the angle (or phase) in radians.

PA-5 Evaluate the following, and give the answer in both Cartesian and polar form. In all cases, assume that the complex numbers are $s_1 = 3 + j3$ and $s_2 = e^{-j(3\pi/4)}$.

(a)	s_1^*	(f)	$s_1 s_2$
(b)	js_2	(g)	$s_1 + s_2^*$
(c)	s_2/s_1	(h)	$\|s_2\|^2 = s_2 s_2^*$
(d)	s_2^2	(i)	$s_2 + s_2^*$
(e)	$s_1^{-1} = 1/s_1$		

Note: s^* means the "conjugate" of s. Part (h) is the *magnitude-squared*.

PA-6 Simplify the following, and give the answer in polar form. Make a plot of all the vectors involved in the complex addition.

(a) $s_a = e^{-j\pi/4} + e^{j\pi/4}$

(b) $s_b = 1 + e^{-j2\pi/3} + e^{j2\pi/3}$

PA-7 Simplify the following complex-valued expressions. Give your answer in either Cartesian or polar form, whichever is more convenient. In parts (a)–(e), assume that A, α, and ϕ are positive real numbers. Your answers may be in terms of these quantities.

(a) For $s = Ae^{j\pi/6}$, find a simple expression for $\Re\{s^*\}$.

(b) For $s = Ae^{-j2\pi/3}$, find a simple expression for $s - s^*$.

(c) For $s = 3e^{j\phi}$, find a simple expression for $\Im\{-js\}$.

(d) For $s = -\alpha\sqrt{3} + j\alpha$, find a simple expression for s in polar form.

(e) For $s = Ae^{j2\pi/3}$, find a simple expression for $|s|/s^*$ in polar form.

PA-8

(a) State the values of the real variables x and y if

$$(2 + j)x + (3 - j2)y = 3 + j5.$$

(b) State the value of the complex variable s if

$$(1 - j2)s + (3 - j) = 2 - j4.$$

PA-9 Express each of the following complex numbers in the form $\sigma + j\omega$, where σ and ω are real.

(a) $(3 + j7) - (2 - j3)$

(b) $(3 + j2)(2 - j3)$

(c) $(1 + j2)^3$

(d) $\dfrac{1 - j}{1 + j}$

PA-10

(a) Sketch each of the following complex numbers in the complex plane. Indicate the real and imaginary part of each number on your sketch, and find the polar representation of each complex number.
    (i) $j5$
    (ii) -3
    (iii) $3 - j4$
    (iv) $-1 + j$

(b) Sketch each of the following complex numbers in the complex plane, and state the Cartesian representation of each complex number.
    (i) $2e^{j\pi/2}$
    (ii) $\frac{1}{2}e^{-j4\pi/3}$
    (iii) $5e^{j\pi}$
    (iv) $2e^{j\pi/6}$

(c) Express each of the following complex numbers in polar form.
    (i) $e^{j\pi/6} + e^{-j\pi/6}$
    (ii) $\dfrac{1 - j\sqrt{3}}{1 + j\sqrt{3}}$
    (iii) $(\sqrt{3} + j)\sqrt{2}e^{-j\pi/4}$
    (iv) $(1 - j)^5$

A-6   Basic Problems

PA-11 Let s be a complex variable, and let σ and ω be its real and imaginary parts, respectively; that is, $s = \sigma + j\omega$. Express each of the following functions in the form $u(\sigma, \omega) + jv(\sigma, \omega)$ where $u(\sigma, \omega)$ and $v(\sigma, \omega)$ are real.

(a) $1 + js^2$

(b) $\dfrac{1 + s}{1 - s}$

PA-12 Using Euler's relation, and without appealing to the law of exponents, show that, if θ_1 and θ_2 are real, then

$$e^{j\theta_1} \cdot e^{j\theta_2} = e^{j(\theta_1 + \theta_2)}.$$

PA-13 Use the definition of the complex exponential or Euler's relations to demonstrate the following (familiar?) trigonometric identities:

(a) $\cos(2\theta) = \cos^2\theta - \sin^2\theta$

(b) $\sin(2\theta) = 2\sin\theta\cos\theta$

(c) $\cos^2\theta = \frac{1}{2}(1 + \cos 2\theta)$

(d) $(\cos\theta)(\cos\phi) = \frac{1}{2}\cos(\theta - \phi) + \frac{1}{2}\cos(\theta + \phi)$

PA-14 This problem derives a number of identities concerning the complex conjugate.

(a) Using either the Cartesian or polar representation for each complex variable, show that
    (i) $(s_1 + s_2)^* = s_1^* + s_2^*$.
    (ii) $(as_1)^* = as_1^*$, where a is any real number.
    (iii) $(s_1 s_2)^* = s_1^* s_2^*$.
    (iv) $(e^s)^* = e^{s^*}$.
    (v) $(s^a)^* = (s^*)^a$, where a is any real number.
    (vi) $\left(\dfrac{s_1}{s_2}\right)^* = \dfrac{s_1^*}{s_2^*}$.

(b) Using the results that you derived in (a), show that

$$[P(s)]^* = P(s^*),$$

where

$$P(s) = a_n s^n + a_{n-1}s^{n-1} + \cdots + a_1 s + a_0$$

is a polynomial of degree n in s with real coefficients.

(c) Using the results of (a) and (b), show that

$$[Q(s)]^* = Q(s^*),$$

where

$$Q(s) = \frac{b_m s^m + b_{m-1} s^{m-1} + \cdots + b_1 s + b_0}{a_n s^n + a_{n-1} s^{n-1} + \cdots + a_1 s + a_0}$$

is the ratio of two polynomials with real coefficients.

(d) Using the results of (b), show that the complex roots of a polynomial with real coefficients occur in complex-conjugate pairs (i.e., show that if $P(s_1) = 0$, where $P(s)$ is a polynomial with real coefficients, then $P(s_1^*) = 0$).

PA-15 Let $A = re^{j\theta}$ and $s = \sigma + j\omega$. Define a complex exponential time function $x(t)$ as

$$x(t) = Ae^{st}.$$

(a) Let $\sigma = 0$, and describe the locus of the point $x(t)$ in the complex plane as t increases from zero to large positive values. Describe the effect of θ and ω on the sketch.

(b) Repeat part (a) for some value of $\sigma < 0$.

(c) Repeat part (a) for some value of $\sigma > 0$.

APPENDIX B:
ANSWERS TO SELECTED BASIC PROBLEMS

Here we include answers for up to ten of the basic problems for each chapter.

P1-15 (a) $P_{\text{inst}}(t) = \begin{cases} 1 - e^{-3t}, & 0 \le t < 1 \\ 0, & \text{otherwise} \end{cases}$

(b) $E = \frac{2}{3} + \frac{1}{3}e^{-3}$

P1-19 (a) $v_r(t) = \begin{cases} 25\cos(50t) & t > 0 \\ 0 & t < 0 \end{cases}$

(b) $v_\ell(t) = -500\sin(50t)$, $t > 0$. The inductor voltage $v_\ell(t) = 0$ for $t < 0$.

(c) $v_c(t) = \frac{1}{30}\sin(50t)$. The capacitor voltage is zero for $t < 0$.

P1-21 (a) a

(i) b $i_c(t) = \begin{cases} \frac{\pi}{2}\sin\pi t, & 0 < t < 2 \\ 0, & \text{otherwise} \end{cases}$

(ii) $E_c(t) = \begin{cases} \frac{1}{8}(1 - \cos\pi t)^2, & 0 < t < 2 \\ 0, & \text{otherwise} \end{cases}$

(iii) $1 < t < 2$

(iv) $0 < t < 1$

(b) (i) $i_\ell(t) = \begin{cases} 0, & t < 0 \\ \frac{1}{2}t + \frac{1}{2\pi}\sin\pi t, & 0 < t < 2 \\ 1, & 2 < t \end{cases}$

(ii) $\frac{1}{2}i_\ell^2(t)$

(iii) $0 < t < \infty$

(iv) There are no values of t for which the inductor is supplying power.

P1-23 $i(t) = \begin{cases} 0, & t < 0 \\ -20/3, & 0 < t < 3 \\ 20/3, & 3 < t < 6 \\ 0, & 6 < t \end{cases}$

P1-25 (a) $-i_a(t) - i_2(t) - i_1(t) = 0$
$i_2(t) - i_3(t) - i_4(t) = 0$
$i_3(t) + i_s(t) + i_1(t) = 0$
$i_a(t) + i_4(t) - i_s(t) = 0$

(b) $-v_s(t) + v_2(t) + v_4(t) = 0$
$-v_4(t) + v_3(t) - v_b(t) = 0$
$v_1(t) - v_3(t) - v_2(t) = 0$

P1-27 $i_1(t) = \frac{3}{8}v_s(t) - \frac{1}{8}i_s(t)$
$i_2(t) = \frac{1}{8}v_s(t) - \frac{3}{8}i_s(t)$
$i_3(t) = -\frac{1}{4}v_s(t) - \frac{1}{4}i_s(t)$
$i_4(t) = -\frac{1}{8}v_s(t) - \frac{5}{8}i_s(t)$

P1-29 $v_1(t) = 2.632\,i_s(t)$
$v_2(t) = 2.368\,i_s(t)$
$v(t) = v_1(t) = 2.632\,i_s(t)$

P1-31 $i_1 = -2/15$ $v_1 = -4/3$
$i_2 = -13/15$ $v_2 = -13/3$

P1-33 $v_1(t) = \frac{5}{2}\sin 30t$, $v_3(t) = \frac{3}{2}\sin 30t$

P1-35 (a) $i(t) - \frac{1}{9}v_s(t)$

(b) $i(t) = \frac{1}{15}v_s(t)$

433

P2-17 Any three of the following four KCL equations along with the two KVL equations are sufficient.

KCL1: $i_1(t) + i_2(t) = i_s(t)$

KCL2: $i_2(t) - i_3(t) - i_4(t) = 0$

KCL3: $i_1(t) + i_4(t) - i_5(t) = 0$

KCL4: $i_3(t) + i_5(t) = i_s(t)$

KVL1: $3i_1(t) - 2i_2(t) - i_4(t) = 0$

KVL2: $-i_3(t) + i_4(t) + i_5(t) = v_s(t)$

P2-19 (a) $i_1(t) = -0.2v_s(t) + 0.3i_s(t)$
$i_2(t) = 0.2v_s(t) + 0.2i_s(t)$
$i_3(t) = -0.4v_s(t) + 0.1i_s(t)$
$i_4(t) = 0.6v_s(t) + 0.1i_s(t)$

P2-21 (d) $v_1(t) = 0.6667v_{s_1}(t) + 0.3333v_{s_2}(t)$
$v_2(t) = 0.3333v_{s_1}(t) - 0.3333v_{s_2}(t)$
$v_3(t) = 0.3333v_{s_1}(t) + 0.6667v_{s_2}(t)$
$i_1(t) = 0.6667v_{s_1}(t) + 0.3333v_{s_2}(t)$
$i_2(t) = 0.3333v_{s_1}(t) - 0.3333v_{s_2}(t)$
$i_3(t) = 0.3333v_{s_1}(t) + 0.6667v_{s_2}(t)$

P2-23 $i_1(t) = 0.4118 i_s(t) + 0.0588 v_s(t)$
$i_2(t) = 0.5882 i_s(t) - 0.0588 v_s(t)$
$i_3(t) = 0.5294 i_s(t) - 0.3529 v_s(t)$
$i_4(t) = 0.0588 i_s(t) + 0.2941 v_s(t)$
$i_5(t) = 0.4706 i_s(t) + 0.3529 v_s(t)$

P2-27 $v(t) = 3.1111 i_{s_1}(t) + 1.6667 i_{s_2}(t)$

P2-31 $i_1 = 2$, $i_2 = 1/6$, $i_3 = 11/6$, $i_4 = -7/6$

P2-33 $\left[\frac{1}{R_1} + \frac{1}{R_2} + \frac{1}{R_3}\right] e_a(t) - \frac{1}{R_3}e_b(t) = i_{s_1}(t)$
$-\frac{1}{R_3}e_a(t) + \left[\frac{1}{R_3} + \frac{1}{R_4} + \frac{1}{R_5}\right] e_b(t)$
$\quad -\frac{1}{R_5}e_c(t) = i_{s_2}(t)$
$-\frac{1}{R_5}e_b(t) + \left[\frac{1}{R_5} + \frac{1}{R_6}\right] e_c(t) = -i_{s_1}(t)$

P2-35 $e_a(t) = -0.4286 v_s(t) + 0.0952 i_s(t)$
$e_b(t) = -0.2857 v_s(t) - 0.2857 i_s(t)$
$e_c(t) = -0.4286 v_s(t) - 0.2381 i_s(t)$
$e_d(t) = -v_s(t)$

P2-39 $v(t) = \frac{8}{25}v_s(t)$

P2-41 (a) $i_1(t) = -\frac{1}{2}i_\alpha(t)$
(b) $i_1(t) = -\frac{1}{3}v_s(t) + i_s(t)$

P3-10 (a) $R_{eq} = 2\Omega$
(b) $R_{eq} = 2\Omega$

P3-12 $v_{out}(t) = -2.73 + 13.09k$

P3-14 (a) $i_1(t) = \frac{1}{2}v_s(t)$
(b) $i_2(t) = \frac{1}{16}v_s(t)$

P3-15 (a) 3 W
(b) 3 W

P3-16 (a) $L_{eq} = L_1 + L_2$
(b) $v_1(t) = \frac{L_1}{L_1+L_2}v(t)$
$v_2(t) = \frac{L_2}{L_1+L_2}v(t)$

P3-18 (a) $\frac{1}{C_{eq}} = \frac{1}{C_1} + \frac{1}{C_2}$
(b) $v_1(t) = \frac{\frac{1}{C_1}}{\frac{1}{C_1}+\frac{1}{C_2}}v(t)$
$v_2(t) = \frac{\frac{1}{C_2}}{\frac{1}{C_1}+\frac{1}{C_2}}v(t)$

P3-20 $i(t) = \left[\frac{1}{R_1} + \frac{1}{R_2}\right]v(t) - \frac{1}{R_1}v_{s_1}(t) - \frac{1}{R_2}v_{s_2}(t)$

P3-22 $R_T = \frac{10}{7}\Omega$
$v_{oc}(t) = \frac{5}{7}v_s(t) + \frac{10}{7}i_s(t)$

P3-24 $R_i = 0.0135\Omega$

P3-30 $P_{inst} = 9.6$ W

P4-2 $v_{\text{out}}(t) = \left(1 + \frac{R_1}{R_2}\right)(v_{\text{in}}(t) + 2)$

P4-4 $v_{\text{out}}(t) = Ri_s(t) + v_s(t)$

P4-6 $v_{\text{out}}(t) = -6v_{\text{in}}(t)$

P4-8 $v_{\text{out}}(t) = -\frac{2}{3}v_s(t)$

P4-10 $v_{\text{out}}(t) = -\frac{R}{L}\int_{t_0}^{t} v_{\text{in}}(\beta)\,d\beta - Ri_\ell(t_0)$

P4-12 $v_{\text{out}}(t) = -\frac{C_1}{C_2}v_{\text{in}}(t) + [v_{\text{out}}(t_0) - v_{\text{in}}(t_0)]$

P4-14 $v_{\text{out}}(t) = -8v_{\text{in}}(t)$

P4-16 $v_{\text{out}}(t) = 20v_{\text{in}}(t)$

P4-18 $v_{\text{out}}(t) = \frac{1}{3}v_{\text{in}}(t)$

P4-20 $i_{\text{out}}(t) = \left[\frac{1}{400} + \frac{1}{10,000}\right]v_{\text{in}}(t) \approx \frac{1}{400}v_{\text{in}}(t)$

P5-6 (a) $x_a(t) = \sqrt{2}e^{-2t}\cos(3t + \frac{\pi}{4})$
    (b) $x_b(t) = 2e^{-t}\cos(2t + \frac{\pi}{6})$
    (c) $x_c(t) = -\frac{2\omega_0}{\sqrt{1+\omega_0^2}}\sin(\omega_0 - \tan^{-1}\omega_0)$

P5-8 (a) $x_a(t) = u(t+2) - u(t-2)$
    (b) $x_b(t) = u(t) + u(t-1) - u(t-2.5) - u(t-4)$
    (c) $x_c(t) = -u(t) + 2\sum_{k=0}^{\infty}(-1)^k u(t-k)$

P5-10 (a) $X_a(s) = \frac{1}{s}\left(1 - e^{-sT}\right)$
    (b) $X_b(s) = \frac{2}{(s+3)^3}$
    (c) $X_c(s) = \frac{5}{s^2+8s+41}$
    (d) $X_d(s) = \frac{1}{s^2}$

P5-12 (a) $X_a(s) = \frac{3}{s}$
    (b) $X_b(s) = \frac{10e^{-6}}{s+2}e^{-3s}$

P5-13 $x(t) = \left(\sqrt{1 + \frac{a^2}{b^2}}\right)e^{-at}\cos(bt + \tan^{-1}\frac{a}{b})$

P5-14 $x(t) = 3 - 4e^{-t} + e^{-2t}, \quad t > 0$

P5-16 (a) $x_a(t) = \frac{5}{2}\delta(t) - \frac{5}{4}e^{-t/2}, \quad t > 0$
    (b) $x_b(t) = \frac{\sqrt{65}}{2}e^{-3t}\cos(2t + \tan^{-1}\frac{7}{4}), \quad t > 0$
    (c) $x_c(t) = e^{-t}\left(\frac{7}{2}t^2 - 14t + 7\right), \quad t > 0$

P5-18 $x(t) = \frac{2}{3}e^{-t} - \frac{2}{3}e^{-t}\cos\sqrt{3}t + \frac{1}{\sqrt{3}}e^{-t}\sin\sqrt{3}t$
    $t > 0$

P5-20 (a) $x_a(t) = \frac{1}{a^2}\left[at - 1 + e^{-at}\right] \quad t > 0$
    (b) $x_b(t) = \delta(t) - te^{-t} + 4e^{-t} - 8e^{-2t}, \quad t > 0$

P5-22 $x(t) = \delta(t) - e^{-t}\left(\frac{1}{4} + \frac{11}{4}\cos 2t - \frac{1}{2}\sin 2t\right)$
    $t \geq 0$

P6-15 $v(t) = v_s(t) - 2i_s(t)$

P6-16 $\frac{1}{2}\frac{dv_{\text{out}}(t)}{dt} + v_{\text{out}}(t) = -10v_{\text{in}}(t)$ At $t = 0$, $i_\ell(t) = 0$, which implies $v_{\text{out}}(t) = 0$ since no current will flow through the feedback resistor.

P6-18 $v(t) = 4e^{-t} - 4e^{-2t}, \quad t \geq 0$

P6-20 $i(t) = -\frac{3}{8}e^{-t} + \frac{3}{2}e^{-2t} - \frac{9}{8}e^{-3t}, \quad t \geq 0$

P6-22 $v_c(t) = -\frac{5}{2}e^{-t} + \frac{9}{2}e^{-3t}, \quad t > 0$

P6-24 $i(t) = \frac{1}{4} - \frac{1}{28}e^{-\frac{12}{17}t}$

P6-26 $v(t) = \frac{1}{2}e^{-t}u(t)$

P6-29 (a) $C_{\text{eq}} = \frac{C_1 C_2}{C_1 + C_2}$
    (b) Two capacitors in series look like a single capacitor with capacitance $C_1 C_2/(C_1 + C_2)$ and initial voltage $v_1(0) + v_2(0)$.

P6-30 (a) $Z_a(s) = R + Ls$
    (b) $Z_b(s) = R + \frac{1}{Cs} = \frac{RCs+1}{Cs}$
    (c) $Z_c(s) = \frac{\frac{R}{Cs}}{R + \frac{1}{Cs}} = \frac{R}{RCs+1}$
    (d) $Z_d(s) = \frac{RLs}{R+Ls}$
    (e) $Z_e(s) = Ls + \frac{1}{Cs} = \frac{LCs^2+1}{Cs}$
    (f) $Z_f(s) = \frac{\frac{L}{C}}{Ls + \frac{1}{Cs}} = \frac{LCs^2+1}{Cs}$
    (g) $Z_g(s) = R + Ls + \frac{1}{Cs} = \frac{LCs^2+RCs+1}{Cs}$
    (h) $Z_h(s) = \frac{1}{\frac{1}{R}+\frac{1}{Ls}+Cs} = \frac{RLs}{RLCs^2+Ls+R}$

P6-33 $v_{\text{out}}(t) = 0.1\sin 1000t + 0.1\cos 1000t, \quad t > 0$

P7-9   (a) $H(s) = \frac{\frac{1}{2}s^2+3}{s^2+5s+6}$

      (b) $i(t) = \left(\frac{1}{2} - \frac{5}{2}e^{-2t} + \frac{5}{2}e^{-3t}\right)u(t)$

P7-10 $H(s) = \frac{1-10^{-5}s^2}{1+10^{-3}s}$

P7-11 $H(s) = \frac{1}{2}[1 - R_1Cs]$

P7-12   (a) $v_a(t) = (1 - e^{-t})u(t)$

      (b) $v_b(t) = \frac{1}{2}\left(1 + \sqrt{2}\cos(t - \frac{3\pi}{4})\right)u(t)$

      (c) $v_c(t) = \left[t - 1 + e^{-t}\right]u(t)$

P7-13 $H(s) = \frac{K \cdot \frac{1}{C_1} \cdot \frac{1}{RC}s}{(s+\frac{1}{RC})(s^2+\frac{1}{RC}s+\frac{1}{LC})}$

P7-14 $v_{out}(t) = \left[\frac{3}{4} + \frac{1}{4}e^{-2t}\right]u(t)$

P7-15   (a) $H(s) = \frac{s-\frac{1}{2}}{s+\frac{1}{2}}$

      (b) $h(t) = \delta(t) - e^{-t/2}u(t)$

P7-18 $i_\ell(t) = (2e^{-t}\sin t)u(t)$

P7-19   (a) $H(s) = \frac{I(s)}{V_{in}(s)} = \frac{1}{3s+6} = \frac{\frac{1}{3}}{s+2}$

      (b) $h(t) = \frac{1}{3}e^{-2t}u(t)$

      (c) $i(t) = \frac{1}{6}\left(1 - e^{-2t}\right)u(t)$
          $- \frac{1}{6}\left(1 - e^{-2(t-1)}\right)u(t - 1)$

P7-20   (a) $h(t) = [6,000e^{-2,000t} - 3,000e^{-1,000t}]u(t)$

      (b) $h(t) = \delta(t) + \frac{1}{RC}u(t)$

P8-14   (a) $v_c(t) = \left(1 - e^{-2t}\right)u(t)$

      (b) $v_c(t) = 2\left(e^{-t} - e^{-2t}\right)u(t)$

      (c) $v_c(t) = (1 - e^{-2t})u(t) + j2(e^{-t} - e^{-2t})$

P8-16   (a) $v_a(t) = \frac{1}{\sqrt{2}}\cos(t + \frac{\pi}{4})$

      (b) $i(t) = \frac{2}{\sqrt{5}}\sin\left(2t - \tan^{-1}\frac{1}{2}\right)$

P8-18 $v(t) = \frac{2\times10^{13}\omega}{\sqrt{(1-10^5\omega^2)^2+(2\times10^{13}\omega)^2}}$

      $\times \sin\left(\omega t - \left(\tan^{-1}\frac{2\times10^{13}\omega}{1-10^5\omega^2}\right)\right)$

P8-20 $v_{out}(t) \approx \cos\left(10t + \frac{\pi}{2} - \tan^{-1}20\right)$

P8-22 $i(t) = 0.074\sin(10t + \tan^{-1}(\frac{10}{9}) - \pi)$

P8-24 $v_c(t) = 5\sin(10t - \frac{\pi}{2}) = -5\cos 10t$

P8-26 $v_{out}(t) = \frac{1}{\sqrt{5}}\cos\left(t - \tan^{-1}2\right)$
        $- \frac{1}{\sqrt{5}}\sin\left(2t + \frac{\pi}{4} - \tan^{-1}2\right)$

P8-28 $v_{out}(t) = \frac{1}{2\sqrt{73}}\sin(4t + \tan^{-1}\frac{3}{8} - \pi)$
        $+ \frac{2}{5}\cos(2t + \tan^{-1}\frac{3}{4} - \pi)$

P8-30 $i_1(t) = \frac{\sqrt{5}}{30}\cos(t + \tan^{-1}\frac{1}{2} - \pi)$

P8-32   (a) $H(s) = \frac{9}{s^2+3s+9}$

      (b) $v_c(t) = -2\cos(3t)$

P9-8 $x(t) = -8 + 2\sin(2t) + 3\cos(4t - \frac{\pi}{3})$

P9-11 $i_c(t) = \sum_{k\ \text{odd}} \frac{2}{(1-4\pi^2k^2)+j2\pi k}e^{j2\pi kt}$

P9-13   (a) Circuit (iii)

      (b) Circuit (ii)

P9-16   (a) Network (i)

      (b) None

P9-17   (b)   (i) $\angle H(j0) = \frac{\pi}{2} - \frac{\pi}{2} - \frac{\pi}{2} + \frac{\pi}{2} = 0$

         (ii) $\angle H(j\infty) = \frac{\pi}{2} + \frac{\pi}{2} - \frac{\pi}{2} - \frac{\pi}{2} = 0$

        (iii) $\angle H(j(a+b)) \approx \frac{\pi}{2} + \frac{\pi}{2} - \frac{\pi}{2} - \frac{\pi}{4} = \frac{\pi}{4}$

P9-19 Network (ii)

P9-20 (a) $\dfrac{V_{out(s)}}{I_s(s)} = \dfrac{50}{(s+\frac{1}{2})^2+\frac{99}{4}}$

(c) $\omega_0 = 5$ rad/s

(d) $\omega_d = \dfrac{\sqrt{99}}{2}$

(e) $\tau = 2$ seconds

(f) $Q = 5$

P9-21 $H(s) = \dfrac{2s^2(1+\frac{s}{10})}{1+\frac{s}{100}}$

$\times \dfrac{1}{1+2\zeta_1(\frac{s}{0.7})+(\frac{s}{0.7})^2}$

$\times \dfrac{1}{\left(1+2\zeta_2(\frac{s}{600})+(\frac{s}{600})^2\right)}$

P9-27 $a = 100, \quad b = 1000, \quad k = 1000$. Note that a and b can be interchanged.

P9-29 (a) $H(s) = \dfrac{s^2}{s^2+20s+100}$

P10-7 (a) $\epsilon_p = 0.1639$

(b) $\epsilon_s = 0.1938$

(c) $\epsilon_p = 0.1170 \quad \epsilon_s = 0.0875$

P10-8 (a) $\omega_p = 0.8342$ radians/second

(b) $\omega_s = 3.1622$ radians/second

(c) $\omega_p = 0.8862, \omega_s = 2.1544$

P10-9 (a) $\dfrac{H(s)}{H(0)} = \dfrac{32\times10^6}{s^2+8000\sqrt{2}s+32\times10^6}$

(b) There are two solutions:

$$L = 0.6043\,H; \qquad C = 0.1036\mu F$$

$$L = 0.1036\,H; \qquad C = 0.6043\mu F$$

P10-11 (a) $H_{hp}(s) = \dfrac{0.8855s^2}{s^2+995.69s+907,030}$

(b) $H_{bp}(s) = \dfrac{982,625s^2}{s^4+1098s^3+(4.5\times10^6)s^2+(2.4705\times10^9)s+(5.0625\times10^{12})}$

P10-12 $H(s) = \dfrac{\frac{1}{RC}ks}{s^2+\frac{3-k}{RC}s+\frac{1}{R^2C^2}}$

PA-11 (a) $1 + j(\sigma + j\omega)^2 = (1 - 2\sigma\omega) + j(\sigma^2 + \omega^2)$

(b) $\dfrac{1+\sigma+j\omega}{1-\sigma-j\omega} = \left(\dfrac{1-\sigma^2-\omega^2}{1+\sigma^2+\omega^2}\right) + j\left(\dfrac{2\omega}{1+\sigma^2+\omega^2}\right)$

Index